SOLID-PHASE EXTRACTION

SOLID-PHASE EXTRACTION

Principles, Techniques, and Applications

edited by

Nigel J. K. Simpson

Varian Associates, Inc.
Harbor City, California

MARCEL DEKKER, INC. NEW YORK • BASEL

ISBN: 0-8247-0021-X

This book is printed on acid-free paper.

Headquarters
Marcel Dekker, Inc.
270 Madison Avenue, New York, NY 10016
tel: 212-696-9000; fax: 212-685-4540

Eastern Hemisphere Distribution
Marcel Dekker AG
Hutgasse 4, Postfach 812, CH-4001 Basel, Switzerland
tel: 41-61-261-8482; fax: 41-61-261-8896

World Wide Web
http://www.dekker.com

The publisher offers discounts on this book when ordered in bulk quantities. For more information, write to Special Sales/Professional Marketing at the headquarters address above.

Current printing (last digit):
10 9 8 7 6 5 4 3 2 1

PRINTED IN THE UNITED STATES OF AMERICA

PREFACE: SPE IN PERSPECTIVE

Few agree on who first performed a solid-phase extraction (SPE), or produced a "SPE" cartridge.[1] Fewer still agree on the date of the first use of SPE for sample preparation. Indeed, the term "solid-phase extraction" was only invented some years after products to perform the technique became available. It is common to find authors of articles in peer-reviewed journals who state model numbers of liquid chromatography pumps, quote buffer pHs to two decimal places, give the address of the distribution company that sold the chromatography column, or even quote the column serial or batch number. And yet, when describing the critical sample preparation step, these same authors claim that "a solid-phase extraction column was used" — no mention of type, size, flow rates, sample work-up, solvents used, let alone discussion of how well it worked! Where are you, editors and referees?

Which brings us to why a book such as this is needed: While most technologies develop from the laboratories of well-placed academic champions, from which theory, applications and acronyms pour forth, this humble technique has had no such birth. Indeed it has had precious little nurturing during the two decades since the first commercially available, pre-packed devices were produced. It is hoped that this book will serve as a catalyst for unified study of this field, as well as a resource for users seeking to optimize their sample preparation step. This book lays a foundation for future users of this powerful and subtle technology. Because expertise is distributed widely in the academic world and because the needs of users vary depending on the type of sample being prepared, it has been necessary to draw from a broad range of scientists in order to present a balanced study of the whole field.

During the preparation of this book I have had many fascinating conversations with contributing authors. One of the contributors, Dr. Martha Wells of Tennessee Technological University, beautifully captured the fun that has been had with this technique: "Today, as we continue to apply and expand SPE applications in our laboratory, I am still amazed at how well this relatively simple procedure can fractionate and concentrate complex chemical mixtures into more manageable subsamples. I realize that I am privileged to have experienced this analytical procedure's growth from a fledgling industry to today's 'workhorse' technique for environmental laboratories as well as for clinical and pharmaceutical applications.

[1] However, it is commonly agreed that the first commercially successful product was the Sep-Pak™ produced by Waters Inc. based on a patent of McDonald, Vivilecchia and Lorenz, bearing the title "Triaxially Compressed Beds" – U.S. Patent #4,211,658 (1980).

"Housed on my office shelves are keepsakes from these fifteen years, devices collected that were used to conduct and market SPE on bonded silicas. A 1979 sample Sep-Pak™ cartridge, the oldest item in my collection, is marked 'Prototype Use Only.' What a long way we have come since then! My collection illustrates that promoters found unique approaches to SPE sales: a game board with holes punched to hold SPE cartridges banded in either black or red for a round of checkers, or SPE columns packaged with Kool-Aid™ and instructions for separating red and blue dyes from this drink."

I am extremely grateful to everyone who contributed to this book. Several authors soldiered on through broken bones, new babies and, in one case, the near loss of his house in a brush fire! The whole process has been a wonderful learning experience for me. I am also indebted to the many who would-have-liked-to-but-couldn't, as well as to those who offered advice and encouragement.

I would especially like to thank Paul Wynne for going far beyond his initial duties by proof-reading several chapters and offering advice and insight on several others. Thanks are due to the staff at Marcel Dekker Inc., especially Brian Black, Anita Lekhwani and Linda Schonberg for their patience and expertise, and to Arnie, Dave, Doreen, Gwen, Kevin, Hung, Rob and the Helpdesk groups at Harbor City and Middelburg -- a pleasure doing business with you and thanks for your comments on the manuscript. Lastly, to my wife and Odette (who was still a theoretical consideration when this book was conceived), whose support and love make it all worthwhile, thank you.

Nigel Simpson

CONTENTS

CONTRIBUTORS

Steven A. Barker School of Veterinary Medicine, Louisiana State University, Baton Rouge, Louisiana

Rokus A. de Zeeuw Groningen Institute for Drug Studies, University Centre for Pharmacy, Groningen, The Netherlands

Jan Piet Franke Groningen Institute for Drug Studies, University Centre for Pharmacy, Groningen, The Netherlands

Michael Henry Beckman Coulter, Inc. Fullerton, California

Steen H. Ingwersen Novo Nordisk A/S, Maaloev, Denmark

Lynn Jordan Zymark Corp., Hopkinton, Massachusetts

Brian Law Zeneca Pharmaceuticals, Mereside, Alderley Park, Macclesfield, Cheshire, United Kingdon

P. Martin Zeneca Pharmaceuticals, Mereside, Alderley Park, Macclesfield, Cheshire, United Kingdom

Maria T. Matyska Department of Chemistry, San Jose State University, San Jose, California

Joseph J. Pesek Department of Chemistry, San Jose State University, San Jose, California

Colin F. Poole Department of Chemistry, Wayne State University, Detroit, Michigan

Salwa K. Poole Department of Chemistry, Wayne State University, Detroit, Michigan

Badrul Amini Abdul Rashid University of Surrey, Guildford, Surrey, United Kingdom

Seyyed Jamaleddin Shahtaheri University of Surrey, Guildford, Surrey, United Kingdom

Nigel J. K. Simpson Varian Associates Inc., Harbor City, California

Derek Stevenson University of Surrey, Guildford, Surrey, United Kingdom

Martha J. M. Wells Tennessee Technological University, Cookeville, Tennessee

I.D. Wilson Zeneca Pharmaceuticals, Mereside, Alderley Park, Macclesfield, Cheshire, United Kingdom

Paul M. Wynne Racing Analytical Services Ltd., Flemington, Victoria, Australia

1

INTRODUCTION TO SOLID-PHASE EXTRACTION

Nigel J. K. Simpson Varian Associates Inc., Harbor City, California
Martha J. M. Wells, Tennessee Technological University, Cookeville, Tennessee

I. THE SAMPLE PREPARATION PROBLEM

We have a sample in front of us. It is of unknown composition, but we know that it is complex, containing anywhere from a few hundred to many thousand chemical components. It is of uncertain physical form — perhaps it looks like a liquid, but it has solid particles floating in it. We think that a particular compound is present in that sample, somewhere and in some form, but we don't know how much is present, nor what form it takes.

And we have in front of us an attractive chromatogram of the compound of interest, our analyte, with another related compound that we will use as our measuring stick. We want to use the conditions given in the chromatogram, but the caption tells us that these compounds were injected in pure organic solvent at a concentration far higher than they would exist in our sample. How do we get our sample to this desirable state?

II. SOLID-PHASE EXTRACTION: WHAT IT IS AND WHAT IT DOES

Solid-phase extraction (SPE) is one of various techniques available to an analyst to bridge the gap that exists between the sample collection and the analysis step. Filtration, homogenization, precipitation, chemical reaction, solvent exchange, concentration, matrix removal, solubilization — these are just a few of the available tools that may be used individually or in combination to get the sample into a form compatible with the analytical instrument required for analysis. Solid-phase extraction is seldom used without other sample preparation steps, such as dilution or pH adjustment. However, as you will see in this and subsequent chapters, the action of performing SPE often simultaneously completes several other preparation goals. Moreover, SPE has, in several creative ways, been coupled with an analytical technique or another preparation method to enhance the benefits of each separate technique.

A. THE BASIC STEPS OF A SOLID-PHASE EXTRACTION

The simple and familiar practice of liquid/liquid extraction (LLE) is an excellent starting point from which to interpret SPE. In LLE the sample is agitated in the presence of an extracting solvent that is not miscible with the sample. When the sample/solvent mixture has settled after agitation, two layers of liquids are visible, one of which will contain most of the compound we are extracting. The shaking action has ensured that all parts of the sample come into contact with the extracting solvent. Compounds from the sample may pass into this extracting solvent and, given time, an equilibrium will be established between the two liquid layers. The equilibrium is described by the partition coefficient for the analyte, which is simply the ratio of concentrations for the analyte in the two liquids. A very high partition coefficient means essentially all the compound of interest will migrate into the extracting phase; a low coefficient means very little of the compound of interest has moved into the extracting phase. For most liquid/liquid extractions, properly chosen conditions will result in most of the analyte being found in the extracting solvent, implying that the partition coefficient has been maximized. This occurs when the analyte interacts better with the extracting solvent than with the sample matrix. In other words, the extracting solvent provides a better environment for the analyte. To complete our LLE we now separate the two liquid layers and keep one for further manipulation, such as concentration, using a rotary evaporator or a stream of dry nitrogen.

1. Retention

In place of an extraction solvent we shall substitute a solid surface. When our compound distributes between the liquid sample and the solid surface, either by simple adsorption to the surface or through penetration of the outer layer of molecules on that surface, an equilibrium is set up, just as it was for LLE. We can define that distribution by a coefficient, K_D, which indicates to us what fraction of the analyte has remained in solution and what fraction has adsorbed on or entered the solid phase. Strictly speaking, this distribution coefficient should be defined in terms of activities of the analyte in either phase. However, convenience dictates that concentrations are used and therefore

$$K_D = [\text{analyte}]_{sorbent}/[\text{analyte}]_{sample} \quad (1)$$

If this process occurs in a column packed with a sorbent into whose outer layer the compound distributes, then we are dealing with a system that is no longer a "batch" partition like LLE. Instead the process more closely parallels distillation and Equation 1 converts to

$$k = 1/(1+k') \text{ or } V_0/V_R \quad (2)$$

where V_0 and V_R are the void volume or empty space in the column and the retention volume respectively. If our compound is to be entirely trapped on this solid surface the distribution coefficient will be very large. So large, in fact, that chromatographic extraction, in contrast to elution chromatography, is best described as a pseudo-equilibrium process. We call the process whereby the analyte is completely adsorbed on the solid surface RETENTION. Chapter 13 in this book, titled "Matrix Solid Phase Dispersion," describes a novel and highly effective technique for getting compounds in a solid sample to retain on a solid surface. However, most Solid-Phase Extractions simply require a liquid sample to be passed through a bed containing sorbent particles onto which the analytes will retain.

2. Elution

Unless we can find a technique for identifying and quantifying the compound of interest while it is retained on the solid surface, we must find a way to remove and collect it. Part II of this text deals with some advanced and unusual techniques for removal from the sorbent and collection of our analytes, such as using a stream of supercritical gas to desorb the analytes. But in a typical solid-phase extraction a simple liquid does the job very well. When a liquid provides a more desirable environment for the analyte than the solid phase does, then the compound of interest is desorbed and can be collected in the liquid as it exits the SPE device. This is called ELUTION. It is characterized by a k' between the concentration of the analyte on the solid surface and in the eluting liquid that is very small.

One way to view the solid phase is as an intermediary between the sample and the elution solvent and this highlights a very important difference between LLE and SPE. An elution solvent may be used which is miscible with the sample in a solid-phase extraction, because the elution solvent and the sample never come into direct contact. Thus, our sample may be aqueous but our SPE elution solvent may be methanol, which is miscible in all proportions with water. Such a scheme, impossible in a LLE, is not only possible with SPE — it accounts for the majority of all solid-phase extractions!

3. Rinsing or Washing

During the retention step, many compounds in our complex sample may have been retained on the solid surface at the same time as our compound of interest. Likewise, at elution it is likely that some of these co-retained compounds will be eluted with our compound of interest. To minimize the interferences these undesirable compounds will create during the analysis stage, we may add one or more wash steps between retention and elution, to attempt to remove or rinse them out. Each wash step involves another distribution between the analyte and the co-retained species, the solid surface and the liquid that is passing over it. You will control each step by careful selection of the wash, elution, and sample loading conditions.

4. A Complete Solid-Phase Extraction

The entire process of the solid-phase extraction is represented in Figure 1. You can now begin to appreciate the potential complexity but also the power of SPE. Each of the steps shown can be controlled. Thus, you can select the sorbent type (the solid phase that does the extraction); you can manipulate the sample to enhance retention of one chemical species over another; you can select an elution liquid that has properties that are not just desirable to the compound of interest, but which may be convenient for your method of analysis or for subsequent sample handling; and in between

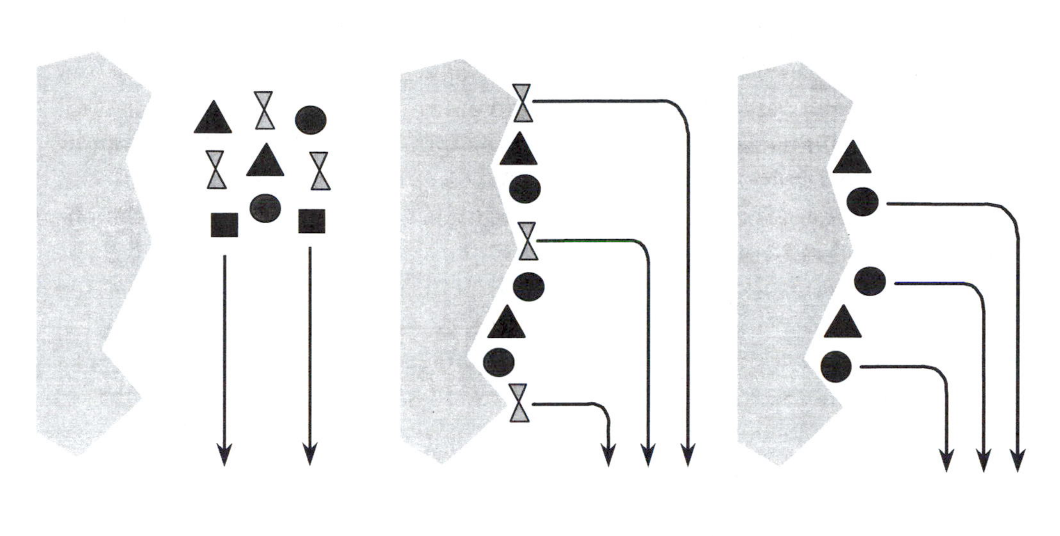

Figure 1. The three steps of a solid-phase extraction of a compound represented by ● after conditioning of the sorbent. A schematic view of what happens during sample loading, sorbent washing, and analyte elution. At each step some compounds will retain and others will be washed off or eluted.

you may use any number of wash steps to ensure that the final extract is of the desired purity.

III. A BRIEF HISTORY OF SOLID-PHASE EXTRACTION

Solid-phase extraction has been in use for thousands of years. Some scientists claim the first literature reference is to be found in the Bible, Exodus, Chapter 15, verses 24 and 25 to be exact (Riemon and Walton, 1970) even if the users were unaware of the science behind what they were doing! McDonald (1998) points out that the term is actually applied inaccurately since the extraction is not really performed by a solid phase but by a solid surface (adsorption), or at most a "meta-surface" like the organic layer of a C18 bonded silica. According to this strict interpretation, the extraction of volatile fragrances from rose petals by medieval laborers in Grasse, France, who performed the feat by embedding the petals in paraffin wax, counts as a solid-phase extraction. The first *modern* use of SPE in the colloquial sense probably employed animal charcoal (later use would involve diatomaceous earth and zeolites) to remove pigments from chemical reaction mixtures. In such cases the charcoal was filtered out of the mixture and discarded along with the compounds it had absorbed (remember that this was more a batch extraction than a column process, though).

Our goal, however, is not to discard compounds of interest but to collect them, preferably by concentrating them from a sample and removing those compounds we do not wish to analyze. Specifically, we want to strip the sample away from the analyte (compound to be analyzed) and to put that analyte into a small volume of a different liquid. In summary, we intend to achieve any or all of three things:

1) Concentration
2) Removal of unwanted molecules from the sample (clean-up)
3) Removal of the sample matrix/solvent exchange

In this sense SPE did not really become a scientific technique until the 1970s. The course of development as a sample preparation technique progressed from initial latency (prior to 1968) through three subsequent phases (1968-1977, 1977-1989, 1989-present). These have been distinguished by exponential increases in popularity, and advances in technology have, in turn, initiated changes in use of sorbent types and product formats. SPE was practiced for at least two decades before 1968, when applications using synthetic polymers (such as styrene-divinylbenzene resins) were first published in the literature.

The introduction of prepackaged, disposable cartridges/columns containing bonded silica sorbents, in October 1977, certainly made the procedure more convenient and initiated another phase of development. In May 1978, the technique was featured on the cover of *Laboratory Equipment*. Also that same year the first article using SPE on a bonded phase silica was published (Subden et al., 1978), which described the use of a Sep Pak™[1] C18 "cell" for the cleanup of histamines from wines. The introduction of stable, covalently bonded chromatographic sorbents and especially reversed-phase ones that improved recovery from aqueous solutions, opened up applications in the environmental, clinical, and pharmaceutical markets.

These improvements played a pivotal role in advancing SPE from a laboratory novelty into a widely used and accepted form of chromatographic extraction. In 1989, SPE discs (also called "disks" or "membranes") were introduced, initiating another phase in the development of Solid-Phase Extraction. These disc devices typically hold a sorbent en-

---

[1] Sep Pak is a trademark of Waters Corporation, Milford, MA.

meshed within a matrix of teflon or glass fiber, or between pads of glass fiber. The result of this design is, relatively speaking, a very short, very fat SPE cartridge.

Prepacked cartridges, columns or discs are now used in tens of thousands of laboratories around the world. Over the years, designs for housing SPE sorbents have ranged from "pipette tip" styles to plastic or glass mini-columns with polymer or steel or teflon frits. Some SPE devices are designed for the sample to be "pushed" through the sorbent while others are designed for vacuum use, enabling the sample to be "pulled" through. Clinical and pharmaceutical researchers working with the small sample volumes to which they are usually limited, drove the initial product development — small bed masses of sorbent, packed in cartridges or columns. Since the late 1980s, however, extraction membranes or discs, and "mega" columns containing several grams of sorbent, have grown in popularity, as a consequence of the increased environmental market for SPE.

Growth of SPE applications is directly attributable to the ease of use of SPE products. Automation and high-throughput via parallel sample processing using large manifolds or centrifuges have been critical to further growth in the 1990s.

A. THE OBJECTIVES OF SOLID-PHASE EXTRACTION

1. Concentration

In order to be able to measure the quantity of a compound accurately we need to concentrate it as much as possible. This will ensure the largest response from our detection system and will minimize errors in precision caused by background noise. Looking at our scheme shown in Figure 2, we can identify three ways in which SPE can help us.

1) Pass a large volume of sample through the smallest bed of sorbent that will completely retain all of our compound of interest.
2) Elute compounds of interest in the smallest volume of solvent possible.
3) Elute compounds of interest in a solvent that permits easy concentration, such as a volatile organic solvent.

In order to optimize each of these processes we need to know more about the capacity of a solid-phase sorbent and how that relates to sample type, sample pH, ionic strength, and so on. We shall also need to know how the properties of the different sorbent types affect retention and elution steps for our compound of interest. This will, in turn, allow us to identify

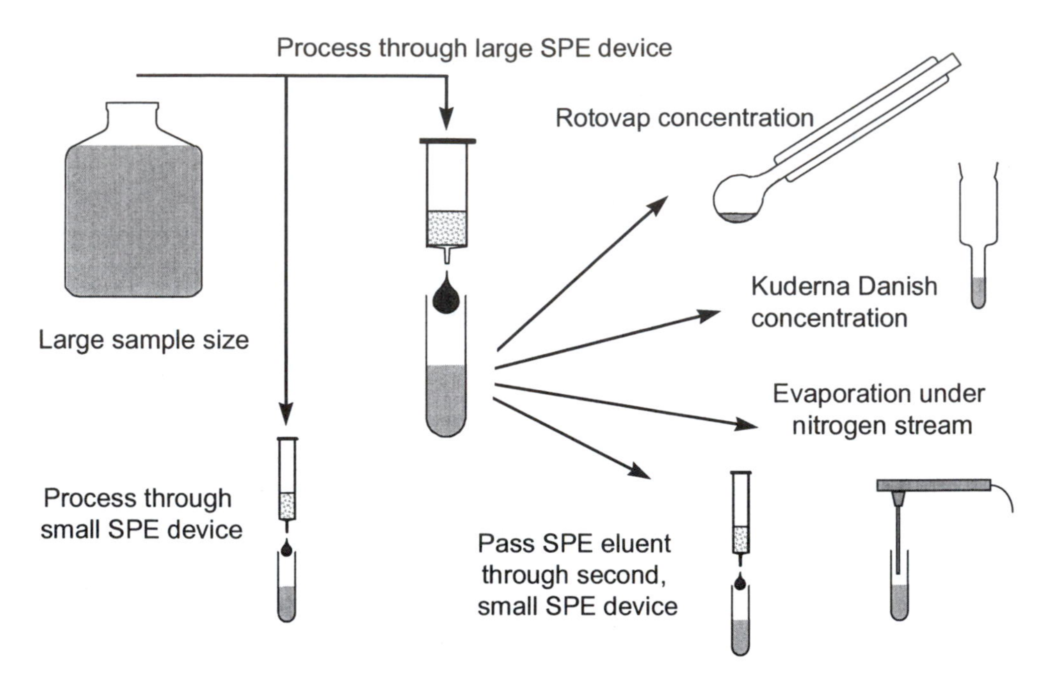

Figure 2. Illustrations of common ways to enhance concentration of an extraction that uses SPE. In many cases it is possible to achieve adequate concentration during just the SPE step. When sample size is not limited concentrations of several hundred fold have been achieved during the solid-phase extraction.

the solvents that give excellent elution while permitting further concentration if this is required. These topics will be dealt with in the chapter on method development and SPE theory.

Note that concentration requires the analyte to be retained on the sorbent bed — occasional methods utilizing SPE simply retain sample interferences on the sorbent bed and do not retain the analytes, thus precluding concentration during the SPE process.

2. Clean-Up

Concentration of an analyte is pointless if we cannot measure the analyte in a final concentrated solution. The most common reason for this is that the extracted sample contains interfering compounds. These are components of the sample that mask the analyte during the analysis (for example, when two or more compounds co-elute in a chromatographic experiment like gas or liquid chromatography). Sample preparation permits removal of such interferences before the analytical/separation step. Figure 3 shows a hypothetical example of how a chromatogram would look before and after a SPE clean-up. The cleaned-up extract gives clearly identifiable signals from the extracted components in the sample. Such a chromatogram is readily inter-

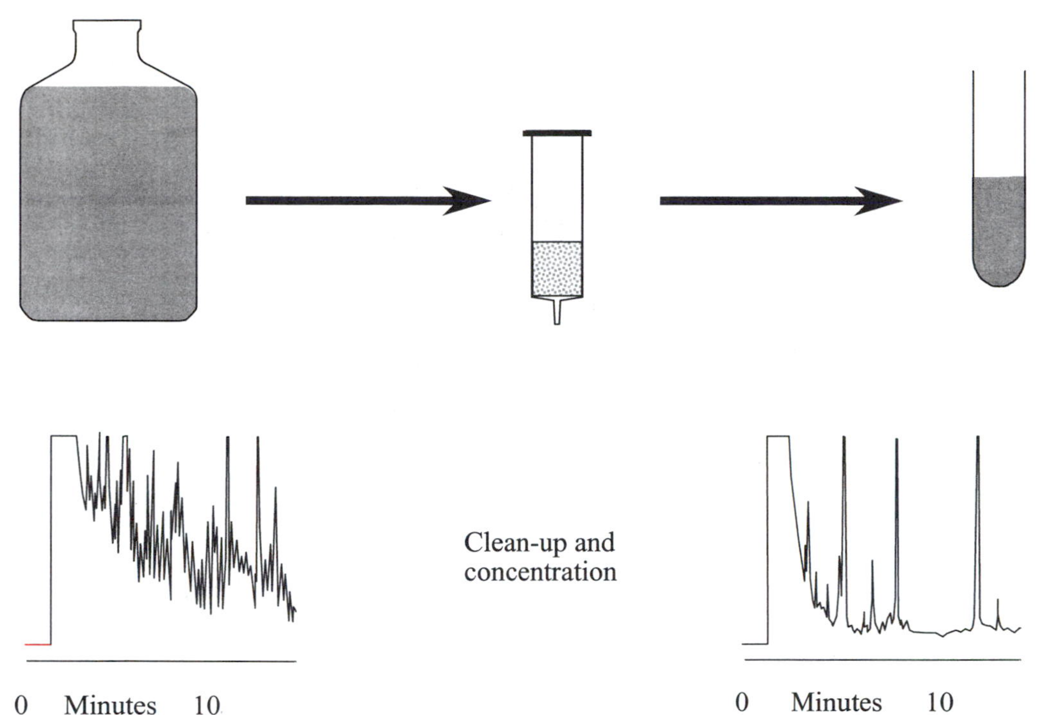

Figure 3. The aim of clean-up during the sample preparation step. A hypothetical pair of chromatograms are shown, "before SPE" and "after SPE."

preted and the quantities of desired analytes present in that sample are easily measured.

Remember that clean-up may be achieved either by retaining the analyte on a solid phase sorbent and washing out interferences or by retaining the interferences and washing out the analyte.

3. Sample Matrix Removal/Solvent Exchange

Many analytical instruments (e.g., gas or liquid chromatographs, nuclear magnetic resonance or infrared spectrometers) require that the sample to be analyzed is in a specific environment. For example, injection of an aqueous sample onto a gas chromatograph would ruin the delicate instrument. You would also, coincidentally, be lucky to see any chromatogram at the end of the experiment. In such cases, whether your sample is drinking water or whole blood, urine or face cream, you will need to remove the *sample matrix,* and convert your sample into a form compatible with the instrument to be used.

A significant advantage of SPE over LLE is that solvents that are miscible with the sample matrix, may be used to elute the analytes. Thus, a human plasma sample containing compounds that are to be analyzed by reversed-phase high-performance liquid chromatography (i.e., using an or-

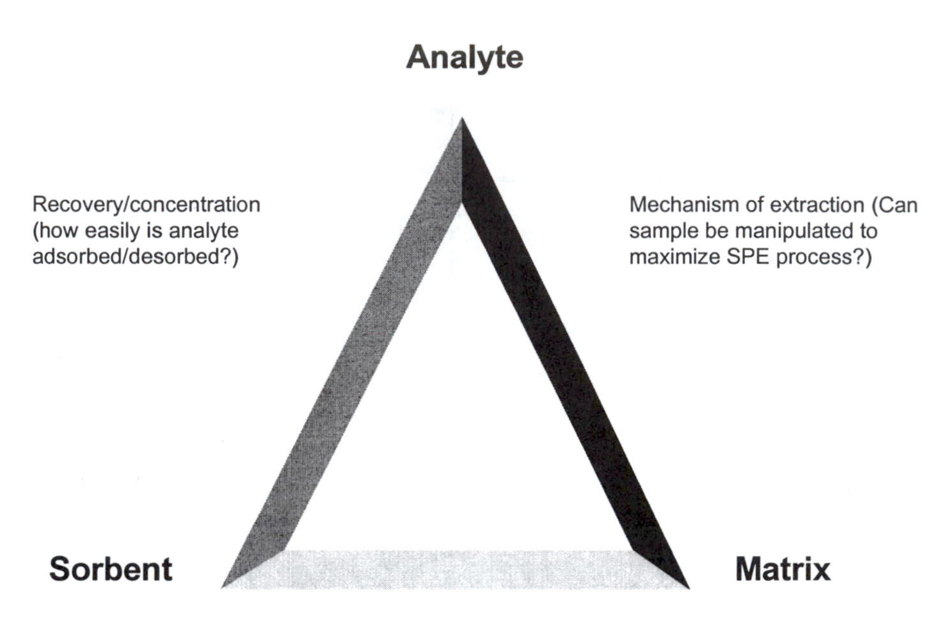

Figure 4. This triangle diagram illustrates the constraints imposed upon an SPE extraction. For example, an ideal choice of sorbent to maximize recovery/concentration may not be compatible with the sample type. Optimum clean-up may only be achievable on a sorbent that is incompatible with the sample matrix. We will address these concerns in later chapters on method development.

ganic/aqueous mobile phase) may be retained onto a SPE sorbent from the water-based sample, and can then be eluted with a water/organic mixture (e.g., H_2O/Methanol). This eluent can be injected directly into a reversed-phase HPLC system. The corresponding LLE commonly leaves the analytes in a water-immiscible solvent that must be dried down and the residue reconstituted in a suitable solvent before analysis may begin.

B. THE CONSTRAINTS THAT OPERATE DURING A SOLID-PHASE EXTRACTION

The three aspects of a sample preparation using SPE can be represented by a triangular diagram as shown in Figure 4. It may be possible to achieve all three goals: satisfactory clean-up, sufficient concentration, and efficient matrix removal in one simple SPE extraction. However, it is more common to have to compromise. To understand why, consider the following.

For a given sample matrix there is an optimum sorbent that will give excellent retention and excellent elution for one specific analyte. If we try to extract more than one analyte at a time (for example, a screen of drinking water for a range of environmental pollutants, or of urine for a parent drug

Table 1. Sorbent types, acronyms and extraction properties of common solid-phase extraction sorbents based on bonded silicas.

Bonded Phase	Acronym	Primary Properties
Octadecyl	C18, ODS	Non-polar
Octyl	C8	Non-polar
Ethyl	C2	Non-polar
Phenyl	PH	Non-polar
Cyclohexyl	CH	Non-polar
Cyanopropyl	CN (Cyano)	Non-polar/polar
Propanediol	2OH (Diol)	Polar/non-polar
Silica (unbonded)	SI	Polar
Alumina (unbonded)	AL	Polar
Florisil (unbonded)	FL	Polar
Diethylaminoethyl	DEA	Weak anion exchange/polar
Aminopropyl	NH2	Weak anion exchange/polar
Carboxyethyl	CBA	Weak cation exchange
Propylsulfonic acid	PRS	Strong cation exchange
Ethylbenzene sulfonic acid	SCX	Strong cation exchange
Trimethylammonium propyl	SAX	Strong anion exchange

and various metabolites of that drug) it is unlikely that one sorbent will be the best choice for every one of these compounds. The wider the range of analytes, the better the chance that several may retain or elute poorly from a given sorbent.

Another explanation can be found if we consider the process of retention and elution. To achieve good concentration we want to pass a large volume of sample through the extracting sorbent. We must have strong retention under these conditions to ensure all the analyte is retained — no analyte "breaks through" the sorbent bed. Strong interaction implies, however, that elution will not be easy, so a larger-than-desired volume of elution solvent may be required to fully desorb the analyte.

To help the analyst overcome such problems, manufacturers of SPE devices have developed an extensive range of sorbents — ones that utilize varying strengths of Van der Waals (non-polar), hydrogen bonding or dipolar (polar) or coulombic (ion exchange) interactions. A select range is shown in Table 1.

The variety of sorbent types adds complexity to SPE but it also adds power. Several chapters of this book are dedicated to understanding the properties and uses of these sorbents and making the best selection for your specific sample preparation need.

Table 2. Tracking SPE synonyms: A list of terms used

All-or-Nothing	Tiselius (1955)
	Morris and Morris (1976)
Extraction chromatography	Braun and Ghersini (1975)
Adsorption Trapping	Ogan et al. (1978)
Solid-Phase Extraction	Zief et al. (1982)
Liquid/Solid, Retention/Elution, Adsorption/Desorption, Stop/Go or Digital Chromatography	Yago (1984)
On/Off Chromatography	Wankat (1986)

IV. TERMINOLOGY

You may know SPE by another of its many synonyms (Table 2). For several years after the introduction of prepacked SPE devices, it was not even called "solid-phase extraction." To get around a lack of a name for the technique, authors had to resort to lengthy descriptions in the titles of their papers. In the early 1980s the terms "adsorption trapping" or "extraction chromatography" were commonly used, and they are good descriptions of the technique. Some authors avoided the use of the term solid-phase extraction for another reason, however — it was a trademark of one of the manufacturers of SPE devices. By the mid 1980s, "solid-phase extraction" was so well-established that authors were being advised by journal editors to use the term before attempting to publish an article.

Today, "solid-phase extraction" is the most widely used term. Using a standard term is certainly best, since the variety of terminology makes tracking the early literature on the subject difficult. Many of the terms in Table 2, (for example, digital chromatography) refer to the nature of SPE, in which liquid chromatographic mechanisms are "taken to their extreme" (Baker, 1984), i.e., $k' = 0$ or ∞.

V. LITERATURE AND APPLICATIONS

Many thousands of applications (Varian, 1996; Waters, 1995) have been published. Yet, however many publications exist, the analytical chemist will find there are never enough. For example, compound X may have been extracted from drinking water before, but for analysis by gas chromatography and not for HPLC; compound Y may have been extracted before from urine, but not from whole blood; compound Z may never have been extracted from any sample type.

Guidebooks for SPE, covering the basic elements of method development, exist (Baker, 1988; Waters, 1995). Two user's manuals, more detailed and comprehensive than these resources, have been written (Varian, 1993; Thurman and Mills, 1997), which describe the properties of bonded silicas, explain method development, and provide resources for optimization of SPE. These are excellent primers but they do not address theory and are not complete reference sources.

Many publications on the application of SPE on bonded silicas published since 1978 do not refer to the now commonly accepted term "solid-phase extraction." However, review of the technique's development is most easily conducted by a search of one of the many electronic databases such as are gathered by *Chemical Abstracts*, for those articles that do contain this term in their titles. Growth in the number of citations that meet this criterion has increased steadily since the first use of the term "solid-phase extraction" by the employees of the J.T. Baker Chemical Company (Zief et al., 1982 a, b; Wachob, 1983 a, b). The preponderance of SPE publications in the mid- to late-1980s dealt with clinical/pharmaceutical applications, as shown in Figure 5. Most applications in this field pertain to pharmacology but there are also plenty of applications from the toxicological field, specifically drugs of abuse confirmation, and many analytical methods for pharmaceuticals, biochemicals and mammalian hormones.

Publication of environmental applications of SPE has exploded during the 1990s (Figure 5). Environmental applications focus primarily on the analysis of water samples (i.e., drinking water, surface waters, groundwater, wastewater, and seawater). Additional environmental SPE publications are categorized in the areas of air pollution and industrial hygiene; waste treatment and disposal; inorganic or organic analytical chemistry; agrochemical bioregulators; and fertilizers, soils and plant nutrition. Applications of SPE for food and feed chemistry are also increasing. Tracking the field of SPE citations can be confusing since they are sometimes classified according to the analyte and sometimes by the matrix. Some manufacturers have introduced applications bibliographies (Varian, 1996; Waters, 1995) which permit an analyst to search for references that extract a specific compound, though it should be kept in mind that no manufacturer to date has been philanthropic enough to reference methods that use other manufacturers' products. The recent creation of easily accessible and fully searchable databases on computer disk, CD-ROM or on the World Wide Web will alleviate this problem.

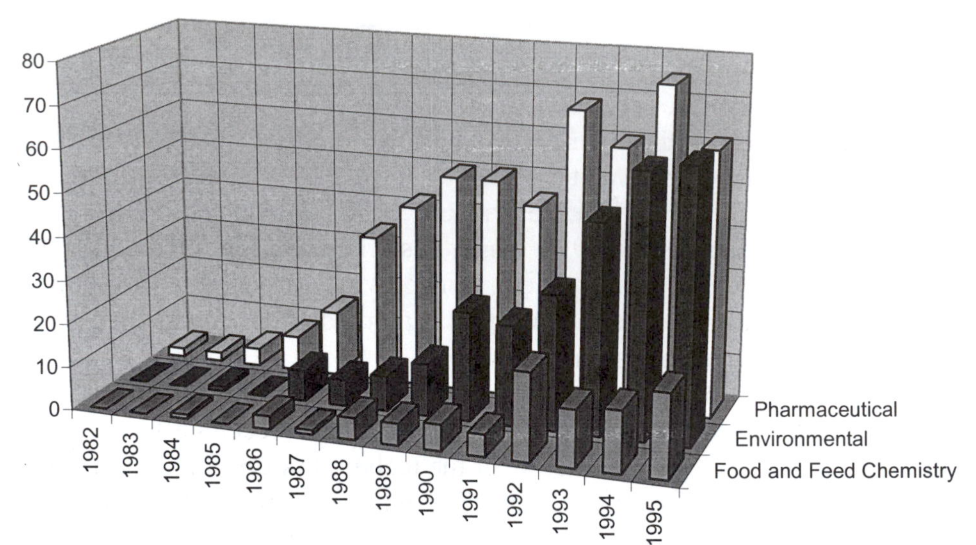

Figure 5. The number of citations having the term "solid-phase extraction" in the title of the article, during the time period 1982 through 1995, according to whether the type of application was pharmaceutical, environmental, or food and feed chemistry.

VI. AN OVERVIEW OF THIS BOOK

This text will provide you with answers to such questions as — "What is solid-phase extraction?" "What techniques exist for using SPE?" "What analytical instruments have been used with SPE?" This knowledge will allow you to develop your own solid-phase extractions successfully, and to troubleshoot and make that extraction rugged and reliable. It will also give you a thorough review of current SPE theory and models, and current SPE techniques and practices over a wide range of analytes, sample types, and extraction mechanisms. To ensure that this book is up-to-date and comprehensive, many experts have contributed sections or chapters on their specialized fields.

To allow this book to be relevant to the chemist at the laboratory bench and to the graduate student in analytical chemistry, it is structured in three parts. The first section gives the user the tools he or she needs to understand and successfully develop methods using SPE, with separate chapters that deal with clean-up and concentration — the two most important goals of a typical sample preparation step. The second part introduces theoretical and advanced practical concepts that allow the researcher to explore the underlying principles and the subtleties of SPE. The final part comprises "stand-alone" modules that deal with specialized aspects of SPE such as its application to drug-of-abuse screening or its application to therapeutic drug development. In addition, this section explores more general topics which, nonetheless, will not be relevant to every user of the technology, such as the coupling of SPE to a particular analytical technique, or the benefits and problems connected with automation of SPE applications.

In order to allow each chapter to be read separately, there is necessarily some overlap, most often in the introductory paragraphs. You are urged to read as much of this book as possible, however. The lack of unifying literature, or of a central "clearing house" like an industry panel in this field means that users, and therefore authors, have developed highly personal approaches to understanding and describing the use of SPE. We believe you will find something relevant in every chapter and section, even if the main topic of that chapter does not appear to directly connect with your field of study.

One other aspect of this book requires explanation. There is no chapter that explicitly covers method development. The reasons for this are two-fold. First, primers already exist (Varian, 1993; Baker, 1988) that cover generalized approaches to method development in a clear and concise manner. Second, the vast differences between sample types and the final analytical goals of each assay necessitate, in the authors' opinions, very different approaches to the SPE process. Thus, method development is introduced in terms of the goals (clean-up, concentration, final sample state) and the context (Chapter 8: systematic screening for abused drugs. Chapter 9: application of SPE to animal doping and metabolic studies) of the extraction process. Chapter 14, which describes automation of SPE methods, is the only chapter that prescribes general method development tools. We believe that enough universal elements of a sample preparation exist in this field, irrespective of the sample preparation goals and context.

Certain data and supporting information have been relegated to appendices to allow for a smoother flow of ideas. Thus, you will find a list of suppliers of automation equipment in an appendix, rather than in the body of the text.

VII. CONCLUSIONS

In this introduction we have identified what we wish to achieve by using SPE: sample matrix clean-up (in every case that SPE is used); analyte concentration; solvent exchange; and matrix removal. Trends in the use of SPE devices and their construction were explored and the history and terminology of SPE were reviewed. This background will allow us to perform literature searches more effectively. All topics covered here will be explored more fully throughout the book.

Solid-phase extraction is a fascinating technology and one that is central to the modern analytical laboratory. We hope you gain as much benefit and fun reading and working from this book as the editor and contributors had writing it.

REFERENCES

Baker (1984) Baker-10 SPE Applications Guide, Vol. 2, J. T. Baker Chemical Company, Phillipsburg, New Jersey, pp. 96-222.

Baker (1988) Solid Phase Extraction for Sample Preparation. J. T. Baker Chemical Company, Phillipsburg, New Jersey.

Braun, T., and Ghersini, G. (1975), Extraction Chromatography, Journal of Chromatography Library, Vol. 2, Elsevier, Amsterdam, The Netherlands.

Mc. Donald, P.D. (1998) SPE: Citius, Altius, Fortius, ISC '98, Rome. Poster # 112.

Morris, C.J.O.R., and Morris, P. (1976) Separation Methods in Biochemistry, 2nd edn., John Wiley and Sons, New York, p. 87.

Ogan, K., Katz, E., and Slavin, W. (1978) Concentration and Determination of Trace Amounts of Several Polycyclic Aromatic Hydrocarbons in Aqueous Samples. J. Chromatogr. Sci. 16: 517.

Riemon, W. And Walton, H.F., (1970) Ion Exchange in Analytical Chemistry, Pergamon Press, Oxford, UK.

Subden, R.E., Brown, R.G., and Noble, A.C. (1978) Determination of Histamines in Wines and Musts by Reversed-Phase High-Performance Liquid Chromatography. J. Chromatogr. 166: 310-312.

Thurman, E.M. and Mills, M.S. (1997) Solid Phase Extraction; Principles and Practice. John Wiley and Sons, Inc., New York.

Tiselius, A. (1955) Elektorphorese und Chromatographie von Eiweibkorpem und Polypeptiden. Angew. Chem. 67: 245.

Varian, (1993) Handbook of Sorbent Extraction Technology, 2nd edn., (Eds. Simpson and van Horne) Varian Associates, Inc., Harbor City, California.

Varian, (1996) Applications Bibliography, Varian Associates Inc., Harbor City, California.

Wachob, G.D. (1983a) The Solid Phase Extraction of Water-Soluble Vitamins from Multivitamin and Mineral Tablets. LC, Liq. Chromatogr. HPLC Mag. 1(2): 110-12.

Wachob, G.D. (1983b) Simultaneous Solid Phase Extraction of Vitamins A, D2, and E from Multivitamin Tablets. LC, Liq. Chromatogr. HPLC Mag. 1(7): 428-30.

Wankat, P.C. (1986) Large-Scale Adsorption and Chromatography, Vol. 2, CRC, Boca Raton, Florida, pp. 18-20, 34-39.

Waters, (1995) Solid Phase Extraction Applications Guide and Bibliography, 2nd edn., Waters Corporation, Milford, Massachusetts.

Yago, L. S. (1984) Bonded Phase Sample Preparation Technology: A Growing Trend. Am. Lab., 16: 4.

Zief, M., Crane, L.J., and Horvath, J. (1982a) Preparation of Steroid Samples by Solid-Phase Extraction. Am. Lab. 14(5): 120, 122, 125-6, 128, 130.
Zief, M., Crane, L.J., and Horvath, J. (1982b) Preparation of Steroid Samples by Solid-Phase Extraction. Int. Lab. 12(5): 102,104-9,111.

2

SPE SORBENTS AND FORMATS

Joseph J. Pesek and Maria T. Matyska, San Jose State University, San Jose, California

I. INTRODUCTION

As explained in the previous chapter, solid-phase extraction (SPE) is based upon principles which are common to chromatographic processes. Therefore, the sorbent materials used in SPE are in most cases very similar to those commonly found in typical HPLC columns, even though the format and the physical properties may differ to some extent between an HPLC column and an SPE cartridge. Since the mechanisms of retention are a consequence of the extracting sorbent, we will be better-placed to develop and troubleshoot an extraction if we know and appreciate the underlying properties of the sorbent. Most sorbents used for SPE are based on silica as the support material and therefore an understanding of the properties of silica will provide an insight into the selection of an appropriate SPE sorbent.

The essential features of porous silica will be described as well as the methods used to chemically modify the surface in order to achieve the properties desired for a particular extraction. A similar but briefer description will be provided for other types of sorbent materials that one might encounter in SPE extraction methods. Once the choice of sorbent material has been made, then the appropriate format for holding the solid sorbent in place must be selected (although in practice it is usually the format that is selected first). The factors affecting this choice will also be discussed in this chapter with emphasis on the intended type of application.

II. SORBENT MATERIALS

A. CHEMICAL AND PHYSICAL PROPERTIES OF SILICA

The silica which is used in most chromatographic applications and in typical SPE procedures is an amorphous porous solid with a surface area of between 50–500 m^2/g and pore diameters of 50–500 Å. The primary advantages of silica are its availability in a wide range of well-defined surface areas and pore sizes as well as its relatively low cost. While a very specific surface area and pore size (narrow distribution) are important in most HPLC applications, such stringent specifications are not necessary in SPE. These reduced requirements make the cost of the silica even lower and account for its popularity as a sorbent material in SPE.

The surface chemistry of silica (Anderson, 1995; Cox, 1993; Nawrocki, 1991) is dominated by the presence of hydroxide groups commonly referred to as silanols. Silica is an inorganic polymer with the general structural formula of $(SiO_2)_x$. It may be for-med by the polymerization of silicic acid usually made by acidification of sodium silicate. Another approach is to polymerize a tetra-alkyl orthosilicate in the presence of water and acid. This leads to the desired polymer and control of the experimental conditions determines particle size, pore size, and surface area. The ends of the polymer chain are thus hydroxyl groups that appear at the surface and are responsible for most of the chemical properties associated with silica. In a

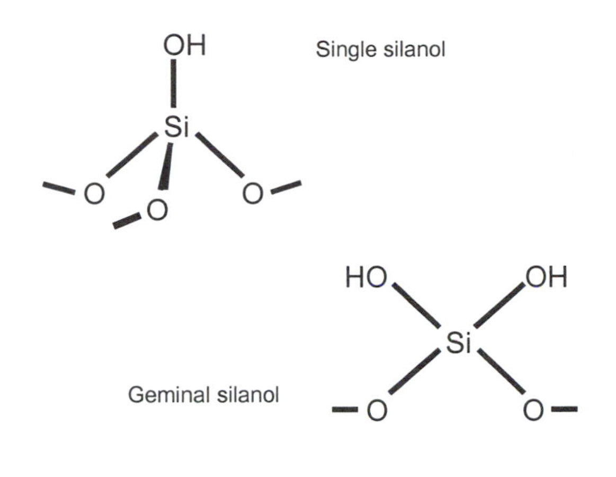

Figure 1. Isolated surface features of silica.

few instances, the polymerization leaves two hydroxyl groups on the same silicon atom. These sites are referred to as geminal silanols (Figure 1).

Overall the surface of silica is even more complex or heterogeneous because of several other factors. The silicon atoms which have a single hydroxyl group are found in two different types of structures on the surface. The first is the isolated silanol in which no other OH groups are physically nearby. The other possibility is the case in which two OH groups on adjacent silicon atoms are oriented in such a way to facilitate hydrogen bonding. These groups are usually referred to as associated silanols and are represented in Figure 2.

Finally because of the highly polar nature of the surface (silanols + siloxane linkages) there is a very strong tendency for silica to adsorb water. Therefore, the complete structure of the surface is a complex mixture of isolated, geminal, and associated silanols as well as a variable amount of adsorbed water. In general, the exact proportion of each of the various silanol groups depends on the method and conditions of the polymerization process.

Another variable which must be considered in describing any silica material is the presence of impurities. For silica made from sodium silicate, metals such as sodium, calcium, magnesium, potassium, iron, and aluminum are often found. If the concentration of aluminum is high, then appreciable alumina characteristics are part of the properties observed in the sorbent. Some of these same impurities, but usually at much

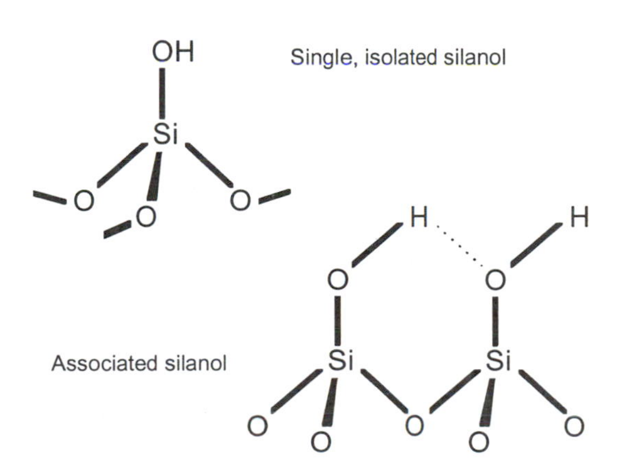

Figure 2. Isolated and associated silanols.

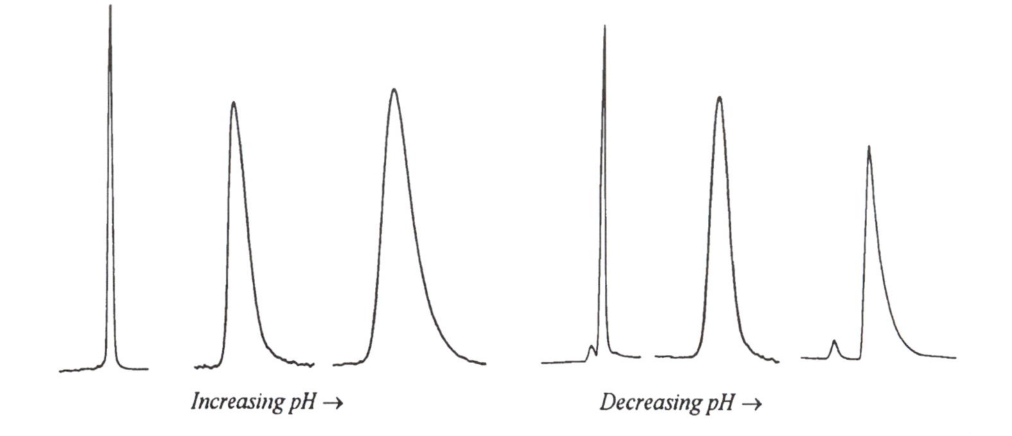

Figure 3. The effect of pH upon silanol activity during a liquid chromatographic experiment for an acidic (right) and basic (left) analyte.

lower levels, can even be found in silica made from the tetra-alkyl orthosilicates. The effects of impurities can be observed experimentally in a number of different ways. For example, the surface hydration, adsorption of water, and pH depend on the nature of the active sites. Most impurities in silica increase the amount of water that is found on the surface. Measurement of sample weight as a function of temperature usually shows that the gradient is higher in silicas with higher levels of impurities. A number of studies have shown a large variation in the apparent pH of silicas from different commercial sources. The values range from 3.8 to 9.5 (Nawrocki, 1991). It is believed that the variations are mainly due to the different synthetic methods of the manufacturers which leads to a wide range and level of impurities. The presence of trace impurities can also be seen in some simple chromatographic experiments. For acidic solutes on bare silica, peaks become broader as the pH is raised, and for basic solutes peaks become broader and tail when the pH is lowered. This is illustrated in Figure 3.

The exact value at which peak broadening and tailing appears, for both the acidic and basic solutes, depends on the type and level of impurity present in the silica being tested.

Overall, the underlying silica can have an important effect on a process like SPE. Masking of this silica surface by reacting it with another chemical species, a process known as bonding, will diminish the effects of the problems described above and will add the desired selectivity for either an HPLC separation or an SPE isolation. Table 1 gives a summary of some of the commonly recognized methods for the chemical modification of a silica surface. Method 1 is classified as an esterification process which involves

Table 1. Reactions for the chemical modification of silica.

REACTION TYPE	REACTION	SURFACE LINKAGE
1) Esterification	Si-OH + R-OH → Si-OR + H_2O	Si-O-C
2) Organosilanization	a) Si-OH + X-SiR$'_2$R → Si-O-SiR$'_2$R + HX b) Si-OH + X_3-Si-R → Si-O-Si$(OY)_2$-R + 3HX Y = Si or H	Si-O-Si-C
3) Chlorination followed by reaction of grignard reagents and organolithium compounds	Si-OH + $SOCl_2$ → Si-Cl + SO_2 + HCl a) Si-Cl + BrMgR → Si-R + MgClBr or b) Si-Cl + Li-R → Si-R + LiCl	Si-C
4) TES-silanization	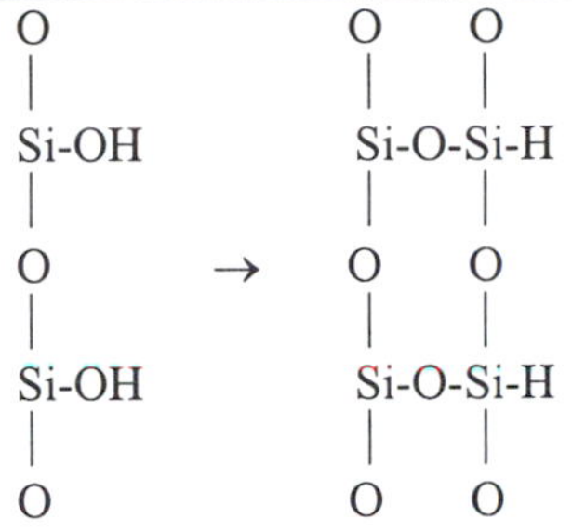 O–Si(OH)–O–Si(OH)–O → O–Si(O-Si-H)–O–Si(O-Si-H)–O	a) Si-H monolayer
then Hydrosylation	Si-H + SH_2==CH-R →Si-CH_2-CH_2-R	b) Si-C

reaction of a surface silanol with an alcohol. The resulting bonded moiety is attached to the surface via a Si-O-Si linkage.

While the reaction is simple, the resulting bonded material is hydrolytically unstable — it will decompose in even gentle aqueous buffers. Method 2 is the most common reaction for the modification of silica and is used in virtually all commercial sorbents for SPE and stationary phases for HPLC. The method is referred to as organosilanization and two types of reagents are generally used. The first type of organosilane contains one reactive group (X = Cl, OMe, or OEt), two small alkyls (R' = Me), and another organic group (R) possessing the desired properties of the modified sorbent. These materials are classified as monomeric-bonded sorbents because there is a single point of attachment to the surface. The other con-figuration for the organo-silane consists of three reactive groups (X) and the desired organic moiety (R). The product of the reaction of this reagent with silica is not only attachment to the surface but extensive cross-linking between adjacent bonded moieties. These materials are classified as polymeric bonded sorbents. The basic difference in structure between these two types of bonded materials is shown in Figure 4.

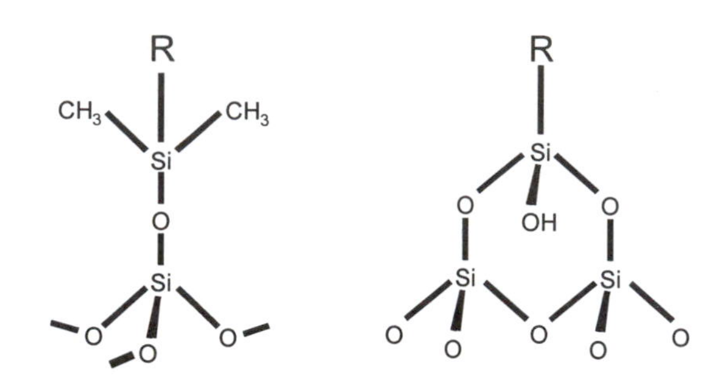

Figure 4. Representation of two approaches to bonding of silica surfaces.

The other methods listed in Table 1 result in monomeric-type bonded materials with a direct Si-C linkage between the surface and the organic moiety (R). Method 3 involves first chlorinating the surface with thionyl chloride followed by reaction with an organometallic such as a Grignard or organolithium compound. While the product possesses a highly stable Si-C linkage, the intermediate is very unstable requiring extreme care in keeping the system absolutely dry. Method 4 results in the same type of final product as Method 3, but utilizes reagents that not only tolerate water but actually require a small amount of it, thus greatly simplifying the synthetic procedure. Neither Method 3 or 4 are in common commercial use at the present time. Whatever reaction sequence is used to modify silica, it is the attached organic moiety (R) that gives the sorbent the desired properties for a particular extraction in SPE or separation in HPLC. Therefore the final sorbent material could be hydrophobic (R is only alkyl), hydrophilic (R contains polar functional groups like hydroxyl, cyano, or amine), or ionic (sulfonic acid, carboxylic acid, or amine).

Because of both the method of synthesis and the amorphous nature of porous silica materials, the product from any of the above reactions will always contain some unreacted silanol groups. As silica particles are formed, the material develops two types of pores. The first are those which constitute the main structure of typical sorbent silica and are usually referred to by the manufacturer when describing the physical properties of the material, which is typically in the range of 50-500 Å as stated above. These are often given the term mesopores. However, there are also present in all silicas pores with diameter of a few Å or less. These are generally referred to as micropores. The mesopores are much more readily accessible to reacting chemical species than the micropores. In addition, the geometry of the surface is such that as the size of the modifying reagent gets larger, steric interaction between adjacent organic moieties increases. Therefore, even in mesopores of large diameter, it is impossible to completely bond all of the free silanols which are present. These situations are shown in Figure 5. Fortunately, the silanols in the micropores are generally inaccessible to the

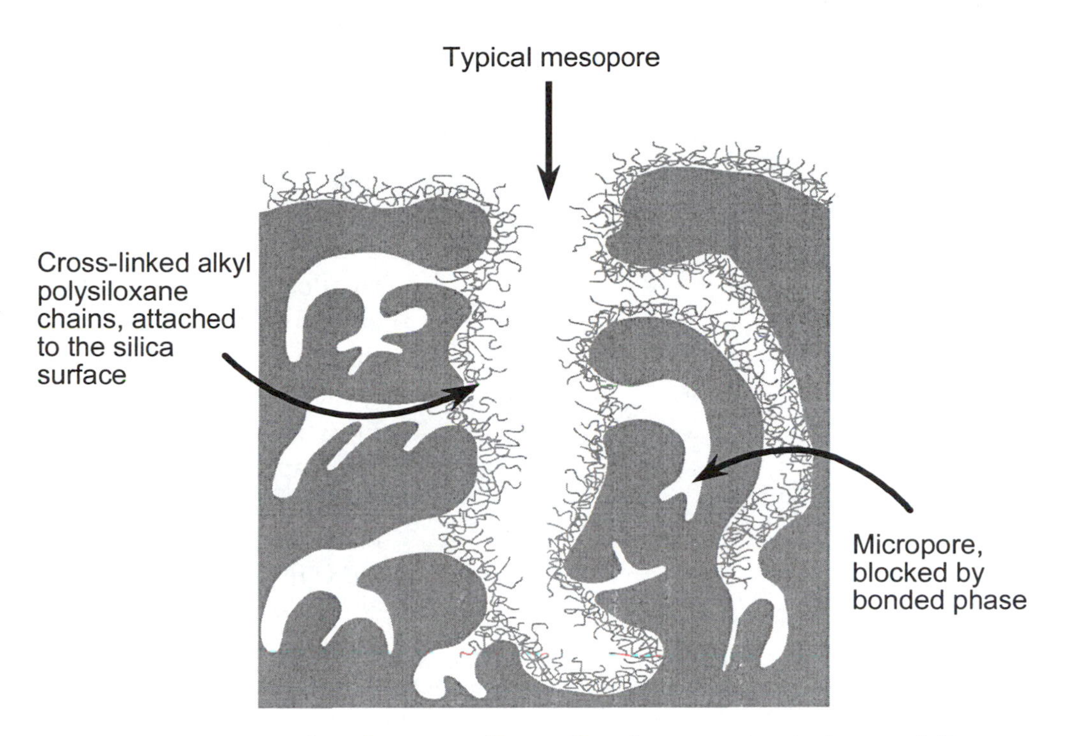

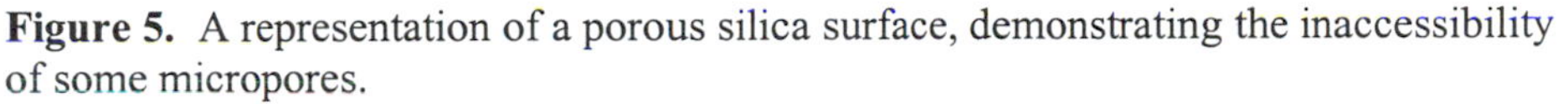
Figure 5. A representation of a porous silica surface, demonstrating the inaccessibility of some micropores.

analytes we wish to extract from a sample, as well as the bonding reagents. The problem of steric interference from the bonded moiety which prevents complete or nearly complete removal of all accessible surface silanols is illustrated in Figure 6.

The bonded organic group is somewhat mobile except for the attachment at one end. For a small R group the movement is very restricted so access to the nearby surface (silanols) is not hindered to any extent. However if R is large like octadecyl (C-18), then the movement of the alkyl chain prevents other organic moieties from reaching the surface, particularly if they are large also.

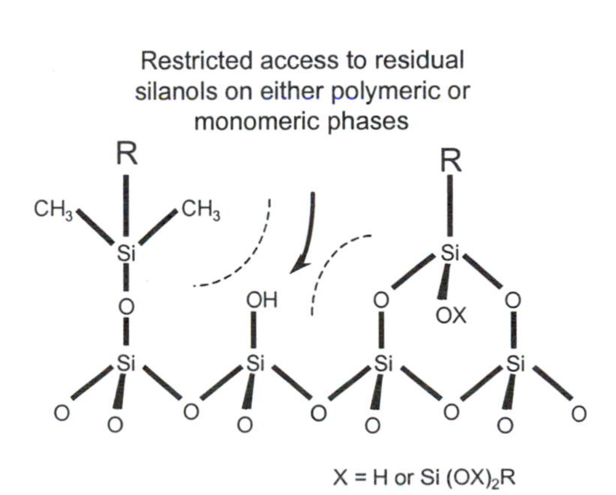

Figure 6. Illustration of the blocking effect of neighboring bonded-phase groups on adjacent silanols.

When larger bonded organic moieties (R) such as octadecyl (C-18) are attached

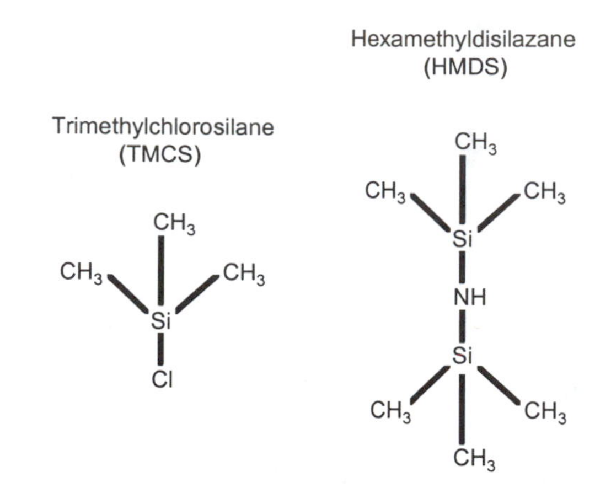

Figure 7. The structures of two common endcapping reagents.

to the surface, there will be a considerable number of free silanols remaining. In order to minimize the Si-OH groups, a secondary reaction with a smaller silane-type molecule follows bonding of the larger primary organic group. This reaction is referred to as "endcapping." The structures of two typical endcapping reagents are show in Figure 7.

In each case, these smaller entities are assumed to be able to reach some of the unreacted silanols between the larger bonded groups. HMDS hydrolyzes so that the bonding moiety is similar to that of TMS ($-O-Si(CH_3)_3$). However, no endcapping reagent is small enough to be able to reach all of the remaining silanols, as can be seen from Figure 6. When a manufacturer tells you they completely endcap their sorbents, don't believe them; "thorough" or "consistent" endcapping is possible — "complete" or "full" endcapping is not!

The remaining silanols are the same types as described above, i.e., isolated and associated. It is the isolated silanols that are by far the most adsorptive with respect to the analytes in SPE. They can strongly retain compounds with polar functional groups and make elution much more difficult. Bases in particular are most strongly retained since the silanol is an acidic group and acid/base interactions are similar to ion exchange (a more powerful interaction). While the absolute number of unreacted silanols is a primary factor in determining the retention of compounds such as bases by a modified silica sorbent, the type of silanol also plays an important role. Since the ratio between isolated and associated silanols varies with different sources, a material that has few isolated SiOH groups before modification is likely to have even fewer after bonding of the organic moiety and endcapping. Therefore, its behavior toward compounds with polar functional groups and in particular bases is generally more desirable. Why is this?

The problem that bases cause with silica-based sorbents in SPE can be best described by a chromatographic ex-ample. In HPLC, the presence of interactions between the solute and surface silanols is manifested by the appearance of tailing peaks as shown in Figure 8.

In Chromatography, the presence of tailing peaks limits resolution and causes poor quantitation. In SPE, a larger volume of eluting solvent is necessary to obtain good recovery of the analyte. This means that sensitivity is poorer because the analyte is more dilute or that an additional concentration step must be added to improve the lower limit of detection. The latter procedure results in more time to complete the analysis and adds additional sources of error to the method. This has implications for the user, since clearly one manufacturer's "endcapped C18" may have very different properties from another "endcapped C18." One significant difference between SPE and other chromatographic processes is that more techniques may be used to eliminate secondary interactions. As you will see in subsequent chapters of this book, secondary interactions can, if they are understood, be used to permit some impressive separations.

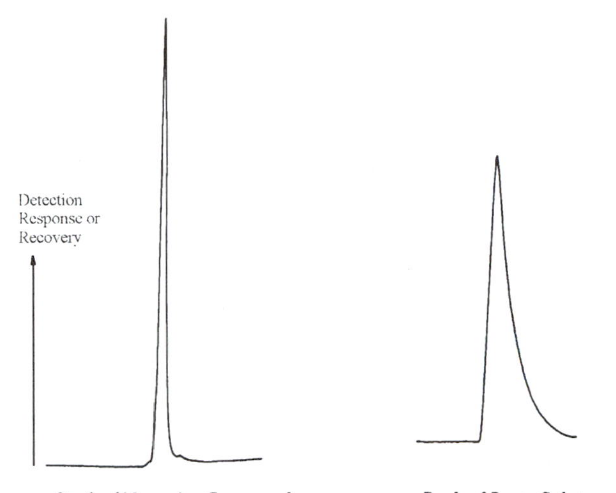

Figure 8. Tailing peaks – a consequence of silanol activity.

1. Conditioning of Porous Silica Sorbents

The presence of an appropriate bonded material with sufficient bonding density on the silica surface is the first requirement for being able to perform efficient solid-phase extractions. However, additional treatment of the material by the analyst is necessary in order to obtain high solute recoveries. The sorbent must undergo proper **conditioning** before the sample passes through the SPE device. Many reasons for this could be cited so to preserve clarity we shall focus on the most common conditioning process, where we condition a C18 sorbent prior to application of an aqueous or largely aqueous sample such as urine or drinking water. In its dry form, the C18 bonded-phase material is completely randomly oriented on the surface. If the sorbent is placed into contact with a completely aqueous sample, the environment surrounding the bonded organic moiety would be highly polar. Such an environment would be entirely incompatible with a C18 bonded phase. This situation would be equivalent to trying to mix water with a hydrocarbon; these materials are completely immiscible with each other and the result would be that separate layers of the two liquids form. The situation for bonded solid phase sorbents is illustrated in Part A of Figure 9. The environment around the bonded hydrocarbon moiety is polar (e.g., aqueous) in this case, so the organic group tries to minimize its exposure to the high polarity medium. It can accomplish this by forming clusters or aggregates among bonded groups that are close to each other. In such a configuration, the organic surface that is exposed to the solute as it comes in contact with the sorbent is very small. This arrangement of the bonded organic groups will greatly diminish the efficiency of the SPE process.

In addition to this, the majority of the pores of a dry C18 sorbent particle would be inaccessible to the aqueous sample because the sample is excluded by water's surface tension and viscosity. The pressure at which the sample could be forced into the pores can be calculated (Bouvier et al., 1997) if a number of assumptions are made. The conclusion, under all conditions, however, is that the surface tension of the sample would prevent penetration of the pores under normal conditions of use.

The situation can be remedied by use of a conditioning step before the sorbent is exposed to the sample. In this process, the non-polar SPE sorbent is first treated with an organic solvent such as methanol. While methanol is more polar than a typical hydrocarbon such as the octadecyl moiety, it is considerably less polar than water. With an environment of methanol surrounding the bonded organic moiety the need to minimize the surface area of the nonpolar group is less and the configuration becomes more like Part B. Under these conditions, the sorbent is more open and available for interaction with the solute.

The extent of conditioning (opening of the clustered structure in Part A) is a combination of the polarity of the conditioning solvent and of the bonded organic group. The ideal situation occurs (Part C) when the sorbent is completely open so that maximum interaction can take place between the solute and the bonded organic moiety. This configuration can result from use of a less polar conditioning solvent than methanol such as acetonitrile or even tetrahydrofuran. However, water miscibility is a critical requirement of the conditioning solvent when the sample to be applied is aqueous. Otherwise, when the sample is applied no penetration of the bonded phase or the pores will result — as if the sorbent had not been conditioned. Thus, good conditioning solvents in addition to those mentioned above are acetone, isopropanol, and in rare cases even ethyl acetate, whose limited water miscibility (3.3%) can permit penetration of the pore structure.

It is easier to achieve the structure illustrated in Part C if the bonded organic moiety is more polar than a hydrocarbon group such as octyl or octadecyl. Examples of more polar phases would be cyano, diol, or amino. In

A

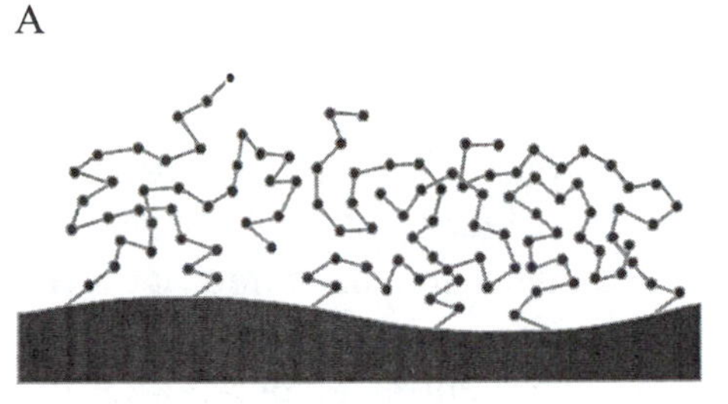

B

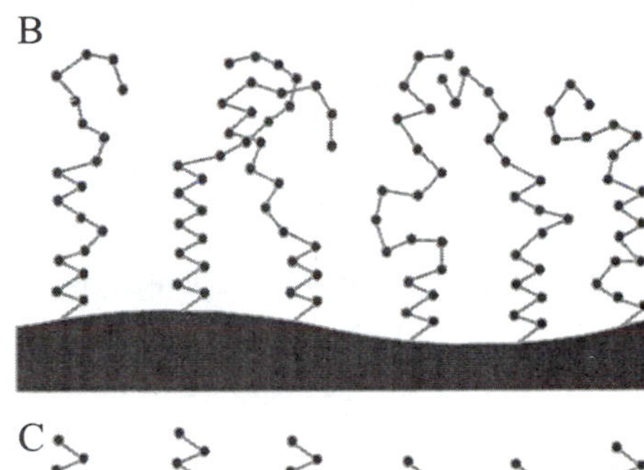

C

Figure 9. Three models for an octadecyl bonded silica, A) Without conditioning; B) Partially conditioned and; C) Fully conditioned.

this case, it would be necessary to decide which bonded group is most suitable for the extraction itself. Then, the conditioning solvent can be selected to achieve the most open structure possible (either Part B or C).

Another consideration is the extent and effect of unreacted silanols. While it is possible to reduce these to a very small number, no bonding and/or subsequent end-capping processes can eliminate all of these groups. Therefore, it is likely that some solutes may interact with these remaining silanols. The effect of conditioning on the activity of these Si-OH groups is not completely understood. However, as the surface becomes more open, these groups become more accessible to the analyte. For solutes with a moderately polar functional entity (hydroxyls, carbonyls, etc.) the Si-OH groups may serve as an additional point of interaction. This would result in a combination of hydrophobic and hydrophilic interactions as a means of retaining the solute on the sorbent. While the presence of silanols can be detrimental to the efficient extraction of solutes with strongly basic groups like amines, they can prove beneficial for molecules with less polar groups as a source of a secondary interaction. This mixed-mode of retention is often used in chromatography to enhance the separating ability of one class of compounds with respect to another. Consequences of the conditioning step are discussed in a later chapter, since the mode of conditioning will influence the role of silanols and hence the mode of retention and elution of an analyte.

A final comment on conditioning: Excess conditioning solvent, held between sorbent particles and in other void spaces in the SPE device, is undesirable. It is therefore common practice to apply a solvent similar to the sample (for example, if the sample is largely aqueous, an aqueous buffer

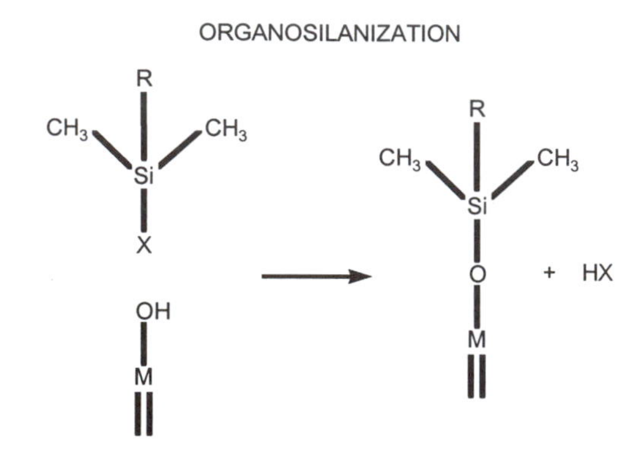

Figure 10. Reaction scheme for functionalizing a metal oxide surface like alumina.

can be supplied). The function of this liquid is to push excess conditioning solvent out of the SPE device and, in some applications, to equilibrate the sorbent at a specific pH or ionic strength.

B. OTHER OXIDE MATERIALS

The active sites for adsorption on silica are the hydroxide groups (silanols) which are present on the surface. Since other oxides also possess such active hydroxy groups on the surface, they have the same potential to act as a sorbent in SPE (Pesek, 1994). Oxides such as alumina, magnesia, zirconia, titania, and thoria have been tested as sorbents in HPLC and to some extent have been evaluated in SPE (Gillespie et al., 1992; Buser et al., 1992). Until recently, the main difficulty in using any of these oxides except alumina and high magnesia silica, for either HPLC or SPE, has been the commercial availability of the materials in suitable particle and pore sizes at a reasonable cost. Therefore, their ability to compete with silica as a sorbent is severely hampered in SPE where the cost of the materials must be considerably lower than in HPLC. One of the advantages that all of the above oxides possess over silica is their higher hydrolytic stability. While this property can have considerable importance in HPLC, the single use most SPE devices are subjected to makes hydrolytic stability a relatively unimportant factor and price a substantially more important one.

Because of the active hydroxides on the surface, the bare oxides can be used as a sorbent in normal phase type extractions. In order to increase the variety of applications for any of these other oxides, it is necessary to chemically modify the surface in a manner similar to that done for silica. The easiest and most direct method for the modification of all of the above oxides is via organosilanization. A general reaction for any oxide is shown in Figure 10.

By proper choice of the R group, the same range of hydrophobic/ hydrophilic interactions found on silica materials can be provided by these sorbents. Likewise, the presence of residual hydroxides on the surfaces leads to secondary interactions in these types of sorbents. In many cases, particularly alumina, these interactions are even stronger than those found on silica-based materials. Therefore, recovery can often be poor for strongly polar compounds, especially bases. This results in decreased recovery when compared to SPE on a silica sorbent. While in the future new advantages or solutions to some of the problems may be found, at present the use of other oxides or their modified counterparts is still limited.

C. GRAPHITIZED CARBON

Graphite is another type of amorphous solid which has been considered as a sorbent for SPE (Anderson, 1995). It is hydrophobic in nature because the predominant structure is planar sheets of carbon atoms with a hexagonal structure. The commercial process for making porous graphite materials results in particles with dimensions similar to silica materials. The surface can also have some functionality (presence of hydroxyl, carbonyl and acidic groups have been identified) so that it can have adsorptive properties or potentially be modified by bonding reactions. However, most of the uses to date have been based on the hydrophobic nature of graphitized carbon. As with silica, proper choice of solvents determine to a great extent the success of extracting a particular type of compound. One particular advantage of graphite is its improved performance with respect to basic compounds. In general, if the eluting solvent has a pH below the pK_a of the analyte the positive charge on the analyte helps to reduce the strong hydrophobic interactions and good recovery is obtained since there are no strong adsorption sites similar to those found on silica and other oxides. Similarly, it is effective for small, polar, acidic species, such as phenoxy acid herbicides (Di Corcia, Marchese and Samperi, 1989). In this case a basic eluent (pH above the pK_a of the analyte) is most effective. The disadvantages of graphite center on its strong hydrophobic nature. Poor recovery is usually found for high molecular weight and other hydrophobic analytes as well as those that are considered electron-rich such as aromatic species. The range of applications for graphite-based materials in SPE has not been fully explored but certainly as with oxides other than silica some specific analyses will be developed based on the unique properties of the material.

D. POLYMER-BASED SORBENTS

One approach to eliminating the problems of the highly active sites found on silica and other oxides is to use a completely organic material (Anderson, 1995; Hosoya, 1993, 1994, 1995). This can be accomplished by polymerizing a compound such as styrene or methyl methacrylate. In order to make the polymer useful for either chromatographic or SPE applications, it must be cross-linked with another olefinic compound. Some examples of cross-linking agents include divinylbenzene and ethylene dimethacrylate. Some examples of monomers and cross-linkers are shown in Figure 11.

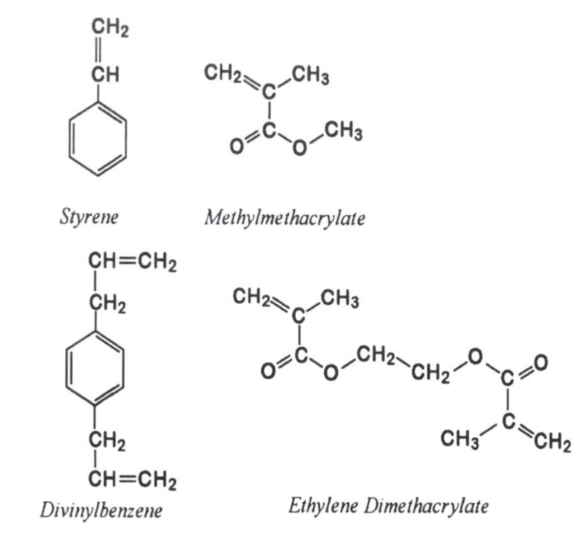

Figure 11. Examples of the monomeric species and their common cross-linking partners.

The result of cross-linking and control of other experimental parameters results in spherical beads which are suitable for SPE. The high degree of hydrophobicity of these polymeric materials generally gives them a large capacity. The proper choice of the main monomer and/or cross-linking agent can moderate the overall hydrophobicity of the completely organic sorbent. Another method of controlling the relative hydrophobicity is to mix the monomer (co-monomers) to achieve the desired properties. However, monomers must be closely matched in reactivity in order to ensure adequate incorporation of both molecules into the polymer structure. Mismatching can result in either physically and/or chemically heterogeneous materials.

At present the main advantage of polymeric materials lie in their ability to withstand pH extremes not achievable with silica-based sorbents. Their cost is generally higher than comparable silica materials. In addition, the range of selectivity for polymer-based sorbents is typically not as great as can be produced by chemically modifying silica surfaces. However, the types of polymer sorbents are rapidly increasing and new synthetic strategies will result in a greater range of applications.

Novel functionalized polymer materials have been introduced from various vendors since 1995 (for example, Oasis™[2] and Bond Elut™ PPL[3]) which offer modified non-polar properties. Claims made against these polymers are that they are less sensitive to drying out after conditioning (Oasis) and that they show enhanced retention of highly polar analytes such as phenols (Bond Elut PPL). Other manufacturers have introduced standard stryrene divinyl benzene polymers with extremely high surface areas (for example, LiChrolut™ EN[4]. These sorbents achieve, through maximizing the available retention volume of the sorbent, what Bond Elut PPL achieves through functionalization. Polymeric SPE sorbent use is expected to grow

[2] Oasis is a trademark of Waters Inc. Milford, Massachusetts.
[3] Bond Elut is a trademark of Varian Associates, Palo Alto, California.
[4] LiChrolut is a trademark of E. Merck, Darmstadt, Germany.

substantially as novel monomers, cross-linking species and the resulting three-dimensional sorbent structures are explored.

III. SOLID-PHASE EXTRACTION APPARATUS

The equipment required for SPE is generally very simple (Ho, 1989). There are two basic formats, cartridges and discs, which are described below in terms of their construction. The resulting chromatographic/SPE properties from each format are discussed in detail in Chapter 5. In some cases when very small sample volumes are required, the SPE system has been adapted to microscale operation.

A. CARTRIDGES

The standard SPE cartridge, also called a column (confusingly, in our opinion, since "column" is used in HPLC to describe a device with very different properties), is fabricated using a typical syringe barrel or a variant on the basic syringe barrel shape. Some examples are shown in Figure 12. In the most common form the sorbent bed occupies about one-third of the sy-

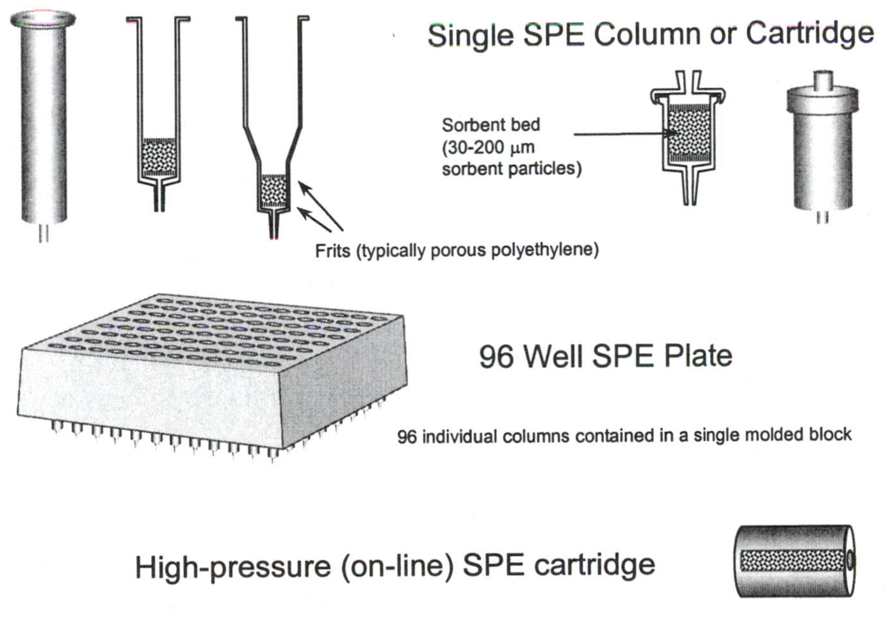

Figure 12. Examples of common and new formats for SPE devices.

ringe barrel volume. It is kept in place by porous discs, usually of polyethylene but occasionally of stainless steel or teflon. The remainder of the cartridge volume is used as the sample or solvent reservoir. The size of the cartridge depends on the amount of sorbent, which in turn is controlled by the expected mass of the analyte or of the contaminants in the sample that will also be extracted.

Cartridges are available with sorbent bed masses of 10 mg to 10 g or more. A usual approximation for determining the appropriate cartridge size is that the amount of analyte should be no more than about 5% of the sorbent weight — but as you will see in later chapters, this figure should be seen as an extreme approximation at best. Theoretical treatment of cartridge capacity is provided in Chapter 6. Practical approaches to optimizing bed mass are given in several of the following chapters.

Another factor that is important in any sample clean-up procedure involves the volume of solution that is necessary to detect a particular concentration of analyte. There are several things to consider, including the sensitivity of the analytical method and the amount of eluting solvent necessary to remove the solute from the sorbent. In some cases, the amount of solution containing the analyte may be limited, such as physiological fluids. In other instances such as wastewater or drinking water there is generally no problem getting as much sample as is needed. The sample is passed through the column using a vacuum to force the liquid over the sorbent bed.

For small columns (containing a few hundred milligrams of sorbent) it may take an hour or more to pass a liter of liquid with a vacuum of about 20" of Hg. For larger columns with several grams of sorbent the same volume may require 20 minutes or less, but a larger column will also require more eluting solvent. The final sample solution may have to be evaporated to bring the analyte into the concentration range detectable by the analytical method. So while a larger column can save time in the extraction step due to higher flow rates, some or all of this time could be lost if evaporation is necessary following elution. Therefore, the analyst will always have to exercise some judgment in developing an SPE method by evaluating sample size limitations and the amount (concentration) of analyte necessary when choosing the size (capacity) of sorbent bed.

B. DISCS

Three types of construction have been developed into commercial products: 1) the sorbent is contained between porous discs which are inert with respect to the solvent extraction process (essentially a very thin, wide SPE cartridge); 2) the sorbent is enmeshed into a web of teflon or some other inert polymer; 3) the sorbent is trapped in a glass fiber or paper filter. Since the latter configuration allows almost any type of sorbent to be housed, then the same choices of materials are available as are found in cartridge devices — silica, modified silica, other oxides and modified oxides, and polymers. However, commercial production of the wide variety of potential disc phases has lagged behind that of cartridges by a long way.

Disc devices can be placed into a syringe barrel or other housing, and the resulting hybrid "disc cartridge" may be used much like a regular cartridge, as shown in Figure 13. Other approaches to providing either low bed mass or fast sample throughput of a disc have included the laminating of a thin layer of small sorbent particles between two glass fiber pads, creating a disc-like packed bed and the incorporation of a high flow, large particle size sorbent with in-line filter in a wide-bodied cartridge.

The original benefit identified for extraction discs was the faster flow rate in the extraction step per unit mass of sorbent. The wider diameter of the apparatus results in a faster flow but still preserves good contact between the sample solution and the sorbent bed. At a vacuum of 20" of mercury one liter of solution can pass through the system in less than 10 min. This represents a significant saving of time particularly for laboratories that are required to perform hundreds or even thousands of analyses per week. However, recent developments of "high-flow" cartridges have reduced the advantage this format has in sample processing times (Pocci et al., 1997). The benefit of smaller elution volumes drove the development of the disc cartridge. Again, recent packed bed developments, most notably in the de-

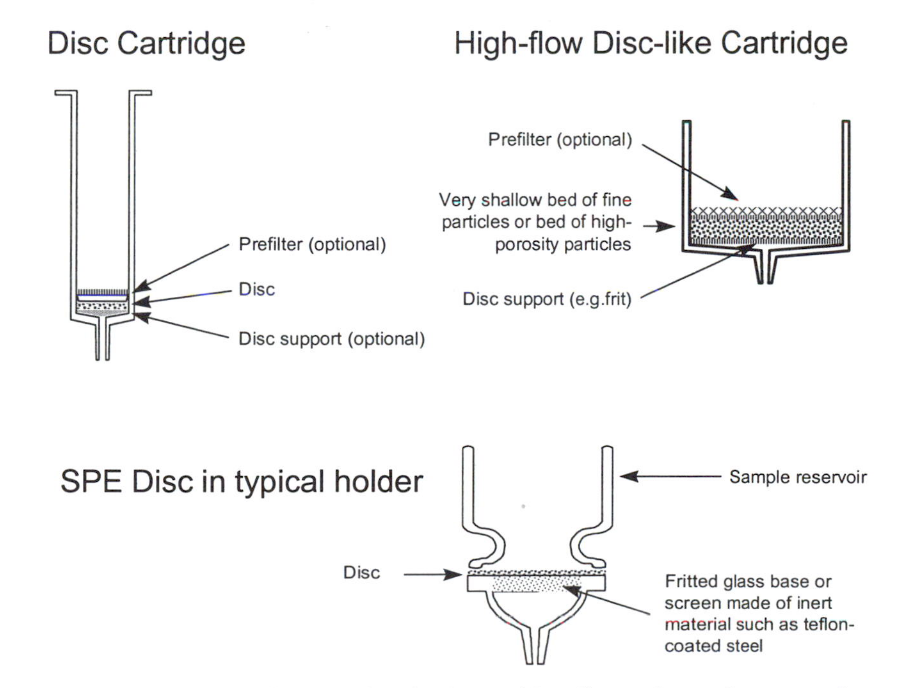

Figure 13. Disc devices, illustrated by both cartridge disc and membrane or glass fiber disc in a typical holder.

velopment of 96-well plate technology (see Figure 12) has reduced, though not eliminated this benefit.

C. MICROSYSTEMS

One of the primary reasons for the development of SPE as a replacement to ordinary LLE is the need to reduce solvent consumption. Miniaturization of SPE apparatus can reduce the use of solvents even further (Arthur, 1990; Louch, 1992). In addition, since many analytical techniques that are used for the samples prepared by SPE require very small volumes (HPLC, GC/MS, capillary electrophoresis, etc.), then the further reduction in sample size is compatible with the instrumental requirements.

Solid-phase microextraction (SPME) utilizes a fine fused silica fiber that can either be uncoated or modified with a thin layer of a selected coating (Arthur et al., 1992). This approach parallels ordinary SPE, which normally uses bare silica or a chemically modified silica. In theory both normal-phase (with bare silica or a polar modifier on the surface) or reversed-phase (with a non-polar modifier on the surface) extraction can be done. In all cases the fiber surface acts as the extraction sorbent. The main drawback to this approach is seen for extractions of aqueous samples, which account for the vast majority of applications encountered in SPME. Under aqueous conditions, a thin static layer of water will surround the fiber. Even vigorous stirring or sonication does not completely remove the thin water film. The presence of the water layer causes many solutes to diffuse slowly to the adsorbent surface. This can result in either longer extraction times or poorer extraction efficiency. One solution to the problem has been to sample the vapor phase above the sample rather than the aqueous phase directly (Zhang, 1993). If the sample solution is well stirred, then new solute material is continuously supplied to the headspace where rapid diffusion to the fiber results in efficient extraction. SPME has the potential to have high efficiency and precision, low-cost, and portability. This technique differs from SPE in various ways, most notably because it is based upon an equilibrium partition between the solid phase and the sample matrix. SPE is based upon disequilibrium. SPME will therefore be introduced only occasionally in this text, to draw distinction between it and regular SPE. A good reference for the SPME technique is provided by Zhang et al., (1994).

IV. CONCLUSIONS

Solid-phase extraction is still a developing field. Many of the advances in sorbent technology parallel those that are made in HPLC. For example, the types of sorbents are similar to stationary phases in HPLC. So where appropriate, the technology developed for HPLC is transferred to SPE for specific applications. Because the mode of operation for SPE is single use, sorbent ruggedness is often not as crucial a factor as it is in HPLC. It is also probable that more on-line applications and miniaturization are likely in the development of new SPE methods and sorbents. In this chapter we have looked at the base materials for sorbents, some of the many schemes used to manufacture bonded silicas, and we have gained an appreciation of how, despite comprehensive surface modification, we can still expect to see "secondary" interactions from the base material. We have examined common formats for SPE devices and summarized the effects these will have upon performance in an SPE experiment. The following chapter will address the implications of the formats discussed here, to enable the analyst to make best selection of the available sorbents and formats when developing an SPE application.

REFERENCES

Anderson, D.J., (1995) High Performance Liquid Chromatography (Advances in Packing Materials). Anal.Chem, 67: 475R.

Arthur, C.L. and Pawliszyn, J., (1990) Solid Phase Microextraction with Thermal Desorption Using Fused Silica Optical Fibers. Anal.Chem., 62: 2145.

Arthur, C.L., Potter, D.W., Buchholz, K.D., Motlagh, S., Pawliszyn, J., (1992) Solid-Phase Microextraction for the Direct Analysis of Water: Theory and Practice. LC.GC. 10 (9): 656-661.

Bouvier, E.S.P., Martin, D.M., Iraneta, P.C., Capparella, M., Cheng, Y-F. and Phillips, D.J., (1997) A Novel Polymeric Reversed-Phase Sorbent for Solid Phase Extraction. LC.GC 15: 152-158.

Buser, H.-R., Muller, M.D., et al., (1992) Enantioselective Determination of Chlordane Components using Chiral High-Resolution Gas Chromatography-Mass Spectrometry with Application to Environmental Samples. Environ.Sci.Technol., 26:1533-1540.

Cox, G. B., (1993) The Influence of Silica Surface Structure on Reversed-Phase Retention. J.Chromatogr. A, 656: 353.

Di Corcia, A., Marchese, M, and Samperi, R., (1989) Extraction and Isolation of Phenoxy Acid Herbicides in Environmental Waters Using Two Sorbents in One Minicartridge. Anal.Chem., 61: 1363-1367.

Gillespie A.M., et al., (1992) Isolation and Determination of Organo-Phosphorus and Organo-Chlorine Residues from Vegetable Oils and Fats using Solid Phase Extraction. Proc. California Pesticide Residue Workshop, Rancko Cordova, California.

Ho, J.S., Hodaklevic, P., Bellar, T.A., (1989) Solid Phase Extraction. American Lab., July 14.

Hosoya, K. and Frechet, J.M.J., (1993) Reversed-Phase Chromatographic Properties of Macroporous Particles of Poly(Styrene-divinylbenzene) Prepared by a Multi-Step Swelling and Polymerization Method. J.Liq.Chromatogr., 16: 353.

Hosoya, K., Kageyama, Y., Yoshizako, K., Kimata, K., Araki, T. and Tanaka, N., (1995) Uniformed-Sized Polymer-Based Separation Media Prepared Using Vinyl Methacrylate as a Cross-Linking Agent Possible Powerful Adsorbent for Solid-Phase Extraction of Halogenated Organic Solvents in an Aqueous Environment. J.Chromatogr. A., 711: 247.

Hosoya, K., Sawada, E., Kimata, K., Araki, T. and Tanaka, T., (1994) Preparation and Chromatographic Properties of Uniform Size Cross-linked Macroporous Poly(Vinyl p-tert-butylbenzoate) Beads. J.Chromatogr. A., 662: 37.

Louch, D., Motlagh, S. and Pawliszyn, J., (1992) Dynamics of Organic Compound Extraction from Water Using Liquid-Coated Fused Silica Fibers. Anal.Chem., 64: 1187.

Nawrocki, J., (1991) Silica Surface Controversies, Strong Adsorption Sites, their Blockage and Removal. Part I. Chromatographia, 31: 177.

Nawrocki, J., (1991) Silica Surface Controversies, Strong Adsorption Sites, their Blockage and Removal. Part II. Chromatographia, 31: 193.

Pesek, J.J., Tang, V., (1994) The Modification of Alumina, Zirconia, Thoria and Titania for Potential Use as HPLC Stationary Phases Through a Silanization/Hydrosilation Reaction Scheme. Chromatographia, 39: 649.

Pocci, R., Dixon, A., Nau, D.R., Nguyen, H and Constantine, F, (1997) Optimization and Redesign of Solid Phase Extraction Columns for the Determination of n-Hexane Extractable Materials (Oil and Grease) for EPA Method 1664. 1997 Pittsburgh Conference Book of Abstracts, Presentation number 648.

Zhang, Z. and Pawliszyn, J., (1993) Headspace Solid-Phase Microextraction. Anal.Chem., 65: 1843.

Zhang, Z., Yang, M.J. and Pawliszyn, J.B., (1994) Solid-Phase Microextraction. Anal.Chem., 66: 844A.

3

THE SAMPLE MATRIX AND ITS INFLUENCE ON METHOD DEVELOPMENT

Nigel J. K. Simpson Varian Associates Inc., Harbor City, California
Paul M. Wynne, Racing Analytical Services Ltd., Flemington, Victoria, Australia

I. INTRODUCTION

The greatest challenge facing the analyst who must develop a solid-phase extraction (SPE) procedure is to resolve the problems posed by the sample matrix. As shown in previous chapters, identification of the molecular characteristics of the analyte such as its potential for ion exchange or non-polar extraction, is quick and simple. The analyte's predominant liquid chromatographic properties can be estimated simply by looking at its structure and solubility in various solvents. For complex molecules, particularly those that have several functional groups that may interact with a sorbent surface, additional information such as previously reported HPLC conditions may also be valuable. Often such conditions have been derived by empirical means and cannot be readily predicted on purely theoretical grounds. If chromatographic properties were the only input variable affecting the outcome of an extraction, a simple decision tree would help us to quickly get to an optimum choice of sorbent and solvents (Figure 1).

A problem arises in that this approach fails to take into account a potentially bigger variable. Intersample matrix variations, even when constrained within a very narrow definition, may be of such significance to extraction efficiency or method performance that unacceptably high variations in recovery may result. To a lesser extent, sample volume also influences the extraction performance. The combined effect of sample type and volume should therefore influence the choice of extraction mechanism and all

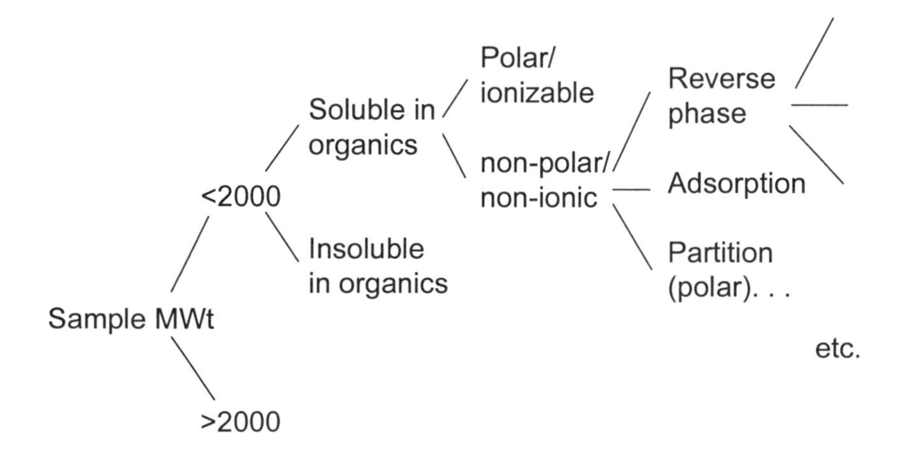

Figure 1. A traditional decision tree approach to method development.

aspects of the SPE process, including the selection of a sorbent and washing and elution solvents. The consequence of this can be appreciated by a quick survey of the application bibliographies published by the leading SPE manufacturers[5] where the diversity of sample type partially explains why twenty or more applications may have been developed for the same analyte. This is particularly the case for compounds such as therapeutic agents and agrochemicals, for which analyses may have been reported in a wide variety of matrices. Importantly, not only does the matrix vary from pure compound through the dose form or commercial product to residual levels in a body fluid, plants, wastewater, soil or other media; each example reflects different concentrations of the target analyte in the presence of vastly different matrix interferences.

The goal of this chapter is to explore and systematize the effects of the sample matrix upon the choice of extraction mechanism and the resulting application.

II. ANALYSIS OF THE MATRIX

A. EFFECT OF MATRIX TYPE AND CONTENT

In this section a single illustrative application has been chosen. The ideas introduced here will underpin the decisions made with respect to extraction mechanisms and conditions in all the subsequent sections and referenced applications.

In the example, our goal is to extract atrazine from each of three different matrices. These are, respectively, seawater, corn oil, and soy beans. Inspection of atrazine's structure, solubility data, and acid/base properties (much of this information may be found in references such as the *Merck Index* or the *U.S. Pharmacopoeia*) suggest that it can be extracted by non-polar, polar, or ion exchange mechanisms. Indeed, this is the case and applications demonstrating each of these principle mechanisms are shown in Table 1.

Two of these samples require some pre-extraction work-up or manipulation — a topic that is discussed in more detail in Section III. The immediate question is; "Why is one mechanism favored over another, resulting in three different applications for the same analyte."

---

[5] For example, application lists provided by Varian Inc. (Palo Alto, CA) or Waters Inc. (Milford, MA).

Table 1. Three contrasting matrix-dependent extractions of the same analyte.

APPLICATION	METHOD
Atrazine from sea water using non-polar (C18, 100 mg) extraction	1. MeOH and water conditioning 2. Apply 20 mL of seawater 3. Wash with 2 mL water 4. Elute with 1 mL MeOH
Atrazine from soy beans using ion exchange (SCX, 100 mg)	1. Homogenize soy beans in AcCN. Filter homogenate 2. Dilute 5:1 with water/1% AcOH 3. MeOH and water/1% AcOH conditioning 4. Apply sample 5. Wash with 1 mL water/1% AcOH, 1 mL AcCN, 1 mL 0.1M K_2HPO_4 6. Elute with 1 mL AcCN/0.1M K_2HPO_4 1:1 (v/v)
Atrazine from corn oil using polar (2OH, 100 mg) extraction	1. Dilute corn oil 1:10 with hexane 2. Hexane conditioning 3. Apply sample 4. Wash with 1 mL hexane 5. Elute with 1 mL MeOH

Atrazine has a solubility in pure water of 70 ppm (parts per million) at 25°C. In a seawater sample at ambient temperature this solubility is expected to be lower still. Thus, atrazine is unlikely to be found in excess of a maximum concentration of a few micrograms per milliliter. In the same sample we would find inorganic salts present at thousands of times this level. Let us assume we have tried to extract atrazine by ion exchange. Even if atrazine were to possess selectivity high enough to compete with inorganic ions for the ion exchange sites on the sorbent (unlikely, given the far greater numbers of inorganic ions) the final eluent would hardly be clean. Some of these co-retained species would survive even a vigorous washing procedure to appear in the extract and interfere with the analysis. The polar nature of seawater also precludes the use of a polar extraction sorbent under normal conditions. Therefore, we are left with the non-polar extraction as the only viable mechanism, since non-polar sorbents are both compatible with the matrix and show little or no affinity for the major matrix impurities.

Similarly, it would be inappropriate to apply non-polar or polar mechanisms to a soy bean matrix as soy beans contain a high level of fatty (non-polar) species and carbohydrate (polar) species in addition to a variable water content, depending upon the sample's state of hydration. Corn oil, consisting mainly of triglycerides and some free fatty acids, would also be a matrix that would be inappropriate for non-polar extraction. As our objective in selecting a SPE sorbent is to maximize the retentive differences between the analyte and other matrix components, we may be tempted to again consider ion exchange. However, in the oily matrix, which requires dilution in an organic solvent to reduce its viscosity, atrazine is in a neutral form. Thus a 2OH sorbent (employing a polar extraction mechanism) is found to provide the best medium for separation while showing little or no affinity for the triglycerides that comprise the bulk of the sample.

These concepts will be more thoroughly and explicitly developed in later sections of this chapter.

B. A COMPARISON OF TWO MATRIX TYPES

The sample matrix components in the above example were considered in a simplistic manner to explain, *a postiori*, the choice of extraction mechanism. The differences between sample types were also emphasized. In each case the matrix was either simple (corn oil or seawater) or easily characterized or well defined (soy beans). To see how we can select, *a priori*, an extraction mechanism or even "borrow" an entire extraction from another application for the same analyte residing in a different matrix, consider the difference between two very different matrix types: green plants and urine.

Urine is a filtrate generated by the mammalian renal system, which by its very nature is comprised of water and water-soluble species. It contains metabolic waste products or energy exhaustion compounds (which are generally highly oxygenated species) and compounds that have been excreted because they were recognized as foreign to the body. Examples of the former class are the aryl-alkyl carboxylates and ureates while typical examples of the latter include drugs, pesticides, and low energy plant components derived from the diet.

Urine frequently forms precipitates on standing, but these are not the result of an initial solids content. Rather, they are the result of changes in the pH of the urine, hydrolysis of soluble conjugates or the result of bacterial putrefaction that yields solids directly or causes a change in pH. Gelatinous solids may also be formed as a result of a change in temperature as the urine cools from body temperature to storage temperature.

Plant material is a solid structure comprised of a fibrous framework of cellulose and some fatty material, supporting an aqueous matrix which provides both structural and transport properties to the plant. In adapting SPE methods developed for urine to the extraction of functionally similar compounds from plants, the plant material must first be transformed into an aqueous phase that may be considered as analogous to the urine without removing the target analytes with any separated solid material.

C. EFFECT OF THE SAMPLE VOLUME

Unlike LLE or other separations based upon achieving an equilibrium partition of analyte between two phases, an increase in sample volume does not automatically result in the need for a larger sorbent bed. However, the practical constraint of time (the desire being to complete the extraction as quickly as possible) usually leads the analyst to reach for a larger, faster SPE device when the volume of sample to be extracted is increased. The presence of particulates in the sample and the risk of co-extracted species exceeding the sorbent bed's capacity also drive analysts towards larger bed mass SPE tubes or discs. Theoretical values of minimum bed masses required for complete analyte extraction and even experiments conducted on spiked surrogate samples, such as those by Hennion et al. (1996), may be misleading and should not be used without careful testing. This is particularly true when blank samples are selected from the routine sample stream. Inevitably, the blank sample is picked because of its good flow characteristics and lack of interferences and suspended material. In such circumstances, the blank sample is rarely a good representation of the worst samples to be extracted although it is these that are likely to cause matrix related failure of the method.

III. NEGATION OF ANALYTE/MATRIX INTERACTIONS

While this section is not an exhaustive account of all the techniques in use or available for releasing a molecule associated with another matrix component, it does cover the most commonly encountered problems and techniques for solving them. Later chapters will deal with some of these topics such as protein binding, adherence of analytes to cellular debris or environmental particulate matter, more specifically and in greater detail.

A. PROTEIN BINDING AND BIOLOGICAL PARTICULATES

Protein binding is invoked as the cause of low recovery when an analyte that recovers well from simple, surrogate samples during method development, yields poor results when spiked into real matrix blanks. A review of SPE methods published in the early 1980s (refer to Chapter 1) shows that a high proportion of these extractions were carried out on biological matrices (primarily urine, serum, or plasma). The problem of protein binding was therefore identified early in the evolution of the SPE field and strategies were developed to mitigate or eliminate its effects. For example, the first edition of *The Handbook of Sorbent Extraction Technology* (Analytichem International, 1985) lists techniques such as precipitation or protein denaturing. Whichever technique is used two goals must be achieved:

1) To completely break up the analyte/protein interaction(s).
2) To release the analyte into the liquid part of the sample.

The first requirement is necessary because a bound molecule will not show the same properties towards adsorption as a free molecule. In fact, if the target analyte is bound to a much larger species then any attempt to identify an extraction mechanism for the free molecule may be totally wasted. This is usually the case when the analyte is a small drug and the binding molecule is a protein (Figure 2). The second requirement is especially important if the sample is to be filtered or centrifuged before it is extracted to ensure that the target analyte is not lost from the sample. Even if the entire sample, including the precipitate or suspended material, is to be loaded onto the SPE cartridge, it is not desirable to have the target com-

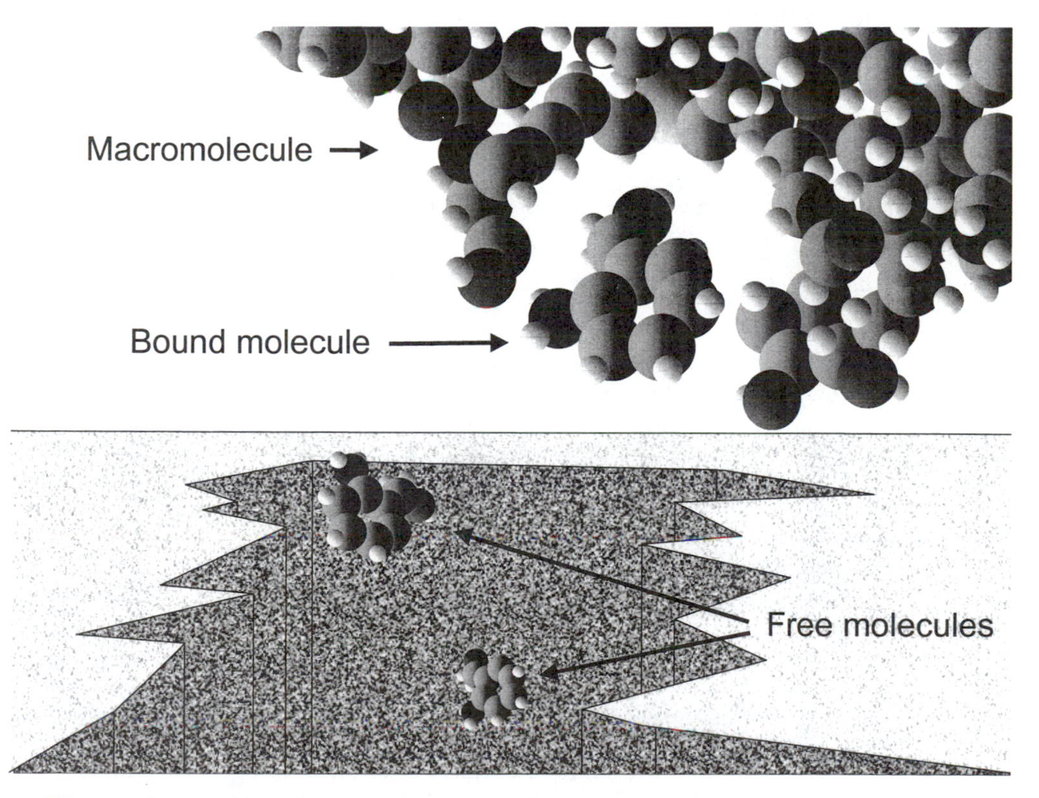

Figure 2. Illustration of the mechanism by which proteins may exclude small molecules bound to them from the pores of a sorbent.

pound associated with a precipitated protein, as the sorbing mechanism is unable to operate in such circumstances and poor recovery may result. Thus, techniques like protein precipitation using lead acetate or zinc sulfate, or even simple pH adjustment, should be used carefully to avoid the co-precipitation of an analyte.

Addition of chaotropic agents like guanidine hydrochloride or urea could also be problematic if these species, present in high concentrations, can interfere with the extraction mechanism. Even the addition of a denaturing or precipitating solvent like acetone, acetonitrile, or methanol must be done with consideration for the extraction mechanism, especially if non-polar sorbents are to be used.

One of the few attempts to compare these techniques (Chen et al., 1992) demonstrated that for a range of acidic, basic, and neutral drugs extracted on a combination non-polar/cation exchange cartridge, recovery was generally highest when the sample — whole blood — was diluted in a buffer and subjected to a physical denaturing (sonication) rather than a chemical one. The presence of cellular material in this case complicates an interpretation of the role protein binding may have played in reducing drug recovery from the untreated sample. Select data from these extractions are shown in Figure 3. The lower recoveries of analytes from samples prepared using tech-

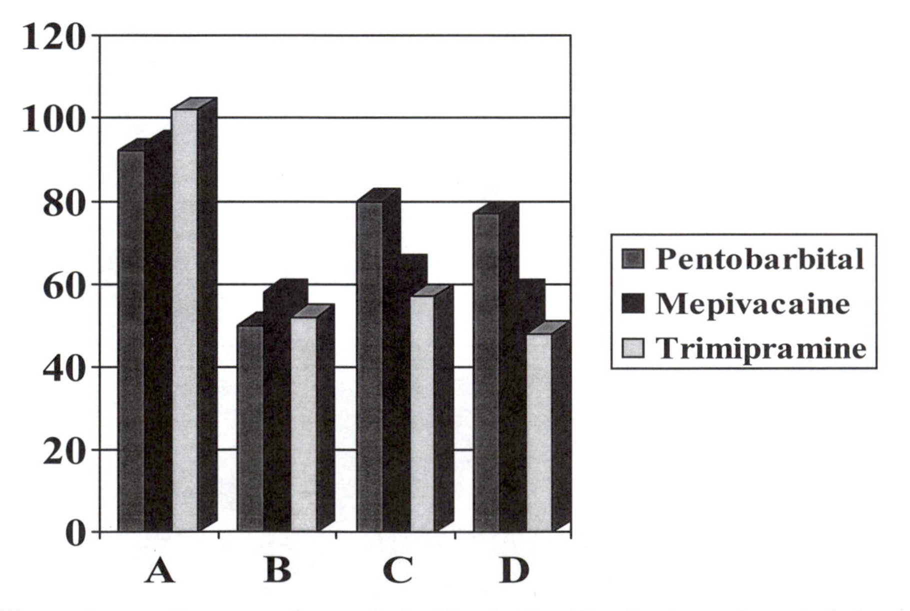

Figure 3. Recovery from whole blood of acidic, basic, and neutral drugs, using four preparations techniques. Technique A uses dilution in a buffer, sonication and loading of the entire sonicated sample onto the cartridge. Technique B uses inorganic precipitation (zinc sulfate/methanol). Techniques C and D use precipitation by AcCN and MeOH respectively.

niques B, C, and D were attributed either to co-precipitation of the analyte with the proteins or cellular material or to poorer non-polar extraction. Sample additives that release a non-polar drug by making it more soluble in the liquid phase of a sample also reduce the strength of retention on a non-polar sorbent.

One possible approach to elimination of protein binding may be drawn by analogy to biochemical receptor inhibition. By adding to the sample an excess of a compound which competitively displaces the analyte from the binding sites on the protein, the effect of the protein on extraction efficiency should be reduced. The difficulty with this approach is that the nature of the analyte/protein interaction must be accurately known or estimated and the added compound must not interfere with the extraction or analysis. As it is likely that a competitive displacing agent will be structurally similar to the analyte that it displaces, it will also compete for the sorbing sites during extraction. As such, the practicality of the technique in all but a few specialized applications remains doubtful. However, the authors are not aware of a published article that uses this technique.

Finally, there is some evidence (Ingwersen, 1994) that some or perhaps most claimed instances of protein binding in the application of SPE to biological matrices may have been incorrectly interpreted. Not all, or perhaps not even the majority of claimed observations can in fact be attributed to protein binding. Retention of molecules with strongly polar moieties and especially basic functional groups can be strongly assisted by secondary interactions (Ruane and Wilson, 1987), an effect that may be reduced or rendered ineffective by other species in the sample.

B. HUMIC ACIDS AND ENVIRONMENTAL PARTICULATES

The presence of humic acids and other environmental particulates often leads to reduced recovery of analytes from environmental matrices. The drop in recovery that they cause is often identified during method development when an analyst switches from spiked surrogates to spiked sample blanks. As for biological samples, the interactions between the matrix components and the target analytes may be disrupted by the addition of a solubilizing species such as an organic solvent or a change in pH.

Despite the desirability of rapid sample processing, due to unimpeded flow of sample through the cartridge that results when suspended solids are filtered out of solution, sample filtration is not a common or appropriate technique for environmental samples. Instead, SPE devices that are more tolerant to the build-up of matrix solids (through the use of in-line filters, high-flow frits, or large particle size beds) should be used.

Suspended solids, when they build up in the extraction device in this way, may also affect elution. Review of data in papers describing extrac-

Table 2. The effect of surfactant concentration on recovery of the drug paclitaxel at 5000 ng/mL, on a CN SPE cartridge using a wash solution of 1:4 mixture of methanol/0.01 ammonium acetate.

WASH VOLUME	SURFACTANT CONCENTRATION	RECOVERY IN FINAL ELUENT
1mL	0%	94
2mL	0%	86
1mL	0.2%	86
2mL	0.2%	84
1mL	1%	78
2mL	1%	68

tion from particulate-laden samples seems to confirm this though sadly the authors know of no definitive study of particle type, quantity, and size versus recovery. A simple expedient in one case (Markell, 1994) is to boost the polarity of the elution solvent (for example, by adding methanol to methylene chloride). In this case an initial method, developed on spiked clean water samples and using methylene chloride elution, gave poor and variable recoveries when applied to real-life samples. The addition of methanol to the eluent substantially improved recovery and consistency of method performance over a wide range of samples.

This can be rationalized on two levels. We may speculate that the suspended matter — silts, clays and biological debris — is of a polar nature and therefore elution of any analyte bound to it may require more polar solvents to disrupt this interaction. A second, and more feasible explanation may be that the ability of the retained particulates to occlude water will prevent adequate drying of the sorbent prior to elution with the aprotic dichloromethane. The addition of methanol to the solvent increases its miscibility with the partially hydrated sorbent and therefore increases the efficiency of analyte desorption.

C. SOLUBILITY OF THE ANALYTE IN THE MATRIX

Any matrix which contains surfactants, emulsifiers, or other species that enhance solubility, will need to be treated carefully. The effect of these species on recovery becomes particularly severe when the analyte is weakly retained even under ideal conditions. Few quantitative analyses have been published, unfortunately, so the effect of interferences such as surfactants usually must be estimated. One exception is provided for the drug paclitaxel in the presence of a pharmaceutical "vehicle" or surfactant, which is necessary in the formulation of the drug due to the poor solubility of the

Table 3. The effect of pH adjustment and salt addition to the sample on non-polar extraction efficiencies (Markell and Hagen, 1991). This work does not attempt to separate the effects of salt addition or pH adjustment. Compounds are listed in order of increasing hydrophobicity.

COMPOUND	C18, no NaCl	C18, 25% NaCl, pH 2	CH, no NaCl	CH, 25% NaCl, pH 2	pKa	Sol. in water (g/L)
Phenol	4%	23%	5%	14%	9.98	667
Cresol	20%	94%	9%	99%	≈10.2	≈240
2-nitrophenol	35%	90%	23%	69%	7.22	2
2,4-dichlorophenol	106%	92%	102%	88%	≈4.5	≈1

drug in common preparation solvents (Huizing et al., 1998). Illustrative data is given in Table 2. This paper demonstrates not only an increase in premature elution during the wash step, presumably due in part to the increased solubility of the analyte in the presence of the surfactant. The data also shows a concentration dependence of recovery that varies with surfactant concentration. This can be interpreted two ways: Either the analyte shows lower capacity toward the sorbent when in the presence of a high concentration of surfactant, or the surfactant reduces the available capacity of the sorbent by binding itself. The actual cause is probably a combination of the two.

The influence of a surfactant on the SPE process is not necessarily counterproductive in all cases. For example, the addition of an ion-pairing reagent may increase the affinity of a non-polar sorbent for a species that remains charged in solution. The overall effect of the surfactant is also highly dependent on whether it is introduced as a component of the sample or as a conditioning reagent in the preparation of the sorbent.

The pretreatment of biomedical PTFE and polyesters with quaternary ammonium compounds has been shown to alter the affinity of these materials for a range of drugs. Golding and Wynne (1988) investigated the effectiveness of mono-, di-, and tri-dodecyl-methylammonium compounds in modifying PTFE surfaces to achieve the retention of anionic drugs. While the pro-inflammatory response elicited by the surfactants resulted in their discontinued clinical use, the uptake and release of the diclofenac from treated materials was investigated. The authors speculated that the number of long chains on the surfactant played a role not only in the extent to which it was adsorbed onto the polymer, but also significantly influenced the extent of steric shielding of the ion-pair formed between the surfactant and diclofenac anion. The protection afforded by the steric shielding by more highly substituted surfactants was shown to reduce the percentage of bound

Table 4. Illustration of the pH/solubility relationship on retention/recovery for select acidic herbicides, extracted by non-polar mechanisms from natural waters (Data from Pichon et al., 1996).

COMPOUND	pKa	pH2	pH3	pH7
Dicamba	2.5-3.5?	89%	46%	0%
Ioxynil		98%	83%	31%
Bentazone		100%	100%	6%
2,4 DB	4.5?	98%	92%	38%
2,4,5 TP	4-5?	100%	78%	10%
Dinoterb		72%	49%	30%

drug that could be washed from the surface by aqueous buffers. Therefore, surfactant type may be seen not only to influence the sorption process but also to impact upon the desorption process and any desorption selectivity.

Taguchi et al. (1996) investigated the contributions of both non-polar interactions and the formation of ion-pairs on the overall retention of a li-

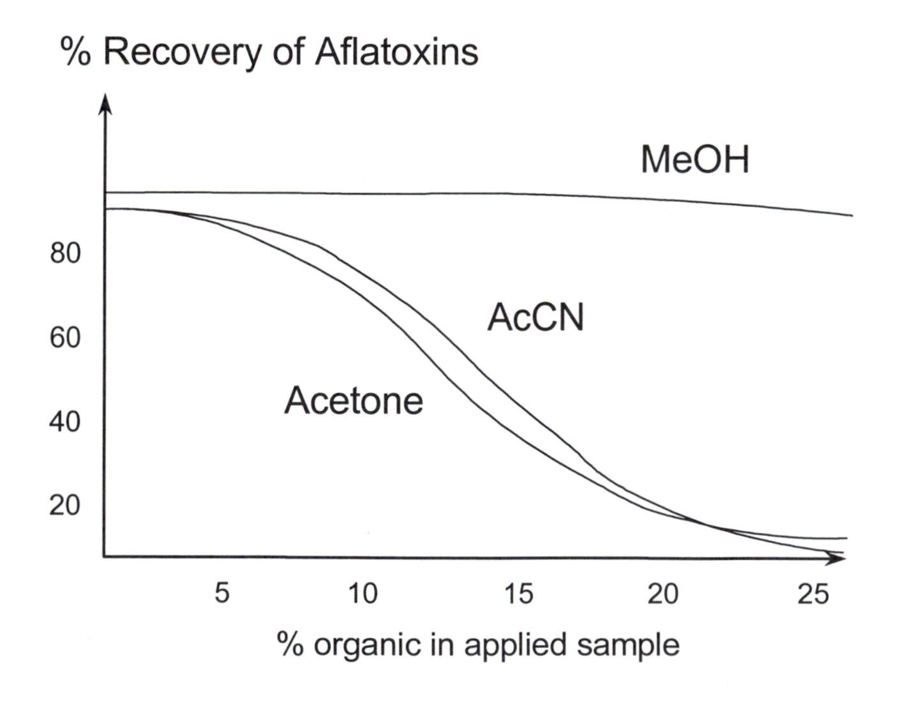

Figure 4. The effect of organic solvent percent composition and organic solvent type on the efficiency of extraction/SPE retention of aflatoxins from a maize slurry, when trapped on a PH sorbent.

gated cobalt (III) species and crystal violet cations. The study employed a variety of both anionic surfactants and sorbents. Li and Lee (1997) have adapted the pre-adsorption of cetyltrimethylammonium bromide on a C18 sorbent as an alternative to mixed-mode SPE for the recovery of ionogenic compounds from environmental water samples.

D. MATRIX MANIPULATIONS

As mentioned earlier, few published studies attempt to systematize or compare the sample work-up options for a single extraction. In most published procedures, little or no explanation is given as to the effectiveness of, or the reasons for, a particular choice of sample manipulation. The most commonly observed technique is dilution with a solvent, a buffer, or pure water. Reduction in the viscosity of samples, such as creams or gels, by the diluent, or control of pH and hence ionic state of the analyte or of potential interferences, by the addition of a buffer are benefits, but such issues are seldom explained or even mentioned.

In developing an SPE method, the work-up used is usually one that the analyst is familiar with from experience of other sample preparation techniques such as LLE, or because of a requirement unrelated to the SPE step. Thus, it is common in performing extractions of pollutants from drinking water, to adjust the sample pH to approximately 1-2. While this pH adjustment may enhance retention of weak acids on non-polar phases (see Tables 3 and 4), it is done primarily to suppress bacterial growth in the sample during its storage and transportation, rather than to enhance the recovery of analytes. Salting out is another technique familiar to those using LLE. In this technique the analyte, often moderately water-soluble, is driven into the organic phase by the addition of salt — an action that reduces the analyte's solubility in the aqueous layer.

Examples of ball-milling, sonication, filtration, centrifugation, precipitation, desalting, salting out, gel permeation, dilution, sample polarity change, LLE, and many others may be found in the literature as work-ups for SPE.

In a typical LLE, the dilution of a sample often necessitates an increase in the volume of extraction solvent, which increases both processing time and costs. In such extractions, sample diluents are desirable. However, modest sample dilution has little impact on SPE sorbent capacity for well-retained analytes and may improve extraction efficiency by improving the liquid handling characteristics of the sample. Therefore, it is regarded far more favorably for routine use in SPE.

CREAM	SI		NH2	
	Recovery	%RSD	Recovery	%RSD
Desonide	99.4	1.0	100	4.6
Methyl hydroxybenzoate	99	0.7	99.3	1.9
Propyl hydroxybenzoate	101.8	1.1	100	3.4
OINTMENT				
Desonide	100.2	1.1	100.5	1.5

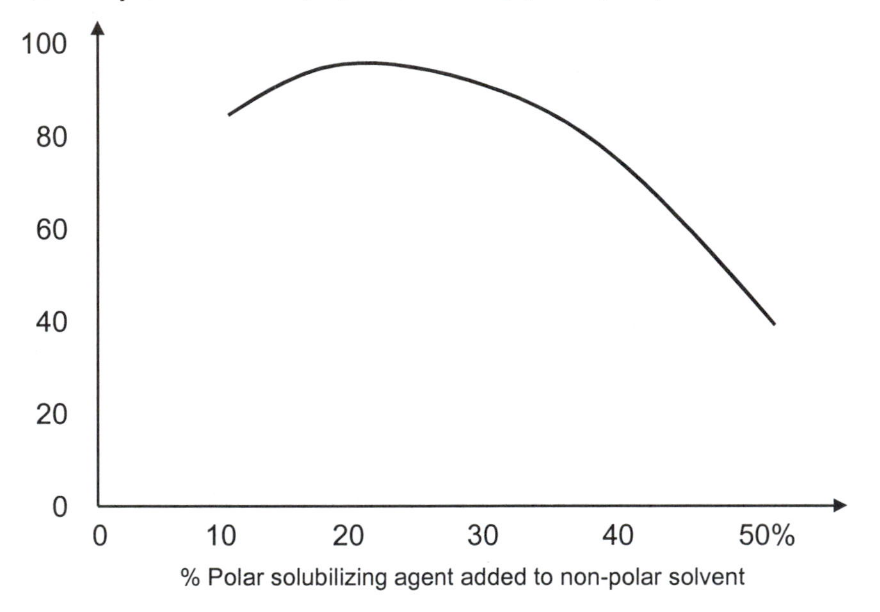

Figure 5. Recovery data on two polar sorbents, and dependence of recovery upon extraction parameters for desonide and parabens from pharmaceutical preparations.

Good examples that not only use several of these techniques but attempt to compare their effectiveness are provided by Bradburn et al. (1990) and Tomlins et al. (1989). In both cases the sample (corn maize) was ground, sieved, and blended in water, before being extracted by a range of aqueous/organic mixtures, and then loaded onto an SPE cartridge in order to identify optimum extraction conditions. Typical results are presented in Figure 4.

These studies explored a variety of bed masses and sorbent types for compatibility with the solvent systems that were found to be most effective at removing aflatoxins from the slurried maize matrix. Another example (Figure 5) in which the use of a sample manipulation was analyzed and optimized is given by Nguyen et al. (1986). Here, they explored the extraction of a corticosteroid and various antimicrobial additives in a viscous, hydrophobic cream or lipid-based ointment. A simple polar mechanism was found to be successful for the solid-phase extraction step, although the sample required dilution to reduce the viscosity such that it would flow through the sorbent.

The authors noted that hexane failed as a diluent because it was poor at solubilizing the matrix, despite the fact that its use should have ensured stronger retention of polar analytes onto a polar sorbent. Dilution with chloroform, on the other hand, failed to ensure adequate retention of the desonide. The authors speculate that chloroform, despite being only very weakly polar, was still polar enough to interfere with weak dipolar interactions. An interpretation the authors prefer is that chloroform's residual acidity (a variable determined by its purity) is the true cause of its elution strength in this application. Whatever the cause, a solution was found; a compromise of 80% hexane/20% chloroform was determined to be the most effective diluent mixture.

E. COMBINED OR MIXED-MODE EXTRACTIONS

The example of aflatoxin extraction given above is actually a hybrid liquid-liquid/solid-phase extraction. Such applications are common when the sample is a solid because cartridge or column SPE is not amenable to the extraction of solids. A clearer example is to be found in the work of Monkman et al., (1994). In this work, non-polar SPE using various sorbent types and elution schemes is compared with multiple LLE and single-step toluene extraction followed by polar SPE to remove interferences co-extracted by the toluene. The non-polar SPE extraction either gave poor clean-up or poor recovery. Protein binding was cited as the cause, for a very weak wash solvent was required to avoid washing out the weakly retained proteins that bound the analytes. The two-stage LLE was abandoned because initial recovery from plasma was good only for two solvents and the extraction was found to be pH-insensitive so a back-extraction was not feasible. An initial LLE in toluene followed by passage of this extract through an NH2 cartridge gave satisfactory clean-up and recovery.

A fine review of how this kind of hybrid technique may evolve is given by Luke (1995) who charts the development of an extraction for pesticide residues from high moisture content fruits and vegetables over the course of thirty years. What began as a simple LLE followed by Florisil[6] column clean-up and thin-layer chromatography evolved into a complex blending/slurrying of the matrix followed by non-polar, polar, and anion ex-

[6] Florisil is a registered trademark of the Floridin Company.

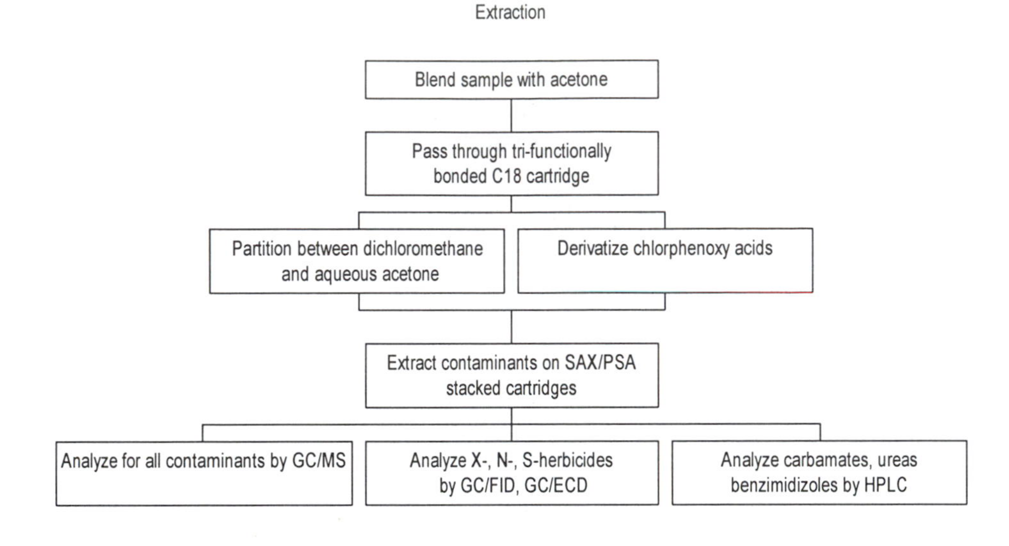

Figure 6. A Schematic of the Luke Procedure (status of method in 1994 revision).

change SPE steps for matrix removal and various liquid-handling steps such as dilution and evaporation.

The resulting method, though now much more complex than the original, is applicable to a wide range of sample types, yields an improved sensitivity and has met the demands for quantitation of a far wider panel of pesticides than were required when the first procedure was developed. A summary of the method is given in Figure 6. The SPE steps in each case contribute a clean-up of matrix components (for example, plant sugars and acids are retained on SAX and PSA sorbents) while the analytes pass through unretained. Note that chlorphenoxyacids must be methylated before passage through the SAX/PSA combination cartridge or they will be retained. The derivatization step in this case is carried out by conventional means although SPE-mediated derivatization may have also been applied.

IV. A SURVEY OF SAMPLE TYPES

While it is impossible to cover adequately every possible sample type, this section attempts to demonstrate extractions on the most common ones, while providing reference sources for some of the more unusual examples. Because the behavior of a matrix will be influenced by the manipulations it is subjected to — for our purpose that depends primarily on which mechanism of solid-phase extraction is being used — the examples given may not be applicable to every extraction from that matrix. For this reason a summary of matrix properties is given in Appendix 1.

Sample components are the major contributor to interferences in a chromatogram. However, an analyst should also keep in mind the materials the sample has come into contact with, since the leaching of compounds from the sample containers such as phthalates from plastics and vulcanizing agents from rubber stoppers, for example, will also affect extract purity.

A. COMMON BIOLOGICAL FLUIDS

The most widely encountered human biofluids are urine, plasma, serum, and whole blood. An early review of their SPE extraction properties was given by Harkey and Stolowitz (1984). These matrices are well characterized and the primary constituents have been widely reported in clinical chemistry primers or collections such as Geigy (1956). However, variability between samples may be caused by many factors such as the diet, state of health, or degree of hydration of the patient providing the sample. Sample-to-sample variations may also be caused, or amplified by the conditions of storage. Thus, plasma or blood samples may develop protein clots, and urine samples may form precipitates if stored for long periods or subjected to repeated freeze-thaw cycles.

One other factor, often overlooked, is the nature of the preservative added to the sample. For example, an anion exchange extraction on a sample preserved with citrate ion may give very different results from one stored in a heparinized tube.

1. Human Urine

Urine is by far the most common matrix analyzed in the screening or confirmation of drugs of abuse in humans and animals. It will be encountered extensively in Chapters 8 and 9, so a detailed study is not provided here. Table 5 summarizes the components of human urine. The relatively high and variable level of inorganic electrolytes is seldom a problem for a solid-phase extraction, although it explains the success of a combined non-polar/ion-exchange approach to sample clean-up. The first stage, a non-polar extraction, is relatively insensitive to variable ionic strength, while the second, an ion-exchange step, which is much more sensitive, occurs after the majority of inorganic ions have been eliminated.

One other feature of urine samples is that many xenobiotics — drugs of abuse, therapeutic agents, and other foreign compounds — are excreted by the kidney only after hepatic metabolism. The metabolic conversions include hydroxy-lation, oxidative de-alkylation, and the attachment of polar functional groups to form glucuronides, sulfates, and other conjugates (after

Table 5. Properties and composition of human urine. Composed by the authors from data listed in a variety of textbooks and other references. Note that the composition of human urine is highly variable and is affected by diet, degree of hydration and diuresis and clinical abnormalities. This table employs the standard representation of g/1200 mL which is equal to g/day for the normal adult.

PROPERTY	NORMAL RANGE
pH	4.8-7.4
viscosity (cP)	0.010-0.013
as a multiple of water	1-1.3
dry residue	55-70
Nitrogen compounds	
urea	20-35
uric acid	0.1-2.0
ammonia	~0.8
creatinine	1-1.5
amino acids - free	~0.5
-total	~1.1
Other organics	
hippuric acid	~0.7
total sugars	0.5-1.5
acetone bodies	0.02-0.05
lipids	0
free fatty acids	0.8-0.05
citrates	0.15-0.3
pigments	0.01-0.13
Inorganics	
chloride	6.0-9.0
phosphate	1.0-5.0
bicarbonate	0-3.0
sulphate	1.4-3.3
sodium	4.0-6.0
potassium	2.5-3.5
calcium	0.01-0.3
magnesium	0.17-0.28

the liver has attached a polar functional group such as a glucuronide or a sulfate to the molecule or has removed an alkyl chain). A decision must be made by the analyst whether to hydrolyze the sample, thereby cleaving selected conjugates, or to extract the metabolites in their excreted form. For example, enzymatic, acid, or base hydrolysis has been used to free parent opiates from their glucuronide-bound forms. Enzymatic treatment is often less convenient but offers the advantage of cleaving only those bonds with the stereochemistry upon which the enzyme is active. This specificity yields cleaner extracts compared with acid or base catalyzed hydrolysis, by generating fewer small molecules from unnecessary fragmentation of the other matrix components. Further, by avoiding extremes of pH, sensitive species that would otherwise be destroyed may be recovered. For example, enzymatic hydrolysis must be used when looking for 6-monoacetyl morphine since acid hydrolysis, as well as cleaving glucuronide groups bound to the opiates, will cleave the acetyl group to yield morphine. Both are metabolites of heroin (diacetylmorphine), which is rapidly metabolized after administration and is not observed in the urine. The 6-monoacetyl derivative is significant in that its presence in a urine sample establishes the administration of heroin rather than or in addition to the administration of morphine.

2. Plasma and Serum

Plasma and, to a lesser extent, serum are encountered in clinical analysis and metabolism, toxicological, pharmaco-kinetic, and bio-availability studies during drug development. Table 6 shows the major constituents of human blood. The biggest component is the protein fraction, the presence of which leads to increased sample viscosity and the formation of clots or precipitates. Centrifugation may be applied to remove agglutinated material, but care must be taken to ensure that the analytes are not also spun down out of solution. Proteins may still build up in the SPE cartridge (which is probably good) or find their way into the analytical instrument (bad!). If proteins are retained on the SPE cartridge during sample loading, assuming a non-polar mechanism is used, then water washes will not dislodge them. Washes containing protein-denaturing agents such as a low level of methanol, sodium dodecyl sulfate, acids, or base may help reduce the final protein level in the eluent.

Proteins are big molecules. The most common SPE sorbents have pore sizes of less than 100 Ångströms, which effectively exclude molecules with a molecular weight greater than approximately 20,000 Daltons. The extent of this exclusion is dependent at least in part on the primary, secondary, and tertiary structure of the protein as it exists in the sample matrix. Naturally, the structure and conformation of the protein are major determinants of its surface chemistry and therefore the extent of its interaction with any sorbent surface. If the protein is excluded from the pores, it is unable to access the vast majority of the retentive surface of the sorbent and thus will show a small breakthrough volume and low retention.

This is usually desirable, so very few manufacturers include wide-pore sorbents in their product offerings. However, a trend to larger pore size (and larger particle size, which results in a sorbent bed less prone to blocking by protein clots) has been observed during the early 1990s. This trend may accelerate as the biotechnology industry develops an interest in larger therapeutic agents, such as peptides.

When selecting a suitable SPE device for an application, it is important to remember that while larger particle sized sorbents may offer reduced incidence of blockage, this gain may come at the expense of a reduced efficiency in some other step in the extraction process. The choice of particle size is therefore a variable that must be optimized during method development.

Table 6. The composition of human whole blood, plasma, and serum. The major difference between plasma and serum is the presence of clotting proteins such as fibrinogen in plasma. Whole blood composition can differ depending on the cell disruption technique employed. The standard way of listing blood constituents (mass/100 mL) is used.

PROPERTY	NORMAL RANGE	
	whole blood	**plasma/serum**
pH		7.3-7.4
viscosity (cP)	3.5-5.4	plasma 1.9-2.3 serum 1.6-2.2
solids	18-25% w/v	8.5-10 % w/v
Nitrogen compounds	(mg/100 mL)	
urea	15-40	10-40
uric acid	0.3-4	2.6-7.0
creatine	2.9-4.0	0.7-1.3
creatinine	1.2-1.5	0.7-1.29
amino acids - free	-	35-65
-total	-	5800-8000
glutathione	28-52	0
Other organics		
polysaccharides	-	73-140
glucosamine	-	50-90
glucose	75-92	61-130
glucuronic acid	4.1-9.3	0.4-1.4
acetone bodies	1.0-3.0	3.0-8.0
lipids	0.2-2	0.45-1.2
free fatty acids	0.29-0.42	0.19-0.64
cholesterol	100-250	100-350
phenols	2.0-8.0	1.0-2.0
bile acids	2.5-6.0	5.0-12.0
lactates	5.0-35	6.1-17.0
citrate	1.3-2.5	1.6-3.2
steroids	-	120-320
Inorganics		
chloride	280-320	340-380
phosphate	6.0-15.0	~12
bicarbonate	-	~120
sulphate	0.8-2.0	1.5-3.0
sodium	170-230	325-330
potassium	150-250	18-22
calcium	5-11	9-11
magnesium	2.0-4.0	1.6-2.2
iron complexes	45-55	0.06-0.22

One study (Chen et al., 1994) investigated the relative extraction properties of plasma and human urine for a range of small drugs molecules, us-

Figure 7. Comparison of the extracts of three mammalian urines, including human urine, demonstrating the wide variety of extractable materials observed when different samples are subjected to identical sample preparation techniques. The effect of diet is also illustrated. Data produced by Wynne.

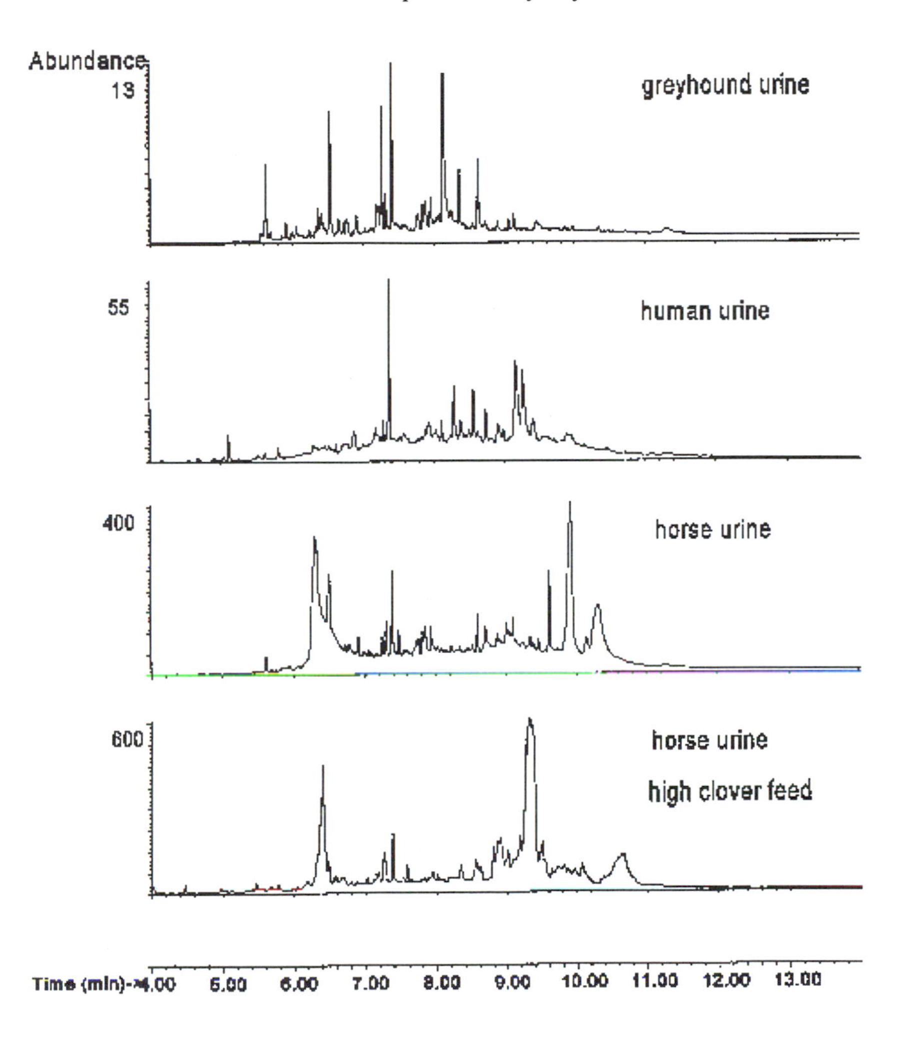

ing GC/flame-ionization detection (GC/FID) and GC/nitrogen-phosphorus detection (GC/NPD) detection.

The principal finding was that urine samples contained many more co-extractives than plasma. Thus, a urine sample, extracted using a method developed for plasma samples on a C8/SCX mixed-mode SPE cartridge, showed a large number of early-eluting interfering peaks on the GC chromatogram. However, this conclusion is not unequivocal, because the interferences, identified as probably nitrogen-containing species (neutral or acidic since they appear in Fraction A) were largely eliminated by the addition of a wash step using water/acetonitrile (4:1 v/v) after the sample load and water wash step.

Similar methodology applied to urine collected from horses produced even higher levels of the interfering species which could be readily identified by GC/mass spectrometry (GC/MS) as hippuric acid and glycine amide conjugates of other simple aryl and aralkyl carboxylates (Batty et al., 1994). Typical chromatograms for extracts of human, greyhound, and equine urine are illustrated in Figure 7 of this chapter and reappear in Figure 2 of Chapter 9 (where they will be discussed in connection with horse doping analysis). In addition to removing a large portion of the glycine conjugates, the relatively gentle wash step (water/acetonitrile 4:1) presumably removed weakly bound, small, highly oxygenated species like glucuronide-bound molecules, phosphates, ureates, and sugars and the glycine amide conjugates of other simple aryl and aralkyl carboxylates. Unfortunately, such investigations comparing performance of a single method on multiple sample types are rare.

3. Whole Blood

Many studies that attempt to measure substances in blood use serum or plasma in preference to whole blood. This is because the separation of cellular material prior to further extraction yields samples which present fewer extraction difficulties. The particulates present in whole blood (see Table 6) represent a problem for SPE since they may either independently block up the SPE cartridge or they may coagulate the entire sample. However, in some applications, whole blood is analyzed either because it is the only fluid available (for example in forensic toxicology) or in instances where the distribution ratio of the target analyte(s) between the plasma and erythrocytes is known or suspected to be significant. In such cases, the SPE of a sample must overcome the two obstacles of suspended particulates such as cells, cellular debris, or clots, and the binding of target analytes to proteins or other matrix components. Thus, the issues concerning SPE of whole blood relate not to the SPE process itself, but to the preparation of the sample with a sufficiently low viscosity and particulates content that it is able to flow through the sorbent.

Protein binding and biological precipitates have been discussed in Section III, above. The techniques used in the preparation of whole blood for SPE were identified as pH adjustment, chemical denaturing of the sample with organic solvents or inorganic salts, chemical hemolysis, addition of chaotropic agents, chemical displacement, dilution, sonication, and centrifugation either alone or in combination. The use of high flow columns to avoid blockages by blood particulates has also been proposed (Applied Separations, 1994; Moore et al., 1993). However, these sorbents must be used with care as cellular debris is sufficiently large that it is excluded from the pores in the sorbent and thus any portion of the sample that is trapped in this debris is not able to interact with the sorbent. Further, the use of high flow sorbents does not negate the need to overcome protein binding or the importance of cell lysis to release protein bound analytes.

The use of hemolytic surfactants such as Triton X, although effective in disrupting cellular material, is inadvisable for non-ion exchange SPE as the surface-active properties of the surfactant may have a profound and often undesirable effect on the interactions between a target analyte and the sorbent.

As demonstrated earlier in this chapter, it is still possible to process blood samples by using techniques like sonication and dilution (Chen et al., 1992), which destroy blood cells and prevent proteins from clotting. It is likely that the addition of a diluent buffer and sonication of the sample effected a change in the physical properties of the blood sample, improving its flow characteristics. This manipulation will also have altered the ionic strength of the sample such that many polar interactions between proteins and bound drugs were disrupted. High concentrations of salts such as sodium or ammonium sulfate, and polyvalent cationic salts such as zinc sulfate and lead acetate are also well-known protein denaturants. However, while their use has little negative impact on reversed-phase SPE (and may in some cases improve the retention of analytes), high salt concentrations interfere with polar and ion-exchange SPE processes and should therefore be avoided for such applications. Careful consideration must also be given to the possibility that inorganic precipitation of proteins may lead to loss of target analytes either by direct precipitation by the additive, or by entrapment in the denatured protein precipitate.

Other examples of chemical displacement of drugs from protein binding sites have been reported. One older example is the displacement of basic drugs from protein binding sites by the plasticizer tris-(2-butoxyethyl) phosphate which leached from the rubber stoppers in the blood collection tubes (Moffat, 1986)

Commonly, blood proteins are denatured by the addition of a water-soluble organic solvent such as methanol, ethanol, acetonitrile, or acetone. Thus, cocaine and its metabolites were analyzed in whole blood by Abusada et al. (1993), following the addition of methanol to the sample. The proteinaceous material was removed by centrifugation and the methanolic solution evaporated, reconstituted in an aqueous buffer and subjected to SPE. A useful variation to this pre-extraction technique in which the blood layer is frozen in an ethanol bath (-30°C) allows effective separation of immiscible or partly miscible solvents such as ether from samples that may otherwise be prone to emulsification (for example, McIntyre et al., 1993).

The manipulation of the sample to either highly acidic or highly basic pH is also an effective means of denaturing protein. Unlike solvent denaturing however, the adjusting of pH provides for the effective lysis of erythrocytes and the disruption of most protein binding of target analytes. Where necessary, the pH of the matrix may be restored to an appropriate level for extraction once the proteinaceous material and other cellular debris has been removed by centrifugation.

B. OTHER BIOLOGICAL MATRICES

Animal health studies, and especially testing of racing animals for low doses of therapeutic and other illegal substances, have been carried out using SPE. The versatility and wide acceptance of SPE in this field is such that a chapter specifically on this subject is provided later in this book. Some general features of these sample types are given here to permit matrix-to-matrix comparison.

The major limitation to analysis of blood in the equine athlete is the relatively high-packed cell volume of equine blood in comparison to other species, which results in a low yield of plasma for extraction. This limitation is readily overcome by collecting a larger sample of blood, a requirement which usually presents little problem given the animal's size.

Horse urine and urine collected from other grazing animals is comparatively viscous and may in some cases present extreme difficulties in processing through SPE cartridges. Speculation on the possible causes of this increase viscosity has included the role of poly-mucosaccharides (or proteoglycans as they are increasingly referred to) causing clotting on the top frit or top of the sorbent bed either directly or by reducing the efficiency of centrifugation during sample preparation.

Other animal samples are encountered in pharmaceutical development and toxicology testing. Thus, Schad et al. (1992) and Schaefer et al. (1993) demonstrated a simple extraction of 125 μL of mouse urine, diluted 40-fold to 4 mL with water, and applied to SAX and NH2 ion exchangers respectively, in their efforts to elucidate benzene metabolism in mice. Diethyl ether LLE was performed on the eluents but unfortunately little detailed discussion of the matrix extraction properties was given. However, in the earlier paper the extent of sample clean-up was illustrated by chromatograms obtained from the analysis of samples before SPE, after SPE, and after SPE and LLE. In another study, Zhong and Lakings (1989) discussed the selection of appropriate elution solvents to avoid co-elution of dog-serum components with a basic drug, Pioglitazone. However, even brief discussions of this sort are rare and even fewer of the published papers comment on the properties of the matrix compared to human or other animal samples. Of those that do it usually appears that these sample types are dirtier (contain more co-extractive species) and are, especially for small rodents, of very limited volume.

From such qualified comparisons, it quickly becomes clear that methods developed for such samples may be adapted to human biosamples with little risk of interference. However, the reverse adaptation of methods developed on human samples to biosamples collected from other species may require special consideration and careful modification to eliminate potential interferences.

Human cerebro-spinal fluid (CSF) has been extracted using SPE (for example, Venn and Michalkiewicz, 1990) without difficulty, often by using the same procedure that was initially developed for urine or plasma. Organ perfusates, amniotic fluids, and pulmonary surfactants (Egberts and Buiskool, 1988) are also amenable to SPE since these are primarily solutions of simple composition, carrying low levels of organic species. A more complex problem is tackled during the extraction of drugs of abuse from hair. This is a matrix which presents many more preparation challenges than urine or blood, but which is favored because it may be collected non-invasively and because it circumvents sample substitution and dilution problems associated with the collection of urine.

Published applications (Moeller et al., 1992, 1993; Harkey et al., 1991) stress the need to select that part of the hair strand to be analyzed that was actively growing during the period of drug use. They also emphasize the need to wash the hair before digestion, to eliminate external contamination. Chemical digestion has been achieved by the use of a combination of 1M Tris buffer at pH 7.5, 10% sodium dodecyl sulfate solution and dithiothreitol in a sodium acetate buffer, followed by enzymatic digestion using Proteinase K (Harkey et al., 1991). Alternatively, the sample may be prepared by ball-milling followed by suspension in a pH 7.6 phosphate buffer and subjection to a beta-glucuronidase/aryl-sulfatase enzymatic digestion (Moeller et al., 1992). Detection limits of low nanograms of analyte per milligram of hair have been demonstrated for some of the extracted drugs.

Saliva, stomach contents, bile, and feces may all be readily processed by mixing or homogenizing them with a suitable diluent buffer for the application and removing solids by centrifugation or filtration. The supernatant liquid may then be subjected to SPE using methods identical to those developed for urine samples. For example, Scalia (1990) extracted free and conjugated bile acids from gastric juices and bile using a combination of C18 and SAX sorbents. The C18 in this case removed proteins and inorganic anions, which would have interfered with the anion exchange fractionation. In some cases (Omori et al., 1986) the samples may be loaded directly onto the cartridge. In others, slurrying in an organic solvent is a necessary first step. This was the route taken by Abusada et al. (1993), who extracted parent cocaine and various metabolites from meconium — the gut content of fetuses in the womb and a thus, reservoir of compounds to which the fetus was subjected during development, including drugs in the mother's bloodstream. By homogenizing the meconium in methanol, centrifuging the homogenate, drying down the supernate to a volume of less than 1 mL and then diluting this in the same phosphate buffer that would have been used for diluting a urine sample, the sample could be extracted successfully, following the same protocol recommended by the SPE cartridge manufacturer for urine samples. This approach was also demonstrated to be an effective alternative extraction procedure for plasma and whole blood.

As most of these matrices are less complex than urine, it is not unusual for the eluent from an SPE sorbent used for their extraction to be significantly cleaner than is obtained by identical extraction of a urine sample. However, as with all samples, the more that is known about the composition of the sample matrix, the greater the opportunity to develop SPE methods that yield cleaner extracts.

Adipose or muscle tissue is usually analyzed with the intent to determine the health effects of consumption of that tissue and, as such, focuses on identification of pesticides or drug residues. Chapter 13 is dedicated to a non-standard SPE technique called Matrix Solid Phase Dispersion (MSPD), which has proved to be very much simpler and more broadly applicable to this kind of solid sample than pre-extraction work-up followed by cartridge SPE. Traditional SPE requires the sample to be liquefied, or at least that the analytes are solubilized and stripped from the bulk of the matrix solids. A typical example is provided by Horie et al. (1991) in which antibacterial drugs were extracted from fish tissue. In this case the fish tissue was homogenized in a slurrying buffer of metaphosphoric acid/methanol (3:2 v/v) in order to deproteinize the sample. Filtration and roto-evaporation of the homogenate yielded a liquid that was suitable to be applied in a simple procedure on a standard SPE cartridge.

In another application, Ono et al. (1997) extracted tributyl tin and triphenyl tin from fish following hydrolysis of the samples with ethanolic potassium hydroxide solution. The hydrolysate was extracted with petroleum ether and the extract concentrated and extracted in aqueous ethanol (1:1). The organo-tin compounds were extracted on a hydrophobic divinylbenzene methacrylate copolymer and analyzed by GC/MS. SPE is a valuable technique in handling such hydrolysates as the high protein and lipid content of these samples often makes them prone to emulsification during LLE. In a further example, the compounds responsible for so called "boar taint" were extracted from pig fat by Hansen-Moller (1992). The fat was extracted with acetone-tris buffer and the extract defatted by passing it through a C18 column prior to analysis of the extract for a variety of indoles by HPLC.

The special case of animal-derived fatty acids is considered below with other food oils.

C. ENVIRONMENTAL MATRICES

Environmental matrices are explored in Chapter 4, with particular emphasis on large volume aqueous samples. The trace enrichment aspect of SPE lends itself very well to the extraction of liquids, especially clean samples such as drinking water or groundwater. The difficulties associated with handling particulate-laden samples like river water or wastewater have already been alluded to. Two other environmental sample types to which SPE has been applied are soil/sludge and air.

1. Soil and Sludge

Much as for solid biological samples, a prerequisite for the successful extraction of soil or sludge samples is liberation of the analytes from a solid matrix into a liquid one. Soxhlet extraction, homogenization in an extraction buffer or some other physical manipulation may be needed, and careful choice of conditions used at this stage will ensure that a minimum of biological debris and inorganic matrix will be co-extracted (for example, Redondo et al., 1993). Even with careful selection of the primary extraction mechanism, considerable additional clean-up is often required prior to analysis. Historically, SPE using a Florisil cartridge clean-up, has been used to perform only a part of the potential clean-up role of which it is capable.

Thus, in the U.S. Environmental Protection Agency Statement of Work for determination of chlorinated pesticides and other chlorinated organic species in sludge and soil samples (U.S. EPA Contract Laboratory Program, 1990) a Soxhlet extraction is used. The extract is rich in humic and fulvic acids, and may also contain a high level of sulfurous compounds and other inorganics. In order to prepare this extract for GC/ECD analysis a gel permeation step (to eliminate large species such as humic acids), a desulfurization step and other contaminant removal procedures must be applied, in addition to a simple, but obviously only partially effective, SPE extraction on Florisil.

On a philosophical level, environmental chemists must resolve a different dilemma: Should they be looking for the total level of toxicants in a sample, or only for that portion of toxicants that could find its way into the water supply and thence, assuming nobody makes a habit of eating soil, into the human population? If the answer to this question is "Yes," then the extraction of soil samples present far less of a problem. An approach referred to as a Toxicity Characteristic Leaching Procedure (TCLP) simply requires tumbling of the sample in an aqueous medium, or alternatively the passage of water through the sample, and extraction of the resulting leachate using LLE or SPE (Crepeau, 1991). The presence of toxic compounds in particular extract fractions, for example, the fraction that is extracted onto a

C18 sorbent under acidic conditions, may be readily determined using a crude bioassay such as the determination of an LD50 or LD100 level for a population of aquatic organisms such as Daphnia.

2. Air Monitoring

In contrast to solid matrices, air represents a simple and easily processed sample. Airborne contaminants can be gases, dispersed liquids such as aerosols or mists (for example boom spray drift), or fine solids as might be found in combustion waste, such as flue gases and motor vehicle exhaust. Each contaminant type is amenable to SPE, although some variation in sample treatments is found.

Airborne contaminants that are primarily volatile or semi-volatile can be analyzed without the need for sample clean-up. When SPE sorbents are used the mechanism is perhaps better described as solid-phase adsorption rather than solid-phase extraction, since the analytes do not migrate into a stationary phase and the device functions as a simple trap for volatiles.

Porous carbon sorbents formed by the controlled pyrolysis of sugars have been used to concentrate organic compounds of varied polarity from air with the adsorbed species being desorbed off-line either thermally or by elution with solvents (Matisova et al., 1995). C18 sorbents have also been used to concentrate the C10-C20 hydrocarbons found in natural gas. Desorption of these compounds in solvent has been found to give results superior to those obtained using thermal or solvent desorption from more traditional techniques using Tenax[7] or charcoal (David et al., 1989).

Novel SPE techniques have also been employed when concentration or derivatization are required. For example, SPE has been used to trap and derivatize volatile carbonyl compounds from air by using C18 sorbent that has been impregnated with derivatizing reagents such as 2,4-dinitrophenyl hydrazone (Kootstra and Herbold, 1995). The Shiff's base formed on the sorbent may be selectively eluted and analyzed using an appropriate analytical technique. This method offers the advantage of increasing the molecular weight of volatile carbonyl compounds thereby improving the applicability of GC and GCMS to their analysis.

Adsorption of aliphatic alcohols and aldehydes from aqueous solution by immersion of thermally activated silicone rubber has been investigated as an alternative to head-space preconcentration above such samples (Jelinek, 1994). The sorbent, while poorly described, may be considered as displaying surface chemistry similar to an un-endcapped C2 on silica sorbent although the diffusion and transfer processes from solution would be less clear. It was found to be effective for C4-C8 aliphatics with higher sen-

[7] Tenax is a registered trademark of Akzo Research Laboratories, The Netherlands.

sitivity towards less volatile aliphatics than is observed with Tenax head space preconcentration. Desorption of the analytes gave superior recovery of higher molecular weight and more polar compounds but poorer recovery of volatile aromatics when compared to other methods. Differences in recovery may be readily explained by the lower volatility of higher molecular weight and higher polarity compounds, which are less abundant in the head space. The poor van der Waals interaction of volatile aromatics with C2 like sorbents when compared to sorbents such as C8, C18, PH and carbon will also contribute to the net effect.

Volatile aromatic amines have been extracted from the gas phase by ion-exchange on a perfluorosulphonate (Nafion) sorbent without drawing large volumes of the gas phase through the sorbent (Cox et al., 1995). The sorbent was also demonstrated to be effective for basic and neutral compounds in aqueous solution and in the gas phase. The application of the unique electrochemical properties of Nafion allowed the sorbent to be used as a preconcentration medium for an amperometric sensor for such compounds. Desorption of the analytes into aqueous mixtures was discussed by the authors in relation to ion displacement, although the wettability of the teflon surface by the extractant was clearly an important factor in determining the rate of solution desorption.

Other examples of air sampling include passive sampling using solid-phase trapping of volatile and semivolatile species such as propellant or accelerant residues in arson debris (Newman et al., 1996).

Aerosols and mists are comprised of fine droplets of liquids and their extraction from the atmosphere directly onto a sorbent presents problems with the wettability of the sorbent. A more appropriate sampling technique involves passing the airborne droplets through a scrubbing device containing a miscible solvent and, after a suitable sample has been drawn, extracting the scrubbing fluid as for any other liquid sample.

A more complex problem of extraction of species from airborne particles is discussed by Schulze (1984), who looked at polyaromatic hydrocarbons (PAHs) and nitro-PAHs in diesel exhaust particulates. The approach used and the philosophy adopted by these analysts towards particulate sampling was akin to those used for soil analysis by leaching: They were concerned about water-leachable rather than total contaminant levels. A similar problem was tackled by Karlesky et al., (1986) who trapped airborne particulates on a filter. However, they adopted the alternative approach of Soxhlet-extraction of the filter, followed by fractionating the pollutant extract on a PSA cartridge.

A case where the airborne contaminant is less well-defined (gas/liquid/particle) is the extraction of fluoride from arc-welding fumes drawn through a cellulose nitrate filter. The CN fulfilled the roles of a sorbent, a condensing surface and a particulate filter. The filter was eluted and interfering metal ions removed by an in-line SCX pre-column before analysis by ion-chromatography (Vasconcelos et al., 1994).

D. FOOD AND BEVERAGES

SPE applications for food and beverages are usually developed either for quality control purposes (for example, hop acid content in beer, pigments, sugars and acids in wine, micro-nutrients in infant foods), or for the detection or identification of drug and pesticide residues or microbial toxins. The goal is to ensure the safety of the consumer.

Wine requires analysis for volatile and semivolatile species, which are largely responsible for the wine's bouquet, and sugar acids, which contribute to the flavor. Solid-supported LLE and SPE have been extensively used for this purpose (Gelsomini et al., 1990). The advantage of using SPE is that class fractionation into acid, base, and neutral fractions is simple and the opportunity to concentrate the target analytes offers enhanced sensitivity that may facilitate detection. Further, SPE allows extraction under mild conditions of pH, thereby limiting the incidence of decomposition or rearrangement of labile compounds. An example is given by Busto et al. (1994), who derivatized biogenic amines present in wines to yield the dansyl derivatives. These were simply extracted on a C18 cartridge and could be readily analyzed following elution due to the 50-fold concentration effect of the SPE step, which gave isolates with concentrations in the region of linear response for the detector.

The volatile aldehydes produced during some fermentation processes, which are normally studied by head-space GC analysis, may also amenable to trapping and concentration using the Shiff's base derivatization technique described above, under air monitoring.

Examples of the use of SPE on beverages for residue analysis, and hence on policies related to the use of agricultural chemicals, include extraction of daminozide (Alar) in apple juice (Kim et al., 1990) and vinclozolin (an antifungal) on grapes (Gennari et al., 1985). In most cases the role of SPE is to provide matrix removal and concentration of the target analytes.

Milk, butter and cheese present a bigger problem. They have a high fat and protein content and they are quite viscous or completely solid. Papers describing the use of MSPD for these sample types have appeared with increasing frequency. However, traditional SPE should not be considered a superseded technique for all aspects of working with these samples. For example, it has been successfully applied in the determination of pesticide residues using solid-supported LLE columns (diMuccio et al., 1988) or regular SPE cartridges (de Jong and Badings, 1990; Schenk et al., 1993; Manes et al., 1993). The extraction of PCBs (Pico et al., 1995) and aflatoxins (Simonella et al., 1988) from milk on C18 sorbents has also been reported. Extraction and separation of the fats themselves is described in greater detail below. Other examples of the application of SPE to fatty samples are given elsewhere in this chapter (for example, extraction of fish meat, toxicological investigations of blood and tissue, and the extraction of various compounds from plant oils).

A thoughtful approach to SPE of carcinogenic amines in food is provided by Gross (1990) who utilized a variety of solid-phase extraction approaches including C18 and a strong cation exchanger (PRS) to effect separation, attributing initial retention of amines in a non-polar solvent to polar secondary interactions. The method employs a careful transition from the non-polar organic loading conditions to aqueous wash conditions under which ion exchange can occur.

Horie et al. (1990) demonstrate a simpler clean-up of meat samples. A sample was homogenized with aqueous methanol, filtered and evaporated to 20 mL. The evaporated sample was applied to a C18 cartridge, washed with water and the target analytes, in this case sulfamethazine and N4-acetyl sulphamethazine, eluted with water. This paper is notable for its comparison of commercially available C18 sorbents and for its evaluation of the role that bed mass has upon recovery and extract purity. The method is applicable to various meat types.

A good example of recovery dependence on method variables is given by Sydenham et al. (1992) in the study of fumonisin recovery from maize. The method is similar to that described by Tomlins et al. (1989) for the extraction of aflatoxins from maize. Both papers used sample liquefaction (grinding in a mill and extracting with methanol/water mixtures) but unlike aflatoxins, the fumonisins in Sydenham's work are potentially ionic. This allowed the extract to be applied directly to an ion-exchange cartridge (SAX) without the need for dilution of the extract to reduce the methanol content. Sample loading and elution rates and elution solvent strength were each tested to optimize conditions. Reuse of the cartridge was also explored and found to be possible, presumably because fairly gentle loading and elution conditions were used.

In contrast to analysis of samples that require work-ups of this sort, those samples that are largely or completely water soluble present much simpler challenges. Thus, the extraction of insecticides from honey (Taccheo Barbina et al., 1989; Taccheo Barbina, 1990) required simple dilution of the sample with water, followed by a quick non-polar matrix removal step on an extraction cartridge. Similarly, various micro-nutrients such as cyanocobalamin (Iwase and Ono, 1997) have been quantified from water-soluble foods such as soup powders by dissolution, filtration, and clean-up carried out on a C18 cartridge. The filtration was used is in this case to remove excess fatty compounds before loading onto a C18 SPE cartridge.

Beverages, provided the alcohol or sugar/syrup content is not high or variable, are simpler still to process by SPE. Thus, wine has been fractionated into volatile components by employing a solid-supported LLE to trap volatiles in one example and SPE on a CH bonded phase to extract pigments (anthocyanins), leaving sugars in the effluent in another (Gelsomini, 1990). This crude class fractionation could be taken a stage further by passing the wine through an anion exchanger to trap the wine acids. A two-cartridge extraction of this kind is demonstrated by Saito et al. (1989) in the clean-up of soft drinks during the analysis of aspartame degradation products. This procedure, using C8 and SCX, removed caffeine, aspartame, sodium benzoate, and caramel and color acids from the drink.

1. Fruits and Vegetables

The impact of SPE in the analysis of fruit and vegetable produce has been concerned primarily with the extraction of pesticide residues associated with food standards and the determination of putrefactive or other compounds arising from spoilage. Classical SPE, as opposed to MSPD sorbent extraction, has been occasionally applied to fruit and vegetable matrices. The high or variable water and fat contents of citrus fruit, berries and nuts can present capacity problems. The most comprehensive study (Luke, 1995) explores the extraction of a large range of pesticides from these matrices (see Figure 6). Considerable sample manipulation and LLE is required by this method, indicating the origins of the method as a LLE to which other preparation steps and SPE extractions have been added, as the method goals became more stringent.

The extraction of benzimidazole fungicides with acetone from a variety of fruit and vegetables followed by SPE clean-up of the extracts on 2OH bonded sorbents has been reported by Pavoni and Errani (1996). Elimination of co-extracted waxes, which are often problematic in the analysis of fruit such as apples, may sometimes be achieved post-extraction by careful selection of solvents for reconstitution. As the waxes are typically insoluble in methanol, the solvent can be used to redissolve an SPE eluate after evaporation with the resultant precipitation of the waxes.

2. Oils and Other Lipidic Material

Triglycerides, fatty acids, and other lipids are important dietary components that are found in a variety of sources. The nature of these compounds also sees them used in areas as diverse as cosmetics and pharmaceutical preparations to industrial lubricants. Because of their strong association with food and diet, they are described here although their wide distribution does

Solvent Fraction	Composition	P_{oct}	Lipid Class Eluted
A	2:1 chloroform: propan-2-ol	4.07	All neutral lipids
B	2% acetic acid in diethyl ether	2.86	Free fatty acids
C	Methanol	5.1	Phospholipids
D	Hexane	0.001	Cholesteryl ester
E	1% diethyl ether, 10% methylene chloride in hexane	0.437	Triglycerides
F	5% ethyl acetate in hexane	0.316	Cholesterol
G	15% ethyl acetate in hexane	0.616	Diglycerides
H	2:1 chloroform: methanol	4.43	Monoglycerides

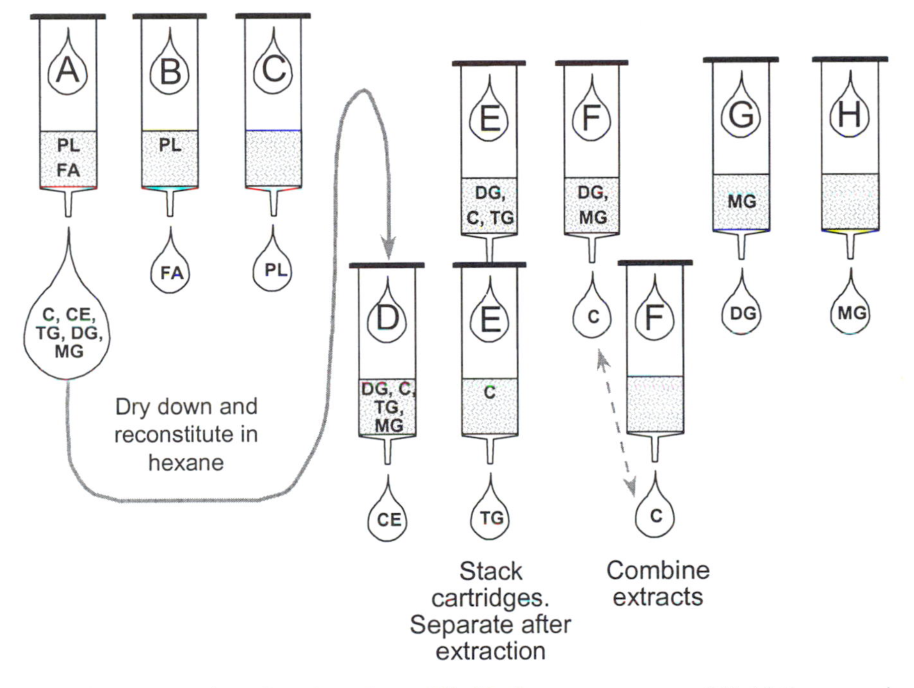

Figure 8. A class fractionation of lipids from an extract of lipid tissue, using a single sorbent type and a complex elution pattern. The elution scheme was developed partly from existing TLC and chromatographic data but also drew upon the authors' experience of prior SPE method development and existing solvent polarity data such as log P_{oct} tables.

not preclude the application of the methods described below to studies in other areas.

The lipid fractions of a sample may be obtained by many different methods including cold pressing of seeds, the heating or rendering of animal tissues, solvent extraction, supercritical fluid extraction and MSPD. Whichever is used, it is important to consider that the method of isolation will determine the distribution of lipid classes that may be recovered and the extent of degradation that may occur to individual compounds or entire classes. SPE techniques have found novel application in the small-scale fractionation of such lipid extracts.

Non-polar lipids have been extracted from non-urea adducting filtrates on a C18 sorbent by Rojo and Perkins (1989). The stepwise elution through a tandem silica column with hexane-dichloromethane (non-polar lipids eluted), dichloromethane (hydrogenated monomeric cyclic fatty acid methyl esters), dichloromethane-methanol (polar lipids) and methanol (very polar lipids) gave a degree of separation of some species on the basis of polarity.

The determination of fatty acids in dairy products was achieved by de Jong and Badings (1990). SPE fractionation was preceded by grinding cheese samples with anhydrous salt and sulfuric acid, followed by an ether/heptane (1:1 v/v) extraction, or by addition of ethanol and sulfuric acid to milk followed by a similar ether/heptane extraction. The organic extracts were then either purified by polar SPE on alumina or by ion exchange/polar SPE on an aminopropyl (NH2) sorbent.

The NH2 extraction draws on a classic application by Kaluzny et al. (1985), a schematic that is shown in Figure 8. This extraction separates a chloroform extract of lipid tissue into seven principle fractions with high efficiency and purity. This parent method has been the starting point for several derivative methods that just use a portion of the original separation scheme. In the case of de Jong and Badings work that part of the scheme was the section relevant to retention and elution of free fatty acids, to which minor modifications have been made. Wilson et al. (1993) have used similar methods to resolve saturated and monounsaturated fatty acid esters from polyunsaturated esters. This procedure permits the characterization of free fatty acids from C2 to C40 with quantitative recoveries. The scheme has also found application for wider lipid class studies. Thus, Kim et al (1990) recovered cholesterol and cholesteryl esters, triglycerides, free fatty acids, spignomyelin, cerebrosides, phospholipids, and acidic phospholipids with a good degree of separation and minimum of loss or degradation.

The analysis of glyceryl esters is important in the characterization of food oils, industrial oils and increasingly for pharmaceutical or therapeutic purposes. The enzymatic lipolysis of the triglycerides in soybean oil was studied by Nett et al. (1992). The hydrolysate was dissolved in hexane, passed through a silica SPE cartridge and eluted sequentially with diethyl ether-hexane (1:9, to elute initially the triglycerides and later the cleaved fatty acids), diethyl ether-hexane-acetic acid (50:50:1, to elute the 1,2 and 2,3-diacylglycerols) and finally with methanol (to elute the monoglycerides).

The separation of diglycerides from olive oil, salami, and cheese on a NH2 sorbent has also been reported (Koprivnjak et al., 1995). However,

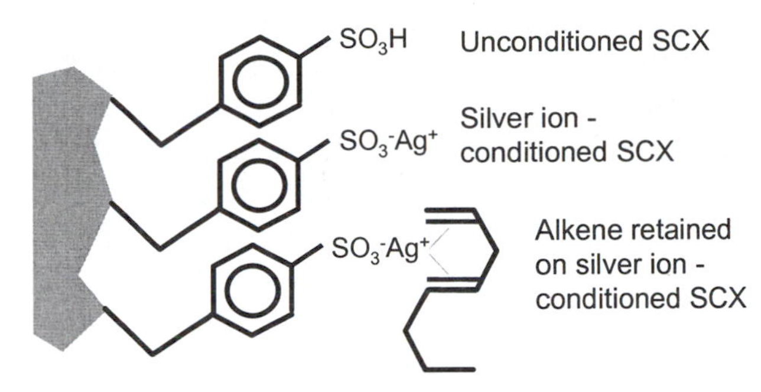

DCM	Acetone	AcCN	Elution Vol.	Fraction
100%			5 mL	Saturated
90%	10%		5 mL	Mono-enes
	100%		5 mL	Di-enes
	97%	3%	10 mL	Tri-enes
	94%	6%	10 mL	Tetra-enes

Figure 9. The creation of an argentation SPE sorbent from SCX and silver ions, for separation of unsaturated fatty acids in classes based upon degree of unsaturation.

later studies of the diglycerides in olive oil and lard demonstrated that isomerization (or migration) of the esters occurred on NH2 sorbents. Extraction of oils on a 2OH phase proved to be fast and reproducible while causing negligible isomerization (Perez-Camino et al. 1996, Coute et al. 1997).

A highly inventive use of SPE to characterize fatty acids is shown in Figure 9. Christie (1989, 1990) applied the technique of argentation chromatography to cocoa butter, palm oil, and sheep adipose tissue to effect a separation of unsaturated and saturated lipids. By conditioning a strong cation exchanger (SCX) with silver ions, a sorbent is created that is able to discriminate between fatty acids with differing numbers of double bonds. This is possible because the silver ion is able to polarize C=C bonds and form a weak complex which, under non-polar conditions, is enough to cause retention of the molecule. The strength of retention increases with the degree of unsaturation of the molecule. Thus, it is possible to separate in successive elution fractions, saturated, mono-ene, di-ene, tri-ene and up to octa-ene species, simply by varying the polarity of the elution solvent. Conveniently, after extraction the silver ions can be completely recovered by passing 1N nitric acid through the cartridge and collecting the effluent. This work incidentally influenced a later publication detailing the extraction of polychlorinated biphenyls (PCBs) and other unsaturated environmental pollutants from hexane extracts of soil or petroleum products.

A further example of how the lipid fractionation method has been modified and applied to other analytical problems is provided by Vaghela and Kilara (1995) who used the building blocks developed by Kaluzny et al., (1985) in the evaluation of lipid classes in whey protein concentrates. The two problems of high lipid load and even higher protein content (protein that often binds to the lipids) were noted by the authors. To circumvent these issues each sample was initially extracted by chloroform/ methanol, roto-evaporated and vacuum-desiccated before the extract was subjected to gel filtration to eliminate the remaining protein. The resulting extracts were dried down and resuspended in chloroform before being applied to a 2g NH2 cartridge. The large sorbent bed is necessary to provide adequate capacity for the lipid load being applied. Lipid content could be checked using this method with a quick gravimetric test and periodic TLC purity evaluations to ensure that the extraction was being performed correctly.

It is important to note that the separation of many crude fatty acid mixtures by argentation-based SPE is limited to only small samples. For larger samples the capacity of the sorbent is invariably exceeded by the saturated fatty acids (which retain by secondary interactions) that are incompletely eluted after sample loading. This limitation may usually be overcome by performing semi-preparative separations of fatty acid mixtures on argentation type HPLC columns on which there are sufficient theoretical plates to effect separation. It is also important to consider that irregularly spaced or dienic double bonds when compared to the abundant "normal" fatty acids often elute unexpectedly, but most commonly in the next most unsaturated fraction. The later elution of such compounds than their normal skipped analogues may be explained on the basis of steric factors. Thus, the greater the spacing between double bonds, the greater the free rotational confirmation changes that the fatty acid can undergo. Such changes present greater opportunity to maximize the number of silver-enic complexes formed and therefore the degree of retention.

E. PLANTS AND PLANT EXTRACTS

While thousands of applications have been published reporting methods for the extraction of biological fluids, far fewer have reported SPE of other matrices for either phytochemical or pharmacognostic investigation. As with food and beverages, the majority of SPE applications reported for the extraction of plant materials relate to the investigation of agrochemical residues for regulatory or food standard purposes. However, as was suggested in Section IIB above, the use of SPE techniques in the analysis of many phytochemicals is limited only by the requirement for the plant material to be processed as a liquid sample. Having achieved this conversion, potential target compounds include fatty acids, lipids, complex lipids/sphignoids, steroids and their related glycosides, carbohydrates, flavonoids, terpenes and terpenoid acids, amino acids and peptides, antibiotic compounds, alkaloids, and vitamins. The matrices from which these compounds may be extracted include woody and soft-stemmed plants, seeds, seed oils, steam distillates of plant matter, algae and molds, bacteria, and broths.

Plants, for all their apparent diversity, can be considered for the purpose of extraction as consisting simply of an aqueous portion, a fatty portion and an insoluble or fibrous portion. When testing for the presence of additives that are not incorporated into the plant tissue (such as anti-oxidants and other protective agents), a simple and common expedient is to suspend a portion of the plant in an appropriate diluent buffer. After allowing the sample to soak for approximately one hour, the residual solids may be removed by centrifugation or filtration and the supernatant liquid processed as though it were a human urine sample.

The aqueous component of green plants consists of both a solution of inorganic ions and soluble organic species. It is therefore often convenient to adopt methodology developed for urine to the extraction of this fraction. This form of adaptation is particularly valuable in the study of the metabolic fate of dietary components such as alkaloids, or pharmacologically active species under investigation for their value as therapeutics. The methods described in many of the examples included in Chapter 9 have been used successfully in the author's laboratory for the isolation of a wide range of basic alkaloids from plant tissues homogenates prepared in phosphate buffer.

Similarly, Strobiecki et al. (1997) employed both SCX and C18 sorbents in tandem for the separation of quinolizidine alkaloids and phenolic compounds in lupin seedlings. While many low molecular weight basic alkaloids are very well retained by a sulfonate SCX phase, the mechanism is apparently less efficient for some of the more complex alkaloids such as strychnine. The poorer recovery of these compounds is particularly evident when the target compound is present at low a concentration in a complex matrix. In some cases, the neutral part of the molecule may be the dominant function controlling sorbing chemistry leading to significant losses during the elution or washing steps. Alternatively, the poor retention of polyhydroxylated species such as aconine on mixed-mode sorbents may be attributed to poor initial retention by the reversed-phase mechanism. Given these anomalous results, it is not surprising that several non-ion exchange extractions have been reported for alkaloids including benzophenanthridine alkaloids (Chauret et al., 1990) and *Catharanthus* alkaloids (Naaranlahti et al., 1990) on C18 sorbents.

Not all phytochemicals are as amenable to SPE as the basic alkaloids. Thus, when SPE was still in its infancy and before the huge range of current applications existed, SPE of plant materials was often used simply as a technique to remove pigments by passage of the lyophilized, soxhlet-extracted or homogenized and extracted sample through a C18 device. The effluent (not the eluent) was then collected (Carmichael, 1982; Meier et al., 1988). Thus, the removal of the insoluble or fibrous material, the aqueous matrix and many of the polar water-soluble components allowed for a rapid processing of samples. However, in recent years there has been increased focus on highly oxygenated, often water-soluble compounds as potential therapeutic agents and important dietary factors. For example, the extraction of bioflavonoids such as rutin from plants on a variety of sorbents has been investigated by Buszewski et al. (1992 and 1993). Flavans were isolated in a phenolic fraction on a C18 sorbent for chemotaxonomic purposes by Katalinic (1997) and non-glycosidic anthraquinones such as alizirin were isolated from plant cell cultures and mushrooms on C8 cartridges (Toth, 1992).

The cytokins are N-substituted derivatives of adenine which control cell division in plants. Zeatin and its riboside have been extracted from plant tissue on a PVPP sorbent prior to ion-exchange chromatography (Cappiello et al., 1990). The recovery of eleven cytokins on C18 and several other sorbents was also reported by Guinn et al. (1990) prior to liquid chromatography on an SCX phase. Other phyto-hormones have been isolated by a variety of SPE techniques.

A general method for the extraction and fractionation of tryptophan derived auxins by on-line SPE and HPLC was reported by Martens and Frankenberger (1991). Acid plant growth regulators were determined in spinach following SPE on DVVP (Polysorb[8] MP-2) and 3-octadecylstyrene-2-sulphodivinylbenzene (Polysorb MP-3) (Patel et al., 1990). Acidic phytochemicals may also be successfully extracted by SAX methods without the high level of co-extractives that is often associated with urine. Thus, phenolic acids in *Echinacea* species were extracted then further purified by SPE on C18 and quaternary ammonium sorbents (Glowniak et al., 1996).

Extraction of fatty acids and other lipid fractions from plants and seeds has been described above in Section IVD. For many phytochemical applications however, the high level of fats in a plant extract are regarded as an impediment to the easy isolation of other classes of compounds. To some extent, these problems can be overcome by traditional wet chemistry techniques. For example, Tsuji et al., (1995) use zinc acetate to coagulate fats in an acetonitrile extract of fruit or vegetables to allow pesticides to be extracted by a simple C18 process. Pigments, too, were reportedly eliminated by this coagulation step. Much of the methodology described above for highly oxygenated neutrals also gives a considerable degree of de-fatting of samples and it is likely that further improvement could be achieved in many cases by adaptations to include sorbents such as silica, C1, and C2.

The extraction of plant sterols is also readily achieved in many cases with the use of SPE. One example is the isolation of ergosterols from lipids isolated from plant tissues colonized by fungi (Gessner and Schmitt, 1996).

---

[8] Polysorb is a trademark of Interaction Chromatography, San Jose, CA.

In this work an alkaline biomass extract was acidified (to neutralize all carboxylic acid groups and increase the affinity of the column for the analyte) and passed through a C18 column. The column was washed with alkaline aqueous methanol and the sterols were eluted with alkaline isopropanol. The isolation of sterols have also been reported from olive oil (Amelio et al., 1992) as has the extraction of other phenolic acids and polar neutrals (Litridou et al., 1997; Papadopoulos and Tsimidou, 1992).

It is not uncommon for phytosterols to be heavily oxygenated and the extraction of several examples containing vicinal diols has been reported. Ecdysteroids (with a 20,22-diol moiety) have been converted to aryl cyclic boronate derivatives to improve their retention during reversed-phase SPE (Pis et al., 1994), presumably by increasing the hydrophilicity of the analyte and therefore its van der Waals attraction to the sorbent. Brassinosteroids (with a 22,23-diol moiety) were effectively retained as their cyclic boronates on an immobilized phenylboronic acid (PBA) sorbent (Gamoh et al., 1994). The method allowed for the effective removal of impurities before the steroids were freed by treatment with an acetonitrile/hydrogen peroxide mixture.

The isolation of acid and neutral fractions has also proved to be an effective method for the recovery of a variety of compounds, which sometimes prove to be useful chemotaxonomic markers. Another example is provided by Lenheer et al. (1984) who, by fractionating iridoid glucosides and flavanoids from plant leaves, allowed a taxonomy of species to be constructed. The lyophilized plant material was ground and extracted with methanol. The methanolic extract was filtered, evaporated and reconstituted in an appropriate buffer before SPE on a C18 sorbent. Similar methodology was reported by Bazylak et al. (1996) for the separation of phenolic acids and anthocyanin like pigments from gluco-iridoids.

Methodology described for the investigation of biological samples, both here and in the scientific literature, may also be used successfully during bioassay guided fractionation of plant material. The advantages of SPE methodology over solvent extraction for such purposes are the same as those described above for the extraction of urine samples. SPE offers the advantage that highly polar compounds such as the flavanoids and other naturally occurring oxygen ring compounds may be isolated from aqueous plant extracts at neutral pH. The ability to minimize the manipulation of pH is of particular importance in the isolation of phenolic and other labile compounds, which may be prone to pH-catalyzed rearrangement or oxidation.

F. PHARMACEUTICAL AND HERBAL PREPARATIONS

Methods for the extraction of drugs are described in other chapters of this book. However, these examples concentrate on the requirements for the extraction of the target analytes. The successful extraction of drugs from pharmaceutical preparations is not always readily adapted from the methods that have been described for biological fluids, plant material or other biological matrices. In comparison with these matrix types, the concentration of active compounds in pharmaceutical formulations is typically high and the number of target analytes is small. Thus, the extraction chemistry, insofar as it relates to the interaction of a target analyte and a sorbent surface, is rarely complex and may be readily defined.

The analysis of pharmaceutical products is complicated by the very high level of matrix components that are present as excipients to the formulation. Such compounds include binding agents, fillers, coating agents and dissolution regulators in tablets, as well as the formulation bases used in the preparation of syrups, elixirs, creams, and pastes. It becomes apparent that while preconcentration of actives is rarely required prior to analysis, some form of matrix removal is often necessary. In cases where the sample solvent must be changed to effect analysis, such as extracting an aqueous preparation for compounds to be analyzed by GC or GC/MS, then it is better to adapt a method in which the target analyte is retained and other matrix components eliminated. In cases where the sample solvent is compatible with the solvent used by the analytical method then retention of matrix impurities and elimination of the target analyte is a simpler option. SPE offers the opportunity to perform this task either completely or in part and is readily illustrated by a simple example. Wisneski et al. (1988) prepared a liquid perfume for analysis by dilution in ethanol/water (9:1 v/v) and passed the resulting solution through a glass column containing a C8 packing. The sorbent adsorbed some interferences while the active agents (cinnamyl alcohols) passed through unretained. Similar applications can, with appropriate consideration to solution chemistry, be adapted to many diluted pharmaceutical preparations.

A greater challenge is found in the development of SPE methodology for the extraction of creams and gels. Many bases are available for the preparation of creams and gels, the choice being governed by factors such as water solubility, buffering capacity, pH, skin permeability, and other mechanical properties. To illustrate the diversity of structure found in such bases, a few better-known examples are given in Table 7. While the structure of the bases is varied, their partial or full miscibility with water is common by virtue of the high degree of hydroxylation. Many topical preparations may also be formulated to include metal oxides as colorants or protectants.

If using SPE to extract active compounds from such bases, we must consider whether the base is to be separated from the target analyte by the initial matrix manipulation or not. If not, then the base will be transferred to the SPE device where its surface-active properties may significantly alter

Table 7. Common polymeric components of pharmaceutical preparations.

POLYMER	SOLUBILITY	USES
Polyvinyl alcohol	Water-soluble	Viscosity-increasing agent
Polysorbate 80	Water-soluble	Surfactant, emulsifier
Alginates	Depends on counter-ion and pH	Thickener, suspending or wetting agent, surfactant
Polyoxyethylene esters	Depends on chain length	Surfactant and wetting agent
Polyoxyethylene alcohols	Depends on chain length	Surfactant and wetting agent
Polyethylene glycol	Water-soluble	Ointment base
Poloxamers	Variable	Gelling agent and emulsifier
Gum tragacanth	Water-soluble	Dispersing agent, emulsifier and thickener
Gum arabic	Water-soluble	Dispersing agent, emulsifier and thickener

the sorbent chemistry that might be anticipated had it not been present. For example, the application of carboxylated species such as alginates may significantly interfere with the performance of a CBA phase in retaining polar organic cations and hydroxylated species such as polyethylene glycols may interfere with the sorbing properties of silica and C1 or C2 phases. The interference that occurs may sometimes be beneficial and yield cleaner extracts by causing the poor retention of co-extractants while not affecting the retention of target analytes. However, such factors must be determined empirically for each combination of sample formulation, SPE sorbent, and solvent regime.

The role of SPE in the extraction of pharmaceutical creams was well illustrated by Nguyen et al. (1986) and Burke et al. (1988). In these cases, either the active component, a buffer, or stabilizer were extracted and quantitated. SPE acted either to eliminate the matrix or the active components to allow faster, easier analysis. The simplicity of the sample was used to advantage by Burke's laboratory in the removal of matrix components from a preparation containing citrate ions, the only significant component not to show a degree of non-polar retention. Thus, even at less than optimum efficiency, passage of the sample through an off-line cartridge or through an on-line, switchable guard column resulted in retention of all interferences while the citrate was found in the effluent fraction of either extraction system.

A similar analytical problem was resolved by DiPietra (1990), who addressed the analysis of common pharmaceutical and dietary sweetener impurities. The samples were powdered (or were already in powdered form), suspended in the HPLC mobile phase and filtered prior to analysis. The authors found however that by passing the filtered solution through a SPE device, impurities like toluene sulfonamides were removed more conveniently than was possible using traditional Pharmacopoeia methods. Dalbacke and Dahlquist (1991) demonstrated a simple SPE procedure as an alternative to an older, less efficient method, for extraction of vitamin B12 in tablets. Here the analyte was retained on a PH sorbent after passage of the sample through a "scrubber" SAX cartridge to which the analyte does not retain. A similar approach, using a SCX sorbent, was applied in the removal of an antimicrobial preservative interference in a cosmetic formulation (Benazzi et al., 1990) during analysis for formaldehyde. This contrasts with a previously published procedure (Benazzi et al., 1989) for cosmetics lacking the preservative where there was no need for the additional stage of extraction and samples were analyzed directly by HPLC.

Applications in which the target analyte is retained by the sorbent have also been reported. For example, 1,4-dioxane was determined in cosmetic products following extraction on silica and C18 sorbents (Scalia et al., 1992). The extraction of cough elixirs for basic drugs is readily achieved by adaptation of methods designed for the extraction of biological fluids on a C8/SCX mixed mode sorbent. The vehicle in this case was a glycerophosphate syrup which has the potential to ion-pair with basic drugs under the conditions required for ion exchange. Use of appropriate dilution with a phosphate buffer and initial retention on the C8 phase eliminated interference by the syrup (Wynne, unpublished results).

The lipid-soluble vitamins A and E have also been determined following the SPE extraction of cosmetic oils and creams (Kountourellis et al., 1992). The extraction and analysis of toothpaste for licorice compounds, which possess mild anti-inflammatory properties, required that the sample be solubilized under alkaline conditions and extracted with hexane before the aqueous solution was applied to a C18 sorbent (Andrisano et al., 1993). Methanol/water wash and elution schemes were optimized for different toothpaste formulations. The raw materials to be used in the finished product were easier to analyze, as demonstrated by Schulz and Albroscheidt (1988). The matrix, though still likely to contain plant sugars, acids and colorants, does not contain the co-formulated excipients and therefore does not present many of the problems associated with them.

This section cannot hope to cover all possible cosmetic or pharmaceutical types to which SPE has been applied. However, the general features of these samples and common strategies for analysis are well covered by the examples given. Whenever the target analyte is present at high levels in a matrix containing difficult components, it is important that the analyst pays particular attention to the nature of these components and designs methods carefully, to eliminate them or at least negate their effect on the sample preparation.

G. INDUSTRIAL MATERIALS AND WASTES

Most industrial processes yield relatively simple product matrices, where the desired compounds are in relatively high concentrations. In such circumstances, it is possible to use in-process monitoring using techniques such as Fourier-transform infrared spectrometry (FTIR), and there is no need for sample clean-up or concentration. Indeed, a sample preparation step is not just unnecessary; often it is impractical for most industrial product streams. Where SPE does have a use it is largely in the areas of product impurity analysis and in the monitoring of wastes from the manufacturing process. Thus, Bitteur and Rosset (1988) describe the extraction of an aroma compound from the waste stream from a food manufacturer. Interestingly this paper explores the issue of breakthrough volumes for analytes for both silica-based C18 and styrene di-vinyl benzene co-polymer sorbents.

Examples of airborne particulate analysis were given in section IV.C. Identical extraction procedures apply to airborne ash residues that are trapped on filters or electrostatically precipitated. Soxhlet extraction of the sample followed by polar SPE fractionation (Kleinveld et al., 1989) provides appropriately clean extracts for GC/MS analysis.

A closely related analytical problem is described by Chen et al., (1994) who perform a Soxhlet organic solvent extraction on powdered sawdust from various wood chips. The acetone extract is then applied to NH2 sorbent cartridges. A slight modification of the lipid fractionation method developed by Kaluzny et al., (1985) provides an appropriate method for analyte fractionation — in this case lipidic materials in the wood chips.

Metals may be successfully extracted from complex mixtures such as adhesives (Mooney et al., 1987) by dissolution of the matrix in an organic solvent. This is then extracted with aqueous acid, non-inorganic extractives are removed on a C18 cartridges, and metal-complexing agents are then added to the C18 effluent. The resulting mixture is then fractionated on a polar SPE cartridge.

A different approach is taken by Frenzel (1998) who uses a complexing agent, 1-5 diphenylcarbazide, immobilized on the sorbent bed by pretreatment with a solution containing this compound. Here the aqueous sample is passed through an extraction disc and Chromium (VI) ions are trapped by the immobilized complexing agent. The analysis (semi-quantitative) was performed using simple colorimetric comparison against discs prepared in an identical manner using solutions prepared at known concentrations. This method was applied to trace level Cr (VI) in water samples. A similar approach is illustrated in Figure 10 (Schwedt and Schunck, 1982).

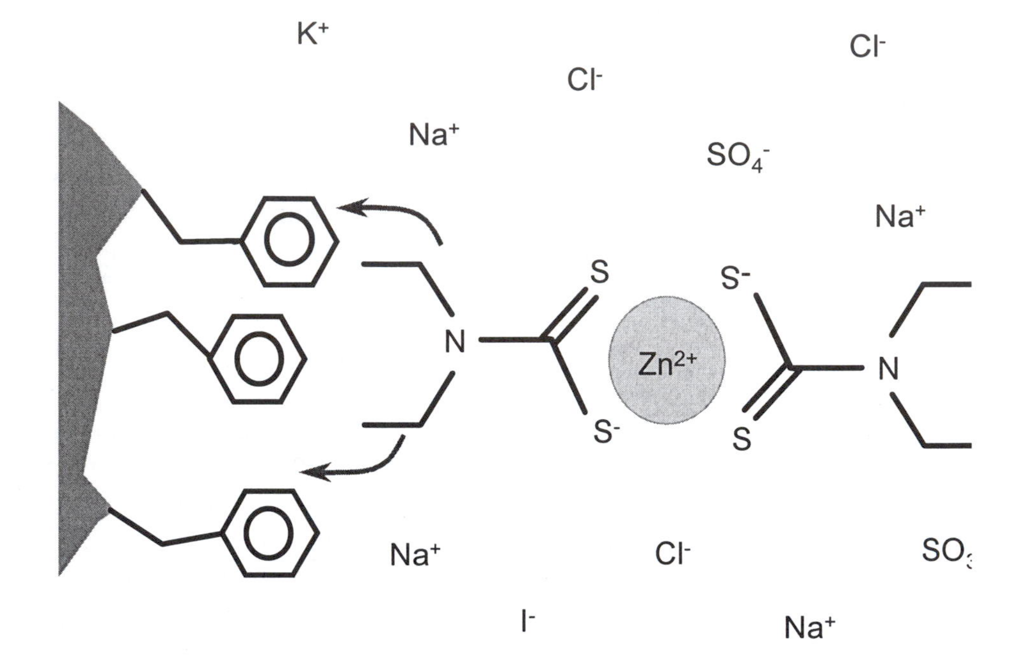

Figure 10. Demonstration of a bulk solution-phase chelation process which permits extraction of certain di- and tri-valent metal ions to be extracted from a sample which contains a large excess of mono-valent metal ions.

Zajicek and Lebo (1992) applied the SPE argentation technique, as pioneered by Christie (1989) and described in Section IVD, to organic solutions containing PCBs. These solutions could model for hexane extracts of petroleum products like transformer oils.

Increasingly, specialised sorbents featuring covalently bound chelating groups are being reported for ion-chromatography and HPLC applications. Kumagai et al. (1998) have prepared iminodiacetate-type chelating resins for the retention of rare earth cations from aqueous solution, the elution of the retained species being achieved by acidification with nitric acid. In a further example, an immobilized crown ether sorbent has been used to extract Pb^{2+} ions from acidic solution and eluted with a competing ligand such as oxalate or citrate. The method was found to be suitable for the extraction of soils and other geological samples following dissolution in nitric acid (Sooksamiti et al., 1996). Such sorbents are suitable for SPE purposes provided that the distribution constant for the metal ion-chelate complex is sufficiently high to ensure adequate retention.

H. REACTION PRODUCTS

One of the most familiar but seldom-considered applications of SPE is the removal of colored compounds from reaction mixtures using animal charcoal. However, it is rare for any investigation of the mechanism of the extraction to be undertaken. This is unfortunate because SPE has a lot to offer the synthetic chemist, especially as synthesis progresses to ever smaller volumes and reactions of ever greater complexity are performed, in the rush to exploit the opportunities opened up by precision liquid-handling robotics and combinatorial chemistry technologies.

Significant applications of modern SPE to reaction products have already been discussed in the previous section on industrial applications, since many of these products are formulations of the products of earlier reactions. Impurities arising from the reaction by-products or raw materials present the most suitable target for SPE since the benefits of trace enrichment can hardly apply to a sample which is already at least 95% analyte. However, structural similarities between reaction product and by-products often mean that one cannot be extracted without the simultaneous extraction of the other. The limited capacity of an SPE device is a problem in such cases and therefore higher capacity extraction techniques like LLE or solid-supported LLE are favored. However, SPE has been successfully performed on reaction products.

Several examples may also be found of SPE applications used in the purification of reaction products following their crude synthesis or isolation. Mixed mode SPE originally developed for the extraction of drugs of abuse from urine has been successfully applied to the extraction of the highly water-soluble compound 6-amino-4-methyl-pyrid-2-one from crude synthesis mixtures (Wynne, unpublished results). Large-scale enantiomeric purification of a variety of β-adrenergic blocking drugs and hyoscyamines has been reported by Chen (1997), using beta-cyclodextrin. The degree of enrichment was found to be strongly dependent on both the polarity and protic nature of the matrix solvent.

The capacity issue and benefits of automated SPE for such work are well illustrated by Lawrence et al., (1996). In this case the analytes were a family of amide drug candidates, which had been synthesized by solution-phase synthesis. Use of SPE permitted solution-phase synthetic reagents to be eliminated easily. Furthermore, the entire process described by the authors — solution-phase synthesis and clean-up — could be performed on one laboratory robot.

I. OTHER MATRICES

The use of a diluent buffer as a carrier fluid is particularly effective for the analysis of pharmaceutical preparations or residues in syringes, vials, nebulizers, and other equipment for drug residues. In such cases the sample is obtained by washing the apparatus with or dissolving the preparation in a buffer suitable for the application and extracting the buffer solution as if it was a urine sample.

When the analyte is impregnated into a material a pre-SPE extraction will be required that is similar to that employed for many solid sample types discussed in this chapter. For example, a dermatological patch, such as would be applied to treat skin injuries or complaints like psoriasis, can be broadly defined as containing elastomer, adhesive, plasticizer, tackifier, anti-oxidants (all more or less hydrophobic) and gelatin, pectin, carboxymethyl cellulose salts and other hydrophobic species. Edwardson and Gardner (1990) describe analysis of such a material in which a liquid-solid extraction was performed, followed by a back LLE, to assure clean-up. Although SPE was used and claimed to be accurate the authors were not happy with its precision (greater than 2% RSD); method optimization could probably improve upon this. A last example of this approach (Zander et al., 1998) uses a simple organic solvent solubilization of chewing gum, to extract nicotine and its oxidation products. By use of molecularly imprinted polymer solid phases, the analytes could be extracted with good efficiency while the matrix components that were also solubilized failed to retain on this specialized polymer. For a treatment of molecularly imprinted solid phases, turn to Chapter 12.

In contrast, an aqueous extraction medium was used by Zwickenpflug and Richter (1987) to extract N-nitroso butylamine from rubber used in pacifiers and baby feeding bottle nipples. This approach was feasible because the goal of the analysis was to estimate the levels of compound extracted by the baby's saliva during normal use of the products — not the total content of analyte in the sample. Hence an aqueous buffer, chosen so as to mimic saliva's solubilizing properties, was used and a simple C18 clean-up completed the sample preparation.

A similar approach could be used to perform the currently fashionable analysis for hormone-mimicking phthalates present in plastics that find their way into babies mouths (in other words, any household item from a baby pacifier to a Barbie™[9] doll to the dog's plastic bone).

[9] Barbie is a trademark of the Mattel Corporation, El Segundo, CA.

J. LARGE-SCALE APPLICATIONS

Application of ion-exchange materials for the medium to large scale desalting of water is well known and will not be further considered here. Charcoal and mineral oxides have also found application for water purification. Similar but considerably more analyte-specific technology has been successfully applied to the removal of heavy metals such as cesium and strontium from nuclear waste (Brown et al., 1995). In this case the extraction device is a disk impregnated with a sorbent containing chelating bonded phase or, less complex, a simple ion exchanger. A similar removal of radioactive species could be achieved with column chromatography using much larger sorbent particles. However, the authors of this paper argue that with the SPE disk they can process larger volumes of liquid and do so faster than permitted by column chromatography. This is because the small sorbent particles in the SPE disk give better extraction kinetics while the short bed depth of the disk ensures low flow resistance.

Another approach to specific removal of metal ions is to apply a known excess of complexing/chelating species to the sample before extraction. Thus, to quantify specific metal ions such as mercury in a large volume water sample, diethyl di-thiocarbamate could be added and the resulting organo-metal complexes could be extracted on a non-polar phase (Schwedt and Schunck, 1982; Frenzel, 1998). The benefit of this approach is that higher capacities result than are found in simple ion exchangers. Hence larger samples can be extracted. Flow rate, because the kinetics of non-polar retention are usually faster than those of ion exchange, can be higher, leading to more efficient sample processing. Finally, extraction can proceed even in the presence of a large number of competing ions — thus, seawater becomes a viable matrix. The reverse process — adding a solution of metal ions to a sample containing an organic species that can bind to the metal ions, followed by extraction of the metal-ion complex — is also feasible.

Most solid-phase extractions could be converted from analytical to process scale. However, SPE offers fewer benefits to the process chemist than to the analytical chemist. Once the analyst has employed the low capacity and high selectivity of SPE to identify that an industrial site is contaminated by PCBs, for example, the engineer may then turn to cruder bulk separation techniques like bubble floatation, foam trapping or microbial action to perform soil remediation. However, a few bulk scale solid-phase extractions (for example, Shannon and Ensley) have been proposed.

V. CONCLUSIONS

In this chapter we have seen examples of the extraction of a wide variety of matrices for an equally diverse set of reasons. Whatever the sample type, SPE has been found to have a useful role to play in the clean-up and concentration steps of the analysis, either by itself or in combination with a LLE or some other manipulation (Soxhlet, grinding, leaching, etc.). In some cases SPE provides little more than a final "polish" after application of a carefully developed set of matrix manipulations, but in others the application of SPE is essential to successful analysis.

In recent years supercritical fluid extraction (SFE) and MSPD have been applied to solid or viscous liquid samples as alternatives to the grinding and liquid-solid separation which constitute the majority of preparations described in this chapter. SFE and MSPD are dealt with later in this book, since they bring with them a distinct set of advantages and difficulties.

This chapter is the last of the general chapters in this text. All subsequent chapters address an area of analytical inquiry, or review technologies which affect, or are affected by SPE. While we cannot claim to have covered all the matrix types and applications in this (long) chapter, the major ones have been addressed. You will have enough information to be well-equipped to apply SPE to your field of research.

A summary of matrix types and properties is given in Appendix 1.

REFERENCES

Abusada, G.M., Abukhalaf, I.K., Alford, D.D., Vinzon-Bautista, I., Pramanik, A.K., Ansari, N.A., Manno, J.E. and Manno, B.R., (1993) Solid Phase Extraction and GC/MS Quantitation of Cocaine, Ecgonine, Benzoylecgonine and Cocaethylene from Meconium, Whole Blood and Plasma. J.Analyt.Toxicol., 17: 353-358.

Amelio, M., Rizzo, R. and Varazini, F., (1992) Determination of Sterols, Erythrodiol, Uvaol and Alkanols in Olive Oils using Solid Phase Extraction, High-Performance Liquid Chromatographic and High-Resolution Gas Chromatographic Techniques. J.Chromatogr., 606(2): 179-185.

Andrisano, V., Cavrini, V. and Bonazzi, D., (1993) HPLC Determination of 18 beta-Glycyrrhetinic and Glycyrrhizinic Acids in Toothpastes after Solid Phase Extraction. Chromatographia, 35(3/4): 167-172.

Applied Separations Inc., (1994) Direct Addition of Whole Blood into SPE Cartridge for Drugs of Abuse Confirmations. J. Analyt. Toxicol., 18: 432.

Bateman, H. G. and Jenkins, T. C., (1997) Method for Extraction and Separation by Solid-Phase Extraction of Neutral Lipid, Free Fatty Acids, and Polar Lipid from Mixed Microbial Cultures. J.Agric.Food Chem., 45(1): 132-134.

Batty, D.C., Wynne, P.M. and Vine, J.H., (1994) Extraction of Acidic and Basic Drugs from Equine and Canine Blood and Urine Using a Single Solid Phase Extraction Cartridge. Proceedings 12th ANZFS Symposium, Wellington, New Zealand, 12-19.

Bazylak, G., Rosiak, A., Shi, C.-Y., (1996) Systematic Analysis of Glucoiridoids from *Penstemon serrulatus* Menz. by High-Performance Liquid Chromatography with Pre-Column Solid-Phase Extraction. J.Chromatogr., A, 725(1), 177-187

Benazzi, C.A., Semenzato, A. and Bettero, A., (1989) High-performance Liquid Chromatographic Determination of Free Formaldehyde in Cosmetics. J.Chromatogr., 464: 387-393.

Benazzi, C.A., Semenzato, A., Zaccaria, F. and Bettero, A., (1990) High-performance Liquid Chromatographic Determination of Free Formaldehyde in Cosmetics Preserved with Dowicil 200. J.Chromatogr., 502: 193-200.

Bitteur S. and Rosset, R., (1998) Use of an Octadecyl-Bonded Silica and a Styrene-Divinylbenzene Copolymer for the Recovery of Blackcurrant Aroma Compounds from a Food Plant Waste Water. J.Food.Sci., 53(1): 141-147.

Bradburn, N., Coker, R.D., Jewers, K. and Tomlins, K.I., (1990) Evaluation of the Ability of Different Concentrations of Aqueous Acetone, Aqueous Methanol and Aqueous Acetone:Methanol (1:1) to Extract Aflatoxins from Naturally Contaminated Maize. Chromatographia, 29(9/10): 435-440.

Brown, G. N., Bray, L. A., Lundholm, C. W., Lewis, L., White, L.R., Kafka, T. M., Decker, Robert, Bruening, R. L., (1995) High Level Radioact. Waste Manage., Proc. 6th Ann. Int. Conf., 1995, 679-681 Publisher: American Nuclear Society, La Grange Park, IL.

Burke, E., Zimmerman, S.R., Brown, D.S. and Jenke, D.R., (1988) On-line Sample Cleanup in the Liquid Chromatographic Analysis of Pharmaceuticals for Citrate Content. J.Chromatogr.Sci., 26: 527-532.

Busto, O., Valero, Y., Guasch, J, and Borrull, F., (1994) Solid Phase Extraction Applied to the Determination of Biogenic Amines in Wines by HPLC. Chromatographia 38 (9/10): 571-578.

Buszewski, B., Kawka, S., Lodkowski, R., Suprynowicz, R. and Wolski, T., (1992) The Influence of Packing Properties on the Isolation of Rutin from Plants and Drugs using Solid-Phase Extraction. J.Liq.Chromatogr., 15(11): 1957-1570.

Buszewski, B., Kawka, S., Suprynowicz, Z. and Wolski, T., (1993) Simultaneous Isolation of Rutin and Esculin from Plant Material and Drugs using Solid-Phase Extraction. J.Pharm.Biomed.Anal., 11(3): 211-215.

Cappiello, P. E. and Kling, G. J., (1990) Determination of Zeatin and Zeatin Riboside in Plant Tissue by Solid-Phase Extraction and Ion-Exchange Chromatography. J.Chromatogr. , 504(1), 197-201.

Carmichael, W.W., (1982) Chemical and Toxicological Studies of the Toxic Freshwater Cyanobacteria *Microcyctis aeruginosa*, *Anabaene flos-aquae* and *Aphanizomenon flos-aquae*. South African J. of Science 78: 367-372.

Chauret, N., Rho, D. and Archambault, J., (1990) Solid-Phase Extraction and Fluorimetric Detection of Benzophenanthridine Alkaloids from *Sanguinaria canadensis* Cell Cultures. J. Chromatogr., 519(1): 99-107.

Chen, S., (1997) Large-scale Enantiomeric Enrichment via Solid State Extraction using Beta-cyclodextrin Crystalline and Anhydrous Acetonitrile as Solid and Liquid Phases. J.Chin.Chem.Soc. (Taipei), 44(6): 629-633.

Chen, T., Breuil, C., Carriere, S. and Hatton, J.V., (1994) Solid-phase Extraction Can Rapidly Separate Lipid Classes from Acetone Extracts of Wood Pulp. Tappi Journal. 77(3): 235-240.

Chen, X-H., Franke, J-P., Wijsbeek, J. and de Zeeuw, R.A., (1994) Determination of Basic Drugs Extracted from Biological Matrices by Means of Solid-phase Extraction and Wide-bore Capillary Gas Chromatography with Nitrogen-Phosphorus Detection. J. Analyt. Toxicol. 18: 150-153.

Chen, X-H., Franke, J-P., Wijsbeek, J. and de Zeeuw, R.A., (1992) Isolation of Acidic, Neutral and Basic Drugs from Whole Blood Using a Single Mixed-mode Solid-phase Extraction Column. J.Anal.Toxicol., 16:352-355.

Christie, W. W., (1989) Silver Ion Chromatography Solid Phase Extraction Columns Packed with a Bonded-sulfonic Acid Phase. J.Lipid Res., 30(9): 1471-1473.

Christie, W. W., (1990) Silver Ion Chromatography of Triacylglycerols on Solid phase Extraction Columns Packed with a Bonded Sulfonic Acid Phase. J.Sci.Food Agric., 52(4): 573-577.

Conte, L. S., Koprivnjak, O., Fiorasi, S., Pizzale, L., (1997) Solid-phase Extraction Applied to Diacylglycerol Determination in Foods. Riv.Ital.Sostanze Grasse, 74(9): 411-414.

Costa Neto, C., Pinto, R. C. P. and Macaira, A. M. P., (1978) Separation and Identification of Aldehydes and Ketones from an Irati Oil Shale Bitumen. Use of the Solid Phase Extraction Technique, Oil Sand Oil Shale Chem., Proc. Symp. (1978), Meeting Date 1977, 345-358. (Editors: Strausz, Otto P., Lown, Elizabeth M.). Publisher: Verlag Chem. Int., New York, N. Y.

Cox, J. A., Alber, K. S., Brockway, C. A., Tess, M. E. and Gorski, W., (1995) Solid-Phase Extraction in Conjunction with Solution or Solid State Voltammetry as a Strategy for the Determination of Neutral Organic Compounds. Anal.Chem., 67(5): 993-998.

Crepeau, K.L., Walker, G. and Winterlin, W., (1991) Extraction of Pesticides from Soil Leachate Using Sorbent Disks. Bull.Environ.Contam.Toxicol., 46: 512-518.

Dalbacke, J. and Dahlquist, I., (1991) Determination of Vitamin B12 in Multivitamin-multimineral Tablets by High-performance Liquid Chromatography after Solid-Phase Extraction. J.Chromatogr., 541: 383-392.

David, F., Nikolai, A. and Sandra, P., (1989) Analysis of C10-C20 hydrocarbons in Natural Gas by Solid-Phase Extraction and CGC [capillary GC], J. High Res. Chromatogr., 12(10): 657-660.

DeJong, C. and Badings, H.T., (1990) Determination of Free Fatty Acids in Milk and Cheese. Procedures for Extraction, Clean-up and Capillary Gas Chromatographic Analysis. J.High Res.Chromatogr., 13: 94-98.

Di Muccio, A., Rizzica, M., Ausili, A., Camoni, I., Dommarco, R. and Vergori, F., (1988) Selective On-column Extraction of Organochlorine Pesticide Residues from Milk. J.Chromatogr., 456: 143-148.

Di Pietra, A.M., Cavrini, V., Bonazzi, D. and Benfenati, L., (1990) HPLC Analysis of Aspartame and Saccharin in Pharmaceutical Dietary Formulations. Chromatographia, 30 (3/4): 215-219.

Edwardson, P.A.D. and Gardner, R.S., (1990) Problems Associated with the Extraction and Analysis of Triamcinolone Acetonide in Dermatological Patches. J.Pharm. & Biomed.Anal., 8(8/12): 935-938.

Egberts, J and Buiskool, R., (1988) Isolation of the Acidic Phospholipid Phosphatidylglycerol from Pulmonary Surfactant by Sorbent Extraction Chromatography. Clin.Chem., 34(1): 163-164.

Frenzel, W., (1998) Highly Selective, Semi-quantitative Field Test for the Determination of Chromium (VI) in Aqueous Samples. Frezenius J.Anal. Chem., 1455: 1-6.

Gamoh, K., Yamaguchi, I. and Takatsuto, S., (1994) Rapid and Selective Sample Preparation for the Chromatographic Determination of Brassinosteroids from Plant Material using Solid-phase Extraction Method, Anal. Sci., 10(6): 913-917.

Geigy, J.R., (1956), Documenta Geigy Scientific Tables, 5th Edition, J.R. Geigy S.A., Basle.

Gelsomini, N., Capozzi, F. and Faggi, C., (1990) Separation and Identification of Volatile and Non-volatile Compounds in Wine by Sorbent Extraction and Capillary Gas Chromatography. J.High Res.Chromatogr., 13: 352-355.

Gennaru, M., Zanini, E., Cignetti, A., Bicchi, C., D'Amato, A., Taccheo, M.B., Spessotto, C., DePaoli, M., Flori, P., Imbroglini, G., Leandri, A and Conte, E., (1985) Vinclozolin Decay on Different Grapevines in Four Differing Italian Areas. J.Agric.& Food Chem. 33: 1232-1237.

Gessner, M. O. and Schmitt, A. L., (1996) Use of Solid-phase Extraction to Determine Ergosterol Concentrations in Plant Tissue Colonized by Fungi. Appl.Environ.Microbiol., 62(2): 415-419.

Glowniak, K., Zgorka, G. and Kozyra, M., (1996) Solid-phase Extraction and Reversed-phase High-performance Liquid Chromatography of Free Phenolic Acids in some *Echinacea* Species. J.Chromatogr., A, 730(1 + 2): 5-29.

Golding, M. and Wynne, P., (1988) Therapeutic devices, Provisional Australian Patent Application PI9321, 15 July 1988 to Biota Scientific Management Pty Ltd.

Gross, G.A., (1990) Simple Methods for Quantifying Mutagenic Heterocyclic Aromatic Amines in Food Products. Carcinogenesis, 11(9): 1597-1603.

Guinn, G. and Brummett, D. L., (1990) Solid-phase Extraction of Cytokinins from Aqueous Solutions with C18 Cartridges and their Use in a Rapid Purification Procedure. Plant Growth Regul., 9(4): 305-314.

Hansen-Moller, J., (1992) Determination of Indolic Compounds in Pig Back Fat by Solid Phase Extraction and Gradient High-performance Liquid Chromatography with Special Emphasis on the Boar Taint Compound Skatole. J.Chromatogr., 624(1-2): 479-490.

Harkey, M.R., Henderson, G.L. and Zhou, C., (1991) Simultaneous Quantitation of Cocaine and its Major Metabolites in Human Hair by Gas Chromatography/Chemical ionization Mass Spectrometry. J.Anal.Toxicol., 15: 260-265.

Harkey, M. R. and Stolowitz, M. L., (1984) Solid-Phase Extraction Techniques for Biological Specimens. Adv.Anal.Toxicol., 1: 255-270.

Hiemstra, M., Joosten, J. A. and de Kok, A., (1995) Fully Automated Solid-phase Extraction Cleanup and Online Liquid Chromatographic Determination of Benzimidazole Fungicides in Fruit and Vegetables. J.AOAC Int, 78(5): 1267-1274.

Hopia, A. I., Piironen, V. I., Koivistoinen, P. E. and Hyvonen, L. E. T., (1992) Analysis of Lipid Classes Solid Phase Extraction and High-performance Size-Exclusion Chromatography. J.Am.Oil Chem.Soc., 69(8): 772-776.

Horie, M., Saito, K., Hoshino, Y., Nose, N., Hamada, N. and Nakazawa, H., (1990) Identification and Determination of Sulphamethazine and N-4-acetylsulphamethazine in Meat by High-performance Liquid Chromatography with Photodiode-array Detection. J.Chromatogr., 502: 371-378.

Horie, M., Saito, K., Nose, N., Nakazawa, H. and Yamane, Y., (1991) Simultaneous Determination of Residual Synthetic Antibacterials in Fish by High-performance Liquid Chromatography. J.Chromatogr., 538: 484-491.

Huizing, M.T., Rosing, H., Koopmans, F.P. and Beijnen, J.H., (1998) Influence of Cremophor EL on the Quantification of Paclitaxel in Plasma Using High-performance Liquid Chromatography with Solid-Phase Extraction as Sample Pretreatment. J. Chromatogr., B., 709;161-165.

Ingwersen, S.H., (1994) A Note on Solid Phase Extraction of the Highly Protein Bound Dopamine Uptake Inhibitor GBR 12909. Sample Preparation for Bio-medical and Environmental Analysis (Eds. D. Stevenson and I.D. Wilson) Plenum Press, NY. pp 139-142.

Iwase, H. and Ono, I., (1997), Determination of Cyanocobalamin in Foods by High-performance Liquid Chromatography with Visible Detection after Solid-phase Extraction and Membrane Filtration for the Precolumn Separation of Lipophilic Species. J. Chromatogr., A, 771(1+2): 127-134

Jelinek, I., (1994), Activated Silicone Rubber as a Sorbent for Solid-phase Extraction of Volatiles from Aqueous Solutions. Chem. Pap., 48(4): 229-235.

Kaluzny, M.A., Duncan, L.A., Merritt, M.V. and Epps, D.E., (1985) Rapid Separations of Lipid Classes in high Yield and Purity Using Bonded Phase Columns. J.Lipid Res., 26: 135-140.

Katalinic, V. (1997), High-Performance Liquid Chromatographic Determination of Flavan Fingerprints in Plant Extracts. J.Chromatogr., A, 775(1 + 2): 359-367.

Kim, H. Y. and Salem, N., Jr., (1990) Separation of Lipid Classes by Solid-phase Extraction, J. Lipid Res., 31(12): 2285-2289.

Kim, I.S., Sasinos, F.I., Stephens, R.D. and Brown, M.A., (1990) Analysis for Daminozide in Apple Juice by Anion-exchange Chromatography--Particle Beam Mass Spectrometry. J.Agric.Food Chem., 38:1223-1226.

Karlesky, D.L., Rollie, M.E., Warner, I.M. and He, C-N., (1986) Sample Cleanup Procedure for Polynuclear Aromatic Compounds in Complex Matrices. Anal.Chem., 58(6): 1187-1192.

Kleinveld, A.H., Verhoeve, P. and Nielen, M.W.F., (1989) Evaluation of Clean-up Methods for the Determination of Chlorinated Aromatic Hydrocarbons in Fly-Ash by Capillary Gas Chromatography and Mass Selective Detection. Chemosphere 18(7/8): 1401-1412.

Kootstra, P. R. and Herbold, H. A., (1995) Automated Solid-phase Extraction and Coupled-column Reversed-phase Liquid Chromatography for the Trace-level Determination of Low-molecular-mass Carbonyl Compounds in Air. J. Chromatogr., A, 697(1 + 2): 203-211.

Koprivnjak, O., Bertacco, G., Boschelle, O. and Conte, L. S., (1995) Determination of Diglycerides in Virgin Olive Oil, Salami and Cheese by Solid-Phase Extraction. Prehrambeno-Tehnol.Biotehnol. Rev., 33(2-3): 97-101.

Kountourellis, J. E., Markopoulou, C. K. and Nathaniel, B. K., (1992 Determination of Vitamin A and E in Therapeutic and Cosmetic Creams Solid Phase Extraction (SPE) and High Performance Liquid Chromatography (HPLC). Chem.Environ.Res., 1(2): 83-87.

Kumagai, H., Inoue, Y., Yokoyama, T., Suzuki, T.M. and Suzuki, T., (1998) Chromatographic Selectivity of Rare Earth Elements on Iminodiacetate-type Chelating Resins Having Spacer Arms of Different Lengths: Importance of Steric Flexibility of Functional Group in a Polymer Chelating Resin. Anal.Chem., 70(19): 4070-4073.

Lawrence, R.M., Fryszman, O, M., Poss, M.A., Biller, S.A. and Weller, H.N., (1996) Automated Preparation and Purification of Amides. Clinical Laboratory, June: 15-20.

Lenheer, A., Meier, B. and Sticher, O., (1984) Modern HPLC as Tool for Chemotaxonomical Investigations: Iridoid Glycosides and Acetylated Flavonoids in the Group of *Stachys Recta*. Planta Medica 5 (Oct.): 365-458.

Li, N. and Lee, H.K., (1997) Trace Enrichment of Phenolic Compounds from Aqueous Samples by Dynamic Ion-Exchange Solid-Phase Extraction. Anal.Chem., 69(24): 5193-5199.

Litridou, M., Linssen, J., Schols, H., Bergmans, M., Posthumus, M., Tsimidou, M. and Boskou, D., (1997) Phenolic Compounds in Virgin Olive Oils: Fractionation by Solid-Phase Extraction and Antioxidant Activity Assessment. J.Sci.Food Agric., 74(2): 169-174.

Luke, M.A., (1995) The Evolution of a Multiresidue Pesticide Method. ACS Conference Proceedings Series, 8th International Congress of Pesticide Chemistry: Options 2000 (Eds. Ragsdale, N.N., Kearney, P.C. and Plimmer, J.R.) pp. 174-182.

Luke, M. A., Yee, S., Nicholson, A. E., Cortese, K. M. and Masumoto, H. T., (1996) Analytical Approach of Multiresidue Analysis of Foods by the Use of Solid-Phase Extraction Technology. Semin.Food Anal., 1(1): 11-26.

Manes, J., Font, G. and Pico, Y., (1993) Evaluation of a Solid-Phase Extraction System for Determining Pesticide Residues in Milk. J.Chromatogr., 642(1-2): 195-204.

Markell, C.G., (1994) Personal communication. C.G. Markell, New Products Division, 3 M Corporation, Minneapolis/St. Paul, Minnesota.

Markell, C.G. and Hagen, D.F., (1991) Extraction of Phenolic Compounds from Water Samples using Styrene-Divinylbenzene Disks. 7th Annual Waste Testing and Quality Assurance Symposium. 2:27-37

Martens, D. A. and Frankenberger, W. T., Jr., (1991) Online Solid-Phase Extraction of Soil Auxins Produced from Exogenously-Applied Tryptophan with Ion-Suppression Reverse-Phase HPLC Analysis. Chromatographia, 32(9-10): 417-422.

Matisova, E., Strakova, M., Skrabakova, S. and Novak, I., (1995) A Novel Porous Carbon and its Perspectives for the Solid Phase Extraction of Volatile Organic Compounds. Fresenius' J.Anal.Chem., 352(7-8): 660-666.

McIntyre, I.M., Syrjanen, M.L., Crump, K, Horomidis, S., Peace, A.W. and Drummer, O., (1993) Simultaneous HPLC Gradient Analysis of 15 Benzodiazepines and Selected Metabolites in Postmortem Blood. J.Analyt.Toxicol., 17: 202-207.

Meier, B., Jalkmann-Tiitto, R, Tahvanainen, J. and Sticher, O., (1988) Comparative High-Performance Liquid and Gas-Liquid Chromatographic Determination of Phenolic Glucosides in *Silicaceae* Species. J.Chromatogr., 442: 175-186.

Moeller, M.R., Fey, P. and Rimbach, S., (1992) Identification and Quantitation of Cocaine and its Metabolites, Benzoylecgonine and Ecgonine Methyl Ester, in Hair of Bolivian Coca Chewers by Gas Chromatography/Mass Spectrometry. J.Anal.Toxicol., 16: 291-296.

Moeller, M.R., Fey, P. and Wennig, R., (1993) Simultaneous Determination of Drugs of Abuse (Opiates, Cocaine and Amphetamine) in Human Hair by GC/MS and its Application to a Methadone Treatment Program. Forensic Sci. Int., 63: 185-206.

Moffat, A.C., Senior Consulting Editor (1986), Clarke's Isolation and Identification of Drugs, 2nd edition., The Pharmaceutical Press (London), 112.

Monagle, M., Hakonson, K. and Roberts, J., (1995) Reduction of Radioactivity in Oils Prior to PCB analysis. A New Use for Solid-Phase Extraction Cartridges, Mixed Waste, Proc. Bienn. Symp., 3rd (1995), 2.8.1-.8.6., Editor(s): Moghissi, A., Love, B. and Blauvelt, R., Publ: Cognizant Communication Corp., Elmsford, N.Y.

Mooney, J.P., Meaney, M., Smyth, M.R., Leonard, R.G. and Wallace, G.G., (1987) Determination of Copper (II) and Iron (III) in Some Anaerobic Adhesive Formulations using High-Performance Liquid Chromatography. Analyst, 112(Nov.): 1555-1558.

Moore, C.M., Brown, S., Negrusz, A., Tebbett, I., Meyer, W. and Jain, L., (1993) Determination of Cocaine and its Major Metabolite, Benzoylecgonine, in Amniotic Fluid, Umbilical Cord Blood, Umbilical Cord Tissue and Neonatal Urine: A Case Study. J.Analyt.Toxicol., 17: 62.

Naaranlahti, T., Nordstrom, M. and Lapinjoki, S. P., (1990) Isolation of *Catharanthus* Alkaloids by Solid-Phase Extraction and Semipreparative HPLC. J.Chromatogr. Sci., 28(4): 173-174.

Neff, W. E., Zeitoun, M. A. M. and Weisleder, D., (1992) Resolution of Lipolysis Mixtures from Soybean Oil by a Solid Phase Extraction Procedure. J.Chromatogr., 589(1-2): 353-357.

Newman, R.T., Dietz, W.R. and Latheridge, K., (1996) The Use of Activated Carbon Strips for Fire Debris Extractions by Passive Diffusion, Part 1: The Effects of Time, Temperature, Strip Size, and Sample Concentration. J.Forensic Sci. 41(3): 361-370.

Nguyen, T.T., Kringstad, R, and Rasmussen, K.E., (1986) Use of Extraction columns for the Isolation of Desonide and Parabens from Cream and Ointments for High-Performance Liquid Chromatographic Analysis. J.Chromatogr., 366: 445-450.

Norman, H. A., Mischke, C. F., Allen, B. and Vincent, J., (1996) Semi-preparative Isolation of Plant Sulfoquinovosyldiacylglycerols by Solid-Phase Extraction and HPLC Procedures. J.Lipid Res. , 37(6): 1372-1376.

Omori, H., Okabe, K., Nakashizuka, T, and Yamazaki, S., (1986) Determination of Prostaglandins in Human Saliva by High-Performance Liquid Chromatography Using a Column-switching Technique. J.High Res.Chromatogr. & Chromatogr. Commun. 9(Aug.): 477-479.

Ono, K., Yamamoto, Y., Kosaka, T. and Takeda, O., (1996) A Simple Method for Determining the Presence of Tributyltin and Triphenyltin in Hatchery Fish. Miyazaki-ken Eisei Kankyo Kenkyusho Nenpo, 8: 76-81.

Papadopoulos, G. K. and Tsimidou, M., (1992) Rapid Method for the Isolation of Phenolic Compounds from Virgin Olive Oil using Solid-Phase Extraction. Bull. Liaison - Groupe Polyphenols, 16(Pt. 2): 192-196.

Patel, R. M., Benson, J. R., Hometchko, D. and Marshall, G., (1990) Polymeric Solid-Phase Extraction of Organic Acids. American Lab. (Fairfield, Conn.), 22(3), 92, 94-9.

Pensabene, J. W., Fiddler, W. and Gates, R. A., (1992) Solid Phase Extraction Method for Volatile N-nitrosamines in Hams Processed with Elastic Rubber Netting. J. AOAC Int., 75(3): 438-442.

Perez-Camino, M. C., Moreda, W. and Cert, A., (1996) Determination of Diacylglycerol Isomers in Vegetable Oils by Solid-Phase Extraction Followed by Gas Ghromatography on a Polar Phase. J.Chromatogr., A 721(2): 305-314.

Pichon, V., Cau Dit Caumes, C., Chen, L and Hennion, M-C., (1996) Solid Phase Extraction, Clean-up and Liquid Chromatography for Routine Multi-residue Analysis of Neutral and Acidic Pesticides in Natural Waters in One Run. Intern.J.Environ. Anal. Chem., 65: 11-25.

Pico, Y., Redondo, M. J., Font, G. and Manes, J., (1995) Solid-Phase Extraction on C18 in the Trace Determination of Selected Polychlorinated Biphenyls in Milk. J.Chromatogr., A, 693(2): 339-346.

Pis, J., Hykl, J., Vaisar, T. and Harmatha, J., (1994) Rapid Determination of 20-hydroxy Ecdysteroids in Complex Mixtures by Solid-Phase Extraction and Mass Spectrometry. J.Chromatogr., 658(1): 77-82.

Redondo, M.J., Ruis, M.J., Boluda, R, and Font, G., (1993) Determination of Pesticides in Soil Samples by Solid Phase Extraction Disks. Chromatographia, 36:187-190.

Rojo, J. A. and Perkins, E. G., (1989) Cyclic Fatty Acid Monomer: Isolation and Purification with Solid Phase Extraction. J.Am.Oil Chem.Soc., 66(11): 1593-1595.

Ruane, R.J. and Wilson, I.D., (1987) The Use of C18 Bonded Silica in the Solid-phase Extraction of Basic Drugs - A Possible Role for Ionic Interactions with Residual Silanols. J.Pharm.Biomed.Anal., 5: 723-727.

Saito, K., Horie, M., Hoshino, Y., Nose, N., Makazawa, H. and Fujita, M., (1989) Determination of Diketopiperazine in Soft Drinks by High Performance Liquid Chromatography. J.Liq.Chromatogr., 14(4): 571-582.

Scalia, S., Testoni, F., Frisina, G. and Guarneri, M., (1992) Assay of 1,4-dioxane in Cosmetic Products; Solid Phase Extraction and GC-MS. J.Soc.Cosmet.Chem., 43(4): 207-213.

Scalia, S., (1990) Group Separation of Free and Conjugated Bile Acids by Pre-packed Anion Exchange Cartridges. J.Pharm. & Biomed. Anal., 8(3): 235-241.

Schad, H., Schaefer, F., Weber, L. and Seidel, H.J., (1992) Determination of Benzene Metabolites in Urine of Mice by Solid-Phase Extraction and High-Performance Liquid Chromatography. J.Chromatogr. 593: 147-151.

Schaefer, F., Schad, H. and Weber, L., (1993) Determination of Phenyl Mercaptouric Acid in Urine of Benzene-exposed BDF-1 Mice. J.Chromatogr. 620: 239-242.

Schwedt, G. and Schunck, W., (1982) Rapid Extraction of Trace Metals from Water with "Baker"-10 SPE Disposable Column. 'Baker'-10 SPE™ Applications Guide, Volume I. J.T. Baker Chemical Company, Phillipsburg, NJ. ©1982.

Schenck, F. J., Calderon, L. and Saudarg, D., (1996) Florisil Solid-Phase Extraction Cartridges for Cleanup of Organochlorine Pesticide Residues in Foods. J.AOAC Int., 79(6): 1454-1458.

Schulz, H. and Albroscheit, G., (1988) High-Performance Liquid Chromatographic Characterization of Some Medical Plant Extracts Used in Cosmetic Formulas. J.Chromatogr., 442: 353-361.

Schulze, J., Hartung, A., Kiess, H., Kraft, J. and Lies, K-H., (1984) Identification of Oxygentated Polycylic Aromatic Hydrocarbons in Diesel Particulate Matter by Capillary Gas Chromatography and Capillary Gas Chromatography/Mass Spectrometry. Chromatographia 19: 391-397.

Shannon, M.J.R. and Ensley, B.D., Solid-Phase Extraction for Removal of Organic Contaminants from Soils. Envirogen, Inc., USSR, U.S., 5 pp. Cont.-in-part of U.S. Ser. No. 744,902, abandoned, US 5242598 A 930907, US 92-936795 920828, US 91-744902 910814.

Shoup, R.E. and Mayer, G.S., (1982) Determination of Environmental Phenols by Liquid Chromatography/Electrochemistry. Anal.Chem., 54: 1164-1169.

Simonella, A., Torreti, L., Filipponi, C., Ferri, N. and Scacchia, M., (1988) Solid-Phase Extraction and Determination by RP-PLC of Aflatoxin M1 from Milk. Riv.Soc.Ital.Sci.Aliment. (1988), 17(1): 55-60.

Sooksamiti, P., Geckeis, H. and Grudpan, K., (1996) Determination of Lead in Soil Samples by In-valve Solid-Phase Extraction-Flow Injection Flame Atomic Absorption Spectrometry. Analyst, 121(10): 1413-1417

Stobiecki, M., Wojtaszek, P. and Gulewicz, K., (1997) Application of Solid-Phase Extraction for Profiling Quinolizidine Alkaloids and Phenolic Compounds in *Lupinus Albus*. Phytochem.Anal., 8(4): 153-158.

Sydenham, E.W., Shepherd, G.S. and Thiel, P.G., (1992) Liquid Chromatographic Determination of Fumonisins B1, B2 and B3 in Food and Feeds. J.AOAC. Int., 75(2): 313-318.

Taccheo Barbina, M., De Paoli, M. and Valentino, A., (1990) Determination of Tau-fluvalinate Residues in Honey. Pesticide Sci., 28: 197-202.

Taccheo Barbina, M., De Paoli, M. and Spessotto, C., (1989) The Determination of Coumaphos Residues in Honey by HPTLC with In-situ Fluorimetry. Pesticide Sci., 25: 11-15.

Taguchi, S., Takayoshi, K., Yotsu, Y. and Kasahara, I., (1996) Ion-pair Solid-Phase Extraction with Membranes: Group Contributions and Steric Effects of Ions on the Extraction Behaviour. Analyst, 121(11): 1621-1625.

Theobald, N., (1988) Rapid Preliminary Separation of Petroleum Hydrocarbons by Solid-Phase Extraction Cartridges. Anal.Chim.Acta, 204(1), 135-134.

Tian, S. and Schwedt, G., (1997) Solid-Phase Extraction of the Chromium (III)-Diphenylcarbazone Complex Prior to Ion-pair Chromatography and Application to Geological Samples. Fresenius' J.Anal.Chem., 354(4), 447-450.

Tomlins, K.I., Jewers, K. and Coker, R.D., (1989) Evaluation of Non-polar Bonded Phases for the Clean-up of Maize Extracts Prior to Aflatoxin Assaying by HPTLC. Chromatographia, 27(14): 67-70.

Toth, Z. A., Raatikainen, O., Naaranlahti, T. and Auriola,, S., (1992) Isolation and Determination of Alizarin in Cell Cultures of *Rubia tinctorum* and Emodin in

Dermocybe sanguinea using Solid-Phase Extraction and High-Performance Liquid Chromatography. J.Chromatogr., 630(1): 423.

Tsuji, M., Takeda, N., Mitsuhasi, T. and Akiyama, Y., (1995) Sample Preparation by Solid-Phase Extraction with Zinc Acetate Coagulation for the Determination of Pesticides in Crops. Anal.Sci.: 11(5): 869-871.

US EPA Contract Laboratory Program (1990) Statement of Work for Organics Analysis. Multi-media, Multi-concentration, Document Number OLM01.0, D; 1 – 59.

Vaghela, M.N. and Kilara, A., (1995) A Rapid Method for Extraction of Total Lipids from Whey Protein Concentrates and Separation of Lipid Classes with Solid Phase Extraction. J.Am.Org.Chem.Soc., 72(10): 1117-1121.

Vasconcelos, M.T.S.D., Gomes, C.A.R. and Machado, A.A.S.C., (1994) Ion Chromatographic Determination of Fluoride in Welding Fumes with Elimination of High Contents of Iron by Solid-Phase Extraction. J.Chromatogr., A., 685(1): 53-60.

Venn, R.F. and Michalkiewicz, A., (1990) Fast, Reliable Assay for Morphine and its Metabolites using High-Performance Liquid Chromatography and Native Fluorescence Detection. J.Chromatogr. 525: 379-388.

Wiegrebe, H. and Wichtl, M., (1992) High-Performance Liquid Chromatographic Determination of Cardenolides in *Digitalis* Leaves after Solid-Phase Extraction. J.Chromatogr., 630(1): 402.

Wilson, R., Henderson, R. J., Burkow, I. C. and Sargent, J. R., (1993) The Enrichment of n-polyunsaturated Fatty Acids using Aminopropyl Solid Phase Extraction Columns. Lipids, 28(1): 51.

Wisneski, H.H., Yates, R.L. and Wenniger, J.A., (1988) Liquid Chromatographic-Fluorometric Determination of Cinnamyl Alcohol in Perfumes. J.Assoc.Off.Anal.Chem., 71(4):821-823.

Zajicek, J.L. and Lebo, J.A., (1992) Silver Ion Complexation Chromatography of Organic Contaminants. Proceedings, 7th Annual Meeting of the Ozark-Prarie Chapter of the Society of Environmenatl Toxicology and Chemistry, May 1992, Carbondale, Illinois.

Zanco, M., Pfister, G., and Kettrup, A., (1992) Determination of Fluazifop-butyl and Fluazifop with the use of Disposable Solid-Phase Extraction Columns for Selective Clean-up and Concentration of Soxhlet Soil Extracts. J.Anal.Chem., 344(7-8): 345-349.

Zander, A., Findlay, P., Renner, T., Sellergren, B. and Swietlow, A., (1998) Analysis of Nicotine and its Oxidation Products in Chewing Gum by a Molecularly Imprinted Solid Phase Extraction. Anal.Chem. 70: 3304-3314.

Zelles, L. and Bai, Q. Y., (1993) Fractionation of Fatty Acids by Solid-Phase Extraction and their Quantitative Analysis by GC-MS. Soil Biol.Biochem., 25(4): 495-507.

Zhong, W.Z. and Lakings, D.B., (1989) Determination of Pioglitazone in Dog Serum using Solid Phase Extraction and High-Performance Liquid Chromatography with Ultra-violet (229nm) Detection. J.Chromatogr., 490: 377-385

Zwickenpflug, W. and Richter, E., (1987) Rapid Methods for the Detection and Quantification of N-Nitrosobutylamine in Rubber Products. J.Chromatogr.Sci., 25(Nov.): 506-509.

4

HANDLING LARGE VOLUME SAMPLES: APPLICATIONS OF SPE TO ENVIRONMENTAL MATRICES

Martha J. M. Wells, Tennessee Technological University, Cookeville, Tennessee

I. INTRODUCTION

Environmental applications of solid-phase extraction on bonded silicas were first developed in the 1980s and have grown rapidly in the 1990s, primarily as an alternative to liquid-liquid extraction. Protocol development mainly focuses on analyses of water although applications in the areas of air pollu tion and industrial hygiene, waste treatment and disposal, inorganic and organic analytical chemistry, agrochemical bio-regulators and fertilizers, soils, and plant nutrition are increasing.

Environmental analyses may present the analyst with some or all of the following challenges:

- large volume samples
- non-homogeneous samples containing particulates and/or dissolved organic matter, and non-aqueous phase liquids
- a wide range of analytes covering both hydrophilic to hydrophobic extremes

Knowing these issues are potential problems is the first step toward overcoming them. This chapter will examine analytical approaches to these challenges during SPE. While it is possible to be unfortunate enough to confront all these challenges in the same sample, for the purpose of this discussion, each point will be addressed individually.

II. LARGE VOLUME SAMPLES

To a clinical chemist, any volume greater than a few tenths of a milliliter seems enormous, but environmental chemists are accustomed to dealing with liters of samples. The reason for this volume difference is not so much that a reservoir of water yields large volumes of samples more willingly than does a human patient. Rather, the sample volume is determined by the typical concentrations of potential analytes and, therefore, the degree to which they will need to be concentrated from the sample prior analysis. Pharmacological levels of drugs in bodily fluids are commonly found in the part-per-million (ppm) or part-per-billion (ppb) levels while trace pollutants are often tracked at the part-per-billion or part-per-trillion (ppt) levels. As the analyte is recovered by extraction, the co-extracted interferences are also recovered, presenting a challenge for the analyst.

Certainly, large volume samples may be obtained from sources of water (i.e. drinking water, surface water, groundwater, wastewater, and seawater), but large samples can also be derived from other matrices. An effective strategy for extracting solid environmental matrices such as soil and plant tissues (dealt with in detail in Chapter 3) is to turn them into a "water" sample before processing. For example, when soil samples are extracted with water-miscible organic/buffer mixtures, several milliliters of sample may result. This volume, in turn, is further diluted with water to reduce the eluotropic strength of the sample.

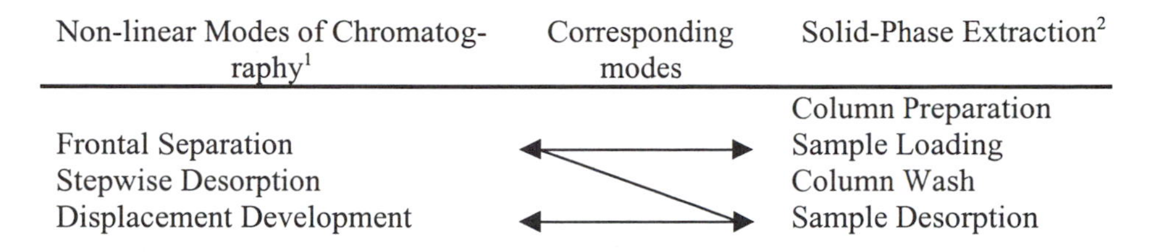

Figure 1. Solid-phase extraction as a combination of non-linear modes of chromatography. 1. A.L. Lee, A.W. Liao and C. Horvath, (1988) J.Chromatogr., 443: 31-43 2. M.J.M. Wells and J.L. Michael, (1987) J.Chromatogr.Sci., 25: 345-350.

A. LIQUID-LIQUID EXTRACTION VERSUS SPE

There are many good *experimental* reasons for using SPE in preference to liquid-liquid extraction (LLE) for extracting large volume environmental samples. For example, exposure to and consumption of large volumes of organic solvents is avoided; operator dedication to manually shaken separatory funnels, which are expensive to purchase, tedious to clean, and subject to breakage, is eliminated; increased production through multiple simultaneous extractions is realized; and the formation of emulsions is reduced. For anyone who has ever experienced the frustration of attempting to "break" an emulsion formed while extracting "real-world" samples by LLE, this advantage alone might make SPE attractive. However, in the author's experience, the primary advantage that SPE has over other separation and preconcentration procedures for large volume samples is the fundamental, *theoretical* difference inherent in the SPE approach — *SPE is a non-equilibrium procedure.*

In contrast, liquid-liquid extraction, batch adsorption, etc., are equilibrium processes. The problem with using an equilibrium process is that you may never know when equilibrium has been reached, and the equilibrium distribution may necessitate multiple extractions. Also, different analytes may exhibit vastly different distribution coefficients between extracting solvents and various matrices because of other contaminants.

LLE procedures depend on the principle of repeated extractions. If a frog jumps 50% of the distance out of a well each time it jumps, it will, theoretically, never succeed in achieving 100% of the distance. Similar to this hapless frog, the extraction process in LLE is driven by the equilibrium distribution or partition coefficient, equivalent to the fractional distance our frog jumps out of the well: Repeated LLEs will yield a recovery closer and closer to 100% without ever achieving complete extraction. Recovery from

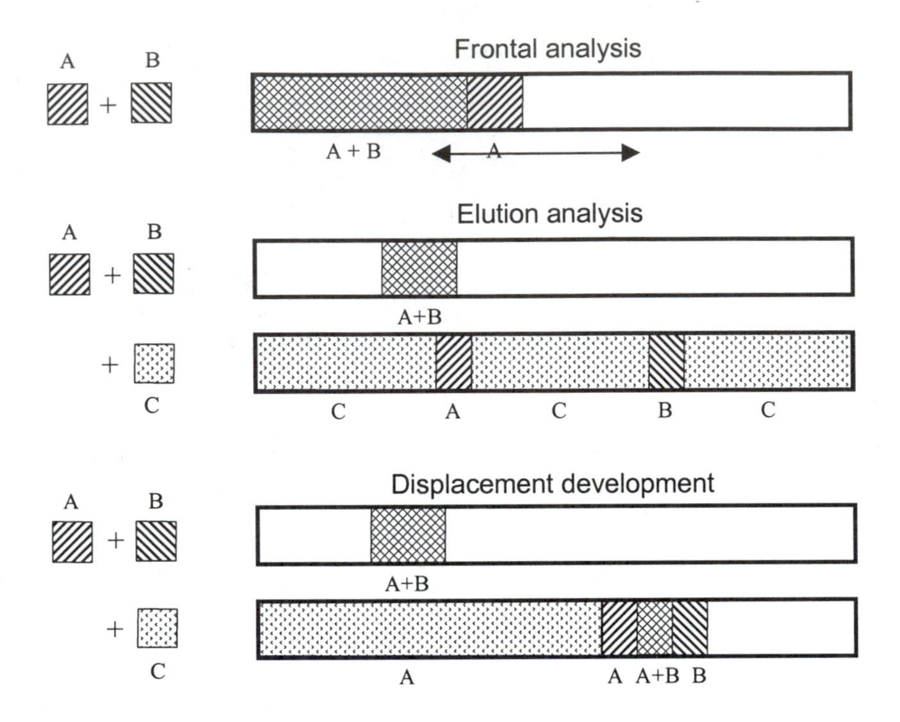

Figure 2. Illustration of the three principal modes of elution chromatography which are relevant to discussion of solid-phase extraction. A and B are the two components to be separated and C is the carrier or displacement agent. Reproduced with permission from Ettre, L.S. and Horvath, C. Foundations of Modern Liquid Chromatography, Anal.Chem. 47:422A. Copyright [1975] American Chemical Society.

samples extracted by separating funnel LLE may be matrix-dependent. The composition of the sample itself, or the presence of contaminating soluble or particulate material in actual samples, may alter the expected distribution coefficient. Problems may result from the inability to adequately predict when virtually complete extraction has been reached. That point will vary from sample to sample as the nature of the matrix varies (Wells et al., 1995). The differences in recovery are not always accurately predicted by synthetically spiked samples. Therefore, equilibrium LLE by a generalized procedure (for example, using a set extraction time or a specified number of extractions) is not always appropriate for environmentally contaminated samples.

SPE is a combination of non-linear modes of chromatography (Figure 1). The sample loading or retention step involves frontal chromatography (frontal analysis or frontal separation), and the sample desorption, or elution step, involves stepwise, or gradient, desorption or displacement development (Lee et al., 1988; Wells et al., 1990). Non-linear chromatography is superior to linear chromatography in column load capacity, reduced mobile phase consumption and simultaneous fractionation and concentration, and it can produce greater concentrations than in the original samples (Lee et al., 1988). Because SPE is a non-equilibrium, non-linear chromatographic procedure, it is able to produce simultaneous purification, fractionation, and concentration.

Large volumes can be handled by SPE sorbents in either the column or disc format. Most often the choice of using a disc or column is a personal preference, although, for very large samples, a column format can be more cost effective. Environmental laboratories already have filtration apparatus including disc holders and side-arm vacuum flasks as standard equipment, so discs or columns may be equally convenient.

Matrix solid-phase dispersion (MSPD) and solid-phase microextraction (SPME) procedures are closely related to SPE. Each is a valuable analytical technique in its own right. However, the analyst should clearly understand that both MSPD (batch adsorption) and SPME (partition-dependent) are equilibrium procedures.

B. BREAKTHROUGH VOLUME

Most quantitative, analytical, chemistry texts for undergraduate college level focus primarily on linear (or infinite dilution) chromatography, as practiced in the pseudo-equilibrium techniques of high-performance liquid chromatography (HPLC) and gas chromatography (GC). Such texts may discuss nonlinear (finite concentration) modes of chromatography such as frontal chromatography or displacement development briefly or not at all (Figure 2). Linear chromatography creates a dilution of the solute because the sample is applied in a small plug at the head of the column followed by the continuous addition of mobile phase (Lee et al., 1988; Denney, 1976). This is represented in Figure 3a. Conversely, in frontal chromatography the sample is continuously added to the column. Only the least sorbed component is obtained in a pure state (Ettre and Horvath, 1975; Denney, 1976). This is represented in Figure 3b.

When dealing with large volume samples, the breakthrough volume of certain analytes may be exceeded. Sample breakthrough is a function of the strength of the interaction between the analyte and sorbent, the sample volume and the mass of sorbent. Sample breakthrough occurs regardless of the type of sorbent. The solute-of-interest has some finite capacity factor in the sample solvent. When the breakthrough volume is exceeded, the solute begins to elute from the end of the column at the same time as sample is still being added to the head of the column (Figure 3b).

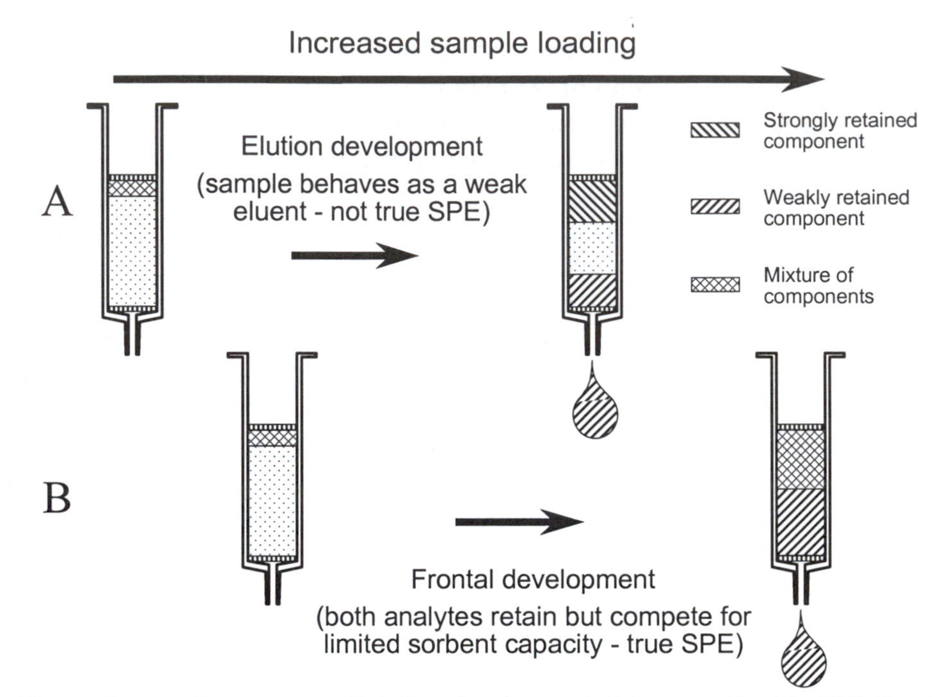

Figure 3. Comparison of elution development (A) as practiced in HPLC and frontal development (B), as they would appear if applied to a SPE cartridge. Note that as the sample size, and hence analyte loading, increases either the analytes both migrate down the column (under conditions of weak retention) or one analyte competes effectively for the sorbent and reduces capacity for the other analyte.

Larrivee and Poole (1994) defined breakthrough volume as "the volume of sample, assumed to have a constant concentration, that can be passed through the SPE device before the concentration of the analyte at the outlet of the device reaches a certain fraction of the concentration of the analyte at the inlet."

Bidlingmeyer (1984) first described theoretical breakthrough volume profiles in SPE. Hypothetical data is graphed (Figure 4) for two compounds having assumed capacity factors of 10 and 30, respectively. With an assumed column volume of 1.0 mL, recovery is theoretically 100% up to and including a loading volume of 11 mL and 31 mL, respectively. That is, a breakthrough volume of 11 column volumes is derived from a capacity factor of 10 plus the volume of solvent in the column when the sample was introduced (1.0 mL), and likewise for a breakthrough volume of 31 mL. Beyond the point where the breakthrough volume is exceeded, recovery efficiency drops nonlinearly (Bidlingmeyer, 1984; Wells, 1985). Actual breakthrough data are well modeled in the literature by theoretical profiles for both column and disc formats for a variety of analytes (Larrivee and

Poole, 1994; Wells, 1982; Wells, 1985, Wells and Michael, 1987a; Wells et al., 1990). The theoretical models used are described in Chapter 6. The theoretical considerations that are particularly applicable to large volume samples are discussed below.

Poole's research group has published a substantial body of research predicting SPE breakthrough volumes by examining the strength of the interaction between the analyte and sorbent. Breakthrough volumes were determined and modeled using solvation or solvatochromic parameters to characterize analyte retention (Larivee and Poole, 1994; Mayer et al., 1995). Solute size is identified as a primary driving force for sorbent retention under SPE conditions with polar interactions favoring retention in the aqueous mobile phase and a decrease in the breakthrough volume (Miller and Poole, 1994). The solvation parameter model also allows the prediction of breakthrough volumes with different sample co-solvents. The selective sorption of the organic solvent by the sorbent changes the stationary-phase volume and system selectivity. This results in changes in up to an order of magnitude in the breakthrough volume when either methanol, propan-1-ol, tetrahydrofuran, or acetonitrile are added to the solvent as an organic modifier at the 1% (v/v) level (Poole and Poole, 1995).

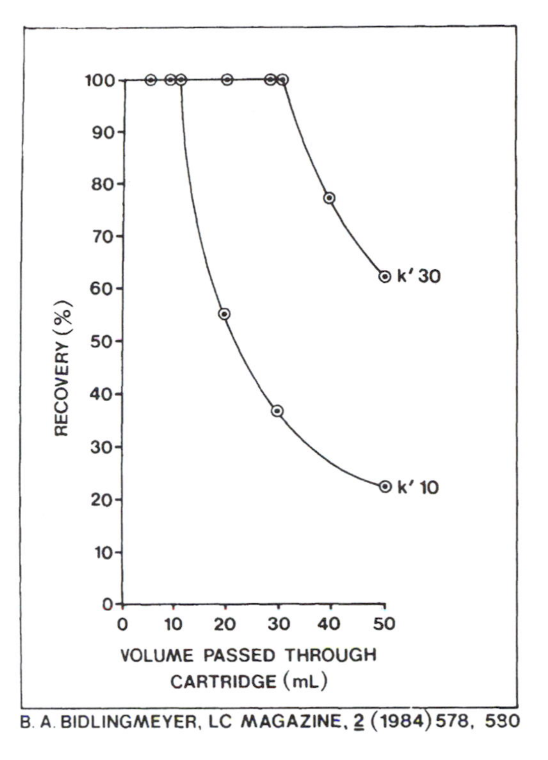

Figure 4. The relationship between recovery, sample loading volume, and capacity factor by solid-phase extraction. Reprinted with permission from Wells, M.J.M. (1985) General Procedures for the Development of Adsorption Trapping Methods Used in Herbicide Residue Analysis. The Second Annual International Symposium, Sample Preparation and Isolation Using Bonded Silicas, Analytichem International, Inc., 63-68.

Hughes and Gunton (1995) applied a model based LLE theory to interpret the elution profile for multi-component SPE data. Rather than a graphical representation of recovery versus volume (as illustrated in Figure 4), they plotted extraction profiles as functions of $[-\ln(1-R_T)]$ where R_T is the total extraction recovery, versus volume, V. The authors illustrated that analytes having extraction profiles with the same slopes exhibited no selectivity i.e., no selective desorption. In these extraction profiles an intercept not significantly different than zero indicated that none of the analyte was irreversibly bound. The slope, intercept and relative shape of extraction profiles are proposed by the authors to be more useful than the more common recovery/elution plots, allowing improved differentiation between various SPE schemes. At the point where the profiles for different analytes converge and became more linear, the effects of sample loading mass are revealed.

Liska et al. (1990) measured breakthrough curves for polar compounds from water. Using breakthrough volumes and widths of the elution curves, theoretical preconcentration factors were calculated for all analyte-aqueous sample-sorbent systems tested. Likewise, Nakamura et al. (1996) concluded that attention to the sample volume is important to successful SPE and reported that, for reversed-phase SPE, the appropriate sample volume could be approximated in relation to the log P_{oct} values (logarithm of the n-octanol/water partition coefficient) of the analytes. They determined that for chemicals with a log P_{oct} value above approximately 3.5, there was no breakthrough from alkyl-bonded, silica sorbents up to a sample volume of one liter. However, the sorbent mass used in these experiments was unspecified.

Bidlingmeyer (1984) reported that recovery is dependent on flow rate through the SPE device because breakthrough volume is decreased due to band-broadening at higher flow rates. Mayer and Poole (1994) found that the recovery of analytes by SPE shows significant flow-rate dependence when the sample volume exceeds the breakthrough volume of the analyte.

C. PRACTICAL IMPLICATIONS

The critical feature of an extraction is that it must be able to meet an anticipated detection limit for quantitation of the analyte. In order to meet this requirement, the sample must be concentrated. The degree of concentration is determined by the amount of sample that can be extracted without loss of analyte in the breakthrough volume. For a given reversed-phase sorbent, the breakthrough volume is a function of the hydrophobicity of the solute and the mass of sorbent used. In such cases the dependence of analyte recovery on sample volume by SPE has been demonstrated for the strongly hydrophobic C18 and polymer sorbents — both strongly hydrophobic. For other sorbents, however, the analyte-sorbent interaction may depend on an ion-exchange mechanism, weak van der Waals forces, or combinations of these and other interactions. When the interaction is not a simple one, the threefold relationship between sorbent mass, sample volume, and analyte/sorbent interaction is less well explored.

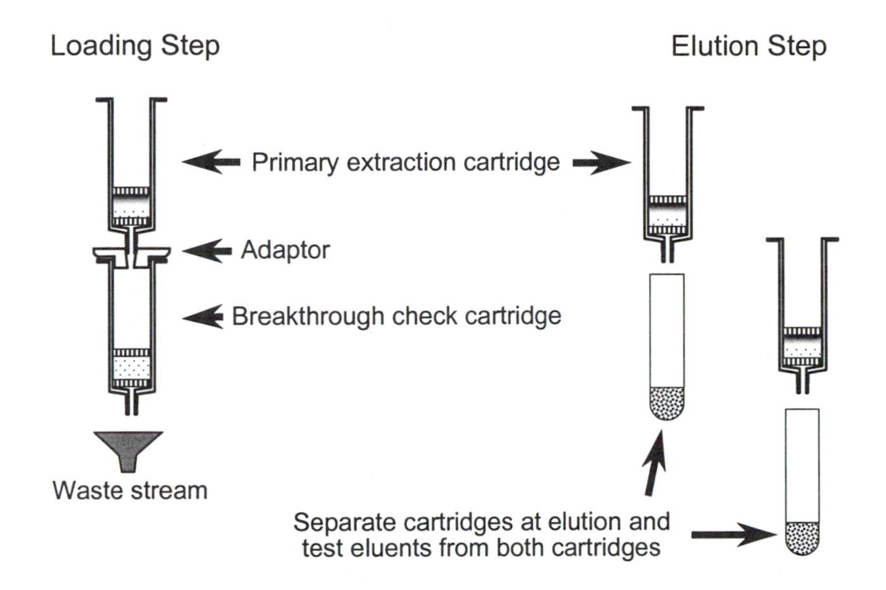

Figure 5. Illustration of how a second cartridge, used in series with the primary extraction cartridge, provides a simple indicator of breakthrough. If any of the analyte appears in the eluent of the second cartridge after sample loading and separate elution of each cartridge, then the sorbent capacity of the first cartridge has clearly been exceeded.

When the analyte is not tightly adsorbed on the sorbent, the mass of sorbent must be increased to maintain the desired sample volume. As the strength of the interaction between the sorbent and the analyte increases, the required sorbent mass decreases, and the sample volume that can be passed over the sorbent before reaching breakthrough increases.

Our goal is to maximize the ratio of sample volume to sorbent mass, since concentration achieved by an extraction depends not just on the volume of sample loaded but on the volume of the elution solvent required to desorb the analyte. While breakthrough volume is roughly directly proportional to bed mass, elution volume is roughly inversely proportional. The loading capacity, or the total amount of material which the sorbent is physically capable of adsorbing, is not generally an issue in trace analyses. However, for analyses of weakly retained species or where poor analyte detectivity requires that high microgram or low milligram quantities must be extracted, then the use of an additional column in series with the primary SPE column is recommended as a backup to monitor for breakthrough (Figure 5).

Pfaab and Jork (1994) determined that the ratio of sorbent and sample volume necessary to avoid breakthrough of phenylurea herbicides during SPE from drinking water should be 1 g of reversed-phase octadecyl bonded sorbent per liter of water for the range of concentrations typically encountered. In our work (Wells and Michael, 1987a; Wells et al., 1994a) a 1 g mass of sorbent has become a standard starting point for SPE method development using reversed-phase sorbents, as well as for cation and anion exchange sorbents. In addition to a sorbent mass of 1 g, a sample volume of 100-200 mL and solute concentration of 100 ppb are reasonable starting values when attempting to establish a new method. Many compounds will be retained from a liter of water by a sorbent mass of 1 g.

Foreman et al. (1993) isolated multiple classes of pesticides from 10L water samples using octadecyl bonded sorbent cartridges containing 10 g of sorbent. For a wastewater containing dyes, surfactants, and dye carrier components, a 1 g-to-100 mL ratio was established to prevent breakthrough of the most hydrophilic, colored components of the effluent (Wells, et al., 1994b). These parameters were successfully applied to the recovery of organics from wastewater by passing 9 L of sample through 90 g of reversed-phase sorbent purchased as the bulk phase and packed into a chromatography column.

III. NATURAL ORGANIC MATTER

Natural organic matter (NOM), comprising both particulate matter (PM) and dissolved matter (often referred to as dissolved organic matter, DOM, or dissolved organic carbon, DOC) is implicated in the environmental fate and transport of chemicals. The presence of NOM is relevant to this discussion, because it can complicate environmental analyses for several reasons.

In environmental matrices, analytes can exist in free form, or complexed with particulate or dissolved organic matter. For example, NOM is known to bind both metals and hydrophobic organic pollutants. The influence of this "associated" state on the transport of contaminants, their ultimate degradation, or their bioavailability to the food chain, is unclear. However, it clearly changes the retention properties of the bound analyte compared to the free analyte. In the presence of sorbents used for SPE, organic matter and analytes complexed with organic matter, can also become adsorbed on the sorbent, complicating analyses by changing the properties of the extracting sorbent as the NOM binds to its surface.

A. PARTICULATES

Environmental samples may contain inorganic, organic and/or biological particulates. As for DOC, pollutants can be reversibly or irreversibly bound to PM. Unlike DOC, PM can more successfully be removed from the sample prior to analysis by SPE.

For the SPE determination of atrazine and simazine, Durand and Barcelo (1993) prefiltered seawater samples in a step-wise manner through glass-fiber filters at 0.7 µm, followed by filtration with 0.45 µm glass-fiber filters to trap particulate matter. Other researchers (Schuette et al., 1990; Dirksen et al., 1993; Kuriyama and Kashihara, 1995; Taylor et al., 1995; Albanis et al., 1997) have tested a variety of filter aids including glass wool and glass beads, either in combination with pre-filtering or in an attempt to alleviate the need for prefiltering. This research has been most intensive when SPE extraction discs were used for analysis of natural water or wastewater. The fact that cartridges have received less attention may be partly because they can function as a depth filter in themselves.

When the sample matrix is a sediment or soil extract, centrifugation and/or centrifugation followed by filtration (Redondo et al., 1996; Sabik et al., 1995) reduced plugging SPE discs. In our laboratory (Wells et al., 1995) a depth filter consisting of diatomaceous earth (Hydromatrix™[10]) was found to be preferable to a nylon depth filter for SPE of non-homogeneous oil and grease samples. When filter aids are used, care must be taken to elute the analyte from the filter's surface – a step which may have the disadvantage of increasing overall elution volume.

B. DISSOLVED ORGANIC MATTER

Early in the development of SPE as a technique for environmental applications, analysts became aware of both a potential pitfall and a benefit to be derived from the interaction between DOM and sorbents used for SPE, particularly the reversed-phase bonded silica sorbents (Landrum et al., 1984). The pitfall arises because organic pollutants and metals are known to bind to DOM such as humic or fulvic acids. If we wish to measure the unbound pollutant, yet the associated humic acid-pollutant complex binds to the sorbent (and it may bind differently under various analytical conditions) and is partially or completely eluted from the sorbent, how can the concentration of the pollutant in the free, unassociated form be confirmed? The analytical optimist will view this anomaly as an opportunity to use the SPE approach to determine binding constants for DOM-pollutant interactions.

[10] Hydromatrix is a trademark of Varian Inc., Palo Alto, California.

The reference *Standard Methods for the Examination of Water and Wastewater* (American Public Health Association, 1995) is well accepted as a source of protocol for environmental analyses. *Standard Methods* and other sources (Adams, 1990) operationally DOC as the fraction of total organic carbon that passes through a 0.45 μm filter. However, Owen et al. (1993) consider different fractionations of NOM to be appropriate, defined in this case as particulate organic matter (≥ 1.0 μm), colloidal organic matter (between 0.22 μm and 1.0 μm), and dissolved organic matter (< 0.22 μm).

Macromolecular DOM is found in the microgram-per-liter range in groundwater and in the milligram-per-liter range in surface freshwater (Aiken, 1985). DOM is composed of a variety of organic compounds in various oxidation states including humic and fulvic acids. DOM may be of aquatic or terrestrial origin, and compositional differences among soil, stream, and marine sources of DOM exist. Humic acids are water soluble at high pH while fulvic acids are water soluble at all pHs. A third type of humic substance, humin, is insoluble in water at any pH.

Depending on the SPE sorbent used, the DOM extracted from an aqueous sample can be observed as a tight yellow-brown band at the top of a column or disc. If this band is eluted with the sample it may interfere with subsequent analyses. Theoretically, based only on pH considerations, fulvics would not be retained on reversed-phase sorbents at any pH, while humics should be strongly retained as the sample pH drops below the pK_a. The pK_a of humic acids is generally reported in the literature to be between 5 and 6. However, this represents an average value, since humic acids have a wide variation of carboxyl, phenolic hydroxyl, alcoholic hydroxyl, and carbonyl functional groups that may individually vary from 2.5 to 8.5 in pK_a. At pHs above the pK_a, the functional groups are negatively charged, and DOM is assumed not to be attracted to reversed-phase sorbents.

Landrum et al. (1984) applied this concept in a study based on the assumption that contaminants associated with humic materials in water will pass through reversed-phase materials at pH > 5, whereas unassociated hydrophobic contaminants will be retained by the column. However, using a guideline of pH 5 may be too simplistic: Humic substances are amphiphiles, composed of both hydrophobic and hydrophilic components, and they can form micelles (Manahan, 1994). These molecular assemblies are so hydrophobic that, even charged, they may be retained on SPE sorbents.

Johnson et al. (1991) suggested that DOM may be capable of saturating sorptive sites on various sorbents, thus giving rise to low recoveries of the desired analyte. Also, the authors claim that pesticides may associate themselves with the DOM, rendering them non-extractable by the sorbent and thereby decreasing the amount recoverable by the SPE process. Their experiments with radiolabeled diazinon and parathion suggest the formation of a pesticide-humic acid complex that was not efficiently extracted with octadecyl bonded silica cartridges (Johnson et al., 1991).

Whatever the real cause, however, several researchers have confirmed that the presence of humic substances and other surface-active materials such as surfactants, influences analyte recovery in a detrimental manner (Johnson et al., 1991; Wang et al., 1995; Nakamura et al., 1996; Albanis et al., 1997). Johnson et al. (1991) tested the recovery of pesticides in the presence of 10 ppm DOC (1 ppm is defined as 1 mg of material per Liter) and found that recoveries from water, adjusted with humic acid to simulate natural water with high DOC content, were generally lower than recoveries obtained from pure water. Albanis et al. (1997) studied the recovery of pesticides on SPE discs from water having DOC concentrations of 0-100 ppm. The recovery of all pesticides tested, representing several classes of compounds, seemed to be reduced as the concentration of DOC in water increased.

Senseman et al. (1995) investigated the influence of dissolved humic acid and Ca-montmorillonite clay on the extraction efficiency of 12 pesticides from water using SPE discs. Humic concentrations ranged from 0 to 25 ppm DOC and clay concentrations from 0-1000 ppm. Aqueous samples were prepared at pH 6.0 and 8.0 and adjusted to an ionic strength of 3×10^{-3} M. Lower pesticide recoveries, determined at pH 8 when the concentration of clay was ≥ 100 ppm, were attributed to greater dispersion, increased surface area, and subsequent adsorption. Effects of humic acid on reduced recovery of pesticides, greater at pH 6 than at pH 8, were postulated to be due to increased neutral character of the DOC as pH is lowered. Senseman et al. (1995) reported that pesticides within chemical families reacted similarly.

Dealing with DOC in aqueous samples is a problem, but its presence is ubiquitous and may be at even higher concentrations in sediment and soil extracts. One approach used in our laboratory (Sutherland, 1994) to confirm the presence or absence of complications to recovery during SPE of soil extracts of the pesticides simazine and 2,4-D, was to compare spiked extracts of field-weathered soils collected prior to pesticide treatment with pure water spikes. To determine if DOM present in soil samples had any adverse effects on the SPE extraction and subsequent derivatization of 2,4-D, blank, pretreatment soil extracts were spiked, extracted by SPE on octadecyl bonded silica sorbents, and derivatized. Similarly, simazine was spiked into blank soil extracts. In this instance, no detrimental effect of the presence of DOM upon recovery of 2,4-D or simazine was observed. Of course, the analyst does not always have the luxury of having available pretreatment samples that are the perfect match for soil type, and DOC composition and concentration, for use in paired analyses.

Nakamura et al. (1996) studied the influences of humic acid and surfactants on the SPE behavior of 25 aromatic compounds and 20 agricultural chemicals. Their research established potential guidelines for the effects of humic acid and a benzenesulfonate surfactant on SPE. The recovery of those chemicals having a log P_{oct} below about 4 when alkylbonded silicas were used for SPE, or below 3 when a polystyrene sorbent was used for SPE, were not influenced by the presence of 1 ppm of humic acid. The recoveries of analytes having higher partition coefficients decreased, although the decrease was less remarkable for the polystyrene sorbent. These guidelines are consistent with the observation from our research: (Sutherland, 1994) no detrimental effect due to the presence of DOM on recovery of 2,4-D or simazine was noted. This is supposedly because the log P_{oct} value is 1.94 for simazine and approximately 2.6-2.8 for 2,4-D (Rao and Davidson, 1981).

In addition to reducing recovery, DOC may also hamper quantitation by co-extracting with the analytes and by enhancing the "matrix effect" that results in spuriously high signals. Altenbach and Giger (1995) adopted an approach of using strongly positively charged, graphitized carbon black for determination of benzene and naphthalene sulfonates in wastewater. By this technique, negatively charged humic substances were permanently retained and found to be almost absent in the final extracts. Bonifazi et al. (1994) approached the problem by destroying humic acids prior to SPE to obtain good recovery of polychlorinated dibenzodioxins (PCDDs) and polychlorinated dibenzofurans (PCDFs) from particle-free water containing DOC. Samples acidified with sulfuric acid were treated with potassium permanganate to oxidize humic substances. Excess permanganate was reduced with hydrogen peroxide and the final pH was adjusted with sodium hydroxide.

C. SIGNIFICANCE OF MATRIX VARIATION

Personnel working in environmental analytical laboratories are familiar with filtering aqueous samples for the determination of total suspended solids (Method 2540D) using procedures detailed in the *Standard Methods for the Examination of Water and Wastewater* (American Public Health Association, 1995). Such a sample, filtered for the determination of solids, may then appropriately be processed by SPE. Prefiltering samples prior to SPE in a standard manner to remove suspended solids is recommended. Glass-fiber filter discs (0.45 μm) having no organic binders should be used. The analytes of interest should be tested for their adsorption potential on the filter selected. Certainly, the filtered particulates may be analyzed further for adsorbed contaminants. Refer to the *Standard Methods* procedures for

Determination of Solids (Method 2540) and for Total Organic Carbon (Method 5310) for more detail.

Filtering will reduce the procedural clogging problems presented by the presence of particulates, but the problems of DOC are not as easily resolved. Analytes exist in the free form or complexed with DOM. In SPE, the complexed and noncomplexed analytes may or may not be retained on the sorbent. The presence of DOC poses a greater problem as the concentration of DOC increases (Albanis et al., 1997) and as the hydrophobicity of the analyte increases (Nakamura et al., 1996). In comparison, during LLE using strong organic solvents, the free, unassociated pollutant as well as the pollutant associated with the organic matter, are probably extracted.

Appropriate "blank" samples devoid of analyte, yet containing representative organic matter simulating background levels of environmentally contaminated samples, are not always easy to generate in the laboratory. Furthermore, there is plenty of argument in the scientific community that commercially available products are not representative of natural organic matter. However, these commercial humic substances are currently the best substitute for natural organic matter available to the analyst for use in performing controlled studies to examine potential extraction problems of the analyte of interest.

With the limited information currently available about the complications of using SPE in the presence of DOC, the following recommendations seem appropriate:

- Determine the level of DOC in the actual samples to be analyzed, and be aware that some of the effects reported in the literature may have derived from unrealistically high levels of DOC.
- Apply the guidelines for log P_{oct} proposed by Nakamura et al. (1996). That is, be concerned that DOC could have a detrimental effect if the analyte of interest has a log P_{oct} above 4 when alkyl-bonded silicas are used for SPE, or above 3 when polystyrene sorbents are used for SPE.
- Test with spiked "real-world" blanks whenever possible.
- Compare SPE and LLE results on split-samples.
- Control composition of additives to the sample and elution solvent.
- When all else fails, use the procedure to determine humic-pollutant association constants!

IV. HYDROPHILIC/HYDROPHOBIC EXTREMES

Multiresidue/multiclass environmental samples may contain analytes that range from hydrophilic to hydrophobic extremes, making it difficult to recover all components effectively from a single extraction. Hydrophilic compounds may not be tightly retained, resulting in the loading of sample volumes that are too small to produce the necessary degree of sample concentration before the breakthrough volume is reached. Conversely, hydrophobic compounds may be so strongly adsorbed that they are difficult to recover from the sorbent. Furthermore, the recovery patterns observed may not depend solely on the hydrophobicity of the solute but also upon the size and planarity of the hydrophobic part of the molecule (Nakamura et al., 1996). This situation is often encountered in environmental studies that require simultaneous analyses of parent molecules and their degradation products. Most commonly, but not always, metabolites are more polar than the parent compound. Consequently, poor recoveries of polar degradation products (Hakkinen, 1991) may be observed under conditions appropriate for recovery of the parent molecule.

Techniques reported to improve the recovery of extremely hydrophilic and hydrophobic components using SPE have included choosing sorbents that should be less retentive toward hydrophobic components or more retentive toward hydrophilic compounds and optimizing sorbent type and quantity. Other approaches have included adding organic modifiers to the sample itself, varying the sample matrix pH, ionic strength, character, and volume and modifying the elution solvent eluotropic strength, pH, and volume. These approaches will be reviewed.

A. SORBENT EFFECTS

The influence of the quantity of sorbent on the breakthrough volume has already been discussed. The quantity of sorbent can be increased to better retain hydrophilic compounds, or decreased to be less retentive for hydrophobic compounds. The type of sorbent selected can also have a profound effect on recovery of multiple analytes ranging widely in hydrophobicity. Sorbent efficiency may vary between manufacturers and even between lots from a single manufacturer, although lot-to-lot reproducibility has greatly improved while our understanding of SPE has matured. The analyst should be aware that the extent of endcapping on bonded phases, and variation in the degree of bonded phase coverage can affect the recovery of solutes.

The character of a bonded phase can be inferred from its chromatographic properties. Fernando et al. (1993) calculated the hydrophobicity index and silanophilicity index for the Empore™[11] octadecylsilica membrane, based on indices developed by Kimata et al. (1989) and Walters (1987). The hydrophobicity index indicates the concentration of organic ligands bonded to silica while the silanophilicity index indicates the concentration of accessible silanol groups. The conclusions from this research were that the Empore C18 membranes exhibited both a high silanophilicity and hydrophobicity. The high hydrophobicity was considered desirable for trace enrichment while the high silanophilic index was considered to imply that irreversible sorption of hydrogen-bonding bases could occur.

Sometimes, sorbents comparatively less retentive than the most commonly used octadecyl, or C18 phase, should be used. C2 (ethyl) or C8 (octyl) sorbents are less hydrophobic because they have fewer methylene groups in the bonded phase chain. The use of phases that contain polar functional groups, such as the hydroxyl groups in the diol sorbent, may also improve recovery of highly hydrophobic compounds.

Analysis for atrazine and its various degradation and metabolic products found in the environment provides a good example of the difficulty of determining multiresidue samples whose analytes vary in hydrophobicity. Pichon et al. (1995) successfully achieved SPE for atrazine and its metabolites by retaining the breakdown products on different sorbents according to the hydrophobicity of the analytes. Highly polar degradation products of atrazine were retained on porous graphitic carbon, whereas derivatives that were less polar were retained by C18 or styrene-divinylbenzene copolymer. Meyer et al. (1993) demonstrated that two triazine metabolites were not fully retained on a C18 SPE cartridge and that retention and flow rate are inversely related. This supports the earlier assertion that once the breakthrough volume is exceeded, recovery is highly flow-rate dependent.

B. SAMPLE MATRIX ADDITIVES

1. Ion Suppression

The ionic character of the sample can be manipulated by adjustments in pH. Acid/base equilibria play an important role in successful recovery of the analytes by SPE. If the analyte is to be retained on the bonded phase solely by a hydrophobic or van der Waals interaction, then the sample pH must be adjusted to produce the un-ionized form of the analyte (Wells, 1985). In the case of highly polar compounds like phenol or picloram it is impossible to get any retention on a hydrophobic sorbent if the analyte is in an ionized form. Conversely, the ionic form of the analyte may be exploited to take advantage of ion exchange sorbents. As discussed earlier, sample pH may

[11] Empore is a trademark of 3M Corporation, Minneapolis/St. Paul, Minnesota

be a consideration if dissolved organic matter is present in the sample. The sorbent should be prepared with conditioning solvents that match the expected pH of the sample.

2. Ion Pairing

Sample matrix additives that form ion pairs with the analytes can improve recovery by SPE. In an analogous manner to which ion pair reagents are used in reversed-phase HPLC these matrix additives increase retention of hydrophilic compounds. Pocurull et al. (1994, 1995) used the addition of an ion pair reagent during SPE to increase the breakthrough volume of phenolic compounds in river water. Taguchi et al. (1995) found that a low concentration of organic solvent (nitrobenzene, o-xylene, or benzene) dissolved in the aqueous phase, rather remarkably, actually enhanced ion pair extraction onto a solid phase.

3. Ionic Strength

Just as the "salting out" effect has been used in LLE to reduce the solubility of analytes in the aqueous phase and favor extraction into the organic phase, so too has this approach been used in SPE to increase retention of water-soluble components. Salts, commonly sodium chloride, can be added to increase ionic strength. Adding salt may also swamp charged sites remaining on the sorbent surface, thereby minimizing secondary interaction effects.

Increased sample salinity has been examined as a parameter in a number of studies, especially for pesticide analyses (Chee et al., 1995; Schuette et al., 1990; Wells et al., 1994a). Kano et al. (1993) reported that recovery of 24 pesticides by reversed-phase SPE was good except for trichlorfon, which could be extracted well when NaCl was added. Bengtsson and Ramberg (1995) studied river water fortified with 62 pesticides of varying polarities. When acidified water was saturated with salt (35%) prior to extraction, mean recoveries of some polar compounds improved.

On the other hand, the addition of salt can have detrimental effects on the SPE recovery of hydrophobic sample components, possibly by increasing an already strong degree of sorption. Kohri et al. (1994) found that the recovery of organotin compounds was nearly quantitative from pure water but decreased to about 50-60% when in simulated seawater (2.45% NaCl solution, equivalent to seawater). The problem was resolved by adding MeOH to a level of 50% (v/v) in the artificial seawater samples. Conversely, Wells et al. (1994a) demonstrated that the detrimental effect on recovery of the herbicide metribuzin, caused by addition of methanol (20%) to the sample, was overcome by the simultaneous addition of sodium chloride (17.4%).

These effects were also observed when Albanis et al. (1997) studied the recovery of pesticides from aqueous solutions ranging from 0-3.5% sodium chloride. For the more lipophilic compounds, increased salinity negatively influenced recovery, while a positive salting-out effect was observed for the more polar pesticides studied.

4. Organic Solvents

Organic modifiers are added to environmental sample matrices for a variety of reasons. Some researchers add low levels of solvent to aqueous samples to promote acceptable and even flow through sorbent columns or discs, or to promote sustained wetting of the stationary-phase surface during loading of large sample volumes (Hannah et al., 1987; Poole and Poole, 1995).

Organic modifiers may be added to prevent sample loss by adsorption on sample containers and tubing. Organic modifiers also are added to samples during protocol development as a part of the preparation of spiking solutions used to fortify test samples. Soil or plant tissue extracts may already contain organic solvents, and dilution with water may reduce the eluotropic strength of the sample solvent itself. Addition of 10-50% (v/v) of organic modifier to a sample is done in attempts to improve recovery of hydrophobic analytes into smaller volumes of elution solvents — an action that appears with increasing frequency in the literature.

For example, publications describing retention on HPLC stationary phases have already firmly established that the addition of even the smallest amount of organic modifier to the sample alters the character of the bonded phase sorbent. The solvent added can be selectively adsorbed on the sorbent relative to water. Poole and Poole (1995), and Seibert and Poole (1995) recently reported changes in sorbent selectivity during SPE due to selective uptake of the processing solvent. The addition of organic solvents (methanol, propan-1-ol, tetrahydrofuran, or acetonitrile) to water at levels as low as 1% (v/v), resulted in changes of up to an order of magnitude in observed breakthrough volume. Nakamura et al. (1996) reported that upon the addition of methanol (40% v/v) to samples containing chemicals having high P_{oct} values (values of 4.17, 4.58, and 5.38), recoveries were increased to about 100% by elution with only 5 mL of methanol. The recovery of metolachlor (log P_{oct} = 2.9) was significantly improved (Wells et al., 1994a) by the addition of methanol (20%) to the sample, while the recovery of metribuzin (log P_{oct} = 1.65) decreased, and atrazine recovery (log P_{oct} = 2.68) was unaffected.

Symons and Crick (1983) demonstrated that recovery of PAHs from water using a reversed-phase C18 sorbent was acceptable for 2- and 3-ring PAHs, but that recoveries became increasingly less quantitative as the number of rings increased. Recoveries were improved by adding 20% (v/v) methanol to the sample, the most drastic improvements being observed for the recovery of highly hydrophobic PAH congeners: benz(a)anthracene from 58 to 90% recovery; benz(a)pyrene from 53 to 89%; perylene from 58 to 89%; and dibenz(a,h)anthracene from 45 to 85%. At that time, the authors attributed poor recoveries for the larger PAHs to increased adsorption onto container surfaces.

Doubtless surface sample adsorption contributes to reduced recoveries by all analytical procedures. However, it would not be unreasonable to suggest that the increased eluotropic strength of the sample, due to the addition of 20% methanol, was a major contributor to the improved recovery observed. If we apply partition coefficient values to Symons' and Crick's data it is apparent that benz(a)anthracene, benz(a)pyrene, perylene, and dibenz(a,h)anthracene (log P_{oct} values are 5.91, 6.0, 6.5, and 7.19, respectively) are highly hydrophobic compounds. For the PAH data reported (Symons and Crick, 1983), recoveries were adequate up to log P_{oct} values of approximately 5. For larger PAHs having log P_{oct} values of about 6 and above, recovery was improved by the addition of organic modifier to the sample.

Nakamura et al. (1996) compared log P_{oct} with SPE percent recovery for groups of aromatic compounds, benzoates, and agricultural chemicals. A simple relationship across families of compounds was not detected, whereas a close relationship between log P_{oct} and the recovery of compounds of similar structure within families of compounds was observed, depending on the sorbent used. Sorption behavior depended not only on hydrophobicity but also on other physical and/or chemical characteristics such as the size and planarity of the hydrophobic part of the chemicals. Our work has also shown that hydrophobicity-dependent SPE recovery may be predictable within a chemical family but can vary widely when different types of chemicals are compared. However, as a rule of thumb, SPE recovery of analytes having log P_{oct} values above about 4 or 5 may benefit from the addition of organic modifier to the sample.

The effects of adding organic modifiers appear to be threefold:

- even at low levels of added organic modifier, the stationary phase volume and its selectivity are altered (Poole and Poole, 1995).
- hydrophobic components are relatively more soluble in the mobile phase when the levels of organic modifier are increased.
- addition of even small amounts of organic modifiers to aqueous samples drastically reduces the surface tension of the sample relative to that of water, improving flow and sorbent penetration.

These effects cause the solute to be less attracted to the stationary phase, and therefore move farther down the SPE column (or membrane) bed. Breakthrough volume is thereby reduced, and solutes can be eluted more easily with smaller amounts of elution volume.

C. ELUTION CONSIDERATIONS

Elution is most successfully accomplished with a solvent having the highest eluotropic strength toward the sorbent being used, thereby minimizing the total elution volume and maximizing the concentration effect of SPE. However, other considerations, such as solvent compatibility with the analytical instrumentation to be used for final determination, or the desire to reduce the need for further sample handling such as solvent exchange, may dictate solvent selection.

Many solvents have been reported as SPE eluents. In increasing order of expected eluotropic strength on reversed-phase sorbents, based simply on their polarity, these include: acetic acid, methanol, acetonitrile, acetone, ethyl acetate, diethyl ether, methyl ^{t}butyl ether, methylene chloride, benzene, and hexane. However, a strong, nonpolar solvent such as hexane may not effect elution at all if a layer of adsorbed water exists on the SPE sorbent surface (Wells, 1985). Polar solvents or solvents capable of hydrogen-bonding with adsorbed water may be more effective. Miscible solvent mixtures can be used to achieve a hydrophilic-hydrophobic balance. Further, secondary interactions between polar functional groups on the molecule and the sorbent upset this order, and analyte solubility in the elution solvent will also play a part.

Elution by selective desorption, used to fractionate components into hydrophilic and hydrophobic components, can be advantageous (Wells and Michael, 1987b; Wells and Stearman, 1996). When picloram and 2,4-D are sorbed onto a reversed-phase octadecyl SPE column, the two compounds can be completely separated into distinct fractions by eluting the more hydrophilic picloram with acetic acid (25%) in water, followed by elution of the more hydrophobic herbicide 2,4-D with methanol.

Selective desorption has also been used to produce fractionation during SPE aquatic toxicity identification evaluations developed by the U.S. Environmental Protection Agency. Each fraction is subsequently tested for its contribution to the overall toxicity of the wastewater sample. However, for the fractionation of samples containing very hydrophobic (high log P_{oct}) toxicants, the MeOH/H_2O elution sequence requires modification to be effective for compounds with log P_{oct} values in the range of 2.5-7 (Durhan et al., 1993). A modified elution system using MeOH-water and MeOH-CH_2Cl_2 mixtures has been suggested by Lau and Stenstrom (1993) to successfully elute non-polar polycyclic aromatic hydrocarbons.

Addition of salts or organic modifiers to the sample or use of stronger elution solvents may not always be enough to improve recovery of very hydrophobic compounds. In certain instances, elution by gravity rather than vacuum can improve recovery (Wells et al., 1994a; Kohri et al., 1994). Elution by gravity may overcome reduced recovery due to the slow mass transfer of highly hydrophobic compounds from the stationary phase into the mobile phase during vacuum elution. Another elution issue to consider is that if the components to be analyzed have a high vapor pressure, sample drying and/or elution by vacuum may be inappropriate.

D. RECOMMENDATIONS TO IMPROVE RECOVERY

The hydrophilic/hydrophobic problem can be regarded as a case in which the degree of separation of the components is too good. When organic solvent is added to the sample, the degree of separation of the hydrophilic/hydrophobic components is reduced. With chromatographic insight, it should be possible to balance the sorbent, sample matrix, and elution solvent concerns, thus reducing the strength of retention of the most hydrophobic components without drastically lowering the breakthrough volume of the hydrophilic components.

The use of organic solvents, surfactants, and neutral inorganic salts as adjuvant sample processing aids can enhance sample flow rates and recovery (Mayer and Poole, 1994). In our laboratory, the addition of salt in combination with organic solvent additives is routinely tested during SPE protocol development to improve recoveries of analytes ranging widely in their hydrophilic/hydrophobic natures.

V. CONCLUSIONS

My own conversion from LLE to SPE methods occurred in 1981 when I began to realize there has to be a better way. In the environmental laboratory where I worked at that time we were extracting liter-sized volumes of agricultural run-off samples for determination of the herbicide picloram. Unfortunately, diethyl ether was the only organic solvent suitable for removing picloram from water by LLE. We were using this solvent in 55 gallon drum quantities. The fire hazard alone was cause for concern.

Even though the personnel in the laboratory were sensitized to the presence of so much hazardous solvent, I was initially in disbelief when I received a report that the ladies' restroom, located nearly 100 meters from the laboratory, was filled with the odor of diethyl ether. The reason was that the discarded water contained a fraction — 6.89% at 20°C, according to the Burdick and Jackson solvent handbook (1984) — of diethyl ether. Each extraction resulted in a liter of water, cleaned of pesticide residues but saturated with this organic solvent. It was this diethyl ether that was off-gassing in the sewer line.

This experience left me ready to embrace a new technique that avoided exposure to and use of large volumes of organic solvents, eliminated emulsion problems, reduced operator involvement with the extraction and the amount of glassware required, and enabled multiple, simultaneous extractions to be performed. These are all good, practical reasons for using SPE in preference to LLE.

As with any technique, there are also disadvantages to SPE. There can be impurities present in the sorbent or in the housing or frits associated with the packaging. The association of large molecular aggregates from the sample matrix with the sorbent can cause problems and should be assessed for a given type of sample.

The need for a greater understanding of the requirements of an SPE extraction have resulted in the approach explored by Hannah et al., (1987). In one example a 2^4 factorial design — 4 variables at 2 levels — was used for the simultaneous investigation of sample pH, non-polar SPE strength, polar SPE strength, and conditioning solvent concentration in the sample. Another example, using a 2^5 factorial design (Wells et al., 1994a), has been applied to investigation of the SPE parameters including sample pH, elution solvent strength, ionic strength, addition of organic modifiers, and elution by gravity or vacuum. These are viable means for assessing the myriad of parameters that individually and collectively influence SPE recovery of environmentally interesting solutes. Meanwhile, a growing body of publications and official methods are available to the analyst, ensuring this technique's viability and use for environmental sample preparation for many years to come.

REFERENCES

Adams, V.D., (1990) Water and Wastewater Examination Manual, Lewis Publishers, Inc., Chelsea, MI, pp. 17, 41.

Aiken, G.R., (1985) Isolation and Concentration Techniques for Aquatic Humic Substances. In: Humic Substances in Soil, Sediment, and Water. G.R. Aiken, D.M. McKnight, R.L. Wershaw, and P. MacCarthy, Eds., John Wiley & Sons, New York, pp. 363-385.

Albanis, T.A., Hela, D.G., Sakellarides, Th.M. and Konstantinou, I., (1997) Influence of Salinity and Dissolved Humic Acids on Pesticides Extraction from Water using Solid-Phase Extraction Disks. Natl. Meet. - Am. Chem. Soc., Div. Environ. Chem. 37(1):13-15.

Altenbach, B. and Giger, W., (1995) Determination of Benzene- and Naphthalene-sulfonates in Wastewater by Solid-Phase Extraction with Graphitized Carbon Black and Ion-Pair Liquid Chromatography with UV Detection. Anal. Chem. 67(14):2325-2333.

American Public Health Association, American Water Works Association and Water Environment Federation, (1995) Standard Methods for the Examination of Water and Wastewater, American Public Health Association, Washington, DC, 19th edn., pp. 2-56, 5-16 - 5-21).

Bengtsson, S. and Ramberg, A., (1995) Solid-Phase Extraction of Pesticides from Surface Water Using Bulk Sorbents. J. Chromatogr. Sci. 33(10):554-556.

Bidlingmeyer, B.A., (1984) Guidelines for Proper Usage of Solid-phase Extraction Devices. LC Magazine 2:578.

Bonifazi, P., Mastrogiacomo, A.R., Pierini, E. and Bruner, F., (1994) Solid-Phase Extraction of Polychlorodibenzodioxins and polychlorodibenzofurans Dissolved in Particle-Free Water Containing Humic Substances. Int. J. Environ. Anal. Chem. 57(1):21-31.

Chee, K.K, Wong, M.K. and Lee, H.K., (1995) Optimization by Orthogonal Array Design of Solid Phase Extraction of Organochlorine Pesticides from Water. Chromatographia 41(3/4):191-196.

Denney, R.C. (1976) A Dictionary of Chromatography, Wiley, New York, pp. 60, 71, 72.

Dirksen, T.A., Price, S.M. and Mary, S.J., (1993) Solid-Phase Disk Extraction of Particulate-Containing Water Samples. Am. Lab. 25(18):26-27.

Durand, G. and Barcelo, D., (1993) Solid-Phase Extraction Using C_{18} Bonded Silica Disks: Interferences and Analysis of Chlorotriazines in Seawater Samples. Talanta 40(11):1665-1670.

Durhan, E., Lukasewycz, M. and Baker, S., (1993) Alternatives to Methanol-Water Elution of Solid-Phase Extraction Columns for the Fractionation of High Log K_{ow} Organic Compounds in Aqueous Environmental Samples. Chromatographia 629(1):67-74.

Ettre, L.S. and Horvath, Cs., (1975) Foundations of Modern Liquid Chromatography. Anal. Chem. 47: 422A.

Fernando, W.P.N., Larrivee, M.L. and Poole, C.F., (1993) Investigation of the Kinetic Properties of Particle-loaded Membranes for Solid-Phase Extraction by Forced Flow Planar Chromatography. Anal. Chem. 65(5):588-595.

Foreman, W.T., Foster, G.D. and Gates, P.M., (1993) Isolation of Multiple Classes of Pesticides from Water Samples Using Commercial 10-gram C-18 Solid-Phase Extraction Cartridges. Natl. Meet. - Am. Chem. Soc., Div. Environ. Chem. 33(1):436-439.

Hakkinen, V.M.A., (1991) Analysis of Chemical Warfare Agents in Water by Solid-Phase Extraction and Two-channel Capillary Gas Chromatography. J. High Resolut. Chromatogr. 14(12):811-815.

Hannah, R.E., Cunningham, V.L., and McGough, J.P., (1987) Evaluation of Bonded-Phase Extraction Techniques using a Statistical Factorial Experimental Design. In: Organic Pollutants in Water, I.H. Suffet and M. Malaiyandi, Eds., American Chemical Society Advances in Chemistry Series Number 214, pp. 359-379.

Hughes, D.E. and Gunton, K.E., (1995) Representing Isocratic Multicomponent Solid-Phase Extraction Data by an Extension of Liquid-Liquid Extraction Theory. Anal. Chem. 67(7):1191-1196.

Johnson, W.E., Fendinger, N.J. and Plimmer, J.R., (1991) Solid-Phase Extraction of Pesticides from Water: Possible Interferences from Dissolved Organic Material. Anal. Chem. 63(15):1510-1513.

Kano, Y., Nakamura, H., Sugiura, T., Yamada, M. and Funasaka, R., (1993) Development of Pretreatment Method with Solid Phase Extraction for GC/MS and HPLC Analyses of Residual Pesticides. Kankyo Gijutsu 22(3):149-157.

Kimata, K., Iwaguchi, K., Onishi, S., Jinno, K., Eksteen, R., Hosoya, K., Araki, M. and Tanaka, N., (1989) Chromatographic Characterization of Silica C_{18} Packing Materials. Correlation between a Preparation Method and Retention Behavior of Stationary Phase. J.Chromatogr.Sci. 27:721-728.

Kohri, M., Inoue, Y., Ide, K., Sato, K. and Okochi, H., (1994) Solid-Phase Extraction of Organotin Compounds in Seawater. Bunseki Kagaku 43(11):933-938.

Kuriyama, K. and Kashihara, Y. (1995) Application of Disk-Format Solid Phase Extraction to Pesticide Analysis. Kankyo Kagaku 5(4):807-819.

Landrum, P.F., Nihart, S.R., Eadie, B.J., and Gardner, W.S., (1984) Reverse-phase Separation Method for Determining Pollutant Binding to Aldrich Humic Acid and Dissolved Organic Carbon of Natural Waters. Environ. Sci. Technol., 18:187.

Larrivee, M.L. and Poole, C.F., (1994) Solvation Parameter Model for the Prediction of Breakthrough Volumes in Solid-Phase Extraction with Particle-loaded Membranes. Anal. Chem. 66(1):139-146.

Lau, S.-L. and Stenstrom, M.K., (1993) Discussion on "Alternatives to Methanol-Water Elution of Solid-Phase Extraction Columns for the Fractionation of High Log K_{ow} Organic Compounds in Aqueous Environmental Samples." J. Chromatogr. 646(2):439-441.

Lee, A.L., Liao, A.W., and Horvath Cs,. (1988) Tandem Separation Schemes for Preparative High-Performance Liquid Chromatography of Proteins. J. Chromatogr. 443:31.

Liska, I., Kuthan, A. and Krupcik, J., (1990) Comparison of Sorbents for Solid-Phase Extraction of Polar Compounds from Water. J. Chromatogr. 509(1):123-134.

Manahan, S.E. (1994) Environmental Chemistry, Sixth Edition, CRC Press, Inc., Boca Raton, FL, p. 80-82, 449.

Mayer, M.L. and Poole, C.F., (1994) Identification of the Procedural Steps that Affect Recovery of Semi-Volatile Compounds by Solid-Phase Extraction Using Cartridge and Particle-loaded Membrane (disk) Devices. Anal. Chim. Acta 294(2):113-126.

Mayer, M.L., Poole, S.K. and Poole, C.F., (1995) Retention Characteristics of Octadecylsiloxane-bonded Silica and Porous Polymer Particle-loaded Membranes for Solid-Phase Extraction. J. Chromatogr., A 697(1 + 2):89-99.

Meyer, M.T., Mills, M.S. and Thurman, E.M., (1993) Automated Solid-Phase Extraction of Herbicides from Water for Gas Chromatographic-Mass Spectrometric Analysis. J. Chromatogr. 629(1):55-59.

Miller, K.G. and Poole, C.F., (1994) Methodological Approach for Evaluating Operational Parameters and the Characterization of a Popular Sorbent for Solid-

Phase Extraction by High Pressure Liquid Chromatography. J. High Resolut. Chromatogr. 17(3):125-134.

Nakamura, M., Nakamura, M, and Yamada, S., (1996) Conditions for Solid-phase Extraction of Agricultural Chemicals in Waters by Using n-Octanol—Water Partition Coefficients. Analyst 121:469.

Owen, D.M., Amy, G.L. and Chowdhury, Z.K., (1993) Characterization of Natural Organic Matter and Its Relationship to Treatability. AWWA Research Foundation and American Water Works Association, Denver, CO.

Pfaab, G. and Jork, H., (1994) Application of AMD [automated multiple development] for the Determination of Pesticides in Drinking Water. Part 3. Solid-Phase Extraction and Influencing Factors. Acta Hydrochim. Hydrobiol. 22(5):216-223.

Pichon, V., Chen, L., Guenu, S. and, Hennion, M.-C., (1995) Comparison of Sorbents for the Solid-Phase Extraction of the Highly Polar Degradation Products of Atrazine (including ammeline, ammelide and cyanuric acid). J. Chromatogr., A 711(2):257-267.

Pocurull, E., Calul!, M., Marce, R.M. and Borrull, F., (1994) Comparative Study of Solid-Phase Extraction of Phenolic Compounds: Influence of the Ion Pair Reagent. Chromatographia 38(9-10):579-584.

Pocurull, E., Marce, R.M. and Borrull, F., (1995) Improvement of Online Solid-Phase Extraction for Determining Phenolic Compounds in Water. Chromatographia 41(9/10):521-526.

Poole, S.K. and Poole, C.F., (1995) Influence of Solvent Effects on the Breakthrough Volume in Solid-Phase Extraction Using Porous Polymer Particle-loaded Membranes. Analyst 120(6):1733-1738.

Rao, P.S.C. and Davidson, J.M., (1981) Estimation of Pesticide Retention and Transformation Parameters Required in Nonpoint Source Pollution Models, In: Environmental Impact of Nonpoint Source Pollution, M. Overcash and J. Davidson, eds., Ann Arbor Science Publishers, Ann Arbor, MI, pp 23-67.

Redondo, M.J., Ruiz, M.J., Boluda, R. and Font, G., (1996) Optimization of a Solid-phase Extraction Technique for the Extraction of Pesticides from Soil Samples. J. Chromatogr., A 719(1):69.

Sabik, H., Cooper, S., Lafrance, P. and Fournier, J., (1995) Determination of Atrazine, Its Degradation Products and Metolachlor in Runoff Water and Sediments Using Solid-Phase Extraction. Talanta 42(5):717-724.

Schuette, S.A., Smith, R.G., Holden, L.R., and Graham, J.A., (1990) Solid-Phase Extraction of Herbicides from Well Water for Determination by Gas Chromatography--Mass Spectrometry. Analytica Chimica Acta, 236:141-144.

Seibert, D.S. and Poole, C.F., (1995) Influence of Solvent Effects on Retention in Reversed-phase Liquid Chromatography and Solid-Phase Extraction Using a Cyanopropylsiloxane-bonded, Silica-based Sorbent. Chromatographia 41(1/2):51-60.

Senseman, S.A., Lavy, T.L. and Mattice, J.D., (1995) Influence of Dissolved Humic Acid and Ca-Montmorillonite Clay on Pesticide Extraction Efficiency from Water Using Solid-Phase Extraction Disks. Environ. Sci. Technol. 29(10):2647-2653.

Sutherland, D., (1994) Method Development using Solid-Phase Extraction of Agricultural Soil and Runoff Water Samples Containing Simazine and 2,4-D for

Determination by Gas and Liquid Chromatography. Thesis. Tennessee Technological University, Cookeville, TN.

Symons, R.K. and Crick, I., (1983) Determination of Polynuclear Aromatic Hydrocarbons in Refinery Effluent by High-Performance Liquid Chromatography. Analytica Chimica Acta 151:237-243.

Taguchi, S., Goki, T., Hata, N., Kasahara, I. and Goto, K., (1995) Effect of a Low Concentration of Organic Solvent in the Aqueous Phase on Ion-Pair Solid-Phase Extraction. Bunseki Kagaku 44(4):307-309.

Taylor, K.Z., Waddell, D.S., Reiner, E.J. and MacPherson, K.A., (1995) Direct Elution of Solid-Phase Extraction Disks for the Determination of Polychlorinated Dibenzo-p-dioxins and Polychlorinated Dibenzofurans in Effluent Samples. Anal. Chem. 67(7):1186-1190.

Walters, M.J., (1987) Classification of Octadecyl-Bonded Liquid Chromatography Columns. J.Assoc.Off.Anal.Chem. 70:465-469

Wang, S., Santos-Delgado, M.J. and Polo-Diez, L.M., (1995) Determination of Organochlorine Pesticides in Water in Presence of Surfactants and Other Organic Compounds by Solid-Phase Extraction Combined with GC. Quim. Anal. 14(2):84-88.

Wells, M.J.M., (1982) The Effect of Silanol Masking on the Recovery of Picloram and Other Solutes from a Hydrocarbonaceous Pre-Analysis Extraction Column. J. Liquid Chromatogr. 5:2293.

Wells, M.J.M., (1985) General Procedures for the Development of Adsorption Trapping Methods Used in Herbicide Residue Analysis. The Second Annual International Symposium, Sample Preparation and Isolation Using Bonded Silicas, Analytichem International, Inc., 63-68.

Wells, M.J.M. and Michael, J.L., (1987a) Reversed-phase Solid-Phase Extraction for Aqueous Environmental Sample Preparation in Herbicide Residue Analysis, J. Chrom. Sci. 25:345-350.

Wells, M.J.M. and Michael, J.L., (1987b) Recovery of Picloram and 2,4-Dichlorophenoxyacetic Acid from Aqueous Samples by Reversed-Phase Solid-Phase Extraction, Anal. Chem. 59:1739-1742.

Wells, M.J.M., Rossano, Jr., A.J. and Roberts, E.C., (1990) Solid-Phase Extraction for Toxicity Reduction Evaluations of Industrial Wastewater Effluents. Analytica Chimica Acta 236:131.

Wells, M.J.M., Riemer, D.D. and Wells-Knecht, M.C., (1994a) Development and Optimization of a Solid-Phase Extraction Scheme for Determination of the Pesticides Metribuzin, Atrazine, Metolachlor and Esfenvalerate in Agricultural Runoff Water. J. Chromatogr. 659(2):337-348.

Wells, M.J.M., Rossano, Jr., A.J. and Roberts, E.C., (1994b) Textile Wastewater Effluent Toxicity Identification Evaluation. Arch. Environ. Contam. Toxicol. 27:555-560.

Wells, M.J.M., Ferguson, D.M. and Green, J.C., (1995) Determination of Oil and Grease in Waste Water by Solid-Phase Extraction. Analyst 120:1715.

Wells, M.J.M. and Stearman, G.K., (1996) Coordinating Supercritical Fluid and Solid-phase Extraction with Chromatographic and Immunoassay Analysis of Herbicides, In: Herbicide Metabolites in Surface Water and Groundwater, M.T. Meyer and E.M. Thurman, eds. American Chemical Society Symposium Series Number 630, pp. 18-33.

5

SPE Technology — Principles and Practical Consequences

Michael Henry, Beckman Coulter Inc., Fullerton, California

I. INTRODUCTION

In its broadest sense, solid-phase extraction as defined in this book is governed by the same physico-chemical principles that influence a wide variety of sorptive processes. These are present in such diverse technologies as the purification of biologically useful materials, catalysis, toxic substance monitors, organic synthesis, biopolymer blotting, ionic exchange and water and wastewater clean-up. Most of these areas of applied chemistry have several elements in common:

- Transport of molecules or ions through a fluid of given properties to a surface.
- Adsorption onto or partition into a solid phase.
- Selective desorption or partition from this phase into a fluid with the same or different properties.

In SPE the three elements above can be integrated physically in two ways as follows: The sorbent is retained within a device of fixed boundaries and all parts of a representative sample are brought into contact with the solid, or the sample is placed into a container and the sorbent is brought into contact with it.

Examples of the first device include the SPE column, cartridge and disc. Examples of the second approach are the fluidized bed, batch extraction, the coated fiber, and SPE disc. In practice only the SPE column, cartridge, and disc are used in the first technique and coated fiber in the second process.

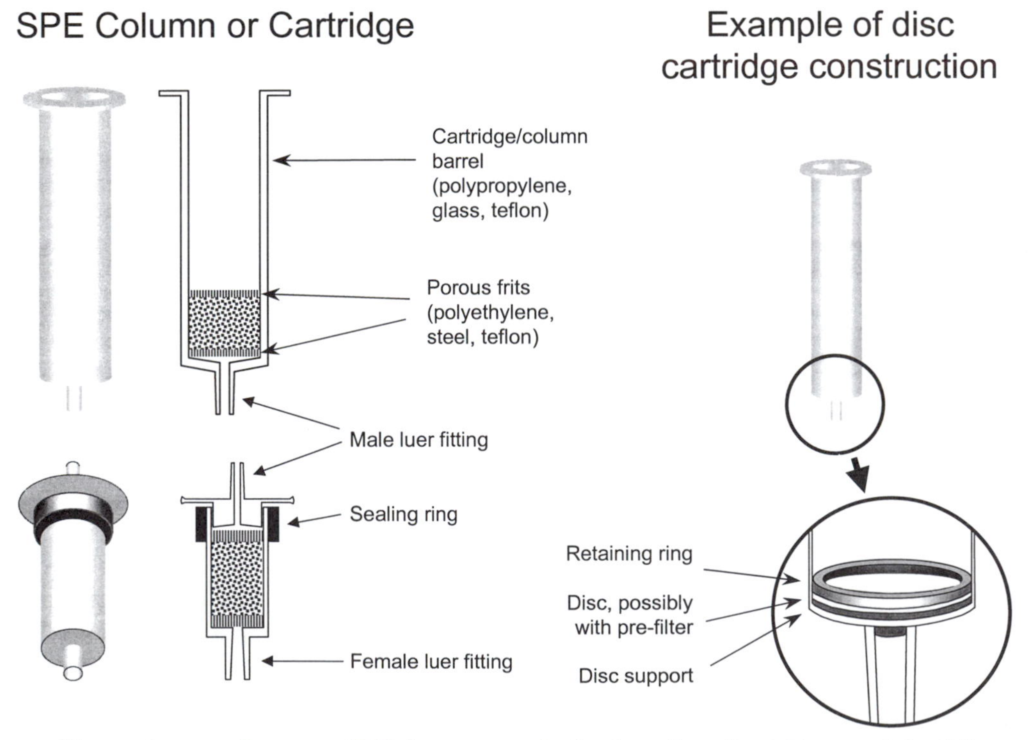

Figure 1. Common SPE formats — the "column" or "cartridge, and the "disc cartridge." The term "column" is only applied to the straight-wall syringe barrel construction. "Cartridge" has been used interchangeably for both formats and is preferred by some users because it avoids confusion with the analytical "column."

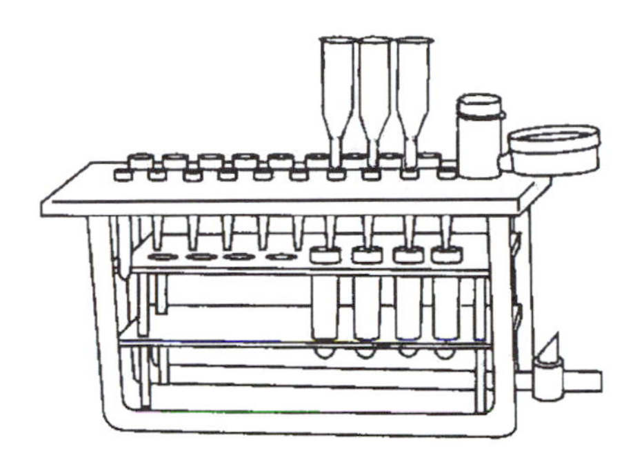

Figure 2. A typical vacuum manifold design (Reproduced with permission from Varian Inc. Sample Preparation Products Catalog, 1996).

A column is defined as an SPE device that can contain both sample and sorbent. The cartridge contains the sorbent only, and the sample must be applied from an external reservoir. An SPE disc is an extraction device in which sorbent particles are enmeshed in a fibrous matrix to form a flat porous solid. Figure 1 is an illustration of the components and structures of each device.

In addition to each device comprising these primary elements of SPE technology (sorbent container and the sorbent), various manual liquid transfer and sample processing systems have been developed. One of the earliest of these was the centrifuge, adapted to process many columns simultaneously under virtually identical conditions. The vacuum box (Figure 2) was adapted for SPE use, enabling several dozen samples to be prepared from a single vacuum source. Individual samples were extracted using a syringe, a vacuum flask or small centrifuge (see Figure 3). All these sample processors can accept cartridge,

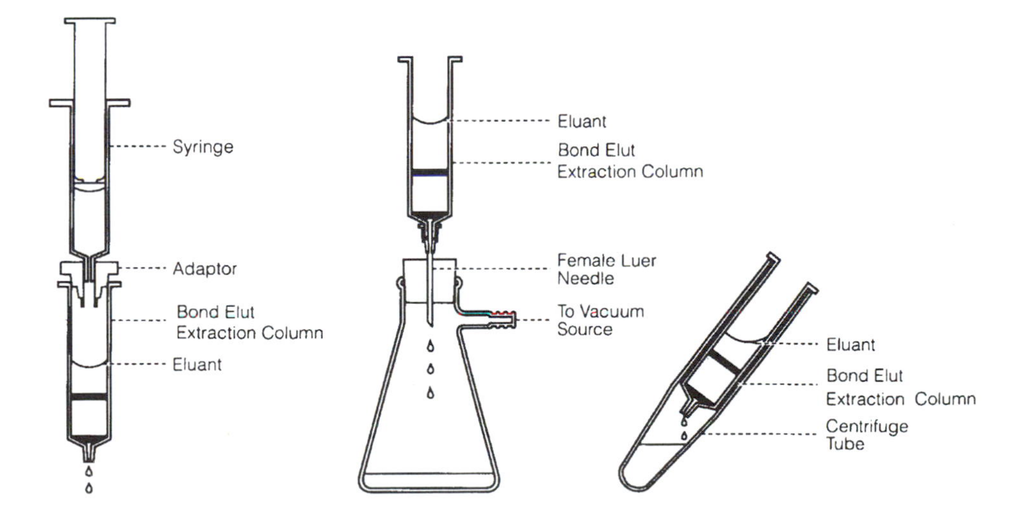

Figure 3. Illustrations of three methods of passing the sample and solvents through the SPE cartridge.

column, or disc configurations.

Large sample volumes can be processed with SPE devices of various sizes, although speed is usually associated with discs of large diameter - typically up to 90 mm. These discs are generally retained in a holder into which the sample liquid is transferred, and moved through the thin sorbent bed under vacuum.

A wide range of accessories has been developed to simplify and improve the processing of single and multiple devices. All basic SPE configurations (cartridge, column, and disc) have been integrated into automatic systems. These are topics covered in Chapter 14 and they will not be covered here.

This chapter will focus on the components of manual systems for single and multiple sample processing in which sub-critical fluids are employed. The objectives of this work are an evaluation of the major performance criteria of the various SPE technologies within the above context and an indication of the practical relevance of these characteristics.

Throughout this chapter many references will be made to components of SPE devices. There are currently over 80 manufacturers of these systems and the reader is referred to reviews and lists of this equipment, given in publications such as *LC.GC* (Majors, 1999) and its *Buyer's Guide* issues (1999) for details.

II. SORPTIVE PROCESSES

A. RETENTION

Analytes and their mixtures with a sample matrix are retained and separated within an SPE adsorption bed by several mechanisms. Here we cover in greater detail the ideas introduced in Chapters 1 and 4.

1. Linear Elution (Equilibrium) Chromatography

In this extraction regime a mixture of two components, say X and Y, is applied to the inlet of the column in a discrete volume, typically less than 1% of the bed volume. The components X and Y undergo a reversible, dynamic sorption, rapidly approaching equilibrium between the mobile and stationary phases. As fresh mobile phase moves through the sorbent bed, it displaces molecules of X and Y from the sorbent into solution, from where they re-adsorb. Depending upon several properties of the system which include the relative adsorptive strengths of the analytes, their solubility in the mobile phase, their concentration and the displacing power of the solvent molecules; components X and Y will tend to separate.

If the volume flow rate of the mobile phase remains constant during this process, then each component must spend the same time, t_o, in the mobile phase — unretained. In chromatography this is known as the unretained peak time. If component X is adsorbed for a time period of t_X and component Y for t_Y, then the total time spent inside the column is $t_0 + t_X$ for component X, and $t_0 + t_Y$ for component Y. The *capacity factors* (k') for the two components are defined in Equations 1 and 2, below:

$$k'_X = t_X/(t_X + t_0) \quad (1)$$

$$k'_Y = t_Y/(t_Y + t_0) \quad (2)$$

The ratio of the k' values for X and Y (equation 3), is known as the selectivity factor, α:

$$\alpha = k'_Y / k'_X \quad (3)$$

The adsorption times t_X and t_Y are known as the adjusted retention times. Ideally k' should be between 1 and 10 in elution chromatography. Note that this is not a k' range normally employed for SPE, where these values are generally in the range of 100-500.

2. Frontal Elution

Here a non-eluting mobile phase containing the components is loaded continuously into the bed. Although the components have appreciable solubility in the mobile phase, they bind strongly to the sorbent and are not significantly displaced by the mobile-phase molecules. The concentration of analytes at the surface rapidly builds up until they saturate the sorbent and breakthrough into the next segment of the bed. The component that saturates first moves down the column ahead of the next component. Here k' is usually much greater than 10, but separation may occur because of saturation.

As with linear elution, this condition is not generally encountered in SPE separations. An exception may be found in attempts to extract phenol from large volumes of water (Coquart and Hennion, 1992).

3. Displacement

A sample containing components X and Y, dissolved in a non-eluting solvent, is loaded onto the head of the column, where it is tightly adsorbed. A so-called displacer component, Z, which binds tightly to the adsorbent, is loaded into the column and displaces component Y. This in turn displaces component X, which begins to elute. Component X will elute from the column first followed by Y, in this example.

In the loading, washing and elution step in SPE, all three of the above mechanisms often operate, as encountered in some ion exchange processes (Helfferich, 1962).

Loading: A discrete volume of liquid sample, usually of the same magnitude as that of the sorbent bed, is applied to the column. This process is essentially that of frontal loading, which is designed so that components of interest are bound to the sorbent and unwanted components are unretained at the surface.

Washing: The solvent is chosen so that it does not displace the components of interest, but weakly bound components are effectively displaced. In addition, the wash step removes unwanted material from the pores and interstices of the packed bed. This process is an elution step in which the wash solvent also acts in some respects as a displacer.

Elution: A stronger solvent displaces the components of interest in a version of the washing step. Solvent molecules or ions (in ion exchange processes) take the place of adsorbed analytes on the surface. The nature and volume of the elution solvent must be such that no proportion of the component of interest remains in the surface or in the pore or interstitial volumes.

4. Mechanisms of Retention in SPE

The processes that occur in the retention step (and in the washing and elution steps) are temporarily dynamic while the sample is being applied. Components bind, displacing solvent molecules that are located at the solid surface in the conditioning step. Strongly bound components may also displace weakly bound ones as well as conditioning solvent molecules. Thus adsorbing and desorbing mechanisms operate in the process of retention, until the entire sample is loaded. The final arrangement of compounds of interest in the sorbent bed depends on the characteristics of the sample matrix, the binding strengths of the analytes of interest and the affinity of solvent molecules for the surface and all molecular components of the sample.

The primary forces that bind an analyte to a surface are hydrophobic, coulombic, dipolar, and electrostatic. In addition to attractive and repulsive forces, changes in entropy determine in part, whether a component will be retained. For example, the ordering of water molecules is powerful driving force towards equilibrium (Tanford, 1973). There are several principles that can be applied in determining whether the analytes of interest in a known matrix will be retained on a given surface. They include having knowledge of solubilities, sample pH and ionic strength, molecular structure and the possible forces that may bind a compound to a surface. It is

Table 1. A guide to the principles of retention in SPE

Analyte	Sample composition	Sorbent choice	Primary interaction mechanism
Hydrocarbons, fat-soluble vitamins, triglycerides, steroids, aflatoxins, phthalates	Water, water/polar organic mixtures	C18, C8, C2, C1, CN-E, CH, PH	Van der Waals, "non-polar", "hydrophobic"
Steroids, dioxins, amides, carbohydrates, esters	Hydrocarbons, chlorinated solvents, non-polar/polar organic mixtures	SI, CN-U, 2OH, Alumina	Dipole-dipole, hydrogen-bonding, "polar"
Anionic: Strong acids, halides	Aqueous buffer, pH 5-8	SAX, NH2, PSA, DEA	Anion exchange, Primarily Coulombic
Cationic: Strong bases, metals	Aqueous buffer, pH 4-8	SCX, PRS, CBA	Cation exchange, Primarily Coulombic
Biopolymers	Proteins, polynucleotides, polysaccharides	C18, C8, C4	Hydrophobic, Van der Waals

also important to know as much as possible about the sorbent surface. Unfortunately this is often difficult to determine without application of sophisticated surface analysis techniques. The simple description of the common silica-based sorbent as C18 covers a range of literally hundreds of commercially available bonded phases (Majors, 1999) with their own, unique sorbent properties. Thousands more have been described in the literature (Regnier, Unger and Majors, 1991; Unger, 1990).

Typical sorbents, analyte classes, sample solvents and major binding forces that operate have been selected in Table 1.

B. ELUTION

The processes of analyte elution from an SPE device are substantially different from the conditions that exist in retention. The objectives in elution are the quantitative recovery of analytes, their resolution from unwanted co-adsorbed components and obtaining them in a concentrated form suitable

Table 2. Sorbent, analyte class and elution conditions

Sorbent	Analyte Class	Examples	Elution Conditions
C18	Organic	Vitamin K	Methanol, acetonitrile
SI	Polar organics	Desonide,	Ethyl acetate, acetone
Ion exchangers	Organic acids	Ibuprofen, wine acids, halides	Low pH, high ionic strength
	Organic bases	Cocaine, pyridostigmine bromide, metals	High pH, high ionic strength, low pH

for analysis. Primarily there is a need to determine the optimum elution solvent; and secondly the appropriate volumes and flow rates.

For example, non-polar analytes adsorbed to a C18 SPE device are typically eluted with a solvent such as methanol or acetonitrile. This material can effectively and simultaneously wet the C18 surface, dissolve the analytes of interest (displacing them from the non-polar surface) and is readily evaporated if necessary. Often the choice of elution solvent that has the above properties is not straightforward, and a certain amount of trial and error is necessary to optimize this step. Section V in this chapter reviews solvent properties that are important in SPE. Table 2 gives a list of solvent type, class of analyte and suggested elution solvents.

The majority of the theory presented in this book and most of the practical applications are based upon non-polar extraction with the common silica-based phases, C18, C8, C2 and the family of divinyl-benzene polymers. Since ion exchange plays a significant part in the extractions described in Chapters 8 through 11, the main features of this extraction mechanism are presented in Appendix 3. More detailed treatment can be found in other texts (Fritz, 1999; Simpson and van Horne, 1993).

III. PERFORMANCE CHARACTERISTICS OF SPE DEVICES

A. THE FOUR ELEMENTS OF SPE TECHNOLOGY

There are many performance criteria for SPE devices, some more important than others, depending upon the objective of the extraction. These characteristics are listed in Table 3, together with their importance and their determinants. Often a performance criterion of a SPE device is determined by the sum of several properties of its components and associated systems.

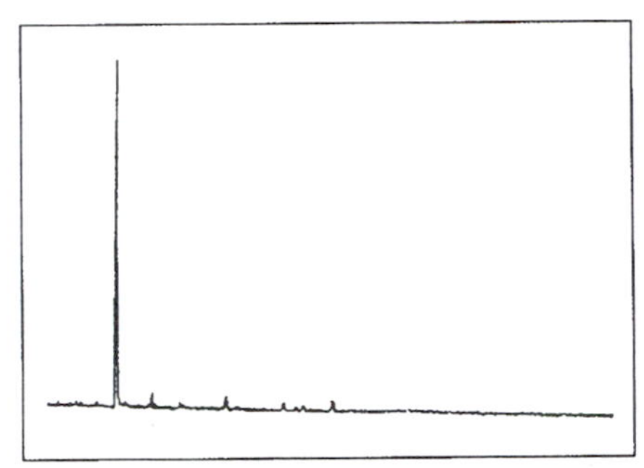

Glass Cartridge with PTFE Frit

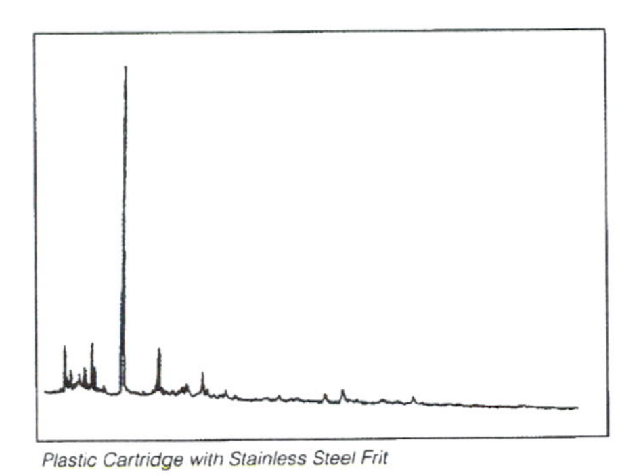

Plastic Cartridge with Stainless Steel Frit

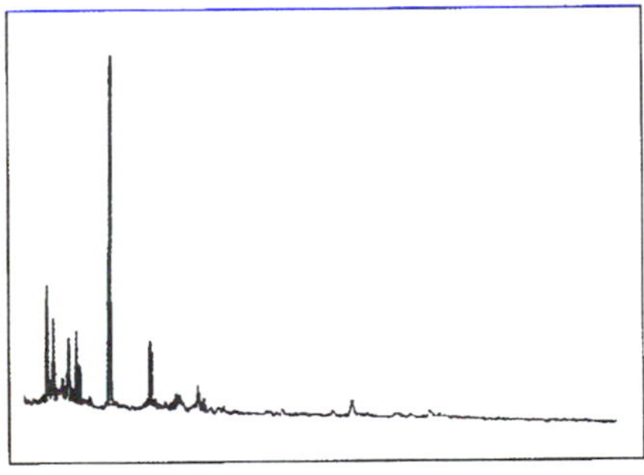

Plastic Cartridge with Polyethylene frit

Figure 4. Examples of methylene chloride extracts of SPE devices, analyzed by GC/FID, demonstrating the different levels of cleanliness to be expected from the use of different construction materials. The obvious conclusion — select the device that gives blanks of adequate cleanliness for the detection technique and detection limits of the analysis (Diagram reproduced with permission from Varian Inc.).

These components and systems, or elements, are the container, the sorbent, the complete SPE device, and the processing system and its accessories. The following discussion will be structured around these four elements as a means of simplifying the performance criteria from Table 3.

1. Element One: Sorbent Container

The sorbent container is defined as the empty cartridge, column, disc substrate and fiber, and associated filters and frits. The nature of these components determines many of the performance criteria in Table 3.

Capacity: Insofar as the container defines the bed shape and size it is naturally a determinant of absolute capacity, an important performance criterion. Specific binding capacity is defined and discussed in more detail in Section III.B and IV.A.2.

Cleanliness: Components of SPE devices such as syringe barrels, frits and filters, often contain impurities such as plasticizers, mold release agents, antioxidants, and monomers. These can leach into extracts and compromise the accuracy and precision of sub-sequent analysis, especially in GC and MS (Junk et al., 1988).

The cleanliness of the device will depend upon the choice of the quality of the initial polymer resins such as polypropylene and high-density polyethylene. Precleaning of the syringe barrel can also substantially reduce extracts. Glass and Teflon syringe barrels are commonly used, and can be cleaned very effectively. Polyethylene frits are a common source of contaminants, and disc technology has an advantage here, since frits are not required. Teflon frits are inert and do not leach additives. Filters that are included in SPE devices, are largely constructed of binder free glass fiber mats, and should not contribute measurable quantities of contaminants to a sample.

The major manufactures of SPE columns have generally high standards of cleanliness, and have published evidence of this; usually in the form of a gas chromatogram of a solvent extract of an SPE device (Figure 4).

Design flexibility, multiple use and automatability: There are several characteristics of SPE containers that influence whether their processing can be automated readily. The original Waters Sep-Pak cartridge had a unique shape designed more for single sample use in conjunction with a syringe. It was readily adaptable to the vacuum manifold (Figure 2), but required a separate reservoir to hold the sample. The use of the mass-produced syringe barrel as a sorbent and sample container and a metal or glass manifold, was pioneered by Analytichem International in the late 1970s, and is now a standard in SPE technology. The syringe barrel design has been easily adapted for single and multiple samples for greater throughput, and shaped for use in robotic systems. The syringe has the advantages of being standardized, cheap, clean and easily filled with sorbent, frits and filters. The use of glass and Teflon syringe barrels in SPE came after the first polypropylene containers. Early packaging of glass SPE columns was inadequate to prevent occasional breakage, but this is now a rarity.

The thin disc configuration, when retained in a syringe barrel, has a void volume that is generally small enough not to require the use of any differential pressure at the conditioning stage. This design is such that this may be carried out rapidly with a large number of SPE columns in a manifold without using vacuum.

Reproducibility: Without this performance characteristic, SPE and any other technology would be of limited value. Consequently column to column variations in several properties of the container must be minimized and known. These properties include physical dimensions, integrity, cleanliness (see above) and chemical inertness. The manufacturers of the raw materials and completed parts are responsible for lot to lot variability. Such information is very rarely published in the SPE literature, although it is often as important as reproducibility of capacity and recovery for example.

Table 3. Performance criteria of SPE devices

Characteristic	Importance	Determinants
Specific Permeability	Determines flow rate and linear velocity and hence processing speed	Particle size and shape distributions, bed structure
Capacity	Predicts maximum sample size	Adsorption isotherms, bed mass, pore properties, temperature, partition coefficients, sorption kinetics, sorbent mass, analyte concentration
Recovery	Accuracy and precision of analysis	Magnitude of undesirable interactions, including no interaction, cross-contamination, pore properties
Bed stability	Loss of sample through channelling	Fluidity of bed components, loss of permeability, bed collapse during processing, particle rigidity
Chemical stability	Change in capacity during use, contamination by bed breakdown products	Bonding chemistry, base support material, temperature, use of aggressive solvent conditions (pH, oxidizing, or reducing species)
Biological stability	Degredation of sorbent through action of microorganisms	Biodegradability of sorbent, temperature, storage conditions
Cleanliness	Interferences in extract	Purity of construction materials, application wash conditions, detection technique used, effectiveness of frits in preventing sorbent leaching into extract, particle size distribution, sorbent impurities (metal ions, monomers)
Reproducibility	Accuracy and precision, lot-to-lot	Manufacturability, quality control and process control in manufacturing, bed fluidization, capacity, (see above)
Selectivity	Suitability of extracted sample for subsequent analysis	Shape and position of adsorption isotherms for individual analytes, temperature, and wash/elution conditions of extraction
Design flexibility and automatability	Small or large volume handling, compatibility with existing or anticipated lab automation	Low void volumes, diverse sorbent chemistries and housings
Simplicity	Minimal training, downtime	Few process operations; little reliance on technique
Throughput	Processing efficiency, costs	Rate of sample application, wash, elution
Cost ($)	Value to laboratory	Supplier quality, efficiency, technical support, experience

Experienced practitioners of SPE however are sensitive to this requirement and make it a matter of routine to subject each new lot to a limited series of tests.

In any manufacturing process the best product is achieved only after multiple production runs. Eventually there is a tendency for quality, especially that manifested in reproducibility, to decline as the product achieves routine manufacturing status. It requires long term commitment to product quality to prevent this deterioration. Unfortunately it is often up to the user to monitor product quality, which in any case is often closely tied to a particular application. The subject of reproducibility is explored further in Sections B, and C.

B. THE SPE DEVICE: CARTRIDGES, COLUMNS, DISCS, AND FIBERS

These are all the components of the SPE device: the container, frits, filters and the sorbent itself. Their combined physical and chemical characteristics determine certain performance criteria listed in Table 3. Discs are included here since they comprise fibrous matrices within which (usually) particles are enmeshed and held in place in the form of a solid.

The characteristics from Table 3 of SPE devices that have the most practical importance are capacity, recovery, bed stability, filterability, design flexibility, reproducibility and resolution.

1. Breakthrough Volumes, Resolving Power, and Capacity

Resolution of wanted from unwanted components in a sample is the primary objective in SPE. The theory of linear elution chromatography tells us that resolution depends upon the width of peaks or zones and their degree of separation or selectivity. Peak width can be expressed in theoretical plates, N and selectivity as the ratio of the capacity factors of two components (Equation 3). In the use of an SPE device during sampling, many unwanted compounds are deliberately allowed to break through; while other, including the wanted analytes, are retained in the sorbent bed, perhaps along with other unwanted components. The analyst will want the last eluted, unwanted component to be resolved from the most weakly retained of the desired components. In other words, there are two important breakthrough volumes, V_B, for these components.

Poole and Poole (Chapter 6) discuss the relationship between V_B and other important sorption characteristics of a sorbent bed. These include the retention volume, V_R; the breakthrough volume, V_B; plate number, N and capacity factor, k'. These authors define V_B as the volume of sample that has passed through a sorbent bed until 1% of the analyte concentration is measured at the outlet. This is often measured as the volume at 1% of the

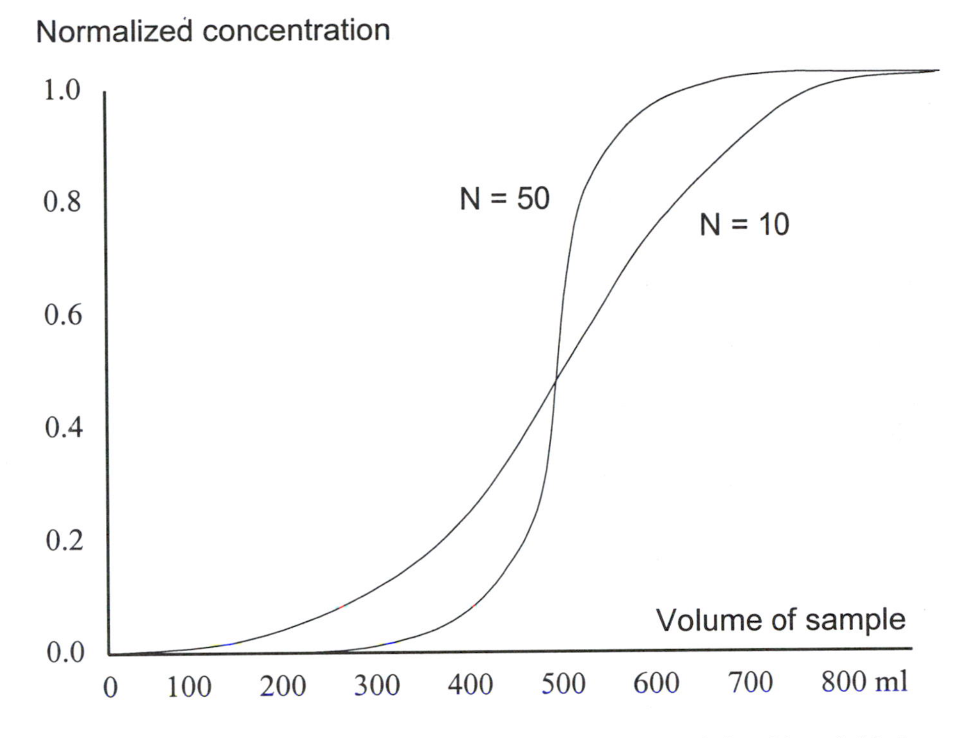

Figure 5. Breakthrough curves for SPE devices containing 50 and 10 theoretical plates (this diagram is based on work by Bouvier, 1994).

height of the breakthrough curve. The value of 1% was chosen so as to reduce the breakthrough losses to a minimum in calculating recovery and ultimately the original analyte concentration in the sample. The retention volume, V_R, of the analyte occurs at the inflexion point on this curve. The interparticle volume, V_0, is the volume between particles; and k' is the capacity factor for the analyte in the sample mobile phase, usually water. N is the plate number for the sorbent bed.

From the theory of frontal chromatography (Werkhoven-Gowie et al, 1981), V_R and V_B are related by Equation 4:

$$V_R = V_B + 2\sigma_V \tag{4}$$

Where σ_V is the standard deviation of the axial dispersion along the sorbent bed, and can be calculated from Equation 5:

$$\sigma_V = V_0(1+k')/\sqrt{N} \tag{5}$$

The standard deviation is readily calculated from measurements of V_0, k' and N for the bed. Hennion and Pichon (1994) have calculated values of V_B for pentachlorophenol through a 100 mg, C18 sorbent bed. Here V_0 is 0.12 mL/100 mg, k' (calculated from $\log P_{oct}$ this is 91,200; N is assumed to be

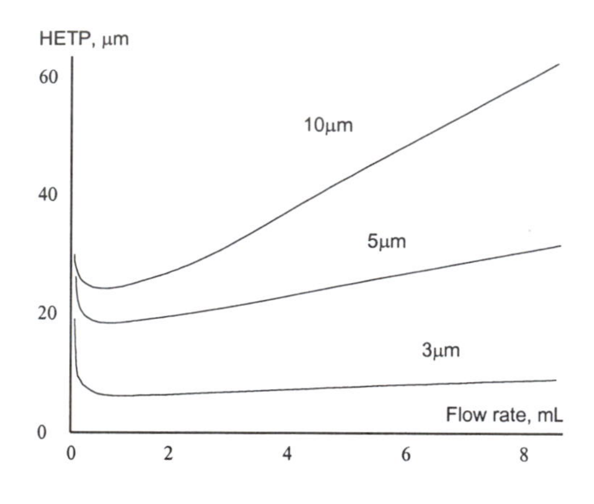

Figure 6. Plots of HETP versus flow rate for columns of different particle sizes.

about 20 plates. These authors found that σ_V was 2.4 liters and V_B was therefore 7 liters. While these calculations and results are interesting, in practice this information is of limited use and rarely obtained.

The shape of the breakthrough curve for the analyte of interest is of crucial importance in obtaining high capacity and high resolution in SPE. Ideally the breakthrough, once it has begun, should be concluded in the minimum sample volume. This means that wanted and unwanted components will be cleanly separated with a minimum of cross-contamination. In terms of dispersion, σ_V should be as small as possible, giving the largest V_B. The standard deviation will be a minimum if N is increased. In other words, if the bed is longer and/or the particles are smaller. Figure 5 (Bouvier, 1994) shows the breakthrough curves for a given analyte for beds of sorbent containing different particle sizes in which N values are 50, 20, and 10 plates. V_B can be as high as 300 mL or as small as 150 mL in the example.

It is well known that higher N values correspond to smaller particles when all else is equal. In other words, for a strict comparison, pore size, surface area and their distributions should remain constant. However even when these properties vary the general relationship holds. The mathematical dependence of N upon particle diameter as a function of flow rate, does not directly include any values of the pore properties mentioned above (for example, see Figure 6 and Engelhardt, 1979).

Axial dispersion, σ_V, along the sorbent bed is smaller with small particles. Hence in principle, SPE columns with finer sorbent particles would be preferred over those with coarser particles, from the point of view of achieving more ideal breakthrough curves. However smaller particles generate high backpressures and have increased tendency to clog with particle laden samples. Amongst the smallest particles used in SPE are those bound into the Empore Disks. These are typically 8-10 μm in diameter, and can only be used efficiently because of the very short path length of the sorbent bed (0.5 mm) in this case.

In practice SPE columns are of such a length and particle size that sample processing under vacuum is rapid enough to generate flow rates between 1 and 100 mL/minute depending upon the diameter of the sorbent bed.

Poole and Poole will describe in detail in Chapter 6 several relevant relations between breakthrough volume, V_B, inter-particle volume, V_o, efficiency, N, and capacity factor, k'. This relationship is shown in the Lovkvist and Jonsson Equations, 6 and 7 (1987), for two components, I and II:

$$V_{I,B} = (a_I + b_I/N + c_I/N^2)^{-1/2} (1+k_I')V_0 \quad (6)$$

$$V_{II,B} = (a_{II} + b_{II}/N + c_{II}/N^2)^{-1/2} (1+ k_{II}')V_0 \quad (7)$$

where a, b and c are constants.

For a retained component, for example I, we want V_B to be > V_0. For an unwanted component, for example II, we want V_B to be < V_0. Poole and Poole correctly point out that V_B is the most important property in determining sorbent suitability. If there are several analytes of interest, we want them all to be retained. Breakthrough volumes of these compounds (I) and those of the matrix (II) must clearly be different. Thus column N, k' and V_0 are the important characteristics that determine V_B and consequently, resolution.

The Lovkvist and Jonsson (1987) equations tell us that larger (favorable) breakthrough volumes are achieved when k' and V_0 are large. In other words, these compounds are strongly retained in a large volume device. Furthermore, a higher N corresponds to a larger V_B. Unfortunately SPE devices in general have smaller N values, lower efficiencies, because particles are large and bed lengths are short. Indeed where N > 4 (Poole and Poole, Chapter 6) the Lovkvist and Jonsson equations hold, but where N < 4, there is premature breakthrough and loss of analyte for kinetic reasons. In other words, molecules travel through and out of the device faster than there is time for them to be adsorbed, regardless of the capacity of the device. This problem could be avoided if flow rates of sample application were reduced. In the limiting case, maximum capacity of the sorbent could be realized if the flow rate was zero.

The important kinetic parameter that needs to be considered here is that of linear velocity, rather than flow rate. High linear velocities may result in premature breakthrough, but flow rates may be increased while keeping linear velocities low by increasing the cross-sectional area of the packed bed. Thus, 47 mm disc-shaped sorbent beds can process large volumes of sample at flow rates up to 200 mL per minute, without breakthrough of desired analytes. Linear velocities in this case are moderate at 1.7 mm per second. This corresponds to a standard HPLC column of dimensions 4.6 x 250 mm,

Table 4. Characteristics of SPE versus LC modes

Characteristic (Isocratic)	SPE	LC
Sample volume (μL)	Large (mL)	Small
Sample solvent	Weak solvents	Intermediate
Quantity of analyte	Large (μ-mg)	Nanograms
Mobile Phase	None	Various
Washing	Stepwise, aliquots	Continuous
Elution	Strong solvents	Continuous
Sorbent clean-up	None	Automatic
Sorbent bed	Fluidized†	Immobile

† Excludes possible monolith beds

in which the linear velocity of the mobile phase at 1 mL per minute is 1.4 mm per second.

Chapter 6 will show that breakthrough volumes are more flow-rate sensitive for cartridge devices than for discs. In part this is due to the smaller particle diameters (for example 10 μm) that can be used in the small bed length membrane. Diffusion of analyte to the particle surface in this case is faster than the cartridge, in which larger particles are generally used.

If the large particle sorbent from a conventional SPE column is packed into standard HPLC column hardware, the resulting device may be operated as a low-pressure chromatography column. Samples may be injected and its components retained and eluted in the dynamic manner characteristic of elution chromatography (see also Section II.A.1). Such a column will have the usual properties of peak efficiency, capacity factors, selectivity factor and symmetry, for example. Indeed an SPE column can readily be configured as a column by using short, fat hardware, with minimal or zero dead volumes above and below the sorbent bed, and also used as a LC column, albeit a short one.

However the operation of this column in the SPE mode will be substantially different from its operation in the LC mode. Table 4 shows several of these features compared for SPE and LC.

Component resolution in LC is achieved by a combination of kinetic (peak width, symmetry) and thermodynamic (selectivity, symmetry) processes. Separation of components of interest in SPE is achieved primarily by making use of selectivity of sorbent with respect to the sample.

By its very nature, resolution in SPE is considerably worse than in low-pressure LC. In SPE large sample volumes, high concentrations, and the tendency to approach the maximum binding capacity of the sorbent, all work against the achievement of high resolution between components on the packed bed. Furthermore, the packed bed in SPE columns is short and

Table 5. Optimum N values for packed beds used in LC and SPE

Separation Device	N (Theoretical plates per column)
HPLC column (5μm, 15cm)	10,000
Low pressure LC column (60μm, 50cm)	500
Low pressure LC column (2cm)	70
Traditional SPE device (60μm, 2cm)	70
SPE disc (10μm, 0.1cm)	10†
SPE disc, (15μm, 0.1cm)	2†

† Compared to cartridges, N values for discs of the same length and particle size are smaller because the fibers that bind the particles together cause significant band broadening.

consequently the plate number, N, which determines resolution, is small. Thus, the value of elution chromatography and its maximizing of efficiency and selectivity in optimizing resolution, are lost in SPE where, for the reasons given above, substantial band-broadening and loss of selectivity due to massive overloading sometimes occurs.

In gradient elution LC, the inlet section of the column may function as an SPE element, with the remainder of the column providing the efficiency and selectivity to resolve compounds in the sample. This has a major drawback in that the head of the column may become contaminated with strongly adsorbed components from the sample. This is less of a problem with internally modified sorbents (Hagestrom and Pinkerton, 1986) where contaminants are excluded from the pores.

Table 4 shows that in isocratic LC sample volumes are small and analyte concentrations must be low. In SPE sample volumes may be several hundred times the bed volume and analyte concentrations can be several orders of magnitude greater than those used in LC.

Maximization of N and α (selectivity) in SPE can be achieved under conditions similar to LC, if small sample volumes and low analyte concentrations are used. But under these conditions SPE would be of limited use. Thus, in SPE the analyst sacrifices N, α, and resolution in order to achieve a gross sample clean-up process, in which the separation of substantially different groups of compounds is achieved. Table 5 lists typical N values of columns used in SPE and LC. These values are obtained under optimal conditions where N can be maximized. Where large volumes and quantities are processed in SPE, these N values are considerably smaller than the values in Table 5.

Disc technology has been in some ways a surprising development in SPE technology. Bed length in a disc is, by definition, very short, typically 1 mm and diameters are as small as 4 mm. Acceptance of discs in joining

Table 6. Values of specific permeability B_0 for SPE devices and HPLC columns

SPE Device/ HPLC Column	B_0 (units, 10^{-14} m^2)
Standard column or cartridge	25
SPEC™ disc	8
Bakerbond Speedisk™	12
Empore™ disk	3
5 micron HPLC column	5

the arsenal of SPE devices was slow in coming, because doubts existed concerning their capacity and perceived assumptions that analyte break-through would occur.

However simple calculations of specific binding capacities of discs versus conventional packed beds clarified at least the capacity issue: The sorbent mass in a conventional 3 mL syringe barrel is typically 500mg. Bed dimensions are 15-mm (depth) and 12 mm diameter. A typical SPE disc in the same 3 mL barrel may have a sorbent mass of 35 mg, contained in a bed which is 12 mm in diameter and 1 mm deep. If the specific breakthrough capacity is assumed to be the same for the sorbent in both SPE devices, say 5% of analyte per gram of sorbent, then the capacity of the conventional bed is 25 mg and that of the disc, is 1.75 mg. Both these capacity values are entirely adequate for most quantities of analyte.

It was originally assumed that analytes applied to disc beds as short as 1 mm would rapidly break through, but the ability to impregnate discs with small particle size sorbents helped to militate against this. Packed-bed SPE columns and cartridges contain sorbents with large particle diameter (40-60 μm). These may channel because the particles are held in essentially a fluidizable bed within the device. Solid, porous discs, in which small particles (10-30 μm) are held close together in a fibrous mesh, show micro-chaneling only. Consequently a much higher proportion of their mass is used effectively to bind sample components. The 3M and Ansys discs are "solid sorbents." There is no tendency for particles to move within the bed and thus channeling is minimal (Hearne and Hall, 1993). The Speedisk™[12] is comprised of particles that are held in a tightly packed bed by means of several screens and filters, and channeling is reported to be minimal (Good and Redmond, 1997).

C. THE MODE OF PROCESSING

The way in which an SPE device is used depends upon several of is performance characteristics.

[12] Speedisk is a trademark of Mallinckrodt Baker, Phillipsburg, NJ.

1. Performance Characteristics

Permeability: An SPE device must be permeable in order to process a sample. The rate at which the sample is processed, for example in mL/min., depends upon several properties of the system. These include the viscosity and particulate content of the sample; and the length, pressure difference (or centripetal g force) and specific permeability, B_o, of the sorbent bed.

The specific permeability parameter is defined for a packed bed, where the fluid is particle-free, in Equation 8 below:

$$B_o = q\eta L/A.\ \Delta P \qquad (8)$$

where q = flow rate, η = fluid viscosity, L = bed length, ΔP = pressure difference across bed, and A = cross-sectional area of bed.

Using SI units throughout, the units of B_o are m^2 and the larger its value, the more permeable is the bed. Table 6 lists various SPE devices and a 5 μm HPLC column for comparison, showing typical values of B_o.

Particle size and particle distribution, and the structure of the packed bed are primary determinants of specific permeability. The presence of a significant proportion of fine particles in the sorbent, although the mean particle diameter may be large (say 50-70μm), will act to reduce permeability. This slows processing speed, which may be important when high throughput is necessary.

For a sample that contains particles that the bed filters out, B_o will decrease as the sample is processed, since the bed structure changes. Ultimately the permeability may decrease to the extent that the SPE device becomes clogged and the flow rate is unacceptably low. An SPE device that has an initial low permeability will clog more rapidly. Most SPE cartridges, columns, and discs contain filters, which are more or less effective at removing particles.

It may be advantageous, however, to remove particles from the sample before it is processed through an SPE device. Particles that reach the sorbent may result in a loss of capacity, so it is more effective to ensure that the sample is filtered before its components can reach the sorbent. Particle removal may be accomplished by centrifugation, off-line filtration, in-line filtration, or integrated filtration.

Off-line centrifugation is the most effective way of removing particles from a sample, and this technique is amenable to parallel sample processing. Conditions of several thousand r.p.m. for a period of a couple of minutes is usually sufficient to give satisfactory clarity.

Evacuation or pressurization can also achieve off-line filtration through a filtration device. There are many accessories available for this process, from standard 1-5 mL SPE columns containing 20 μm frits as filters; to large diameter filters and membranes. However removal of particles must be carried out with care, especially if the analyte has a tendency to adhere to the particles in the sample.

Throughput: Sample throughput is an aspect of SPE technology that is becoming increasingly important. Analytical laboratories are under more pressure to increase output with fewer resources. Total automation, with computer assistance in sampling, sample preparation, analysis, data collection, and intelligent feedback, is a major objective in this drive to efficiency.

However, manual and semi-automatic processors are still being used to prepare multiple samples using SPE. The glass-walled vacuum manifold seems to be a universal tool of the trade. This is one of a number of elements of the SPE process that determines throughput. Table 7 lists the most important contributors to the efficiency of SPE in sample preparation. Several of these contributors are discussed below.

For maximum speed and control of variables, all samples should be subjected to the same process step at the same time. This will entail assembling the sample vials in racks suitable for the use with multiple dispensing systems. Combine reagents as much as possible (for example, acid or base in salt). Choose buffers and organic solvents that cause adequate precipitation, where this step is required. Centrifuge samples to clarify where possible.

For sorbent conditioning use the smallest organic solvent volumes. Small volume SPE devices such as discs require minimal amounts (100-200 μL) of solvents. Furthermore, small beds often do not require a vacuum or pressure at this stage, since the entire bed may be effectively wetted by its wicking action. More polar surfaces such as Mallinkrodt Baker's Light Load™, Waters' Oasis™, or Varian's NEXUS™ may be conditioned with organic solvent in the sample. Optimization of the wash solvent is important in giving the cleanest sample. Bed drying proceeds most rapidly with the smallest bed volume. Analyte elution with the smallest volumes of solvent is rapid and leads to fast evaporation. Re-constitution with the mobile phase may be slow and alternative solvents should be tried. For example, redissolve an acid in a basic solvent.

Disc technology has several advantages over deep packed bed systems, for large sample volumes (>100 mL) although laminar flow cartridges (Speedisk™) and high flow, wide body cartridges (EnvirElut™) can also achieve very high flow rates. This aspect of SPE is covered in Chapter 4.

The appropriate choice of the specific SPE system and elimination of steps in the process can realize large savings in time and reagents.

Table 7. Elements of the sample preparation process that includes an solidf-phase extraction.

Process Step	Details
Pretreatment 1	Adjust pH, ionic strength, organic solvent content.
Pretreatment 2	Precipitation with salt, phosphoric, caprylic or acetic acid, organic solvents.
Filtration/ centrifugation	Off-line, in-line, integrated into the device.
Sorbent conditioning	Usually with an organic solvent, to improve adsorption and flow rate.
Sample addition and transport	Preferably as soon as the conditioning step is finished.
Wash solvents	Matrix removal, elution of weakly bound components.
Bed drying	Necessary when elution solvent is immiscible with wash solvents and sample. Useful as a means of reducing elution volume.
Analyte elution	Smallest volume of solvent, often with volatile solvent to allow dry-down/concentration.
Evaporation	Take precaution against losses on collection vessel walls, oxidation or other decomposition of analyte.
Reconstitution	Choose a solvent that rapidly reconstitutes the analytes and is compatible with the analytical technique. The HPLC mobile phase may be suitable.

2. Processing Systems

This section describes basic manual systems and accessories with which an SPE device is processed. Manual processing systems can be classified into two types. They are:

1) Single sample, single device, serial processing
2) Multiple sample, multiple devices, parallel processing

The first commercialized SPE device from Waters — the Sep-Pak — is an example of a serial-processing cartridge. Generally, the syringe is used as a manual pump to move liquids (solvents, sample) through the stationary phase (see Figure 1). Many companies have introduced products with similar designs. The reader is referred to the *1999 LC.GC Buyer's Guide Issue* for a fairly complete listing of companies who manufacture serial processing devices.

Where a syringe is used in this technique it must be detached from the SPE cartridge in order to retract the plunger. If it is not detached, air may be drawn through the packed bed, destroying the conditioned state of the

Table 8. Advantages and limitations of serial processing

ADVANTAGES	LIMITATIONS
Ideal for learning the basic SPE technique and for determining applicability of SPE to extraction problem	Low throughput
Useful for small sample numbers	Large footprint per sample
Appropriate for large volumes	Expensive
Good control of process	Difficult to adapt to automated systems
Large body of reference literature	
Moderate skill required/little training needed	

sorbent. In order to avoid this, the syringe barrel may be attached to the cartridge or column via an adapter. By this means the syringe may be filled off-line with the solvents and sample, and then reattached to the adapter and processing continued.

The Baker Miser™ (Mallinkrodt Baker, no longer available) was a multidirectional valve, which could remain attached to the SPE column or cartridge during all processing steps. All solvents and sample could be drawn into a syringe barrel sequentially from reservoirs attached to different ports on the valve. These were then pumped through the sorbent bed. A home-made device that achieves the same end as the Baker Miser could be constructed using common laboratory equipment and a switchable liquid-handling valve.

Vacuum: An early serial device used in SPE consisted of a side arm vacuum flask fitted with a stopper. In this set-up the sorbent column or cartridge is attached to a needle, which is passed through the stopper. Flow through the column is driven by applying vacuum to the flask. This is a serial version of the parallel processing vacuum manifold.

Serial processing is often used to prepare large volume samples (>50 mL). High capacity columns or wide discs are employed as the sorbent device. A variety of standard filtration glassware and more specialized systems have been designed for single sample treatment with high velocity disc technology. For example the Accu.prep 7000™[13] single extraction vacuum station can be fitted with 47 or 90 mm discs for processing up to a liter of sample. The station incorporates a unique 3-way valve, which allows the operator to extract and elute with no disassembly. Collection vessels may

[13] Accu.prep is a trademark of Carbon Products, Inc. Santa Rosa, CA.

be changed without breaking the vacuum.

The Diskcover-47™ (Restek) is a simple disc holder that is attached to a source of vacuum via a stopcock. It is suitable for both glass fiber and PTFE-based discs. The versatile Speedisk™ family of laminar extraction discs (Mallinkrodt Baker) is integral with a unique holder. This device may be adapted to remote sampling or to a vacuum manifold.

A creative variation on the single device has been commercialized by Ansys Diagnostics Inc. A small disc is friction-fitted into a pipette tip. The sample is drawn up through the disc so the components of interest bind to the bottom of the disc. Elution occurs in the opposite direction, so that the analytes may be obtained in a very small volume.

Centrifugation: An extraction column can be placed in a centrifuge tube and processed this way. The centrifuge must be stopped when a new solvent or sample is introduced, so this technique is better suited for the preparation of multiple samples. Table 8 lists the Advantages and Limitations of serial processing systems.

Parallel Processing: In this technique multiple samples are processed in a step-wise fashion. In other words, all samples are sequentially or simultaneously subjected to one step of the method, before proceeding to the next step.

Vacuum: Analytichem International (now Varian Sample Preparation Products) introduced systems for parallel processing in 1979. The basic approach required a vacuum manifold to draw solvents and samples through several SPE columns simultaneously (see Figure 2). These manual processing systems are widely available from several manufacturers, and the reader is again referred to the 1999 *LC.GC Buyer's Guide* for a fairly complete listing of companies who manufacture parallel processing devices.

The first manifolds were comprised of a stainless steel bath, fitted with a lid that could be vacuum-sealed with polypropylene or neoprene gasket. The lid contained holes threaded on the bottom to accept steel tubes, 1/16th inch outside diameter, to guide liquids into glass tubes and flasks. The holes on the top of the lid accepted the tips of polypropylene syringe barrels containing sorbent. The bath contained a drain leading through a vacuum gauge to a trap and a source of vacuum. A variety of SPE column sizes were available, and receiving vessels from test tubes to volumetric flasks could be aligned in racks within the manifold.

Table 9. Advantages and limitations of manual parallel processing systems

Advantages	Limitations
High through-put	Limited or complex control over flow rates
Small footprint	Requires moderately high skill level
Appropriate for large numbers of samples	Moderate training required
Adapts to most accessories and is easily automated	
Large legacy of existing applications/literature. Efficient for method development	

Modern vacuum manifolds are constructed of thick-walled glass baths and lids of an inert and rigid polymer like acetal, for example. Fluid pathways in such devices can be made resistant to corrosion, resistant to solvents and free of contaminating impurities. This is usually done using Teflon tubes and valves. The Vac Elut™[14] SPS 24 is a unique vacuum manifold: The major feature of this cylindrical, see-through device is the ability to relocate all 24 columns from a divert-to-waste position to the collection position, by a simple rotation of the cylindrical lid.

A very compact multiple-processing system for large discs is available from Mallinkrodt Baker. Their Diskmate™ II rotary extraction station has a hexagonal, space-saving design. CPI has commercialized a six-station version of the Accu.prep 7000 system. The Resprep-6D system (Restek) accepts multiple Diskcover-47 disc holders. Ansys Diagnostics Inc. has designed a compact system that accepts up to 6 disc holders for 47 or 90 mm size. Many standard vacuum manifolds are available for large disc use and for processing large sample volumes, from companies including 3M, Ansys, Waters, Supelco, Restek, and Varian.

Several companies including Varian, Ansys, 3M, Waters and Whatman have introduced 96-well plates containing SPE sorbent. These are normally used in automated instruments and are discussed in Chapters 2 and 17.

Positive Pressure: One of the first parallel processors to utilize pressure rather than vacuum to mobilize the solvents and sample was commercialized by Applied Separations Inc. in 1987. This instrument was automated, with a simple, logical, programming capability. Automated instruments for carrying out SPE are discussed in more detail in Chapters 14 through 16.

---

[14] Vac Elut is a trademark of Varian Inc., Palo Alto, CA.

Subsequently, several devices that deliver pressure to a vacuum manifold have been introduced (Supelco, Alltech). In 1996, Varian Inc. introduced a manual positive pressure manifold. These offer several advantages over vacuum system, including improved flow control and more rapid drying, as well as the opportunity to use nitrogen gas as an alternative to air, when the analytes are oxygen-sensitive.

Centrifuging: Centrifugation of samples is another form of parallel, manual processing. Many commercially available centrifuges can be used for this method. Typically an array of SPE columns are placed in special holders; solvents and samples are added, and all columns are processed together. This technique has the advantage of being able to subject all samples — usually several dozen — to the same force regardless of its location in the centrifuge. In addition, centrifugiation of samples is a widely-practiced and familiar technique.

Unfortunately, since there is often a variation in permeability from one column to another, the flow rates may differ and the processing time must be tied to the slowest flowing column. Furthermore, the centrifuge must normally be stopped in order to add the next aliquot of solvent or sample. Column drying is also a slow process using this technique. Ansys Diagnostics Inc. have designed a shaped SPE column specifically for centrifuging samples, but this is no longer commercially available. Table 9 lists the advantages and limitations of parallel processing systems.

VI. CHARACTERIZATION OF SPE SORBENTS

The characterization of SPE sorbents is a widely published (Unger, 1990) and important topic. Measurement of key properties of these materials is carried out routinely by researchers, manufacturers, and end-users. The prime objective in sorbent characterization is the determination of its suitability for the pertinent extraction.

A sorbent suitable for SPE must be able to sorb rapidly and reproducibly, defined quantities of sample components of interest. Ideally the mechanism of sorption, manifest by the isotherm for individual analytes of interest in the presence of the sample matrix, is such that there is maximum separation desired from undesired components.

Figure 7 displays illustrative adsorption isotherms (in the form of breakthrough curves, see Section III.B.1 above) of three components on a given sorbent, which by virtue of their different locations along the liquid volume axis, are likely to be fairly cleanly separated on a bed of this material. Such isotherms are the results of an effective characterization

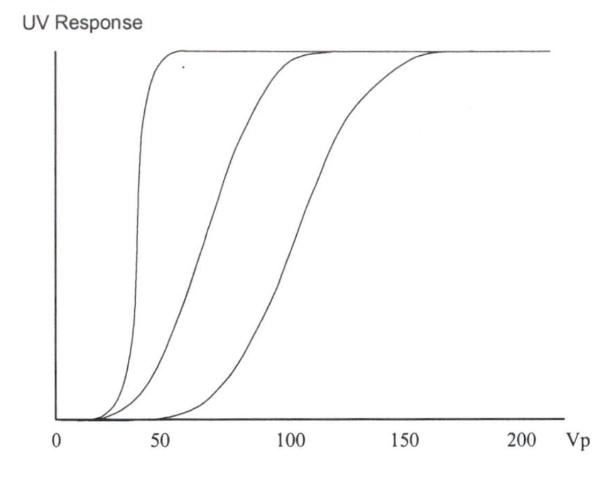

Figure 7. Typical breakthrough curves of three compounds of differing hydrophobicity, recorded on the same C18 sorbent.

technique. They allow the analyst to predict breakthrough quantities of given compounds and the degree of cross-contamination that may be expected. The shape of these curves will vary depending on several factors including linear velocity of the sample solvent as it passes down through the SPE bed, the distribution of concentrations of various analytes and the nature of the sorbent itself.

Unfortunately the determination of a family of isotherms for targeted analytes is usually a very time-consuming process and is only occasionally done by end-users or even some manufacturers. More commonly, specific chromatographic properties of the sorbent are measured and the sorbent's suitability for SPE, including its capacity and selectivity, is inferred from the results. These same chromatographic properties are used extensively to provide evidence of batch-to-batch reproducibility. In addition to chromatographic properties, static measurements of particle size distributions, pore properties, stationary phase loadings and cleanliness are made. In addition, the sorbent must be chemically and physically stable, in order to avoid releasing contaminants into an extract, and it must maintain flow properties (3.C.1) during solid-phase extraction and elution.

The following section describes briefly the most important characterization techniques used for SPE sorbents.

A. CHARACTERIZATION TECHNIQUES

1. Chromatographic Methods

We saw in Section III. B.1 how the loose sorbent of a traditional SPE column could be converted to an LC column. This device will have properties of analyte retention, efficiency, peak symmetry, and pressure. All of these properties will depend upon the choice of sorbent, test compounds, mobile phase, and temperature.

Usually a series of compounds will be chosen to probe the nature of the sorbent surface. It is important to have some understanding of the nature of the surface, since the analytes of interest and possible interferences will all be bound to this surface by one or more types of bonds. In the case of silica-based SPE sorbents the surface may contain a functional group (or ligand), silanol groups of varying acidity, metals, and adsorbed species of various kinds including water, solvents and by-products of silanes used to modify the surface. The dynamic interaction of these probes with the surface can be observed using chromatographic measurements, made under carefully controlled conditions.

For example, the most sensitive conditions for chromatography will include a mobile phase containing few or no ions. These conditions will promote any retention by coulombic forces, if they are present. The use of mobile phases with even moderate ionic strength (25-100 milllimolar) may substantially reduce the number of ionic interactions that an analyte has with a surface. Thus typically a 70/30 v/v mixture of methanol/water may be used to measure chromatographic properties of reversed phases.

Probe analytes may include pyridine and the basic amitryptiline to detect interactions with silanols. If there are strong bonds formed between compounds and the surface, there is likely to be problems in efficient extraction and elution of basic compounds. Phenol and benzoic acid may be employed to detect acidic sites on a surface. If these are present, it may be difficult to obtain good recoveries of organic acids. Chelating probe molecules, such as acetylacetone or 8-hydroxyquinoline, are often used to detect the presence of silica impurities such as iron, by forming complexes with these metal cations. In addition, uracil is used to measure the volume of the packed column that is filled with liquid (void volume). Knowledge of the void volume is useful, as it determines in part the breakthrough volume and the minimum useful elution volume.

The retention of toluene and butylbenzene measure the hydrophobicity of the column sorbent. These can be viewed as representative non-polar hydrocarbons. The logarithm of the chromatographic selectivity between

Table 10. Definitions of hydrophobicity and silanophilicity of non-polar bonded phases.

Group	Hydrophobic Index	Silanophilic Index
Kimata et al[†]	$k'_{(pentylbenzene)} / k'_{(butylbenzene)}$	$k'_{(caffeine)} / k'_{(phenol)}$
Waters[‡]	$k'_{(anthracene)} / k'_{(benzene)}$	$k'_{(N,N\text{-diethyl-m-toluamide})} / k'_{(anthracene)}$

[†] Mobile phase: 4:1 methanol: water

[‡] Mobile phase: 65:35 acetonitrile: water

these compounds is a direct measure of the free energy required to transfer three methylene groups from the mobile phase (aqueous) into the stationary phase. This can be used as an indicator of hydrophobicity of the sorbent.

Kimata et al. (1989) and Walters (1987), as shown in Table 10, have defined more quantitative hydrophobicity indices and silanophilic indices of non-polar phases.

The chromatographic behavior of a selected number of these compounds is frequently measured and the data tabulated, to stand as an overall index of the nature of the sorbent. These properties include retention time, symmetry, and efficiency. Table 11 gives a list of probe analytes, typical chromatographic values and ranges for the Validated™ sorbent (E.G & G, Inc., previously The Perkin-Elmer Corporation).

In general, manufacturers set ranges for these properties,which form some of the specifications for the bulk sorbent. A major drawback of using these chromatographic measurements as indicators of sorbent quality, is that they are made under carefully equilibrated, aqueous/organic conditions. In SPE using C18 phases the sorbent may typically contain only a small amount of organic methanol absorbed into the non-polar stationary phase from the conditioning step. The normally wholly- aqueous sample is then applied under non-equilibrium conditions. Under these circumstances the interactions of analyte with surface are expected to be substantially different from those existing during liquid chromatography of the same probe analytes. Furthermore, elution of analytes will often occur from a dry sorbent bed using a neat organic solvent, rather than from a fully wetted sorbent, as occurs during elution chromatography. This process may confer quite a different selectivity upon the sorbent, than that in a wetted state. This type of characterization of sorbents used in SPE can therefore be used to determine the nature of their chromatographic behavior only (as opposed to their true SPE properties), and to obtain information about reproducibility of this chromatographic behavior.

Table 11. Retentions, capacity factors, selectivities and pyridine symmetry specifications for the Validated™ sorbent†.

Specification	Range
Uracil	1.30-1.55 mins.
Pyridine	1.30-1.55 mins.
Propylparaben	1.85-2.05 mins.
N,N-dimethylaniline	2.55-2.80 mins.
K' toluene	1.70-2.10
Selectivity values	
Toluene/propylparaben	2.30-2.50
Toluene/N,N-dimethylaniline	1.25-1.35
USP tailing factor	
Pyridine	< 2.2

† This sorbent is a 5 micron, 115A material for HPLC use.

Another important property is the extraction performance for a range of specific analytes, such as the Baker Atrazine column, and the Varian TCA and Certify products. Ansys Diagnostics Inc. have published such data as part of their reproducibility claims (Blevins and Henry, 1995).

A more accurate measure of the performance of a sorbent in SPE can be obtained from chromatographic measurements of analyte retentions extrapolated to 100% water (log k'_W), since this is generally the nature of the sample solvent. However Poole and Poole in Chapter 6 explain the drawbacks of this method. At low organic solvent concentrations, retentions are very long and severe band-broadening and tailing reduce the accuracy of measurement. In addition at high aqueous content the rate of change of retention with solvent composition is very rapid and the accuracy of the value of k'_W is low. The method also assumes that extrapolation is a legitimate procedure; in other words, the shape of the curve is consistent right to the point of 100% water.

Log k'_W values may be readily obtained directly from values of log P_{oct} (Braumann, 1986) where P_{oct} is the equilibrium coefficient describing the distribution of the analyte between octanol and water. However this is not a method that is accurate in all cases and the analyst should use this estimate with caution.

Retention Maps: In the absence of rigorous measurements of breakthrough curves for individual and mixed analytes, retention maps are often used as general indicators of sorption characteristics. These are plots showing the relationship between retention of an analyte on a non-polar phase for example, usually expressed as log k', and mobile phase composition, expressed in terms of volume % organic in water. These maps often provide a useful indication of the adsorption capacity of a given sorbent for a series of analytes.

For example Pichon and coworkers (1995) have compared the binding ability of three quite different sorbent classes for the herbicides atrazine and its degradation products using retention maps. This is a particularly useful study as it examines the ability of specific sorbents to bind highly polar compounds, a much-neglected field. The sorbents were C_{18} silica, polystyrene/divinylbenzene (PSDVB) and porous graphitic carbon (PGC).

Those sorbents suitable for extraction of the polar degradation products of atrazine and simazine are shown in Table 12. All sorbents are suitable for the extraction of these less polar parent compounds. PGC was shown to be suitable for all analytes regardless of their polarity. This sorbent showed high values of log k' for polar compounds in the study, even at 10% methanol in the mobile phase. Clearly the use of log P_{oct} as a measure of log k'_W is not appropriate for the carbon based sorbent. On the other hand log P_{oct} values for the most polar compounds in the series were typically less than 1,

Table 12. Suitability of sorbents for binding decomposition products of Atrazine and Simazine.

Compound†	DEA	OHA	DIA	ADE	ANE	ACY	OHDEA	DAA	OHDIA
C18	•	•	•						
PSDVB	•	•	•						
PGC	•	•	•	•	•	•	•	•	•

† DEA = desethylatrazine; OHA = hydroxyatrazine; DIA = desisopropylatrazine; ADE = ammelide; ANE = ammeline; ACY = cyanuric acid; OHDEA = hydroxydesethylatrazine; DAA = desethyl desisopropylatrazine; OHDIA = hydroxy desisopropylatrazine

and showed predictably, weak bonding ability with the C18 and PSDVB sorbents.

It is believed that PGC contains residual acidic and basic functional groups at its surface which are responsible for its ability to bind polar and non-polar compounds. This work emphasizes the importance of experimenting with alternative sorbents when problems of capacity, binding and recovery arise with conventional SPE methods. This is especially true of highly polar compounds whose capacity to bind to non-polar sorbents is frequently very low.

2. Pore and Particle Properties

These include particle diameter, pore diameter, pore volume and surface areas and their distributions. If we assume that all pores are open cylinders that have the same size and shape, the following relationship applies (Equation 9):

$$PD = (4\ PV/SA) \times 10^4 \quad (9)$$

Where PD = pore diameter in Angstroms, PV = pore volume in mL/gram, and SA = surface area in m^2/gram. Typical ranges for PD, PV and SA are given in Table 13 for various sorbents.

Pores in general are not circular, not cylindrical and have a range of sizes and shapes (Rodriguez-Reinoso et al, 1991). Thus values of pore properties that are commonly given in the literature are averages of some kind (Henry, 1991). The range or distribution of these properties is also important in SPE, as we shall see in the following discussion:

Pore diameter: This determines the size of and speed with which a given amount of a species may enter and exit a pore. These materials include surface modification reagents, buffer components and analytes of interest. Species that are too bulky are more or less excluded from different regions

Table 13. Commonly used sorbents for SPE

Sorbent	Pore Properties†	Solubility	Rigidity	Polarity	Sample Solvent
Silica-based bonded phases	60-1000Å, 50-500 m^2/g 0.5 - 2 mL/g	Alkali	High	Various	Aqueous
Silica (SiO_2)	60-1000Å, 100-200 m^2/g 0.5-2 mL/g	Acid	High	Polar	Non-polar/organic
Alumina (Al_2O_3)	30-300Å 50-200 m^2/g 0.5-1 mL/g	Acid	High	Polar	Organic/ Non polar
Florisil ($M_gS_iO_3$)	60Å 1 mL/g	Alkali	High	Polar	Non-polar
Diatomaceous earth/Water	Macroporous 1 mL/g	Alkali	High	Polar	Organic
Polystyrene/Divinylbenzene	30-4000Å, 0.5-1 mL/g	Toluene	Moderate	Various	Aqueous
Polymethacrylates (porous)	100-800Å, 100m^2/g 1 mL/g	Toluene	Low	Various	Aqueous
Porous Carbon	60Å, 800m^2/g 1.0-1.5 mL/g	Insoluble	High		Various
Polysiloxanes/Polyimide (SPME)	(Liquid)	Organic Solvents	(Liquid)	Various	Aqueous
Polydivinylbenzene/vinylpyrrolidone	80Å, 800 m^2/g 1.0-1.5 mL/g	Alkali	Moderate	Non-polar /polar	Aqueous

† Pore diameters in Å ; surface areas in m^2/g ; pore volume in mL/g

of a pore and may not bind. By this means a sorbent will select a proportion of analyte species in a sample.

Small molecules and ions may enter a pore and have access to a large proportion of the binding region. In general, analytes must diffuse into a pore, since there is no convective flow. An exception to this is the group of perfusion sorbents (Afeyan et al, 1990) that contain large through-pores as well as smaller pores. Thus it can be seen that the binding capacity (in mg of analyte per gram of sorbent, for example) for a given molecular species,

Table 14. Capacities of several commercial sorbents

Sorbent Type	Substrate Capacity (per g)	Manufacturer
Strong anion exchanger	PSDVB, 0.5-1.5 meq	Alltech
Strong cation exchanger	PSDVB, 0.5-1.5 meq	Alltech
C18	Silica, 0-50 mg	Varian Inc.
Silica	Silica, 0-30 mg	Varian Inc.
Porous Graphitic Carbon	Carbon	Supelco
Metal Chelation	PSDVB, 1 meq	Alltech
N-Vinyl Pyrollidone	NVPDVB, 30mg	Waters Corp.

will depend upon the total surface area it can access. This in turn depends upon the pore size distribution. A sorbent with a broad pore size distribution from 60-500 Å, for example, may have significant binding capacity for large molecules. In this case such a sorbent may bind up to 10% of its weight of a 50,000 MW compound. Table 14 shows the stated capacities of several commercially available SPE sorbents for small analytes (Alltech, 1996; Larrivee et al 1994).

Pore diameter may also influence recovery of certain classes of analyte. Globular polypeptides for example may enter a pore and upon adsorption inside, denature and increase in size. Desorption may not then be possible, as the enlarged molecule may be unable to diffuse back through the mouth of the pore.

The general reproducibility of a given SPE sorbent is clearly determined in part by a constant pore diameter distribution. The latter will ensure a constant analyte capacity distribution, all else being equal. In other words, for a given sample type containing a variety of molecular species, the binding capacity of each analyte in the presence of the others will be constant over different batches of sorbent. This is especially critical for compounds having low binding capacity. For example the binding of a large protein such as IgM may depend upon the existence of a proportion of very large pores in a given sorbent. Only a small shift in pore size distribution to eliminate these large pores may have the effect of eliminating any binding capacity for IgM at all.

In principle, the larger a pore diameter, the faster will be the transfer of material through it. Thus for rapid sample processing, a larger pore, say >200 Å, will have kinetic advantages over a smaller pore of say 60 Å. The "perfusive" sorbents with pore sizes up to 4000-6000 Å, from Perseptive BioSystems, can bind bio-polymers at very high linear velocities. These materials are generally too expensive for most SPE applications.

Table 15. Surface area at various pore diameters†

Inside all Pores Greater	Pore Area m^2/g	% of Total Surface Area
15 Å	350	100
50 Å	280	80
115 Å	175	50
200 Å	65	19

† Nucleosil, 115 Å silica, 5 micron

Surface areas: Surface areas of sorbents, usually expressed in m^2/g, have been indicated above to be a primary determinant of capacity. Adsorption in SPE usually occurs at a surface, so capacity will be proportional to surface area, since each molecule occupies a specific number of square nanometers. As explained above the key property in adsorption is accessible surface area. This is related to pore size distribution, which in turn is characteristic of the nature of the support. Selected data for a specific silica is summarized in Table 15. It is apparent that significant surface area is present at 200 Å or greater in this silica, although it is nominally given a value of 115 Å.

The size of a molecule that can freely diffuse in and out of a pore of a given diameter depends upon a number of factors. Some of these include the shape of the pore, the dimensions of the molecule (in particular its largest dimension) and its effective diameter, taking into account bound solvent molecules and ions, for example. Taking as an approximation that the pore diameter must be five times the molecular diameter, a 40 Å diameter globular protein for example could access a 200 Å pore. Table 16 lists several proteins whose largest dimension is about 40 Å (Creighton, 1993), and which therefore have access to 65 m^2/g of surface area in the above silica sorbent.

In principle, accessible surface area can influence recovery. If a surface is not constructed properly it may bind more or less permanently, a proportion of a strongly bound analyte. The greater the surface area then, the smaller the recovery of such an analyte. Conversely, a weakly bonded analyte molecular would benefit from a larger surface area. Thus a larger weight of sorbent is necessary in this case to prevent breakthrough.

Adsorption and subsequent reaction of destabilizing chemical species such as hydroxide ions may be increased with a larger surface area. On the other hand, while reaction rates such as hydrolysis may increase, there are more bonds to be broken when the surface area is greater. Depending upon

Table 16. Molecular weight, volume and largest dimension of several proteins.

Protein	MW (kD)	Volume (Å)^3 x 10^{-3}	Largest Dimension (Å)
Cytochrome	11.99	23.1	37
Ribonuclease A	13.60	23.4	38
Lysozyme	14.3	40.5	45
Myoglobin	17.80	48.4	44

factors such as reaction mechanisms and the architecture of the support, the destructive effects of these processes may cancel out.

Pore volume: This is defined as the space within all pores of all particles in a unit mass of sorbent. The measured value however, depends upon the size of the molecule used in the technique. A gas such as nitrogen will have access to a higher proportion of pore volume, than say an organic molecule such as uracil.

During the various steps involved in SPE, where the particle and its interior surface are wetted, the nature of the stationary phase or bonded layer may influence pore volume. For example the C18 chains in the surface of the popular ODS silica sorbent are more or less extended depending upon the nature of the wetting agent. Thus the unretained volume for elution of a small non-interacting molecule may be smaller when isopropanol (a good wetting agent) is the organic modifier, compared to methanol (a poor wetting agent) as the modifier.

In the case of polymeric resin based sorbents, pore volume will depend upon their tendency to swell and shrink (Davankov et al, 1992 and 1994). This varies with different solvents and resins.

Pore volume is important in SPE since this property will contribute significantly to elution volumes. Furthermore, where drying of the sorbent bed is necessary, a larger pore volume will hold a larger amount of water and slow the drying process considerably.

Particle Size Distribution: Another important property of the sorbent used in SPE, is its particle size distribution. This can be given as the number or the volume of particles of a given diameter. These can be interpreted differently, so it is important to be aware of the parameter that is being used in the distribution.

Coarse particles tend to channel more readily than fine particles, but confer high permeability and filtering efficiency upon a bed packed with such material. Fine particles have a tendency to be lost through the outlet filter and contaminate the sample. Furthermore, small particles reduce bed permeability, increasing processing times. In addition sorbents containing fines tend to be clogged more readily by sample particles unless a prefilter is used (see Section III.C.1).

Ideally a bonded phase used in a column should have a narrow particle diameter distribution, with an average around 40-60 microns for easy processing. The sorbents produced by International Sorbent Technology and the Oasis from Waters for example, claim good permeability due to the fact that fines have been removed. Finer particles can be used in disc technology (see Section III.A) because the bed path length is very short.

Particle size distribution is the major sorbent characteristic that determines specific permeability, B_o (see Section III.C.1).

Particle Shape: The large majority of particles used in SPE are irregular in shape, containing some sharp edges. Spherical silicas (from Hypersil or The Separations Group) are not often used because of the increased cost. However spherical beads of porous polymer are embedded in Empore and Nova-Clean™[15] discs (polystyrene/divinylbenzene beads) and Oasis columns (polydivinylbenzene/polyvinylpyrrolidone beads). These synthetic polymer sorbents packed into standard SPE columns are available from an increasing number of sources.

In principle, the structures of beds prepared from particles of different shape are similar enough not to play a role in SPE performance. The major issues in particle shape choice are availability and cost.

3. The Nature of Solid Surfaces

Surfaces and Particles: The nature of the sorbent surface as it pertains to SPE, is a field that is described very thoroughly in the literature (Adamson, 1976). In SPE, solid supports have a surface, most of which is located inside pores. Pores may be of different size, shape and tortuosity, and it is the combination of all of these properties, which determines the important pore-related performance criteria (see Table 3). In simple terms, a sorbent surface conditioned for extraction, will consist of ligands — functional groups capable of interacting with analyte molecules — solvating molecules, and ions that come from a buffer, metal salt, or that are part of the substrate.

Once gaining access to this surface through diffusion or convection in pores, or by convection in the case of non-porous particles, the sample components may adsorb to the surface. In adsorbing, solvent molecules and ions must be displaced by the analyte. The displacing power of an analyte will determine the extent of its binding. Molecules and ions that are good

[15] NovaClean is a trademark of Alltech Inc., Deerfield, IL.

Table 17. Values of carbon coverage* for several C18 sorbents used in SPE

Manufacturer	Product Name	Carbon Coverage (Moles of carbon/m²) †
Mallinkrodt Baker	Light load octadecyl	20.2
Mallinkrodt Baker	Polar Plus™ *(C18)*	27.0
Mallinkrodt Baker	Octadecyl	28.3
Supelco	LC-18	23.3
Alltech	Extract-Clean Hi-load C18	28.3
Machery Nagel	C18	23.3
Ansys	C18 AR	31.9

† Obtained by dividing surface area of the unbonded silica into the %C, and multiplying by 1000/1.2. This is an approximate value of carbon coverage only.

displacers will have good capacities; and conversely weak displacers will have low capacities. The pattern of adsorbed species of different binding strengths and amounts can be described as the selectivity of the sorbent.

Insofar as SPE involves flow, transport phenomena will also play a role. Rapidly diffusing analytes will reach the surface rapidly, whereas slowly diffusing compounds will reach the surface later. Mass transport is a field supported by a comprehensive theoretical and experimental framework (Fernandez and Carta, 1996), and will not be covered in this chapter in any depth.

In certain sorbents used in SPE, there is a distribution of ligands from outside to inside the pores. For example a dual porosity media has been described (Revis and Williams, 1988) that comprises porous silicas in which the outer region of each pore will contain one type of ligand, while the inner region will contain a different ligand. The major objective of such dual regions is to separate larger polymers such as proteins from small analyte molecules such as drugs. The separation mechanism is usually size exclusion.

Spectroscopy: A common method of surface characterization is that using various forms of spectroscopy. This field is very large and the only example we will give in this chapter is the intriguing application of diffuse reflectance infrared Fourier transform (DRIFT) spectroscopy.

Murthy and coworkers (1991) measured DRIFT spectra of several manufacturers' cyano bonded phases. This functional group is potentially unstable due to a tendency to hydrolyze to the corresponding amide and carboxylic acid, as shown in equation 10:

$$—C{\equiv}N + H_2O \rightarrow —C(=O)NH_2 \rightarrow —C(=O)OH \qquad (10)$$

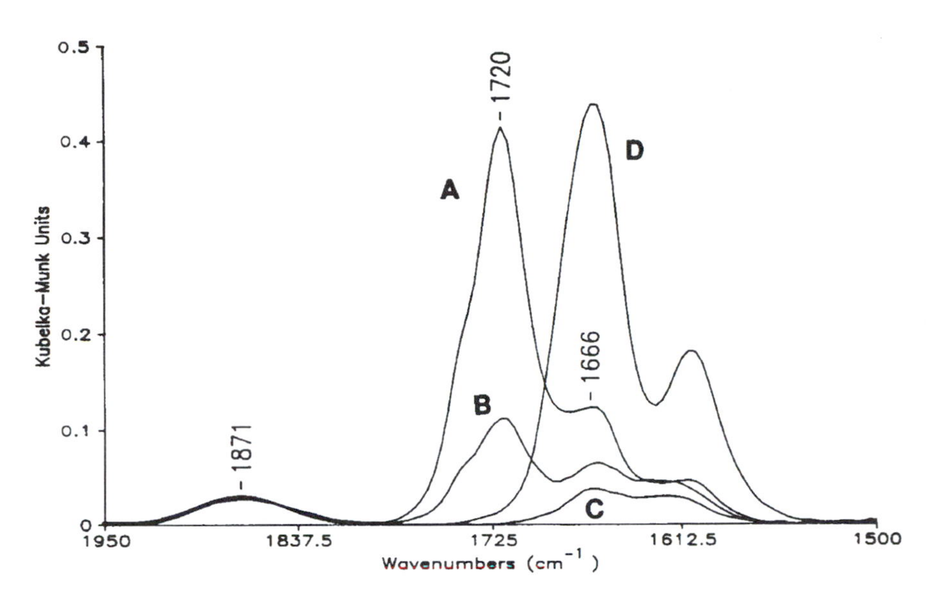

Figure 8. Drift spectra of various CN bonded phases. Diagram reproduced with permission from J.Chromatogr. (Murthy et al., 1991).

DRIFT spectra were measured of several sources of cyano bonded phases used in SPE. The carbonyl of the amide and carboxyl groups gives characteristic infrared absorptions, which can be used to indicate varying degrees of decomposition. Figure 8 shows comparative spectra for sorbents from a number of manufacturers. They indicate the almost complete lack of cyano group in one case to a pure cyano phase in another type of sorbent. Clearly there is decomposition at some point in the preparation of certain phases, for which DRIFT can be used profitably as an indicator. The importance of maintaining the presence of the correct functional group cannot be stressed too much. The cyano group in this case confers a specific type of selectivity upon the SPE sorbent, which will be different with an amide or carboxyl group. Furthermore there is the potential for severe non-reproducibility in this phase if due care is not taken in its manufacture.

Surface Coverage: This property is often used to characterize a silica-based sorbent used in SPE and is commonly defined as the ratio of the number of moles of ligands per gram of sorbent, to the surface area of the sorbent in square meters per gram. Another definition uses the ratio of the moles of carbon (in a hydrocarbon-based bonded phase) per gram to the surface area in meters per gram. Table 17 is a set of data from sorbents of various manufacturers, giving values of carbon coverage for specific C18 sorbents.

The wide variation in coverage indicates several differences in the nature of the sorbent, including surface reactivity, the extent of completion of the surface modification reaction, and the level of endcapping. These variations are not indicators of general suitability for SPE, but simply mean that some sorbents will be better or worse than others will be for given sample types. The coverage variations will certainly indicate differences in chromatographic properties and sorbent selectivity. Usually the surface area used in the calculation is that of the silica substrate, obtained from a measurement of total area with nitrogen. It is unlikely, however, that large modifying reagents such as octadecyl silanes will have access to all this surface area, so the true surface coverage of a C18 ligand for example is likely to be much higher than the calculated one.

For a given set of reaction conditions, constant average pore properties (see above), and constant reactivity of the substrate, coverage should be reproducible. This means that the type of surface that is exposed to the analytes as they adsorb, will be the same from lot to lot of sorbent. Thus taking all these pore properties and coverage measurements into account, these types of characterizations should give an assurance that practical properties of SPE sorbents such as capacity and recovery should be reliable and obtainable reproducibly from different batches of the same sorbent.

It is important, however, to distinguish between the ligand densities of the primary functional group and the endcapping reagent. Both may contribute to %C and it is therefore necessary to know the carbon that originates from the various types of ligands. This is often difficult, as there is an assumption that subsequent reactions do not alter the carbon level of the prior surface. Ultimately ligand coverage as a measure of the nature of a surface; it is an inexact property, and only serves as a rough guide to the reproducibility of the surface.

Chemical Stability: One of the most important properties of the SPE sorbent support is its stability. Ideally the support or substrate should be resistant to reaction with solvents and their components. Silicas react with water, hydroxide ions, and phosphate ions and absorb metals and basic compounds. Polystyrene/divinylbenzenes react with concentrated mineral acids, strong Lewis acids and bases and reactive double bonds. Aluminas and hydroxyapatites are soluble at $pH < 4$. Cross-linked acrylates, urethanes and amides begin to react with water in the presence of hydroxide ions.

The coating or bonding of various types of layers on top of or inside the support surfaces are processes which have been given considerable attention in the form of extensive experimental (Regnier et al, 1991) and commercialized technologies (Majors, 1999).

Chemical Stability and Reactivity of a Solid Surface: Ideally a surface or substrate should not react with any component of the cleaning, conditioning, washing or eluting solvents, or the sample itself. Any reaction will alter the nature of the sorbent or deplete the sample of analytes. Hydrolysis is the most common reaction to consider here, and it is preferred that it occurs negligibly during the processing time of an SPE device. Other reactions may occur between a functional group on a surface and a solvent or sample component. For example, surface amino, hydroxyl, ester and carboxyl groups, may react with carbonyl compounds, amines, or alcohols in solution. These reactions may alter the nature of the surface or more importantly, irreversibly bind analytes.

In principle, we may not expect a large change in the nature of the surface by reaction with analytes. The binding capacity of sorbents used in SPE is very high, typically 10-50 mg per 500 m^2 of surface. When the analyte is present at sub-milligram levels in the sample, only a fraction of the surface may be changed. However, the small active fraction of that surface, such as exposed silanols in the case of silica-based sorbents, may become completely covered with a sample component, such as a nitrogen base. The subsequent binding of similar compounds will then be different. Capacity may decrease and recovery may increase. The reactivity of synthetic polymer-based SPE sorbents also needs to be considered. The familiar polystyrene/divinylbenzene based media will bear unreacted vinyl groups. These are capable of reaction with other active double bonds, creating covalent attachment of certain analyte molecules. These reactions however, usually require initiators or higher temperatures, which are usually not used in SPE processes.

Acrylate-based SPE media such as the NEXUS sorbent contain hydrolyzable ester groups. These are fairly stable to hydrolysis up to pH10; and in any case should cause no problems in SPE where contact time is brief. The Oasis SPE media consist of polyvinylpyrrolidone cross-linked with divinylbenzene. This sorbent contains amide bonds which can be slowly hydrolyzed in acidic or basic solutions. As with NEXUS, however, the extent of reaction during an extraction is negligible.

The ligands bound to a silica surface via siloxane bonds are subject to acid- and base-catalyzed hydrolysis (Kirkland et al, 1997). The velocity of this reaction depends upon the concentration of water (in aqueous/organic solvents), the type of organic solvent, pH, temperature, ion type and the nature of the ligand layer. Kirkland and coworkers have measured the influence of these parameters upon the rate of ligand and silica loss, for sorbents used in HPLC. In all cases ligands were the mono-functional type. In other words, where the silylating reagent contains a single reactive group. These groups include methoxy, ethoxy, and chloro. Kirkland et al. observed that at low pH ligands were cleaved by hydrolysis. Where the silane contained groups such as isopropyl and butyl (Kirkland et al., 1989), these alkyls appear to shield the siloxane bond from hydrolysis and stability was enhanced. These sterically protected ligands also enhance the stability of bonded phases up to a pH value of eleven. The authors also observed that buffer components such as phosphate and carbonate caused more rapid hydrolysis, than acetate, citrate or borate. Although these observations relate to the behavior of small particle sorbents, it seems likely that these properties may be generally characteristic of the large particle, bonded silicas used in SPE.

There have been several studies of the relative stabilities of monofunctional versus polyfunctional surface chemistries (Leister, 1992). In the latter case, the majority of surfaces consist of a more or less polymerized silane bound to the surface at one or more attachment points. These have been shown to be considerably more hydrolytically stable than standard monofunctional ligands layers, in which only a single siloxane bond per ligand is possible. Most sources of commercial SPE sorbents have both mono- and polyfunctional (polymeric) silica sorbents available.

Cleanliness: Sources of contamination of a prepared sample include several components of the SPE holder. Compounds such as plasticizers, mold release agents and antioxidants, may be extracted from the syringe barrel, frits and filters, and find their way into the sample. Contaminants may also come from the sorbent. These may include silanes, monomers and metals. Porous solids that are used commonly in SPE can absorb excess reagents and these must be cleaned from the sorbent by thorough washing. Many companies who manufacture these sorbents will have specifications for their cleanliness. These specifications are usually in the form of the weight of say a methanol extract residue as a fraction of the weight of unextracted sorbent. The availability of such data depends upon the manufacturer, but it is always useful to request it. This property is very important, since its value is not only a measure of the contaminants that may reach a sample, but also a measure of the reproducibility of the sorbent and hence the SPE device itself.

Particles may leak through improperly sealed frits, through the frits themselves, or may be shed from SPE discs. This problem is of concern where the extraction process is not followed by a final particle removal step, such as filtration or centrifugation. Fine particles will cause increased wear in HPLC injectors, increased column backpressure or harm the integrity of surfaces and data in many forms of mass spectrometry. They may also occlude analytes if the eluent is dried down. While a final filtration of a sample prior to injection is always recommend, it is useful to reduce particles in samples at prior stages of their processing. SPE devices should not show evidence of particles that accumulate in packaging, on the walls of columns or above or below the sorbent bed. The SPE columns from Jones Chromatography and the Empore Disk from 3M claim to be particularly clean in this respect. In the former, elutriation or comprehensive particle sieving can be used to reduce fines substantially; and in the latter, the PTFE enmeshing process reduces tendency to shed to a negligible level.

The analyst may wish to wash off the contaminating particles and/or determine whether there is evidence of continuous leakage from the bed itself. In the latter situation, no material should be observed after the evaporation of a second extract of the bed with a solvent such as methanol.

Components of SPE devices such as syringe barrels, frits and filters, often contain impurities such as plasticizers, mold release agents, antioxidants, and monomers. These can leach into extracts and compromise the accuracy and precision of subsequent analysis, especially in GC and MS.

The cleanliness of the device will depend upon the choice of the quality of the initial polymer resins such as polypropylene and high-density polyethylene. Precleaning of the syringe barrel can also substantially reduce extracts. Glass and Teflon syringe barrels are commonly used, and can be cleaned very effectively. Polyethylene frits are a common source of contaminants, and disc technology has an advantage here since frits are not required. Teflon frits are inert and do not leach additives. Filters that are included in SPE devices, are largely constructed of binder free glass fiber mats, and should not contribute measurable quantities of contaminants to a sample.

The major manufactures of SPE columns have generally high standards of cleanliness, and have published evidence of this, usually in the form of a gas chromatogram of a solvent extract of a SPE device (see Junk et al., 1988).

One final note of caution is sounded by Crowley et al. (1995). This research group shows that residue observed in the dried-down extract of a water sample containing inorganic salts was composed of both residue from the sorbent itself and the inorganic salt. This was despite the fact that the sorbent — a C18 material — should show no retentive interactions with the inorganic salt. Further, the inorganic salt was observed in the residue despite extensive washing. The authors speculate that the salt is trapped in the C18 bonded layer during sample loading and that subsequent water washes cause the C18 layer to collapse over the trapped species. These compounds will only be released when a substantially organic solvent that is able to solvate C18 chains is applied — for example, the elution solvent. Thus observation of residue in the extract may be a function of the application rather than the SPE device itself.

B. ASSESSING METHOD ROBUSTNESS

A robust method may be defined as one that can yield a near quantitative recovery, over the entire desired concentration range, of a compound or compounds from a representative sample. The method must be able to do this with an acceptable degree of precision both within the same laboratory and outside it, and both on the same day and over periods of time.

SPE is but one step in the analytical method, which normally consists of sample pretreatment, extraction, elution and analysis. Sample pretreatment and clean up should be designed with the specific analytical technique in mind. Attention paid to the design of the SPE method should certainly be no less than that of the analytical method itself. In general the method of analysis will be used to determine the robustness of critical stages in the SPE method, therefore the analytical technique itself must naturally be predetermined to be robust.

SPE is a technique and therefore depends upon human skill for ruggedness. The definitions of the level of technical skill required and its measurement are very complex tasks and will not be attempted here. However there are several aspects of SPE methodology that are necessary — but not sufficient — for that method to be called robust.

The analyst will look for 100% recoveries, confident that quantitative extraction and elution has occurred in this case. Values greater than 100% may indicate contamination of the sample or extract from a source other than the sample. This means for example that in the analysis of phthalate esters in water, in which plasticizers may be leached from plastic tubing. The origin of recoveries less than 100% are more difficult to pinpoint as such results may indicate poor technique, low binding capacity (breakthrough) non elution or decomposition of the analyte. Determining the origin of and maximizing poor recoveries is a process that is discussed in other chapters of this book.

This section will focus on the primary contributors to differences in sorbents and products from one manufacturer to another, and also the level of reproducibility from a single manufacturer.

1. Manufacture-to-Manufacturer Sorbent and Product Differences

A comprehensive discussion of this topic would include most of the 17 performance criteria of SPE devices listed in Table 3. However from a limited, practical point of view the major inter-manufacturer differences that concern the analyst are analyte recovery, throughput, flow rates, chemical stability, cleanliness and the reproducibility of these criteria.

Table 18. Percent recoveries of PENT and STOB from serum using various manufacturers' C18 SPE cartridges.

Manufacturer	Mallinkrodt Baker	Waters	Supelco	Supelco	Polymer Institute	Machery Nagel
Cartridge volume	1mL	3mL	3mL	1mL	3mL	3mL
PENT	80.6	56.3	94.3	60.9	78.3	97.8
STOB	70.5	97.8	69.3	100.0	98.2	92.1

Recovery: Marko and Radova (1991) have compared serum recoveries of the basic compounds pentacaine (PENT) and stobadin (STOB) using SPE columns from five different manufacturers, using a standard technique. Their results are tabulated in Table 18. These results are very typical when comparing several manufacturers' products. As the authors point out, there is no good or bad sorbent. Marko and Radova ascribed low recoveries to levels of irreversible polar interactions between the drugs and the sorbents.

The experiments carried out by these authors with one manufacturer's product, extracting blank serum as a conditioning solvent and using triethylamine in methanol as an eluting solvent are very interesting. Recovery of both drugs was boosted to 91-92%; but without triethylamine, recovery reached a maximum of 87% (PENTA) and 56% (STOB). These results indicate several creative approaches to surface modification in attempting to increase and unify recoveries of analytes from sorbents of varying quality from different manufacturers. Surface treatment can be attempted at the conditioning, the wash or the elution stages. Such procedures may be included in method development and may also serve to reduce future problems in batch-to-batch reproducibility problems, which are inevitable. Much useful work remains to be done in this area of sorbent performance. A study of this effect is presented in Chapter 7.

In a general in-depth study comparing the Empore Disks to a traditional packed bed product, Wells, Lensmeyer and Wiebe (1995) measured several analytical parameters that characterize the method. These include linearity, recovery, precision, limits of detectability, and especially capacity. The study of the extraction of mexitilene and flecainamide from serum indicated equivalent recoveries (> 90%) for both SPE devices. Unfortunately, as is common in comparisons of this type, the large differences in particle size (8 versus 50 μm), and bed mass (15 mg versus 100 mg) between the two devices preclude the drawing of any rigorous conclusions regarding the intrinsic advantages of one over the other.

The disc format has one fundamental advantage over the conventional packed bed, which is reduced channeling leading to a higher specific ca-

Table 19. Flow rates of various samples with SPE discs.

Product (Particle Diameter μm)	Flow Rate (mL/min.)			Specific Permeability (10^{-14} m^2)
	D.I. Water	Tap Water	Lake Water	
Empore Disk (10)	143	101	100	3
ENVI Disk (15)	182	175	100	8
Speedisk (10)	250	240	240	12

pacity. This is due to the porous solid nature of the disc-based sorbent. Conventional packed beds are in essence fluidizable partly because of the large particle size and partly because of the dry pack manufacturing method. For this reason, they are more likely to contain larger channels.

Studies of sorbent and product differences from one manufacturer to another generally involve comparisons of commercial products, rather than fundamental technology. Many of the general advantages of one SPE device over another are eliminated when the best of aspects of each technology are compared. For example in the study by Wells et al. (1995), the performance of the disc and conventional packed bed would have been much closer if both had the same weight and volume of sorbent. On the other hand, certain sorbents and products present tangible, operational advantages over others due to improvements and changes in engineering, configuration and chemistry.

Flow Rate: Flow rate depends upon several properties of the SPE device, including pressure difference, liquid viscosity, cross-sectional area of the bed, bed length and its structure. It is easier to compare specific permeabilities of packed beds of sorbent, thus eliminating differences in these factors that play a role in determining flow rates. However this approach is rarely used, and it is more common to compare specific commercial products for flow rates of specific sample types under similar pressure differences for example.

Flow rates obtainable from several types of disc SPE products have been compared in literature from Mallinckrodt Baker. The discs from 3M (Empore), Supelco (ENVI Disk) and Baker (Speedisk) were compared. One liter of water from three sources was filtered through approximately 47 mm discs in each case. Some of the data is summarized in Table 19.

There is a clear relationship between bed permeability and flow rate in these results. Unfortunately, the source of data does not make it clear whether all discs were covered with a particle filter. The Speedisk device contains an integral fiberglass filter and flow distributor that will assist in removing particles and giving higher flow rates. This may explain the fact that all flow rates using the Speedisk are essentially the same regardless of the source of water.

Other comparisons of disc and conventional packed beds have been previously described in Sections III.B.1 and III.C.1

2. Performance Variance within a Manufacturer's Product

As explained in section IV.B.1, there are many performance criteria that may be considered in determining variance within a single manufacturer's product. Here again, the most important are analyte recovery, throughput, flow rates, chemical stability, cleanliness and the reproducibility of these criteria. Published reports of variance in recovery are legion, but details of the reproducibility of other characteristics are rarely found in the literature.

A great deal of analytical work is expended by manufacturers in measuring batch physico-chemical properties such as particle and pore properties, flow properties, the extent and nature of surface modification, and the cleanliness of their sorbents and containers. These companies have large data banks in which are stored the results of these tests for hundreds of batches of their products. Occasionally some of this data is published to support reproducibility claims. Sometimes customers can request and obtain lot history information for a specific product. This information will give some degree of confidence that a manufacturer is serious about minimizing variances within its product and how they go about accomplishing this goal.

One of the important performance criteria of SPE sorbents and products that is frequently measured is that of recovery. As it was explained above, the analyst will seek 100% recoveries and perhaps more importantly, reproducibility of this property. Most of the larger manufacturers of SPE devices have published data in support of reproducibility of recovery.

Table 20. Recovery of drugs from plasma, on various lots of sorbent from a single manufacturer.

Lot number	**31204**	**40101**	**30701**	**31212**	**31004**	**Average**
Amobarbital	91.8[a] (2.0[b])	94.1 (2.2)	94.5 (1.9)	92.1 (2.5)	89.6 (2.1)	92.4 (2.9)
	31004	**30701**	**31204**	**31201**	**31202**	
Hydrocortisol	103.6 (3.0)	98.6 (1.7)	100.9 (1.8)	102.5 (0.8)	99.7 (1.7)	101.1 (2.7)
	31204	**31202**	**31004**	**31201**	**30701**	
Mepivacaine	90.5 (1.4)	94.3 (2.0)	91.4 (0.7)	97.2 (2.5)	91.6 (1.6)	93.0 (3.1)

[a] Average of three measurements. [b] Relative standard deviation

Table 21. Recoveries of PENTA and STOB from serum.

Lot number	3738	384	551	552
PENTA	90.1	78.3	35.9	53.4
STOB	86.7	98.2	57.1	34.6

For example Blevins and Henry (1995) have published a study of the recovery of a neutral, basic, and an acidic drug from plasma using a C18 disc. Although the number of drugs investigated was small, they examined the performance of five separate lots of disc material held in an SPE column. The drug concentrations varied from 1 to 5 μg per mL. Table 20 summarizes the results from this study.

The results shown in this Table are generally considered satisfactory. In other words, recoveries of over 90% in all cases and lot-to-lot CV values of less than 3%. It goes unsaid in this study, that such uniform recovery behavior implies a similar range of solute-sorbent interactions in each batch.

This is not the case with the SPE products characterized by Marko, Radova and Novak (1991) in a study of the variations in basic drug extractions on another type of bonded silica. The basic compounds pentacaine (PENTA) and stobadin (STOB) at 1 μg per mL of water or serum were extracted onto 3 mL columns of C18 silica and eluted in methanol. Recoveries for four different batches of columns are shown in Table 21.

These results indicate a typical and complex case of irreproducibility of recovery. There is little consistency in the way recovery changes form one lot to another. The authors conclude in part that differences in surface polarity are the cause of the variation. Furthermore, the differences between the batches lie in their ability to adsorb the basic compounds rather than any differences in their elution patterns.

It is sometimes useful in these cases to examine some of the chemical properties of the sorbent to determine the source of the problem. For example it is relatively simple to measure the percent carbon in each lot and relate it to the specific surface area to calculate ligand coverage, to determine whether there is a wide variation. The experimental results form Marko et al. (1991) are given in Table 22, demonstrating no clear correlation of recovery and ligand coverage. In general this information is inadequate to explain variability in recovery, even though there may be good liquid chromatographic correlations of retention and level of coverage.

Table 22. Recovery and C18 ligand coverage for four lots of C18 SPE cartridges.

Lot number	Coverage (micromols/m^2)	Recovery, PENTA	Recovery, STOB
3738	5.8	90.1	86.7
384	6.2	78.3	98.2
551	5.4	35.9	57.1
552	5.5	53.4	34.6

The data gathered by Marko, Radova and Novak (1991) is insufficient to provide a definite explanation of the observed lack of reproducibility amongst the lots tested. Direct measurements of the nature of the surface are needed, such as solid state NMR or ESCA. Unfortunately such measurements are expensive and require experience in interpretation. It is doubtful whether such expense is warranted in a commercial context, and variable lots like those above should be simply rejected as being outside of specifications.

C. REPRODUCIBILITY

Reproducibility of the sorbent properties was discussed in Section III.A.1, and that of the combined components of an SPE device is no less important. The sorbent mass (capacity, recovery), packing density, flow rate, filterability, cleanliness, and dimensional integrity are the major criteria from which reproducibility is desired.

The automatic machines that pack and assemble SPE cartridges and columns can achieve high precision (< 1% relative standard deviation) in sorbent mass and packing density. Column to column variability in these properties can be observed, and is often due to two factors: Variability in the packing machinery and its operation, and the tendency of some segregation of particles to occur without due care and attention.

Sorbent mass and packing density (whether cartridge, column, or disc) determine measured capacity, recovery, and flow rate of an SPE device. Manufacturers have detailed and large lot history data banks, which evidence reproducibility. Summaries of this information can be found in company literature and technical bulletins. It is rarely published in journals, since this data is considered too commercial. Table 20 contains data showing the lot-to-lot reproducibility of recovery for example, for several lots of commercial SPE columns.

A tightly packed SPE bed will flow more slowly than a loosely packed bed, since permeability is changed (see Section III.C.1). This property of SPE columns still appears to be problematic in reproducibility, although the variability in flow rates that arise from it is a quality parameter that depends upon the manufacturer. The filterability of a sample depends in part upon bed structure, and its reproducibility is very important. Manufacturers constantly assess the reproducibility of dimensional integrity of the SPE device. Most cartridges and columns are made by injection molding. This is a process which, given a moderate degree of competence by the mold maker, is highly precise. Tolerances are generally required in the 2-4 thousandths of an inch range for most dimensions of the cartridge or column.

The reproducibility of capacity and recovery will be enhanced when they are as independent of flow rate as possible. Sorbent beds that are short and wide and that contain adequate excess of bed mass, will be the closest to the ideal in this case. Linear velocities here will be relatively slow, allowing radial diffusion processes to predominate over axial transport (see Section III.B.1). Reproducibility of the SPE device, as in any facet of analysis, is largely in the hands of skilled, observant analysts and operators. The degree of perfection that will be achieved here ultimately depends upon many human traits, rather than scientific principles.

V. A SUMMARY OF SOLVENTS USED IN SPE

A. INTRODUCTION

Solvents are of fundamental importance in most processing steps in SPE. In liquid samples they maintain both the dissolved state of soluble species and the dispersed state of colloidal material. Organic solvents such as acetonitrile may be added to aqueous samples in order to selectively precipitate salts and biopolymers, while maintaining a single liquid phase in the treated sample. In the sorbent conditioning step, an organic solvent is used to wet or solvate a hydrophobic surface. A selective solvent may be used in the wash step to remove unwanted components from the pores and interstices of the packing and from the sorbent surface. A solvent elutes the analytes of interest from the surface and voids and may be evaporated. Finally the solvent-free sample may be re-dissolved prior to analysis.

Solvents used in SPE will be chosen so that their properties of miscibility, solvating power, purity, and chromophoric nature are integrated to give speed, simplicity, effective sample clean-up and quantitative recovery in the relevant process steps. This discussion of the practical decisions that need to be made in solvent selection for SPE will focus on the importance of the above priorities and their effects.

B. SOLVENT PROPERTIES OF IMPORTANCE IN SPE

1. The Eluotropic Series

Table 23 contains a list of the most commonly-used solvents in SPE, arranged in order of eluting strength measured as elutropic value, relative to alumina as the sorbent. The order is very similar for silica, but their magnitude is different. This table is an extract of the more comprehensive list of solvents examined by Snyder (1968). In general it allows the analyst to select solvents that:

- are miscible with each other
- will effectively wet a given surface
- can be effective eluting solvents
- will contribute to overall processing efficiency

Many of these general characteristics are based upon the qualitative principle of "like dissolves like." Solvents that have significantly different elutropic values therefore are likely to be non-miscible, have different surface wetting abilities, different abilities to elute analytes from a surface, and have different abilities to dissolve compounds.

For example, acetonitrile (E_0 0.65) and methanol (E_0 0.95) are commonly used eluting solvents in SPE. Marko and Radova (1991) have shown that the solvents have remarkably different abilities to elute two tertiary nitrogen bases (pentacaine and stobadin) from conventional C18 sorbents. Methanol gives recoveries of up to 98%, whereas acetonitrile appears to have no powers of elution at all. Indeed, these authors have used the comparative elution properties of these solvents as probes to estimate certain surface properties of the sorbents. In this case the results suggest the presence of analyte-sorbent polar interactions, which are interrupted by the more polar methanol. The larger value of E_0 for methanol in the series, coupled with its greater eluting power, imply that there are significant numbers of polar regions in C18 sorbents. These, in turn, are capable of binding basic compounds, a well-known and well-understood phenomenon.

However, the relative eluting power or selectivity of a given solvent or solvent mixture for the members of a family of analytes in a sample is not so well understood. Poole and Poole have developed an impressive approach to this problem in the Chapter 6. These authors have separated the various sorbent-analyte interactions that contribute to the overall breakthrough volume. This breakthrough parameter will be different for different analytes and sorbents, and so the total matrix of breakthrough volumes for a complex mixture can be calculated. Such a comprehensive framework

Table 23. Characteristics of Solvents Useful in SPE.

Solvent	Eo	Viscosity (cp, 20°C)	Boiling Point, °C	Surface Tension, (25°C) mN/m	UV Cut off, nm	Toxicity† LD50, mg/kg
n-Hexane	0	0.41	98	19.65	200	49
iso-Octane	0.01	0.53	98	19.4	200	NA
p-Xylene	0.26	0.65	138	28.01	215	8000
Toluene	0.29	0.59	110	27.93	290	7530
Diethyl ether	0.38	0.23	34	16.65	280	2400
Methylene chloride	0.42	0.44	40	27.2	320	2524
Tetrahydrofuran	0.45		109		225	NA
Triethylamine		0.36	90	20.22	345	460
Acetone	0.56	0.32	56	23.46	330	5800
Dioxane	0.56	1.54	105	32.75	215	7120
Dimethylsulfoxide	0.62	2.24	189	42.92	260	17900
Acetonitrile	0.65	0.37	82	28.66	190	2460
iso-Propanol	0.82	2.37	82	20.93	205	5000
n-Propanol	0.82	2.27	98	23.32	205	4000
Ethanol	0.88	1.2	78	21.97	205	7060
Methanol	0.95	0.6	65	22.07	205	10
Acetic acid	Large	1.26	118	27.1	230	3310
Water			100			NA

† Toxicity measured orally with a rat.

would be invaluable in being able to shed some light upon the underlying parameters that control sorbent selectivity.

Often an eluting solvent may be suitable for direct injection into an HPLC column. In this case the evaporation and reconstitution steps are eliminated. For example, a sample adsorbed on a C18 sorbent may be fully eluted with a 1:1 methanol: 25mM KH_2PO_4, pH 7 solvent/buffer mixture and an aliquot analyzed without evaporation or dilution. This technique normally works best if full recovery can be achieved using small (say 100 µL) volumes of eluting solvent. This maintains the high analyte concentrations required for low detectable quantities. Small elution volumes generally imply small bed volumes.

An alternative to elution with a solvent compatible with the mobile phase is an elution with a small volume of non-aqueous solvent such as acetonitrile containing 0.1% trifluoroacetic acid or 5% triethylamine in methanol. This can then be diluted with a largely aqueous buffer such as 2 mM ammonium acetate containing 0.2% formic acid prior to direct injection.

2. Viscosity

Solvent viscosity has a clearly understood and simple effect in SPE process. The more viscous the sample or solvent, the slower the processing. The relationship between sample processing speed and viscosity can be seen in Equation 8. Flow rate will be inversely proportional to viscosity, if all other parameters are held constant. For example, isopropanol, ethanol, and methanol have viscosities of 2.37, 1.2, and 0.6 centipoise at 20°C, respectively. Methanol will therefore flow twice as fast as ethanol, which will be double that of isopropanol as the processing solvent. On the other hand, the elutropic power of ethanol on alumina (E_0 0.88) is not that much different from isopropanol (E_0 0.82), so that the former solvent may be preferred for speed.

One property of a packed bed that often changes during processing is the specific permeability itself. This may be caused by some accumulation of particles from the sample into the bed, causing a decrease in permeability.

3. Solvating Power

Solvents may be used in the first conditioning step of SPE for a preliminary clean up of the sorbent. A more subtle effect is the role the conditioning solvent has upon the extraction properties of a sorbent. Methylene chloride, toluene, acetone, and methanol are effective in extracting impurities from improperly purified sorbents.

In the sorbent conditioning process a hydrophobic C18 surface for example, must be prepared so that it is energetically favorable for analytes to be adsorbed into the surface from an aqueous sample. In thermodynamic terms this means that the interfacial surface tension must be reduced using a solvent having solubility in both the solid surface and the bulk sample liquid, to allow penetration of the sorbent pores by the aqueous sample, and to improve the kinetics of partitioning.

For example, isopropanol is an effective conditioning solvent for the hydrocarbon-like C18 silica sorbents. Its surface tension (20.93 mN/m) is the closest of the commonly used alcohols to that of hexane (19.65 mN/m) which resembles the sorbent surface in some respects. Isopropanol is completely miscible with hexane, whereas methanol is only partially miscible.

Once the analytes of interest have been adsorbed and the majority of the sample components have hopefully been removed, the washing and eluting solvents then play a crucial role in the selective desorption of unwanted and desired components, respectively. The solvents chosen must have appropriate eluting powers, which means the ability to desorb compounds and also to dissolve them. Solubility of analytes at this stage is rarely given serious thought, as it is more expedient to develop a method in an empirical fashion using a very limited, but generally satisfactory range of solvents. Table 23 in Section V.B.1 contains some of the most commonly used elution solvents.

The solubility of analytes of interest in given processing solvents, is naturally of great importance. This is particularly true of the final step in the SPE process if the eluent is dried down before reconstitution in a small volume of solvent suitable for chromatographic or spectroscopic analysis. This reconstitution may become part of the clean-up process.

The well-known Cohn protein fractionation process (Lehninger, 1975) relies in part upon the relative insolubility of certain blood components in ethanol. Thus, biopolymers such as proteins, polysaccharides and DNA, which have limited solubility in aqueous solutions that are rich in organic solvents such as acetonitrile or ethanol, may be eliminated by suitable choice of reconstitution solvent.

4. Volatility

The requirement for volatility of a solvent used in SPE depends upon the stage in the extraction at which the solvent is applied. The lower volatility of n-propanol (b.pt. 98°C) compared to isopropanol (b.pt. 82°C), may have advantages in the conditioning and wash steps where in general the sorbent needs to remain wetted. More volatile elution solvents, however, are necessary at the sample evaporation stage. Methanol (b.pt. 65°C) has proved to be a useful solvent here, and is the most universal one found in SPE.

However, certain solvents form azeotropic mixtures with water and therefore have the added ability of acting as a drying agent during evaporation. For example acetonitrile (% by weight in azeotrope of 83.7, b.pt. 76.5°C), isopropanol (87.4%, b.pt. 80°C) and n-butanol (57.5%, b.pt. 92.7°C) are in this class.

5. Chromophoric Nature

The importance of ultraviolet and visible radiative transparency in a solvent is well understood in liquid chromatography. A commonly used measure of this transparency is the UV cut-off, defined as the wavelength where absorbance is 1.0 for a 1 cm pathlength of the neat solvent. In SPE this property is a requirement where eluted samples will be prepared for liquid chromatographic analysis, without evaporation. In addition, care must be taken to remove residues of chromophoric solvents such as methylene chloride (UV cut-off 320nm) or acetone (330nm).

Table 23 gives the UV cutoff wavelengths for the solvents most commonly used in SPE.

6. Purity

This is one of the most clearly appreciated solvent properties in SPE. Impurities may bind temporarily to the sorbent, concentrate and be eluted later. The potential for concentration of impurities from the solvent may mean that a compound present but unobservable in a solvent blank, can appear in an SPE extract, giving the appearance that the impurity came from the SPE device. Non-volatile impurities may remain with the analytes after evaporation and interfere with subsequent analyses.

7. Reactivity and Toxicity

Naturally neither solvents nor their impurities should react with analytes or the sorbent. Occasionally solvent impurities may react with the analytes themselves, resulting in inaccurate analyses. Short-term stability or shelf life tests of solutions of the analytes of interest may need to be conducted to ensure that these problems do not arise.

The analyst should always be aware of the adverse health effects of solvents, and take precautions to avoid too much contact with them. Naturally, flammability is an important property of solvents used in SPE, and open flames or sparks should be carefully controlled. The subject of health and safety issues that pertain to solvents is too large to be discussed here and the reader is referred to manufacturers for information such as Material Safety Data Sheets (see also Table 23 for selected toxicity data).

C. SUMMARY

Elutropic strength, miscibility, viscosity, solvating power, volatility, chromophoric nature, purity, reactivity and toxicity are the most important properties of solvents that are used in SPE. It is clearly a complex process to consider every one of these properties during method optimization. Choice of solvent will strongly influence the crucial results of recovery, selectivity, speed, and simplicity. Fortunately, it is possible to produce very satisfactory results for most extractions using just a few common solvents

VI. CONCLUSIONS

The practice of SPE is based upon a wide range of well-understood physical and chemical principles. Among the fundamental fields of science that are important in SPE are chromatographic theory; surface chemistry; surface analysis; solubility; polymer chemistry; flow in packed beds; mass transfer phenomena; binding capacity; product design, quality, and reproducibility; solvent properties; and retention mechanisms. In addition, the analyst needs to have a strong grasp of applied fields such as inorganic chemistry, organic chemistry, and biochemistry. The logical application of these technical areas to practical SPE therefore requires a truly multidisciplinary synthesis of knowledge. In this chapter, I have attempted to describe just a few of the physico-chemical principles that are often used in practical SPE.

The introductory section serves to define SPE in its most simplistic form: a three-step process involving mass transport, adsorption, and desorption. The process generally occurs within a sorbent bed of fixed boundaries. The technique of SPE is applicable to any type, size, and form of sample and is amenable to simple manual or highly automated operations. The results may be the removal of unwanted components, the removal of the desired components, and/or the concentration of analytes of interest.

Section II summarizes the fundamental processes of retention and elution in SPE, extrapolating these processes from the mechanisms that pertain to liquid chromatography. This separation science is familiar to most analysts and therefore it is appropriate to begin a description of SPE from its principles.

Section III described in some detail the practical issues surrounding SPE from the viewpoint of performance characteristics of SPE devices. Table 3 is an exhaustive list of the major performance characteristics of SPE devices, their importance and their determinants. This list formed the basis for discussions of the sorbent container, the SPE device as a whole, and the various modes of processing the device.

The entire SPE device, defined as the container, frits, filters and sorbent, was then considered. The issue of column efficiency versus resolving power versus capacity was discussed, following on from the use of the liquid chromatography phenomena to explain certain processes in SPE. The performance characteristics of permeability and throughput in an SPE device were then described. Processing systems, both serial and parallel, were considered under the headings of pressure, vacuum and centrifugation.

The most important topic in SPE performance is that of sorbent characterization, discussed selectively in Section IV. Under this topic were included adsorption isotherms, chromatographic characterization, particle properties (size, shape, and pore properties), the nature of solid surfaces, chemical stability, and cleanliness. Method robustness, as influenced by sorbent properties, was explored in terms of manufacturer-to-manufacturer differences in recovery and flow rates; and performance variances within a single sorbent type.

The important aspects of solvent choice and their properties, were described in Section V. As solvents are required in all the major processing steps in SPE, their quality and performance characteristics are vitally important in achieving success in this technique. Table 23 gives a list of some of the more important solvents and their relevant properties such as eluotropic value, viscosity, UV cut-off, and toxicity.

There are dozens of physico-chemical principles whose successful practice in the technique of SPE will contribute to its success. This chapter has dealt with only the more important ones.

REFERENCES

Adamson, A.W., (1976) Physical Chemistry of Surfaces. Published by John Wiley and Sons, Inc. New York, pp 385-422.

Afeyan, N.B., Gordon, N.F., Mazsaroff, I., Varady, L., Fulton, S.P., Yang, Y.B., and Regnier, F.E., (1990) Flow-through Particles for HPLC Separations of Biomolecules: Perfusion Chromatography. J.Chromatogr. 519: 1-30.

Alltech, (1996) Factors that Determine the Capacity of an SPE Packing Bed. Separation Science, Bulletin Number 340, pp 8-10.

Blevins, D. D. and Henry, M. P., (1995) Pharmaceutical Applications of Extraction Disc Technology. American Laboratory, May, 32-35.

Bouvier, E.S.P. (1994) Solid Phase Extraction: A Chromatographic Perspective. Waters Column, 5 (1): 1-3.

Brauman, T., (1986) Determination of Hydrophobic Parameters by Reversed-phase Liquid Chromatography; Theory, Experimental Techniques, and Application in Studies on Quantitative Structure-Activity Relationships. J.Chromatogr. 373: 191-225.

Coquart, V. and Hennion, M.-C., (1992) Trace Level Determination of Polar Phenolic Compounds in Aqueous Samples by HPLC and On-Line Preconcentration on Porous Graphitic Carbon. J.Chromatogr. 600: 195-201.

Creighton, T.E., (1993) *Protein Structures and Molecular Properties*. 2nd Edition. Published by W.H.Freeman and Company. New York, pp 264-268.

Crowley, X.W., Murugaiah, V., Naim, A. and Giese, R.W., (1995) Masking as a Mechanism for Evaporative Loss of Trace Analyte, Especially After Solid-Phase Extraction. J.Chromatogr., A, 699: 395-402.

Davankov, V.A. and Tsyurupa, M.P., (1992) In *Synthesis, Characterization and Theory of Polymeric Networks and Gels*. Published by Plenum Press, New York. p 179.

Davankov, V.A., (1994) Analytical and Technological Applications of Hypercrosslinked Polystyrene-Type Sorbents. Proceedings of *Solid Phase Extraction*

Europe, November 28th-29th, 1994, Amsterdam. Published by Advanstar Communications, London, UK, pp 33-38.

Engelhardt, H., (1979) High Performance Liquid Chromatography: Chemical Laboratory Practice. Published by Springer-Verlag, NY, pp 21-24.

Fernandez, A.M and Carta, G., (1996) Characterization of Protein Adsorption by Composite Silica-Polyacrylamide Gel Anion Exchange. Equilibrium and Mass Transfer in Agitated Contactors. J.Chromatogr. A 746: 169-183.

Fritz, J.S., (1999) Analytical Solid Phase Extraction, John Wiley and Sons, Inc. New York, NY.

Good, T.J. and Redmond, A.F. (1997) Microcolumns For Extraction Of Analytes From Liquids. U.S Patent 5,595,653.

Hagestrom, I.H. and Pinkerton, T.C. (1986) Characterization of Internal Surface Reversed Phase Silica Supports for Liquid Chromatography. J.Chromatogr. 368: 77-84.

Hearne, G.M and Hall, D.O., (1993) Advances in Solid Phase Extraction Technology. American Laboratory. 25:28H-28M.

Helfferich, F. (1962) Ion Exchange. Published by McGraw-Hill Book Company Inc., NY, pp 434-442.

Henry, M.P., (1991) Design Requirements of Silica-Based Matrices for Biopolymer Chromatography. J.Chromatogr. 544: 413-443.

Junk, G.A., Avery, M.J. and Richard, J.J., (1988) Interferences in Solid Phase Extraction Using C18 Bonded Porous Silica Cartridges. Anal.Chem. 60: 1347-1350.

Kane, P.F. and Larrabee, G.B., (1976) Characterization of Solid Surfaces. Published by Plenum Press, New York.

Kimata, K., Iwaguchi, K., Onishi, S., Jinno, K., Eksteen, R., Hosoya, K., Araki, M. and Tanaka, N., (1989) Chromatographic Characterization of Silica C18 Packing Materials. Correlation between a Preparative Method and Retention Behavior of Stationary Phase. J.Chromatogr.Sci. 27: 721-728.

Kirkland, J.J., Glajch, J.L. and Farlee, R.D., (1989) Synthesis and Characterization of Highly Stable Bonded Phases for High-Performance Liquid Chromatography Column Packings. Anal.Chem. 61: 2-11.

Kirkland, J.J., Henderson, J.W., DeStefano, J.J., Van Straten, M.A., and Claessons, H.A., (1997) Stability of Silica-based End-capped Columns with pH 7 and 11 Mobile Phases for Reversed Phase HPLC. J.Chromatogr. 762: 97-112.

Larrivee, M.L. and Poole, C.F., (1994) Solvation Parameter Model for the Prediction of Breakthrough Volumes in Solid Phase Extraction with Particle Loaded Membranes. Anal.Chem., 66: 139-146.

LC.GC., (1999) Magazine, Buyers Guide Issue. Published by Advanstar Communications Inc., Duluth, MN. 17 (8).

Lehninger, A.L., (1975) in Biochemistry, 2nd Edition. Published by Worth Publishers Inc., NY, pp 162-163.

Leister, W.H., (1992) The Stability of Monofunctional vs Trifunctional C18 Chromatographic Phases Bonded to Silica on Exposure to Trifluoroacetic Acid. MS Thesis, Lehigh University.

Lovkvist, P. and Jonsson, J. A., (1987) Capacity of Sampling and Preconcentration Columns with a Low Number of Theoretical Plates. Anal.Chem. 59: 818-821.

Majors, R.E., (1999) New Chromatographic Columns and Accessories at the 1999 Pittsburgh Conference, Part II. LC.GC Magazine of Separation Science. 17: 304-315.

Marko, V and Radova, K, (1991) Variations in Solid-Phase Extraction of Basic Drugs Using Bonded Silica. I. Manufacturer-to-Manufacturer Variations. J.Liq.Chrom. 14: 1645-1658.

Marko, V, Radova, K and Novak, I., (1991) Variations in Solid-Phase Extraction of Basic Drugs Using Bonded Silica. II. Batch-to-Batch Variations. J.Liq.Chrom. 14: 1659-1670.

Murthy, R.S.S., Crane, L.J. and Bronniman, C.E., (1991) Characterization of Cyano Bonded Silica Phases from Solid-Phase Extraction Columns. Correlation of Surface Chemistry with Chromatographic Behavior. J.Chromatogr. 542: 205-220.

Perry, S. S. and Somerjai, G. A., (1994) Characterization of Organic Surfaces. Anal. Chem. 66: 403A-415A.

Pichon, V., Chen, L., Guenu, S. and Hennion, M.-C., (1995) Comparison of Sorbents for Solid Phase Extraction of the Highly Polar Degradation Products of Atrazine (Including Ammeline, Ammelide and Cyanuric acid). J.Chromatogr. A 711: 257-267.

Poole, C.F. and Poole, S., Chapter 6.

Regnier, F.E., Unger, K.K. and Majors, R.E., ed., (1991) Liquid Chromatography Packings. J.Chromatogr. 544 (entire issue).

Revis, A. and Williams, D.E., (1988) US Patent Number 4,782,040.

Rodriguez - Reinoso, F., Rouquerol, J., Singh, K.S.W. and Unger, K.K., ed., (1991) Characterization of Porous Solids II. Proceedings of the IUPAC Symposium (COPSII), Alicante, Spain, May 6-9, 1990. Published by Elsevier, Amsterdam.

Simpson, N., and van Horne, K.C., (1993) The Handbook of Sorbent Extraction Technology, 2nd Edition. (Eds. Simpson and van Horne). Published by Varian Inc. Palo Alto, CA.

Singh, K. S. W., (1991) Characterization of Porous Solids II, pages 1-11, Proceedings of the IUPAC Symposium (COPSII), Alicante, Spain, May 6-9, 1990. Edited by Rodriguez - Reinoso, F., Rouquerol, J., Singh, K.S.W. and Unger, K.K. Published by Elsevier, Amsterdam, pp 1-11.

Snyder, L. R., (1968) Principles of Adsorption Chromatography. Published by Marcel Dekker, NY.

Snyder, L.R. and Kirkland, J.J., (1979) Introduction to Modern Liquid Chromatogaphy. Published by John Wiley and Sons, NY, pp 365-370.

Svec, F. and Frechet, J.M.J., (1992) Continuous Rods of Macroporous Polymers as HPLC Separation Media. Anal.Chem., 64: 820-822.

Tanford, C., (1973) The Hydrophobic Effect: Formation of Micelles and Biological Membranes. Published by John Wiley and Sons, NY, pp 16-35.

Unger, K.K. ed., (1990) Packings and Stationary Phases in Chromatography Techniques. Published by Marcel Dekker, New York.

Walters, M. J., (1987) Classification of Octadecyl-Bonded Liquid Chromatography Columns. J. Assoc.Off.Anal.Chem. 70: 465-469.

Werkhoven-Goewie, C.E., Brinkman, U.A.Th. and Frei, R.W., (1981) Trace Enrichment of Polar Compounds on Chemically Bonded and Carbonaceous Sorbents and Application to Chlorophenols. Anal.Chem. 53: 2072-2080.

Wells, D.A., Lensmeyer, G.L. and Weibe, D.A., (1995) Particle-Loaded Membranes as an Alternative to Traditional Packed-Column Sorbents for Drug Extraction: In-Depth Comparative Study. J.Chromatogr.Sci. 33: 386-392.
Wynne, Paul (1997) Personal Communication.

6

THEORY MEETS PRACTICE

Colin F. Poole and Salwa K. Poole, Wayne State University, Detroit, Michigan

I. INTRODUCTION

In this chapter we will explore the advances that have been made in turning SPE into a predictive science, through theory and modeling of extraction behavior. Since the success of a solid-phase extraction is contingent on concentration, we will look at how we can optimize two key parameters of an extraction, the breakthrough volume and the elution volume. The ration of these two, V_B/V_E equals the concentration effected by the extraction. The other aspects of SPE — solvent exchange, sample clean-up, or matrix sim plification — are much more specific to sample and application, and are left for detailed consideration in later chapters.

Solid-phase extraction is used to isolate and concentrate selected analytes from a liquid, fluid, or gas by their interaction (sorption) and transfer to a solid phase. After physical separation of the sorbent and sample medium the analytes are recovered by liquid or fluid elution, or by thermal desorption. In addition, sorbent immobilization provides a mechanism for matrix simplification using selective desorption to remove co-extracted matrix components without displacing the analytes of interest. Provided that the analytes are recovered in a volume significantly smaller than the original sample volume then significant sample concentration is achieved. Certain boundary conditions can be set for the above experiment that are capable of a theoretical treatment.

Until recently solid-phase extraction was performed using short columns or cartridges packed with sorbent particles of a fairly large size (average particle diameter 50 μm) to allow sampling by gravity flow or aided by vacuum (Liska et al., 1989; Poole et al., 1990; Hennion and Pichon, 1994; Berrueta et al., 1995). In the last few years new products for SPE have appeared in the form of particle-loaded membranes (Hagan et al., 1990; Mayer and Poole, 1994), particle-embedded glass fiber discs (Mayer et al., 1995) and polymer-coated fibers for solid-phase micro-extraction or SPME (Zhang et al., 1994). General characteristics of the discs and cartridges were reviewed in Chapters 2 and 3. Only features pertinent to the theoretical modeling will be discussed here.

The particle-loaded membranes in various diameters and a thickness of 0.5 mm, consist of sorbent particles of about 8 μm diameter (90% w/w) immobilized in a web of polytetrafluoroethylene microfibrils. These membranes are flexible and, for general use, are supported on a sintered glass disc (or other support) in a standard filtration apparatus that uses suction to generate the desired flow through the membrane.

Particle-embedded glass fiber discs contain sorbent particles embedded in a supporting matrix of glass fiber. The small diameter glass fiber discs are rigid and self-supporting; larger diameter discs are usually used with a supporting structure. The particle-loaded membranes and the particle-embedded glass fiber discs are generically referred to as disc technology and represent an evolution of the cartridge-type sampling devices with claimed advantages over the latter which will be discussed shortly. The polymer-coated fibers for SPME provide a new approach for sampling headspace or liquid samples. Isolation occurs by sorption into a thin film of polymer (5 to 100 μm thick) immobilized on a silica fiber, followed (usually) by the solventless transfer to a gas chromatograph by thermal desorption in the heated injection port of the instrument. The coated fibers are thin enough to fit inside the needle of a modified microsyringe, which provides a convenient protective housing for manipulating the sampling device.

This is the only chapter in which SPME will be discussed in detail. It is introduced to provide a comparison to SPE theory and to demonstrate that, despite similarities in their names, SPE and SPME are very different techniques.

II. PHYSICAL PROPERTIES OF SOLID-PHASE EXTRACTION DEVICES

General properties characteristic of the sorbent are usually provided by the manufacturer as part of a quality assurance protocol using standardized nonchromatographic techniques (Poole and Poole, 1991). More recently, some crude HPLC-derived capacity and selectivity data has also been supplied on the Certificates of Analysis of some manufacturers. Typical properties include particle size distribution by laser granulometry, specific surface area by the BET method, the surface coverage of chemically bonded phases by combustion, an apparent sorbent pH determined from an aqueous suspension of the sorbent, UV absorbance of a solvent extract, and a nominal mean pore diameter by calculation or size exclusion chromatography.

These values, although useful, do not in themselves tell us how the sorbent will perform in SPE and are more useful for the comparison of sorbents from the same or different manufacturers. For example, a high surface coverage and surface area for a chemically bonded sorbent indicates high retention for those solutes able to explore the internal pore structure of the sorbent. The apparent pH indicates likely difficulties in the recovery of acidic or basic compounds unless methods are modified to accommodate the contribution of ion-exchange type interactions to analyte retention (Martin et al., 1993). Acidic compounds are difficult to elute from basic sorbents by neutral solvents, and vice versa. The indication of a relatively high proportion of particles below average size in the particle size distribution may suggest problems in obtaining an adequate flow during sampling and the occurrence of sorbent particles as contaminants in the recovered extracts.

A. CHROMATOGRAPHIC METHODS

The above properties are capable of a "common sense" interpretation and will not be considered further. Of greater interest are those properties of the sorbent and sampling device that can be determined by chromatographic methods that define the sampling characteristics of the devices. Under typical experimental conditions a sample of fixed concentration enters the sorbent bed with a constant velocity and is quantitatively retained by the sorbent up to the point were the sample volume exceeds the retention capacity of the sorbent. Further sample entering the sorbent bed will not be adequately retained by the sorbent and eventually the sample concentration entering and exiting the sorbent bed will be the same.

This process is an example of frontal chromatography and is illustrated in Figure 1. Figure 1 is generally referred to as a breakthrough curve. The point on the curve at which some arbitrary amount of sample is detected at the outlet of the sorbent bed is the breakthrough volume (V_B). Authors have arbitrarily defined the concentration of sample exiting the sorbent bed at different levels (Josefson et al., 1984). For our purpose we choose 1% of the sample concentration at the entrance to the sorbent bed as reasonable from a measurement point of view and in keeping with the desire to define a maximum sample volume that can be processed with minimal (acceptable) loss of analyte.

The shape of the breakthrough curve is bilogorithmic and a second point, V_M , corresponding to the volume at which the sampling capacity of the sorbent is exceeded and the flux of analyte entering and leaving the sorbent bed are nearly identical can be defined. Based on the definition of the breakthrough volume V_M corresponds to the point on the breakthrough curve where the concentration of the analyte at the exit of the sorbent bed is 99% of its value at the entrance. The point of inflection for the breakthrough curve corresponds to the retention volume of the analyte (V_R) since the differential of the breakthrough curve is a Gaussian distribution similar to the peak response observed during elution chromatography.

The breakthrough volume and the maximum sampling volume are important terms in SPE. The breakthrough volume corresponds to the largest sample volume that can be processed without significant loss of analyte and for which recovery after elution for all sample volumes less than the breakthrough volume will be 100% in the absence of irreversible sorbent interactions. It is the breakthrough volume that is most important in determining the suitability of a

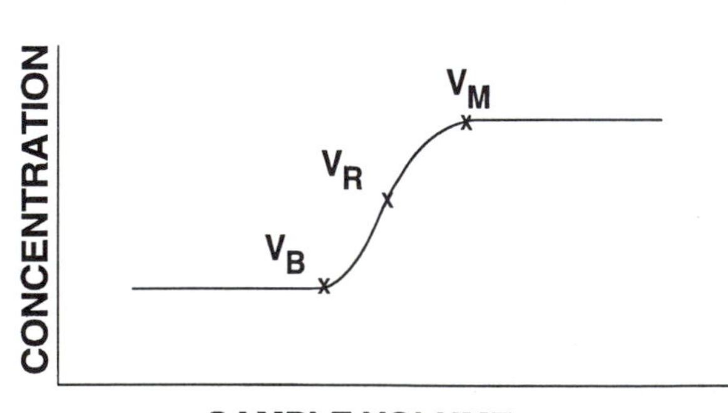

Figure 1. Breakthrough curve for a solid-phase extraction device. V_B is the breakthrough volume, V_R the retention volume, and V_M the maximum sampling volume.

sorbent for a particular isolation procedure. We can define a maximum sampling volume that corresponds to the largest sample volume that can be processed to obtain the largest possible amount of isolated analyte. The amount of sample that is retained on the sorbent is larger than at the breakthrough volume but since some analyte is lost during the sampling process the recovery on elution is always less than 100%.

From the theory of frontal chromatography (Werkhoven-Goewie et al. 1981) the breakthrough volume is simply related to the retention volume by Equation 1:

$$V_R = V_B + 2\sigma_v \qquad (1)$$

where σ_v is the standard deviation depending on the axial dispersion of analyte along the sorbent bed and is evaluated through equation 2:

$$\sigma_v = V_O\,(1 + k) / \sqrt{N} \qquad (2)$$

where V_O is the interparticle volume of the sorbent bed, k the retention factor (capacity factor), and N the number of theoretical plates for the sorbent bed calculated by Equation 3:

$$N = V_R\,(V_R - \sigma_v) / \sigma_v^{\,2} \qquad (3)$$

In principle, it should be possible to determine V_O and N for a sorbent bed and, by measuring V_R, determine the breakthrough volume for any analyte of interest.

The above equations are derived assuming the conditions of linear chromatography apply. Large sample amounts or strongly retained matrix components that result in curved sorption isotherms (Bitteur and Rosset, 1987) or sorbent beds with very low numbers of theoretical plates (Lovkvist and Jonsson, 1987) may result in inaccurate estimates of the breakthrough volume based on the above equations. Still, the boundary conditions for linear chromatography seem realistic for many typical applications of SPE, where trace concentrations of analytes are concentrated from samples with low concentrations of competing matrix components in solvents with weak sorbent interactions.

The properties of cartridges are easily determined by packing the sorbent into a column adapted for use in high-pressure liquid chromatography (Miller and Poole, 1994). A different approach is required for particle-loaded membranes and particle-embedded glass fiber discs. In this case an apparatus for forced flow planar chromatography is used, Figure 2 (Fernando et al., 1993), in a form originally used to characterize the kinetic properties of thin layer chromatography plates (Poole and Fernando, 1993).

The apparatus consists of an overpressure development chamber that can be operated up to a pressure of about 20 atmospheres, a reciprocating piston pump for constant volume solvent delivery, a standard microliter

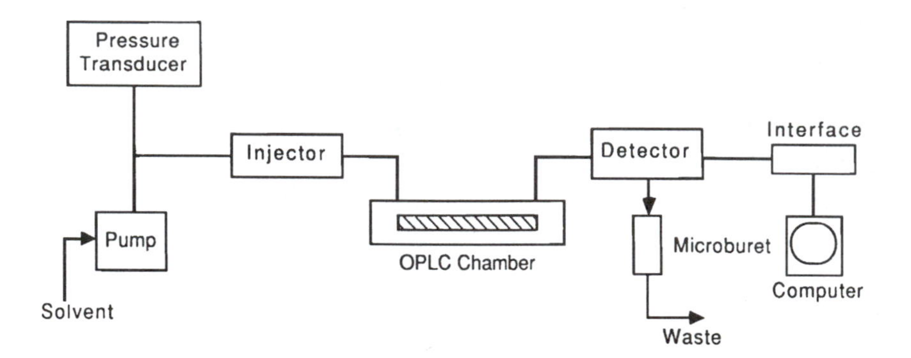

Figure 2. Block diagram of the apparatus used to determine the kinetic properties of particle-loaded membrane and particle-embedded glass fiber discs. From Fernando et al., (1993) with permission.

loop injection valve for solute introduction, on-line UV detector, and a microburette to determine the flow rate. A high-precision pressure transducer is also employed, to determine the pressure drop across the membrane as a function of the solvent velocity. Particle-loaded membranes or particle-embedded glass fiber discs of dimensions 10 x 20 cm are supported on an aluminum sheet of the same size, sealed around the edges with latex glue. Two troughs are cut through the sorbent media, corresponding to the solvent entry and exit positions for the overpressure development chamber. Once inserted in the overpressure development chamber, hydraulic pressure applied to the reverse side of the cushion in contact with the surface of the sorbent media effectively seals it. This allows measurements to be made at selected flow rates and pressure drops analogous to those made with cartridge sorbents using high pressure liquid chromatography (HPLC).

The total porosity of a sorbent medium, ε_t, is defined as the sum of the interparticle porosity, ε_u, and intraparticle porosity, ε_I. The total porosity can be determined from the elution volume of a totally permeating and unretained substance as indicated by Equation 4:

$$\varepsilon_t = F\, t_o / V_c \tag{4}$$

where F is the volumetric flow rate through the sorbent medium, t_o the retention time of the unretained substance, and V_c the volume of the sorbent medium (column or membrane volume). The interparticle porosity is determined in the same way by choosing a solute which is unretained and excluded from the intraparticle volume of the sorbent. The intraparticle porosity is then obtained as the difference between the total and inter-particle porosities. Due to problems with stationary phase sorption and various normal and ionic exclusion mechanisms the choice of unretained solute is generally arbitrary and arrived at by trial and error. For aqueous mobile phases and low polarity sorbents, sodium nitrate at a high concentration (26 mg/ml) is generally effective for determining the total porosity and at a lower concentration (1 mg/ml), where ion exclusion prevents permeation of the pore structure, the inter-particle porosity. From Darcy's law, Equation 5, the specific permeability of the sorbent medium, B, is obtained from the slope of a plot of the mobile phase velocity, u, against $\Delta P / \varepsilon_t L \eta$

$$u = \Delta P\, B / \varepsilon_t L \eta \tag{5}$$

where ΔP is the pressure drop across the sorbent medium, η the viscosity of the solvent used to make the measurements and L is the length of the sorbent bed. From the specific permeability the apparent chromatographic average particle size, d_p, is obtained using the semi-empirical Carman-Kozeny Equation 6:

$$B = (d_p^2 / 180\Psi^2)(\varepsilon_u^3 / [1 - \varepsilon_u]^2) \tag{6}$$

where Ψ^2 is a shape factor, assumed to have a value of 1.7 for irregular porous particles. Also, from the pressure drop across the sorbent media, the flow resistance parameter, ϕ, can be calculated using Equation 7:

$$\phi = \Delta P\, d_p^2\, t_o / L^2 \eta \tag{7}$$

Some typical values for cartridge and disc sorbent media are presented in Table 1 (Poole et al., 1996). For the cartridge devices it is necessary to consider the difference in packing density between the column used for the measurements and the typical values indicated for cartridges. The columns used for the chromatographic measurements are prepared by the tap-and-fill method, so it is reasonable to anticipate that well-packed cartridges would approach these values. That they do not in general is one reason for poor reproducibility of flow characteristics. In addition, breakthrough volumes in SPE are influenced by inadequate sorbent packing density (Miller and Poole, 1994; Poole et al., 1996).

The C18 sorbent cartridges in Table 1 contain almost 20% additional free volume compared to a well-packed column. This obviously affects their kinetic properties, since the sample flow streams will encounter regions of different packing density as they percolate through the sorbent bed. Although it is often claimed in the literature that cartridges are packed with sorbent particles of approximately 40 μm, this is generally untrue and a value of 50-60 μm is more typical of contemporary practice. Absolute particle size is not too critical; more important is the particle size distribution and, in particular, the relative proportion of particles with diameters below average size.

Table 1. Characteristic properties of sorbent media determined chromatographically

Property	Cartridges			
	C-4[1]	C-18[2]	CN[3]	Diol[4]
Cartridge packing density (g/ml)	0.322	0.667	0.599	0.557
HPLC column packing density (g/ml)	0.453	0.726	0.634	0.794
Apparent particle size (μm)	46.8	55.1	30.3	
Total porosity (HPLC column)	0.74	0.47	0.52	0.58
Interparticle porosity	0.60	0.41	0.41	0.41
Intraparticle porosity	0.14	0.06	0.11	0.17
Specific permeability ($10^{12}m^2$)	5.57	2.14	34.1	3.29
Flow resistance parameter	1100	707	2050	
Nominal pore diameter (nm)	25	6	6	6

	Particle-loaded polymer membranes		Particle-embedded glass fiber disc
	C-18[5]	C-18[6]	C-18[7]
Apparent particle size (μm)	7.7	5.8	15.3
Total porosity	0.52	0.54	0.51
Interparticle porosity	0.37	0.48	0.47
Intraparticle porosity	0.15	0.06	0.04
Specific permeability ($10^{14}m^2$)	2.50	2.21	8.38
Flow resistance parameter	1100	1040	970
Nominal pore diameter (nm)	6	6	8

[1] J. T. Baker (Seibert and Poole, unpublished results)
[2] J. T. Baker (Miller and Poole, 1994)
[3] J. T. Baker (Seibert and Poole, 1995)
[4] J. T. Baker (Seibert et al., 1996)
[5] J. T. Baker (Fernando et al., 1993)
[6] Varian Sample Preparation Products (Mayer et al., 1995a)
[7] Ansys (Mayer et al., 1995)

The cyanopropylsiloxane bonded sorbent in Table 1 shows a high value for the flow resistance parameter (normal range for N = 500-1000) and an average particle size that is smaller than would be expected. This is an indication that the sorbent contains an unacceptable proportion of fine particles leading to poor cartridge permeability and reduced efficiency. The total porosity values are indicative of the packing density, which is lower for the cartridges than the liquid chromatographic column, and consequently the interparticle porosity of typical cartridges will be higher than the values shown.

Values for the intraparticle porosity of the chemically bonded silica sorbents are generally quite small. The octadecylsiloxane bonded sorbent in Table 1 is virtually non porous, indicating that in the attempt to obtain sorbents with a high loading of bonded phase a substantial fraction of the pore volume is filled by the chemically bonded phase. An important observation can be made from this — the sorbents are designed to promote *retention* over *kinetic efficiency* and, given their intended use, this is acceptable.

The apparent average particle size of the disc materials (Table 1) is much smaller than that of the cartridges, as is required if a significant efficiency is to be obtained for a bed height of 0.5 mm. The flow resistance parameter is in the normal range, indicating a reasonably homogeneous packing structure free of holes and excessive amount of particles below average size. The pore volume of the sorbents is largely filled with bonded

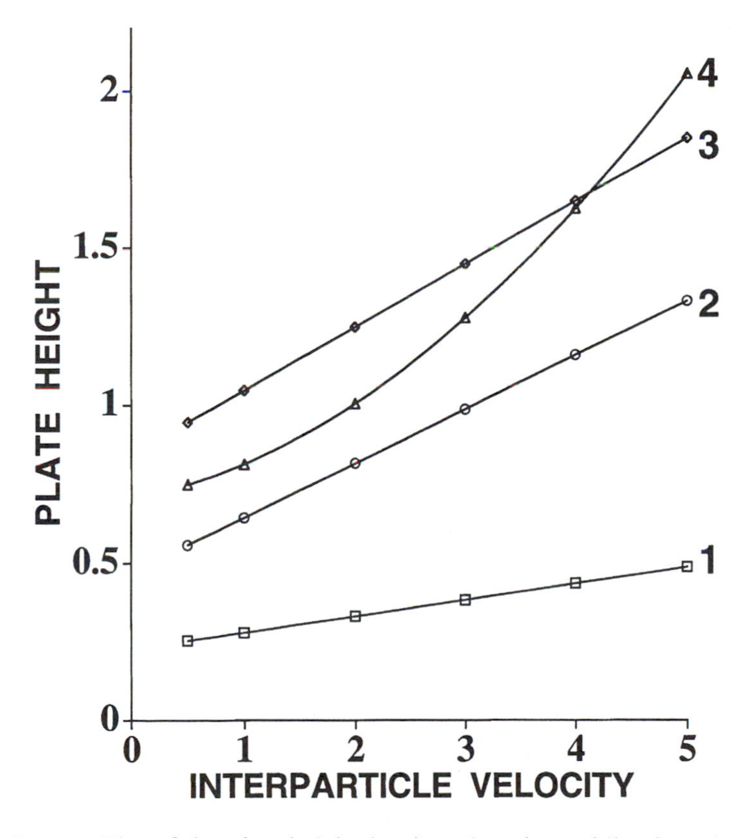

Figure 3. Plot of the plate height (mm) against the mobile phase interparticle velocity (mm/s) for the silica-based cartridge sorbents described in Table 1. 1 = octadecylsiloxane; 2 = spacer-bonded propanediol; 3 = cyanopropylsiloxane; and 4 = butylsiloxane. The test solute is anthracene and the mobile phase methanol/water (60:40).

phase with the same consequences discussed for the cartridge sorbents. Interparticle porosity is similar to the values expected for a well-packed column, indicating an adequate packing density for the sorbent medium.

B. KINETIC PROPERTIES

A plot of the observed plate height as a function of the interparticle mobile phase velocity for the chemically-bonded cartridge sorbents identified in Table 1 is shown in Figure 3. The plate height measurements come from the HPLC data. Note that the column packing density is larger than the cartridge packing density, and thus the values indicated are considered an upper bound; typical cartridges will provide lower values.

Several general features are noticeable. The efficiency of the columns is significantly flow rate dependent in the interparticle velocity range 0.5 - 5.0 mm/s. This corresponds to a flow rate of about 3 to 30 ml/min through a cartridge with an internal diameter of about 1 cm. Although there is curvature in the plot for the butylsiloxane-bonded phase, there is no minimum, and for the other three sorbents the data fit a linear relationship well. The main contributions to the plate height arise from flow anisotropy in the packed bed and resistance to mass transfer. Over the velocity range indicated, the various sorbents provide from 5 - 15 theoretical plates for the butylsiloxane-bonded sorbent, 6 - 10 for the cyanopropylsiloxane sorbent, 10 - 20 for the spacer-bonded propanediol sorbent, and 15 - 40 for the octadecylsiloxane-bonded sorbent per centimeter of packed column.

Additional zone dispersion caused by the lower packing density of the cartridge devices compared to the column packing density suggests that a figure of about 5 - 10 theoretical plates per centimeter is all that can be anticipated for a typical cartridge at normal sample flow rates. This determination is in good agreement with the results of Gelencser et al., (1995), who obtained an average value of 15 theoretical plates for phenols using a 360 mg octadecylsiloxane-bonded silica cartridge and the frontal analysis method based on equation (3).

The relationship between the plate height and the inter-particle mobile phase velocity for a particle-loaded membrane containing octadecylsiloxane -bonded silica particles is shown in Figure 4 (Fernando et al., 1993). The kinetic properties are different to those observed for typical cartridge devices. A minimum value for the plate height corresponding to about seven particle diameters is observed at an optimum interparticle velocity of 0.19 mm/s. This is about three times larger than the plate height predicted for an ideal sorbent bed and is achieved at an interparticle velocity an order of magnitude slower.

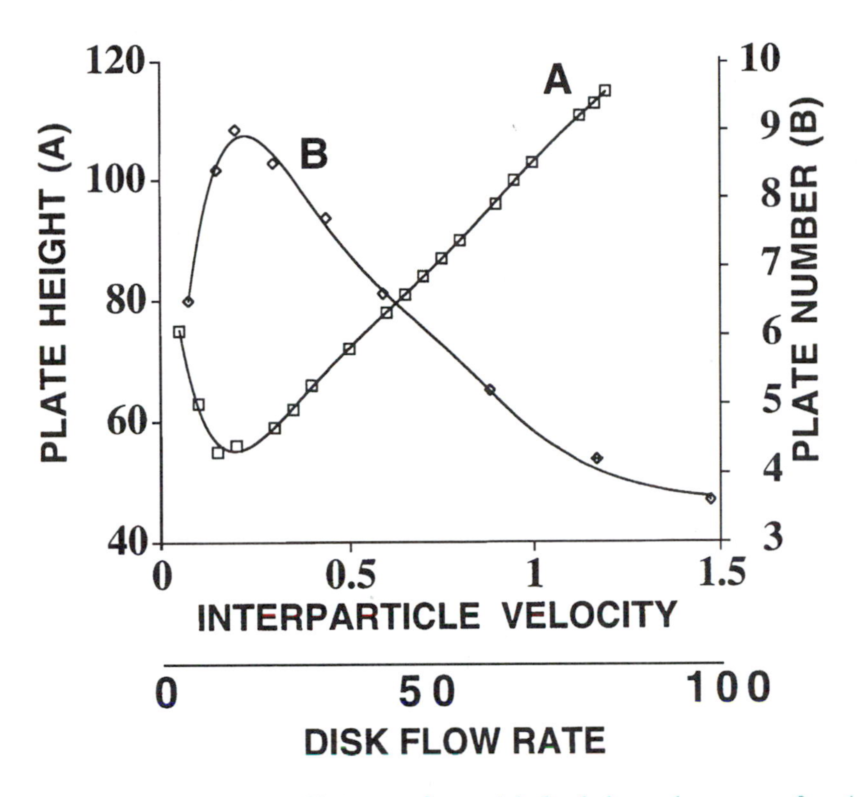

Figure 4. Variation of the efficiency of a particle-loaded membrane as a function of mobile phase velocity. (A) a van Deemter type plot with the plate height (μm) expressed as a function of the interparticle mobile phase velocity (mm/s) and (B) the transformation of (A) to represent the number of theoretical plates (N) achieved for a range of sample flow rates through a 0.5 mm disc with a 38 mm diameter active sampling area (ml/min).

Fitting the data to the Knox model of the relationship between plate height and mobile phase velocity indicates that the contribution of flow anisotropy is about three times greater and the contribution from resistance to mass transfer about an order of magnitude greater than predicted for an ideal sorbent bed. The heterogeneous nature of the membrane structure may contribute to the unfavorable flow anisotropy term, whereas the mass transfer term seems to be influenced by both the heterogeneous nature of the membrane structure and/or the bonding density of the organic ligands to the silica particles.

It is possible that the web of polytetrafluoroethylene microfibrils, responsible for holding the sorbent particles in place, contributes adversely to the kinetic properties of the sorbent bed. It could do this by disrupting the flow streams through the membrane and by trapping a portion of the mobile phase in the inter-particle space, increasing the volume of the stagnant mobile phase and inhibiting favorable mass transfer characteristics. The usual flow rate operating range for a 47 mm particle loaded membrane with a 38 mm active sampling area is about 10 - 100 ml/min. Over this flow range the particle-loaded membrane will provide between about 4 to 9 theoretical plates, the largest value corresponding to the optimum flow rate of about 13 ml/min (Figure 4).

Neither cartridge nor disc SPE devices provide large numbers of theoretical plates but this may not be detrimental to their application since they are used for *isolation* and not *separation*. Lovkvist and Jonsson (1987) have proposed a model described by Equation 8 to characterize the sampling properties of sorbent beds with low plate numbers that is more general than the model obtained by combining equations 1 and 2:

$$V_B = (0.98 + [13.59 / N] + [17.60 / N^2])^{-1/2}(1 + k)\ V_O \qquad (8)$$

The numerical coefficients are selected for a 1% breakthrough level; Lovkvist and Jonsson (1987) provide other coefficients in tabulated form for different breakthrough levels. A general aspect of this model is that it predicts, for well-retained analytes, that breakthrough volumes can be large, despite sampling devices having very low numbers of theoretical plates. It has been demonstrated (Mol et al., 1995) for open tubular traps that when $N < 1.5$ the breakthrough volume can not be increased by increasing retention, indicating that a threshold value or minimum efficiency is required for quantitative trapping which is not predicted by the Lovkvist and Jonsson model. Larrivee and Poole (1994) and Miller and Poole (1994) have shown that the Lovkvist and Jonsson model provides a reasonable interpretation of sampling properties for sorbent traps with more than four theoretical plates.

It has also been suggested (Pankow et al., 1982) that premature breakthrough for sampling devices with low plate numbers occurs because of poor transport. This means that there is a lack of sufficient time for all of the analyte molecules to reach the surface of a particle before exiting the sorbent bed, and is independent of the retention capacity of the sorbent. Thus, there may be a minimum number of theoretical plates required for acceptable sampling performance of a SPE device, but this minimum number is not well defined or easy to define.

The influence of the kinetic performance of the sampling device on the breakthrough volume is contained in the first part of the right-hand side of Equation 8. The homogeneity of the sorbent bed, the quality of the sorbent, and the sample flow rate will influence the breakthrough volume through the dependence of the number of theoretical plates on these parameters. The contribution of sorbent efficiency to the breakthrough volume as predicted by Equation 8 can be evaluated numerically as indicated in Figure 5. At about 100 theoretical plates the quotient reaches about 95% of its asymptotic value.

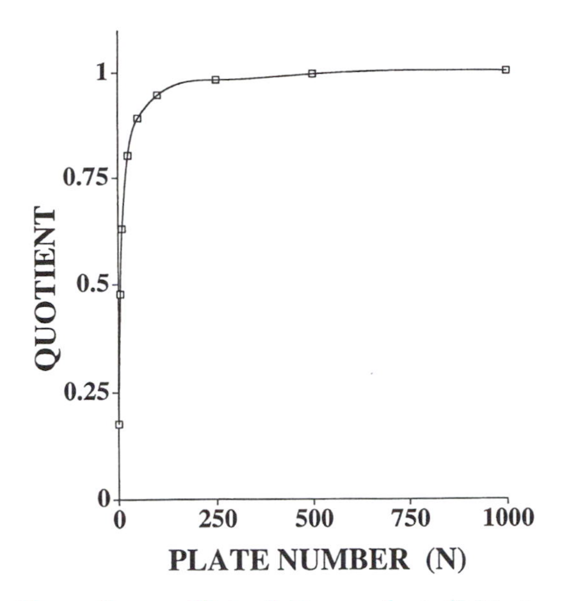

Figure 5. Plot of the quotient $(0.98 + [13.59 / N] + 17.60 / N^2)^{-1/2}$ against the number of theoretical plates N to determine the kinetic contribution to the breakthrough volume at the 1% breakthrough level.

One hundred theoretical plates are practically possible for short precolumns packed with particles of small diameter (3 to 10 µm) used in on-line sampling in HPLC. However, this is significantly larger than typical values observed for sorbent cartridges and particle-loaded membranes, which are generally an order of magnitude smaller. This means that we usually observe a smaller breakthrough volume than is theoretically possible. Further, the sampling process is performed in a region of the plot in Figure 5 where the breakthrough volume is most strongly influenced by kinetic properties. These undesirable features are tolerated in contemporary practice, in order to preserve a simple sample processing procedure that uses suction or gravity flow, and minimizes material costs.

A cartridge packed with a 60 µm sorbent such that its plate height was equal to two to three times the particle diameter need only have a bed length of 1.2 to 1.8 cm to provide 100 theoretical plates. From a hypothetical point of view, higher quality sorbents could be used to obtain improved sampling performance of cartridges. To obtain a more favorable resistance to mass transfer, however, it is likely that the volume of bonded phase would have to be reduced, resulting in lower retention (k value in Equation 8) and a smaller breakthrough volume.

For the isolation of small molecules from aqueous solution retention is probably more important than efficiency, but for larger molecules the opposite is probably the case. In all cases the results would be improved by optimizing the kinetic contribution to the breakthrough volume by maximizing the packing density, the most obvious deficiency of typical commercial products.

Given the low plate numbers of typical SPE devices, it is anticipated that the break-through volume will exhibit flow rate dependence. This dependence is illustrated in Figure 6 for an octadecyl-siloxane-bonded silica

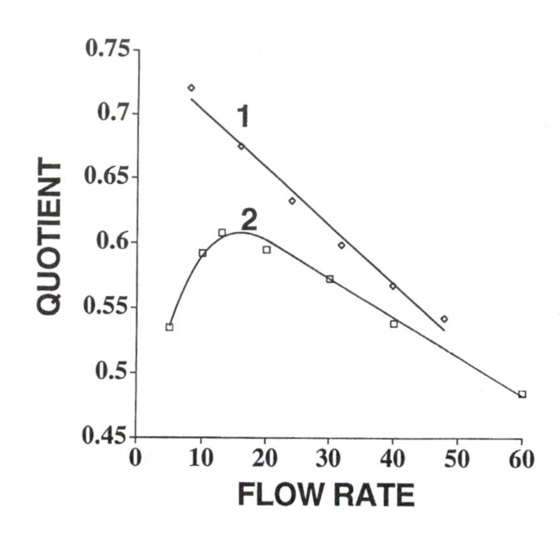

Figure 6. The influence of the kinetic properties of the sampling device on breakthrough volume. (1) an octadecylsiloxane-bonded sorbent particle-loaded membrane with a 38 mm diameter active sampling area. (2) a spacer-bonded propanediol silica sorbent cartridge with a 13 mm internal diameter and 1 cm bed height at a packing density of 0.794 g/ml.

particle-loaded membrane and a spacer-bonded propanediol sorbent cartridge material. The nature of the flow rate dependence for the two devices is different in its characteristics. The cartridge shows a continuous decrease in the breakthrough volume with flow rate, indicating that the slowest practical flow rate is the preferred choice within the indicated flow rate range. For the particle-loaded membrane the breakthrough volume is not strongly affected by sample flow rate between about 10 to 30 ml/min. At both higher and lower flow rates a decrease in the breakthrough volume is expected but the sensitivity to flow rate changes is not very great over the flow rate ranges likely to be used in practice.

Breakthrough volumes are more sensitive to flow rate for cartridge devices than for the particle-loaded membranes and if reproducible results are to be obtained then the flow rate should be controlled within reasonable limits. The data presented in Figure 6 are for a cartridge with a packing density higher than the typical packing density (Table 1) with the outcome that the kinetic contribution is inflated as shown. Although the nature of the relationship will be the same, the position on the y-axis will be lower for a typical cartridge device.

III. DETERMINATION OF BREAKTHROUGH VOLUMES

There are two practical approaches to determine the breakthrough volume of an analyte by experiment (Liska, 1993; Hennion and Pichon, 1994). The simplest is the direct method, the result of which is shown in Figure 1. It is achieved by passing a solution of the analyte at a constant flow rate through the sampling device and monitoring the appearance of the analyte at the exit of the sorbent bed. The analyte concentration should be such that it does not overload the sorbent and the detector should be sufficiently sensitive to record the breakthrough of the analyte at the concentration used in the experiment and be insensitive to the properties of the sample solvent.

Since the breakthrough volume may be flow-rate dependent, the chosen flow rate for the measurement should be similar to that used in practice. In the absence of a suitable method for on-line detection the effluent from the sorbent bed can be collected by aliquots using a fraction collector and each aliquot analyzed separately to determine when breakthrough occurred (Liska et al., 1990). Alternatively, samples of varying concentration can be processed to determine the breakthrough volume (Larrivee and Poole, 1994; Mayer and Poole, 1994; Poole and Poole, 1995). Initially, solutions are screened using decade changes in sample volume to estimate the approximate breakthrough volume, followed by a more systematic experimental design. For compounds with an estimated breakthrough volume between 0 and 50 ml, measurements are made at 2.5 ml volume increments, 50 and 100 ml at 5 ml increments, 100 and 1000 ml at 10 ml-increments, and greater than 1000 ml at 100 ml increments. The data are plotted in the form of the observed recovery for the complete sampling process against the sample volume and the breakthrough volume estimated using the line providing the best fit through the data. Initially the recovery line should be parallel to the volume axis. At the point where breakthrough occurs and beyond, the observed recovery falls below this line.

A similar approach can be used on-line to determine the breakthrough volume for precolumns in HPLC (Subra et al., 1988; Pichon and Hennion, 1994; Pichon et al., 1995). In this case the sampling device is usually a short precolumn connected to the analytical column via a selection valve. This set-up allows the sample to bypass the analytical column during loading and the eluent from the precolumn can be switched to the analytical column for separation at the recovery step. A small sample volume spiked with a known amount of all analytes is percolated through the sampling device (precolumn) and the chromatogram corresponding to the on-line elution is recorded and peak areas are measured. The initial volume is selected to be less than the breakthrough volume. The sample volume is then serially increased, and to each volume a constant amount of analytes are added. Each sample is then processed as before. Initially there will be no change in peak areas up to the breakthrough volume, then in each subsequent volume that exceeds the breakthrough volume there will be a decrease in the peak areas in the chromatogram representing loss of the analytes from the precolumn. Plotting peak area for each analyte against sample volume processed enables the breakthrough volume to be estimated.

The breakthrough volume can also be determined from the retention volume using the equilibrium method (Shoup and Mayer, 1982; Carr and

Harris, 1988; Jandra and Kubat, 1990; and Gelencser et al., 1995 and 1995a). A solution of known volume and concentration of analytes is pumped through the sampling device and returned to the solution reservoir in a cyclic fashion until a steady state is reached. The sorbed amount of each analyte is then determined by any suitable method after recovery from the sampling device. The method is unsuitable for the determination of breakthrough volumes for solutes with large retention factors, because an excessive volume is required to reach equilibrium. In addition, the concentration of sorbed analytes must be below the capacity of the sorbent; that is, the sorption isotherm must conform to the linear chromatography model for the purpose of calculation. The experiment yields the retention factor from which the retention volume is calculated and the breakthrough volume determined using Equations 1 to 3.

IV. PREDICTION METHODS FOR BREAKTHROUGH VOLUMES

Recording breakthrough curves is time-consuming and the exact estimate of the breakthrough volume is difficult and somewhat subjective. For this reason methods that enable the breakthrough volume to be predicted with minimal experimental measurements are particularly attractive. In practice, an approximate value of the breakthrough volume is needed in order to select the appropriate sorbent and the amount of sorbent for sample concentration with SPE devices. Early in the development of solid-phase extraction it was demonstrated that a good correlation existed between the breakthrough volume on porous polymer sorbents and the aqueous solubility of an analyte (Thurman et al., 1978; Josefson et al., 1984; and Przyjazny, 1985). As discussed in Chapter 5, this method, although useful, ignores stationary phase contributions to retention and is limited by the availability and quality of solubility data. This is especially true for sparingly soluble solutes.

The most common method of use of HPLC for the prediction of breakthrough volumes of cartridge SPE devices is to determination the retention factor. Using Equation 9 and an estimated value for the interparticle volume of the sorbent cartridge, the retention volume is obtained and can be used as a surrogate estimate of the breakthrough volume. Alternatively, an estimate of the number of theoretical plates for the cartridge can be applied to Equation 8 to give the breakthrough volume directly (Werkhoven-Goewie, 1981; Hennion and Coquart, 1993; Hennion and Pichon, 1994; Pichon et al., 1995).

$$V_R = V_o (1 + k) \tag{9}$$

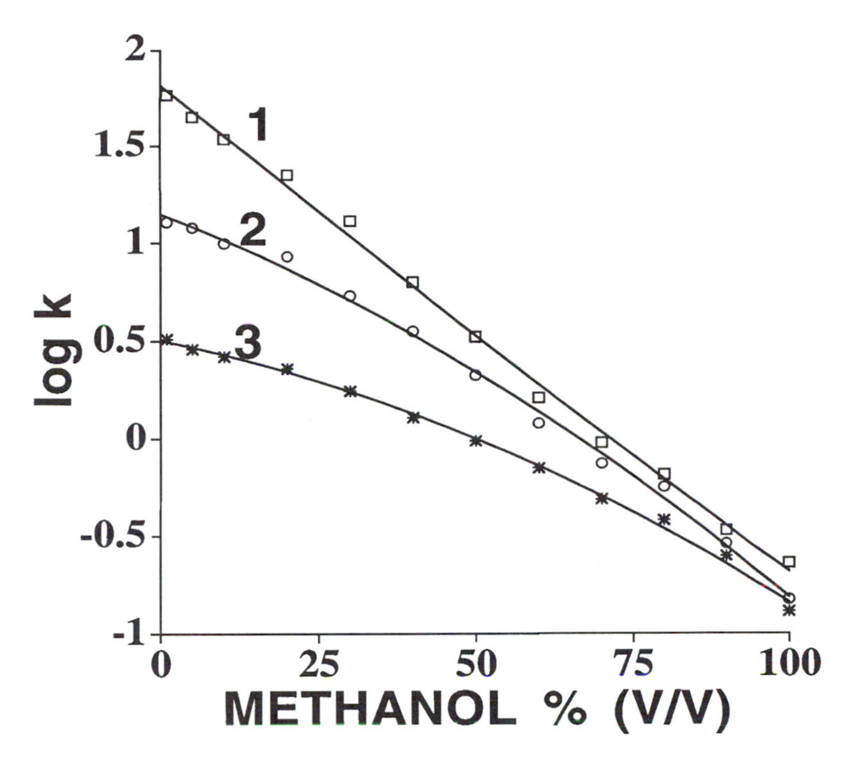

Figure 7. Plot of the retention factor against the volume fraction of methanol in the mobile phase for (1) naphthalene, (2) chlorobenzene and (3) 1-phenylethanol on a cyanopropylsiloxane-bonded silica sorbent.

Application of this method so far has been restricted to reversed-phase chromatography where it is usually necessary to use an extrapolation method to determine the retention factor when the sample solvent is water. Even with short columns it is very difficult to determine the retention factor when using water as the mobile phase. This is because of the often long retention times involved and the very poor efficiency of the chromatographic system under these conditions, resulting in very broad or even unsymmetrical peaks that are difficult to detect (Jandra and Kubat, 1990). The retention factor for pure water, k_w, can be estimated more conveniently by extrapolation of the retention factors observed at higher levels of organic solvent to zero volume fraction of organic solvent with a binary mobile phase containing an organic solvent such as methanol. Either a linear, Equation 10, or quadratic, Equation 11, model is fitted to the experimental data to perform the extrapolation where 2 represents the volume fraction of organic solvent.

$$\text{Log } k = \log k_w + s\theta \quad (10)$$

$$\text{Log } k = \log k_w + a\theta + b\theta^2 \quad (11)$$

The correct form of the extrapolation is both solute-dependent (Figure 7), and solvent-dependent. The accuracy of the estimate of the intercept at zero volume fraction of organic solvent depends on the range of the experimental data recorded (Schoenmakers et al., 1979 and 1982; Jandra and Kubat, 1990; Poole and Poole, 1991). At high organic solvent concentrations data can be gathered rapidly but if a long extrapolation is performed neither the legitimacy of the extrapolation equation used nor the estimate of log k_w can be guaranteed. Considerable time may be required to gather data at low organic solvent compositions but then the estimate of log k_w is improved by the short extrapolation used.

Most literature values of log k_w have been obtained with reversed-phase column packings which not only differ in retention characteristics from one manufacturer to another but also have different design characteristics to materials commonly used as SPE sorbents. Estimates based on results using these materials may provide a poor model for the SPE sorbent and a poor estimate of the appropriate log k_w value.

There is also a conceptual problem in the determination of log k_w by the extrapolation method. Selective uptake of the organic solvent component of the mobile phase results in a more complex stationary phase whose composition varies in a complex manner with the composition of the mobile phase with which it is in equilibrium. At very low amounts of organic solvent the stationary phase contains an appreciable amount of organic solvent compared to expectations based on solvents containing a higher volume fraction of organic solvent. Potentially, this may distort the interpretation of log k_w as the value of the retention factor for water. The sorbent conditioning step commonly used in solid-phase extraction is likely to result in the sample being applied to a sorbent containing an unknown amount of organic (conditioning) solvent, which could also cause deviation from predicted results, particularly when processing small sample sizes. Although log k_w values are easy to determine and many values can be found in the literature, their accuracy is undefined and in the general sense the use of log k_w values to predict breakthrough volumes yields both acceptable and disappointing results depending on circumstances that are not clearly defined.

There is generally a linear relationship between log P_{oct} and log k_w for closely related compounds. The large pool of octanol-water partition coefficients (log P_{oct}) can be used as a surrogate estimate for log k_w for predicting breakthrough volumes for octadecylsiloxane-bonded silica sorbents when experimental values for log k_w are unavailable (Hennion and Pichon, 1994). Literature values for log P_{oct} , however, vary widely with the method of measurement or calculation, adding to the uncertainty when they are used to predict breakthrough volumes.

A. SOLVATION PARAMETER MODEL

The solvation parameter model provides a general framework for interpreting solvent-dependent behavior in terms of fundamental intermolecular interactions starting from a simple cavity model of solution (Abraham, 1993; Carr, 1993). Successful applications of the model include the characterization of the solvent properties of stationary phases and the prediction of retention in gas chromatography, prediction of retention in reversed-phase liquid chromatography, and the response characterization of polymer coated sensors (Poole et al., 1992; Abraham and Roses, 1994; Abraham et al., 1994; McGill et al., 1994; and Poole and Poole, 1995a). The cavity model of solution represents the transfer of a solute from one phase to another as occurring by three independent operations:

- The creation of a cavity in the solvent of a suitable size to accommodate the solute.
- Reorganization of the solvent molecules around the cavity containing the analyte.
- Interaction of the solute molecule with the surrounding solvent molecules.

Transfer from one phase to another occurs with the formation of a cavity in the acceptor phase with the formation of solute-solvent interactions in that phase. This is accompanied by the collapse of a cavity of the same size in the donor phase, with the breaking of the interactions between the solute and the donor solvent. For this to occur there must be a favorable free energy of transfer, represented as the difference in the free energy of solution (or sorption) between the two phases. Under typical extraction conditions the sample solutions can be reasonably assumed to be infinitely dilute such that all interactions that are formed in solution are of the solute-solvent type. These interactions can be characterized as dispersion, orientation, induction, and complexation.

Dispersion interactions are non-selective, and solute-solvent dispersion interactions will be approximately equal in the sample solvent and solvated sorbent and not a significant contributor to the change in free energy accompanying the transfer process. The capacity of a solvent or solvated sorbent for selective polar interactions is more important in solute transfer. Molecules with a permanent dipole moment can interact with each other by the cooperative alignment of their dipoles (orientation interactions) and by their capacity to induce a temporary complementary dipole in a polarizable

Table 2. Solute descriptors used in the solvation parameter model.

Compound	$V_X/100$	R_2	π_2^H	α_2^H	β_2^H	log L^{16}
Naphthalene	1.085	1.340	0.92		0.20	6.014
n-Propylbenzene	1.139	0.604	0.50		0.15	4.230
Benzene	0.716	0.610	0.52		0.14	2.786
Chlorobenzene	0.839	0.718	0.65		0.07	3.657
Bromobenzene	0.891	0.882	0.73		0.09	4.041
Iodobenzene	0.975	1.188	0.82		0.12	4.502
1,2-Dichlorobenzene	0.961	0.872	0.78		0.04	4.518
1,2-Dibromobenzene	1.066	1.190	0.96		0.04	5.456
1,1,2-Trichloroethane	0.758	0.499	0.68	0.13	0.08	3.290
1,1,2,2-Tetrachloroethane	0.880	0.595	0.76	0.16	0.12	3.803
Benzaldehyde	0.873	0.820	1.00		0.39	4.008
Hexanal	0.970	0.146	0.65		0.45	3.357
Benzonitrile	0.871	0.742	1.11		0.33	4.039
Anisole	0.916	0.708	0.75		0.29	3.890
Di-n-butyl ether	1.295	0.000	0.25		0.45	3.924
1,4-Dioxane	0.681	0.329	0.75		0.64	2.892
Acetophenone	1.014	0.818	1.01		0.49	4.501
4-Chloroacetophenone	1.136	0.955	1.09		0.44	
1-Nitrobutane	0.846	0.227	0.95		0.29	3.415
1-Nitropentane	1.057	0.212	0.95		0.29	3.938
Nitrobenzene	0.891	0.871	1.11		0.28	4.557
3-Nitrotoluene	1.032	0.874	1.10		0.25	5.097
Methylbenzoate	1.073	0.733	0.85		0.48	4.704
n-Propyl propanoate	1.028	0.070	0.56		0.45	3.338
2-Hexanone	0.968	0.136	0.68		0.51	3.262
2-Octanone	1.252	0.108	0.68		0.51	4.257
Cyclohexanone	0.861	0.403	0.86		0.56	3.792
Phenol	0.775	0.805	0.89	0.60	0.31	3.766
3-Cresol	0.916	0.820	0.88	0.57	0.34	4.310
4-Cresol	0.916	0.820	0.87	0.57	0.32	4.312
1-Naphthol	1.144	1.520	1.05	0.61	0.37	6.130
2-Chlorophenol	0.897	0.853	0.88	0.32	0.31	4.178
4-Chorophenol	0.898	0.915	1.08	0.67	0.20	4.775
Hexan-1-ol	1.013	0.210	0.42	0.37	0.48	3.610
Heptan-1-ol	1.156	0.211	0.42	0.37	0.48	3.106
Benzyl alcohol	0.916	0.803	0.87	0.33	0.56	4.221
2-Phenylethanol	1.057	0.811	0.91	0.30	0.65	4.630
2-Nitroaniline	0.990	1.180	1.37	0.30	0.36	5.627
Benzamide	0.973	0.990	1.50	0.49	0.67	5.770
Acetanilide	1.114	0.870	1.40	0.50	0.67	

molecule (induction interactions). Other selective polar interactions include complexation interactions involving the sharing of electron density or a hydrogen atom between molecules (hydrogen bonding and charge transfer, for example).

Those properties that affect the transfer of a solute between phases, therefore, include the availability of complementary intermolecular forces to those of the solute in both phases, as well as differences in the cohesion of the phases (e.g., solvent-solvent interactions). This latter contribution determines the effect of cavity formation on the transfer process. The formal description of the solvation parameter model in a form suitable for interpreting processes in liquid chromatography and solid-phase extraction can be set out as follows:

$$SP = c + m\, V_x/100 + rR_2 + s\, \pi_2^H + a\, \alpha_2^H + b\, \beta_2^H \tag{12}$$

where SP is some free energy-related property of the system, typically the retention factor (log k) or distribution constant in liquid chromatography or the breakthrough volume (log V_B) in SPE.

The solute descriptors are identified as the following: the characteristic volume V_x; the excess molar refraction R_2; the ability of the solute to stabilize a neighboring dipole by virtue of its capacity for orientation and induction π_2^H; and α_2^H and β_2^H— parameters that characterize the solute's effective hydrogen-bond acidity and hydrogen-bond basicity, respectively. The subscript "$_2$" indicates that the solute descriptors are values applicable for infinite dilution in a solvent other than themselves. They should therefore be distinguished from compilations of solvent values, which are given a subscript "$_1$" and are appropriate for interactions with like solvent molecules.

These solute descriptors are parameters derived from equilibrium measurements for complexation and partition processes and are available for more than 2000 compounds with others available as estimates by calculation (Abraham, 1993 and 1993a; Abraham et al., 1994a). A small collection of solute descriptors, useful for characterizing the retention properties SPE sorbents, are compiled in Table 2.

The solute descriptor V_x is McGowan's characteristic molecular volume and is calculated by summing the characteristic atomic volumes for each atom in the molecule and then subtracting 6.56 cm^3/mol for each bond, no matter whether single, double, or triple (Abraham and McGowan, 1987). A value for V_x is always available by trivial arithmetic when the structure of a compound is known. The characteristic atomic volumes for some common atoms are summarized in Table 3 along with a representative calculation of the characteristic molecular volume for nitrobenzene. The characteristic molecular volume is divided by 100 to create a scale of somewhat equal proportion to the other solute descriptors indicated in Table 2.

The solute excess molar refraction, R_2, which models polarizability contributions from n and π electrons, is calculated from the refractive index and the characteristic molecular volume. It substitutes as the difference between the molar refraction of the compound and an n-alkane of identical

Table 3. Characteristic atomic volumes and the calculation of the characteristic molecular volume for nitrobenzene.

Atom	Atomic Volume (cm^3/mol)
Carbon	16.35
Hydrogen	8.71
Oxygen	12.43
Nitrogen	14.39
Sulfur	22.91
Phosphorus	24.87
Fluorine	10.48
Chlorine	20.95
Bromine	26.21
Iodine	34.53

$$V_{X(nitrobenzene)} = (6 \times 16.35) + (5 \times 8.71) + (1 \times 14.39) + (2 \times 12.43) - (14 \times 6.56) = 89.06$$

characteristic volume (Abraham et al., 1990). R_2 is tabulated in units of $cm^3/10$. Since there are several methods of estimating the refractive index of compounds from fragmental constants and the relationship between the molar refraction of the n-alkanes and their characteristic volume is well established, R_2 can be estimated by calculation for many compounds. The π_2^H parameter is a measure of a solute's ability to stabilize a neighboring dipole by virtue of its capacity for orientation and induction interactions (Abraham et al., 1991). Normally, it must be determined by experiment, from gas chromatographic data for example, although simple rules have been proposed for its estimation for aliphatic compounds.

The parameters α_2^H and β_2^H are measures of a solute's effective hydrogen-bond acidity and effective hydrogen-bond basicity, respectively. They were originally determined from the equilibrium constants for hydrogen-bond complexation in an inert solvent such as carbon tetrachloride and, more recently, from water/solvent distribution constants. Both scales are normalized such that $\beta_2^H = 1.0$ for hexamethylphosphoric triamide (Abraham, 1993). α_2^H and β_2^H scales are not related to full proton-transfer acidity or basicity scales such as aqueous pK_a, etc. For some specific solutes, mainly anilines, sulfoxides and alkylpyridines, in which a distribution between a predominantly aqueous phase and an organic solvent or stationary phase containing a significant amount of water is considered, it is necessary to replace β_2^H by a new constant β_2^0 (Abraham and Roses, 1994). Such conditions are probably germane to SPE from an aqueous solution. These two scales are identical for most compounds besides those mentioned above. When the purpose is to characterize sorbent retention properties the confusion is easily eliminated by avoiding those compounds that exhibit variable basicity (for example, using the compounds given in Table 2).

The system constants in Equation 12 are unambiguously defined and are determined by the difference in capacity, for specific intermolecular interactions, of the solvent and solvated sorbent.

Table 4. System constants for the fit to Equation 12 of the breakthrough volumes for particle-loaded membranes. The organic modifier (present at 1% v/v) is indicated at the top of each column of data. R = multiple correlation coefficient; SE = standard error; F = F-statistic; *n* = number of solutes. From Poole and Poole, 1995 with permission.

	Porous polymer				Silica C18
	Methanol	Acetonitrile	Tetrahydrofuran	Isopropanol	Methanol
m	5.16 (± 0.11)	5.72 (± 0.10)	5.09 (± 0.09)	7.20 (±0.12)	5.14 (± 0.17)
r	0.81 (± 0.04)	-0.26 (± 0.04)	0.00	0.41 (±0.04)	0.00
s	-0.65 (± 0.05)	-0.35 (± 0.04)	0.00	-0.34 (±0.05)	-0.92 (± 0.08)
a	-1.85 (± 0.04)	-1.17 (± 0.04)	-0.92 (± 0.03)	-1.50 (±0.04)	-1.05 (± 0.08)
b	-2.93 (± 0.08)	-2.81 (± 0.07)	-4.19 (± 0.06)	-4.53 (±0.08)	-2.24 (± 0.15)
c	-0.77 (± 0.10)	-1.19 (± 0.09)	-1.00 (± 0.08)	-2.10 (±0.10)	-1.23 (± 0.17)
	Statistics				
R	0.9987	0.9985	0.9986	0.9990	0.996
SE	0.036	0.041	0.039	0.035	0.070
F	1108	1269	2478	1313	594
n	22	25	25	22	23

- *r* constant refers to the capacity for interaction with solute n or B electrons.
- *s* constant refers to the capacity to take part in dipole-dipole and dipole-induced dipole interactions.
- *a* constant characterizes the hydrogen-bond basicity of the system (because a basic phase will interact with an acidic solute).
- *b* constant characterizes the hydrogen-bond acidity of the system.
- *m* constant describes the relative ease of forming a cavity for the solute in the bulk solvent and solvated sorbent.

The sign of the system constant indicates whether the contribution is favorable (positive coefficient) or unfavorable (negative coefficient) for transfer to the solvated sorbent, enhancing or diminishing retention respectively. System constants for any solvated sorbent and sample solvent are determined by obtaining experimental values for the observed parameter, SP. This must be done for a group of solutes of known properties which are sufficiently varied to define all interactions in Equation 12. The number of values obtained must be of sufficient number to establish the statistical validity of Equation 12, using the computational technique of multiple linear regression analysis. It is usual for upwards of 20 measurements of SP to be used in the regression analysis for a good statistical outcome for the fit to Equation 12. Once the system constants are known then the retention properties in the same system for any solute with known solute descriptors are accessible by calculation through Equation 12.

Table 5. Contribution of different intermolecular interactions to the breakthrough volume for some representative compounds. IPA = isopropanol; MeOH = methanol; AcCN = acetonitrile; and THF = tetrahydrofuran. From Poole and Poole, 1995 with permission.

Organic solvent	Contribution to log V_B					Predicted
	$([mV_x/100]+c)$	$r\,R_2$	$s\,\pi_2^H$	$a\,\alpha_2^H$	$b\,\beta_2^H$	V_B (cm^3)
			Toluene			
IPA	4.070	0.246	-0.177		-0.634	3199
MeOH	3.652	0.487	-0.338		-0.412	2449
AcCN	3.712	-0.156	-0.182		-0.393	957
THF	3.362				-0.587	596
			Cyclohexanone			
IPA	4.099	0.165	-0.292		-2.537	27
MeOH	3.673	0.326	-0.559		-1.641	63
AcCN	3.735	-0.105	-0.301		-1.574	57
THF	3.383				-2.346	11
			Benzonitrile			
IPA	4.171	0.304	-0.377		1.494	402
MeOH	3.724	0.601	-0.722		-0.967	433
AcCN	3.792	-0.193	-0.389		-0.927	192
THF	3.433				-1.383	112
			Heptan-1-ol			
IPA	6.223	0.087	-0.143	-0.555	-2.174	2742
MeOH	5.195	0.171	-0.273	-0.685	-1.406	1005
AcCN	5.422	-0.055	-0.238	-0.433	-1.349	2223
THF	4.884			-0.340	-2.011	341

1. PARTICLE-LOADED MEMBRANES

With the breakthrough volume as the dependent variable, Equation 12 has been used to establish the system constants for an octadecylsiloxane-bonded silica particle-loaded membrane (Larrivee and Poole, 1994). A solution of 1% (v/v) methanol in water was used as the sample solvent, and to evaluate the contribution of solvent effects to sorbent sampling with a porous polymer particle-loaded membrane, 1% (v/v) methanol, acetonitrile, tetrahydrofuran and isopropanol were used as sample solvents (Poole and Poole, 1995). Addition of small quantities of organic solvents with large sample volumes is common practice in SPE to increase the speed of sample processing (Mayer and Poole, 1994). Poly(styrene-divinylbenzene) porous polymer has a greater affinity for the organic solvent than water and is more flexible than silica-based materials, allowing the polymer to swell by the selective sorption of the organic component of the sample solvent. Selective uptake of organic solvent by the porous polymer particle-loaded membrane has a dramatic influence on its sampling properties as indicated by the system constants for the sample processing solvents summarized in Table 4.

Experimentally determined breakthrough volumes were observed to vary by up to an order of magnitude for the different sample processing solvents. The m constant for the solvated porous polymer sorbent covers a wide range, 5.09 to 7.20, indicating that the uptake of organic solvent by the polymer has a considerable influence on the ease of cavity formation. This is probably accompanied by changes in the phase ratio of the sorbent which, along with other factors, is incorporated into the constant term in Equation 12. Thus, as a rough measure of the influence of differences in cohesion of the solvated sorbent on the breakthrough volume the term ($c + mV_x / 100$) is used. An indication of the relative contribution of size and polar interactions to the breakthrough volume of a few varied solutes can be gleaned from Table 5.

Isopropanol and methanol are the preferred sample-processing solvents for toluene because of the more favorable cavity formation in the solvated sorbent (polar interactions are less important since toluene is only weakly polarizable and is a weak hydrogen-bond base). Cyclohexanone is of a similar size to toluene, but is a stronger hydrogen-bond base and has a larger capacity for dipole-type interactions. Since water is a stronger hydrogen-bond acid than any of the solvated sorbents the breakthrough volume for cyclohexanone is much smaller than that for toluene. Methanol and acetonitrile are the preferred sample processing solvents because of their more favorable b constants. However, this advantage would be lost for hydrogen-bond bases larger than cyclohexanone, in which case isopropanol would again be the preferred solvent with the more favorable cavity term offsetting the solvated sorbent's less favorable hydrogen-bond acidity. Benzonitrile is slightly larger than toluene, has a larger capacity for dipole-type interactions and is a weak hydrogen-bond base.

Methanol is the preferred sample processing solvent because the methanol-solvated sorbent is a weaker hydrogen-bond acid than the isopropanol-solvated sorbent. The isopropanol solvated sorbent is less cohesive and is more competitive with water in dipole-type interactions than the methanol-solvated sorbent and would be the preferred choice for compounds with similar properties to benzonitrile but larger in size. Heptan-1-ol is both a hydrogen-bond acid and a hydrogen-bond base with a modest capacity for dipole-type interactions. Acetonitrile and isopropanol are the preferred sample processing solvents. The acetonitrile solvated sorbent is more cohesive than the isopropanol solvated sorbent but is able to offset its unfavorable cavity term by its more favorable hydrogen-bond acid and hydrogen-bond base interactions. For higher molecular weight homologues isopropanol would be the preferred sample processing solvent because the

Table 6. Contribution of different intermolecular interactions to the breakthrough volume of some varied compounds using a octadecylsiloxane-bonded silica (C-18) and porous polymer (PP) sorbent.

Sorbent type	Contribution to log V_B					Predicted
	$([mV_x/100]+c)$	$r\,R_2$	$s\,\pi_2^H$	$a\,\alpha_2^H$	$b\,\beta_2^H$	V_B (cm^3)
			Benzene			
C18	2.450		-0.478		-0.336	43
PP	2.925	0.494	-0.338		-0.440	438
			Naphthalene			
C18	4.347		-0.846		-0.448	1130
PP	4.829	1.085	-0.598		-0.586	53703
			Nitrobenzene			
C18	3.350		-1.021		-0.739	39
PP	3.825	0.706	-0.722		-0.967	700
			2-Hexanone			
C18	3.746		-0.626		-1.142	95
PP	4.225	0.110	-0.442		-1.494	251
			Phenol			
C18	2.754		-0.819	-0.630	-0.694	4
PP	3.229	0.652	-0.579	-1.110	-0.908	19
			Benzamide			
C18	3.771		-1.380	-0.515	-1.501	2
PP	4.251	0.802	-0.975	-0.907	-1.963	16

solvated sorbent is less cohesive; acetonitrile would be a better choice to maximize the breakthrough volume for lower molecular weight homologs because the solvated sorbent is more competitive with water in hydrogen-bond interactions.

Table 4 also provides the necessary information to compare porous polymer and octadecylsiloxane-bonded silica particle-loaded membranes for the isolation of organic compounds with 1% (v/v) methanol as the sample processing solvent. The *m* constants are similar for both sorbents but the *c* term is significantly less favorable for the octadecylsiloxane-bonded silica sorbent, suggesting that either the solvated octadecylsiloxane-bonded sorbent is more cohesive or it provides a less favorable phase ratio for extraction. Solvated octadecylsiloxane-bonded sorbent is less competitive than the solvated porous polymer sorbent for interactions of a dipole-type (*s* constant) but it is a stronger hydrogen-bond acid (*b* constant) and hydrogen-bond base (*a* constant) than the methanol-solvated porous polymer. The porous polymer has an additional capacity for lone-pair interactions (*r* constant) that are favorable for retention.

In determining which sorbent to choose to maximize the breakthrough volume it is necessary to consider size as well as the capacity of the analyte for polar interactions. Some typical results are summarized in Table 6. In all cases the breakthrough volumes are larger for the porous polymer particle-loaded membrane, in some cases by a significant amount. Even for small molecules such as phenol, which is a strong hydrogen-bond acid and base, the balance of properties still favors the porous polymer. Its breakthrough volume is still relatively small, however, since the capacity of water for polar interactions easily exceeds that of the solvated sorbents and the contribution from the size of the solute is unable to compensate sufficiently for these interactions.

The general picture that emerges for samples in predominantly aqueous solution is that the dominant contribution to retention is the ease of cavity formation in the solvated sorbent, with a smaller contribution from electron lone-pair interactions (m constants are always positive, supported by the r constant in some cases). The driving force for retention is expulsion of the analyte from the water network and the characteristic property that distinguishes individual solvated sorbents becomes their ease of cavity formation. For polar analytes orientation and hydrogen-bond interactions reduce retention (s, a, and b constants are negative) since neither solvated octadecylsiloxane-bonded silica nor porous polymer sorbents are able to compete with water for these interactions.

Kamlet et al. (1985) have provided values for the infinite dilution distribution constant for a varied group of solutes between water and activated carbon to which we can apply equation (12). SP is now log K_g, the infinite dilution distribution/zero surface coverage constant. This provides Equation 13:

$$\log K_g = 3.81[V_x/100]+0.59\pi_2^H-3.06\beta_2^H - 1.83 \quad (13)$$

$R = 0.992$; $SE = 0.100$; $F = 555$; $n = 30$. There are some uncertainties which may range from a couple of percent in the term $[V_x/100]$ to as much as 20% on the π_2^H term.

The solvated carbon sorbent is considerably polar, having a larger capacity for dipole-type interactions than water (s constant is positive) and similar basicity and capacity for electron lone-pair interactions (a and r constants are statistically insignificant). It cannot compete with water as a hydrogen-bond acid with the result that the retention of hydrogen-bond bases will be less than that of other polar compounds of a similar size. The solvated carbon is less hydrophobic than the porous polymer and octadecylsiloxane-bonded silica sorbents (m constant) but has complementary properties to these sorbents. More extensive work is required to determine the optimum range of properties desired for solid phase extraction by the many different forms of carbon.

Table 7. Influence of sample processing solvent (1% v/v) on the system constants for three silica-based cartridge sorbents.

System constant	Solvent: Methanol	Isopropanol	Acetonitrile	Tetrahydrofuran
		Butylsiloxane-bonded silica		
m	3.36 (±0.18)	3.27 (±0.12)	3.24 (±0.14)	3.25 (±0.16)
r	0.00	0.00	0.12 (±0.09)	0.25 (±0.11)
s	0.00	0.00	-0.23 (±0.10)	-0.33 (±0.11)
a	-0.46 (±0.09)	-0.40 (±0.07)	-0.42 (±0.08)	-0.13 (±0.09)
b	-1.53 (±0.14)	-1.60 (±0.11)	-1.48 (±0.14)	-1.93 (±0.16)
c	-1.38 (±0.15)	-1.27 (±0.11)	-1.15 (±0.11)	-1.12 (±0.13)
		Cyanopropylsiloxane-bonded silica		
m	2.06 (±0.18)	1.86 (±0.06)	1.95 (±0.10)	1.82 (±0.10)
r	0.53 (±0.07)	0.43 (±0.05)	0.40 (±0.04)	0.41 (±0.05)
s	0.00	0.00	0.00	0.00
a	-0.51 (±0.08)	-0.38 (±0.06)	-0.34 (±0.05)	-0.30 (±0.07)
b	-1.45 (±0.10)	-1.54 (±0.08)	-1.53 (±0.06)	-1.60 (±0.09)
c	-0.88 (±0.16)	-0.68 (±0.08)	-0.77 (±0.09)	-0.68 (±0.09)
		Spacer-bonded propanediol silica		
m	1.57 (±0.11)	1.80 (±0.13)	1.62 (±0.11)	1.54 (±0.11)
r	0.61 (±0.06)	0.44 (±0.07)	0.45 (±0.06)	0.58 (±0.08)
s	0.00	0.00	0.00	-0.25 (±0.11)
a	-0.45 (±0.07)	-0.26 (±0.09)	-0.30 (±0.08)	-0.20 (±0.07)
b	-0.80 (±0.11)	-1.20 (±0.12)	-1.02 (±0.11)	-0.82 (±0.17)
c	-1.05 (±0.09)	-1.03 (±0.12)	-1.02 (±0.11)	-0.95 (±0.10)

2. Cartridge Sorbents

The system constants for butylsiloxane-bonded, cyanopropylsiloxane-bonded (Seibert and Poole, 1995a) and spacer-bonded propanediol (Seibert et al., 1996) silica-based cartridge sorbents with four sample processing solvents are summarized in Table 7. The influence of solvent effects on sorbent retention are less important than those observed for the porous polymer particle-loaded membrane. In those cases were the difference in system constants exceed statistical limits, they are too small to provide meaningful changes in breakthrough volumes. In these cases the data for 1 % (v/v) methanol as the sample processing solvent can be taken as representative of the other sample processing solvents.

Less reliable data, obtained by the extrapolation of log k values, are available for an octadecylsiloxane-bonded silica sorbent with methanol as the sample processing solvent (Miller and Poole, 1994) and can be used for comparison with the methanol-solvated sorbents in Table 7: m = 5.65; r =

Table 8. Contribution of intermolecular interactions to retention of some varied solutes on four chemically bonded cartridge sorbents. C-18 = octadecylsiloxane; C-4 = butylsiloxane; CN = cyanopropylsiloxane; and Diol = spacer-bonded propanediol.

Compound	Sorbent			Contribution to log k				Predicted
		$(mV_x/100)$	$r\,R_2$	$s\,\pi_2^H$	$a\,\alpha_2^H$	$b\,\beta_2^H$	c	Log k
n-Propylbenzene	C18	6.44	0.42	-0.38		-0.49	-1.18	4.81
	C4	3.82				-0.23	-1.38	2.22
	CN	2.20	0.33			-0.22	-0.88	1.44
	Diol	1.79	0.37			-0.12	-1.05	0.99
Benzonitrile	C18	4.92	0.52	-0.84		-1.08	-1.18	2.34
	C4	2.93				-0.51	-1.38	1.04
	CN	1.68	0.40			-0.49	-0.88	0.71
	Diol	1.37	0.45			-0.26	-1.05	0.51
Acetanilide	C18	6.29	0.61	-1.06	-0.20	-2.18	-1.18	2.28
	C4	3.74			-0.23	-1.03	-1.38	1.11
	CN	2.15	0.47		-0.26	-0.98	-0.88	0.50
	Diol	1.75	0.53		-0.23	-0.54	-1.05	0.47
Phenol	C18	4.38	0.56	-0.78	-0.24	-0.98	-1.18	1.76
	C4	2.60			-0.28	-0.47	-1.38	0.47
	CN	1.49	0.43		-0.31	-0.44	-0.88	0.29
	Diol	1.22	0.49		-0.27	-0.24	-1.05	0.15
2-Hexanone	C18	5.47	0.09	-0.52		-1.66	-1.18	2.20
	C4	3.25				-0.78	-1.38	1.09
	CN	1.87	0.07			-0.75	-0.88	0.31
	Diol	1.52	0.08			-0.41	-1.05	0.15

0.70; s = -0.76; a = -0.40; b = -3.26; and c = -1.18. In all cases the driving force for retention is the relative ease of cavity formation in the solvated sorbent. This is augmented, in some cases, by the solvated sorbent's favorable capacity for electron lone-pair interactions. Sorbents with polar-bonded functional groups are more cohesive than those with simple bonded alkyl chains, resulting in less favorable conditions for solute transfer from aqueous solution. The solvated spacer-bonded propanediol and cyanopropylsiloxane-bonded sorbents are more competitive with the sample solvent than the alkyl bonded sorbents for interactions of a dipole-type (s constant) and are stronger hydrogen-bond acids (b constant). The larger capacity of these sorbents for polar interactions cannot offset their higher cohesion so that they are less retentive than the alkyl-bonded sorbents, as shown in Table 8. Because the polar-solvated sorbents remain uncompetitive with water for polar interactions and are more cohesive, they will not provide increased capacity for solute isolation from predominantly aqueous solutions.

There are only a few data points available for the retention properties of polar-bonded phases with non-aqueous sample solvents (Seibert et al., 1996). These are summarized in Table 9. From what results we do have, there is a leveling of the ease of cavity formation in the sample solvent and

Table 9. System constants for normal phase separations (log k) on chemically bonded polar stationary phases with 20 % (v/v) diethyl ether-pentane as the mobile phase. Amino = 3-aminopropylsiloxane-bonded silica sorbent; Cyano = 3-cyanopropyl siloxane-bonded silica sorbent; and Diol = spacer-bonded propanediol silica sorbent. From Seibert et al., 1996 with permission.

Sorbent	System constants						Statistics			
	m	*r*	*s*	*a*	*b*	*c*	R	SE	F	*n*
Amino	0.00	0.35 ±(0.08)	0.00	1.79 ±(0.17)	0.45 ±0.21)	-0.65 ±(0.16)	0.941	0.096	43	21
Cyano	0.00	0.00	0.63 ±(0.06)	0.00	0.28 ±(0.14)	-0.91 ±(0.10)	0.916	0.083	57	25
	0.00	0.00	0.59 ±(0.06)	0.00	0.00	-0.74 ±(0.04)	0.900	0.088	98	25
Diol	0.00	0.00	0.33 ±(0.08)	0.47 ±(0.14)	0.74 ±(0.18)	-0.29 ±(0.12)	0.831	0.107	16	26

the solvated sorbent, in contrast to the results for predominantly aqueous sample solvents, such that the *m* constant is not generally significant and retention is dominated by the polar interactions. As would be expected the dominant characteristic of the solvated aminopropylsiloxane-bonded sorbent is its high basicity (*a* constant). For the solvated cyanopropylsiloxane-bonded sorbent its capacity for dipole-type interactions (*s* constant). For the solvated spacer-bonded propanediol sorbent its capacity for hydrogen-bond type interactions, particularly as a hydrogen-bond acid (*b* constant) combined with a significant capacity for dipole-type interactions (*s* constant). In effect, the reasons for selecting a particular sorbent for extraction from aqueous and non-aqueous solvents are completely different with most of this difference attributed to the rather unique properties of water. There remains, however, a critical need for further studies of the influence of solvent properties on sorbent selectivity for chemically bonded sorbents and non-aqueous sample solvents in SPE.

V. RECOVERY OF SORBENT-IMMOBILIZED ANALYTES

In this section recovery is taken to mean the completeness with which an immobilized analyte is desorbed in the elution step. It should be distinguished from the *overall recovery,* which refers to the outcome of the full extraction process including all operations from sample collection to instrumental determination. Poor recovery can result from the following sources:

- Incomplete elution of the analytes from the sorbent.
- Loss of analytes during the intermediate wash steps due to insufficient retention under the conditions used for matrix clean-up.
- Irreversible binding of a fraction of the analytes to specific sorbent sites.
- Losses during further sample processing such as solvent reduction.

It is well established that octadecylsiloxane-bonded silica sorbents contain a low concentration of silanol groups with sufficient ion-exchange capacity to adsorb basic solutes too strongly for them to be eluted by polar organic solvents in the absence of a competing base added to the elution solvent (Martin et al., 1993; Albert et al., 1994; Martin et al.,1995). These secondary interactions should always be borne in mind as we move to more general models for optimization of the matrix clean-up and elution steps in the recovery of sorbent-immobilized analytes.

The previously introduced Equation 9 indicates that the volume of solvent required for elution is directly proportional to the interparticle volume of the extraction device (V_0) and the retention factor of the analyte with the eluting solvent. These parameters are now in opposition to the requirements for larger breakthrough volumes since the object is to recover and concentrate the analytes in as small a solvent volume as practical. That means both the interparticle volume and the retention factor should be minimized. For efficient elution the general aim is to find a solvent in which the retention factor is less than two. Under this constraint the analytes are recovered in a solvent volume less than about three inter-particle volumes (the figure of two bed volumes is often quoted as a minimum rugged elution volume). By determining the retention factor for the sorbent and estimating or measuring the interparticle volume for the sampling device, a good estimate of the elution volume is obtained.

Optimization of solvent strength for the elution step is achieved by identifying those solvents that provide an acceptable value for the retention factor. Suitable wash solvents are those too weak to cause elution of the analytes while efficiently displacing matrix interferences from the sorbent (or, less frequently, too weak to cause elution of the matrix interferences if the analytes are isolated first). The minimum retention factor required to avoid elution of the analytes in the wash steps depends on how many interparticle volumes of wash solvent are applied for matrix simplification.

Again, from equation 9, if three interparticle volumes of wash solvent are employed then the retention factor must be greater then 2. A margin of safety dictates that it should be significantly greater than the minimum

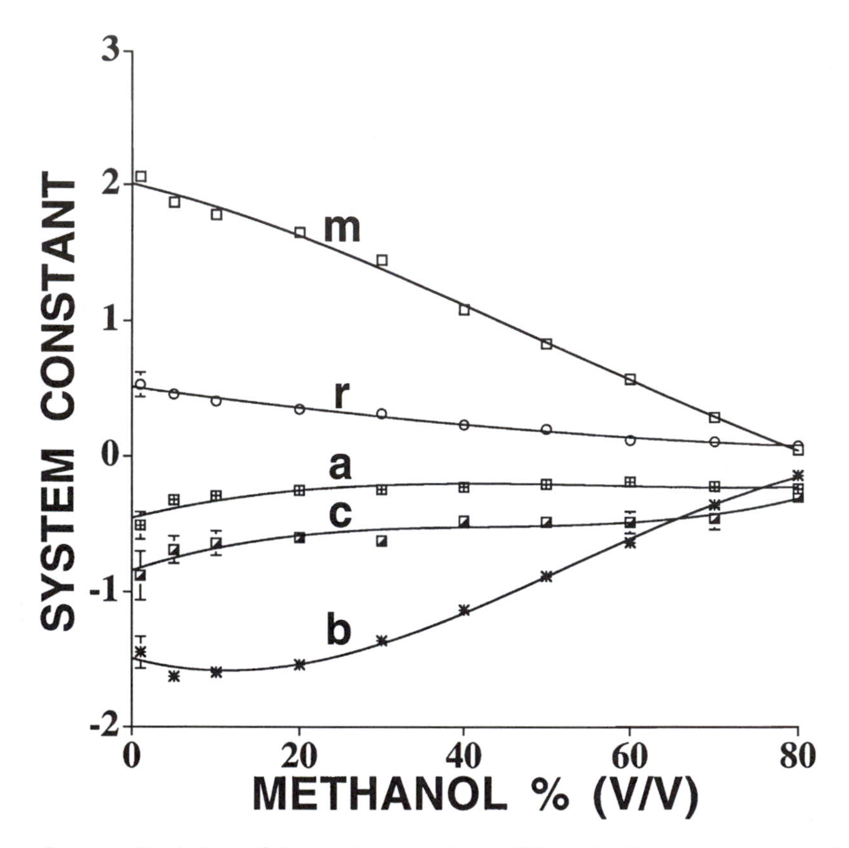

Figure 8. Variation of the system constants of the solvation parameter model with composition of the water/methanol mobile phase for a cyanopropylsiloxane-bonded silica sorbent for solid-phase extraction. From Seibert and Poole, 1995 with permission.

value since several factors affect the validity of the predictions based on Equation 9. The conditions of ideal chromatography are assumed; during sample processing the sample solvent residing in the inter-particle volume (unless removed in an intermediate drying step) will mix with the eluting or wash solvent and change its solvent strength with respect to that of the pure solvent. Further, retention of the analytes occurs predominantly in a zone close to the sample entry position of the sorbent bed. In spite of these objections, the primary information required to select appropriate wash and elution solvents and their volumes is a knowledge of the retention factors of the analytes and the interparticle volume of the sampling device.

Weidolf and Henion (1987) packed liquid chromatographic columns with sorbent from a solid-phase extraction cartridge. They then predicted eluent volumes and compositions for wash and elution solvents from parameters obtained by HPLC and the interparticle volume of the SPE cartridges. The volume and composition of the wash solvent for matrix simplification was determined from the retention factor of the peak front of the analyte, and the composition for elution was calculated from the retention factor of the peak tail in appropriate mobile phases. Casas et al. (1992) determined the elution curve (a plot of recovery against solvent composition) for a sorbent cartridge. This research group then related the inflection points (composition of the strongest binary solvent giving complete sorption and the weakest binary solvent in which the sample was unretained) to the retention factor determined by HPLC on a similar column packing material. Using the relationship between the retention factor and the inflection points of the elution curve, it is possible to predict the most suitable composition of the wash and elution solvents for other compounds that bear a family resemblance to the standards used to determine the relationship between the solid-phase extraction and liquid chromatographic data.

A more flexible approach is the use of retention maps. These plots of the system constants from the solvation parameter model against solvent composition, as indicated in Figures 8 and 9 (Seibert and Poole, 1995a; Seibert et al., 1996). The left-hand side of the figures indicates the system constants required for prediction of retention during sample application. The right-hand side of the figures indicates the range of solvent compositions at which the analytes can be recovered with a low retention factor from the sorbent. Intermediate solvent compositions indicate the retention factor for analytes under conditions used for matrix simplification.

The retention maps and Equation 12 can be used to predict the retention factors in the sampling system for all solutes whose descriptors are either known or available through estimates, since the system constants are (ideally) solute-independent. Thus, once a retention map has been produced, it can be used to estimate the outcome for any analyte whose descriptors are available without resort to additional experimental data. With a sufficient number of retention maps for different sorbents and mobile-phase combinations, it should be possible to optimize the sorbent selection process and identify conditions providing adequate breakthrough volumes, recovery, and matrix simplification through calculation. Once again, the results should always be verified by experiment and iterative variations on the experiment performed to identify whether the predicted sorbent/solvent system really does provide an optimum separation, or only one that is close to optimum.

Hughes and Gunton (1995) have proposed a graphical method to estimate the volume and solvent strength of wash and elution solvents in SPE, based on an extension of the classical theory of LLE. This leads to Equation 14:

$$\ln (1 - R_T) = n \ln (1 - R_o) + \ln (R_{ret}) \tag{14}$$

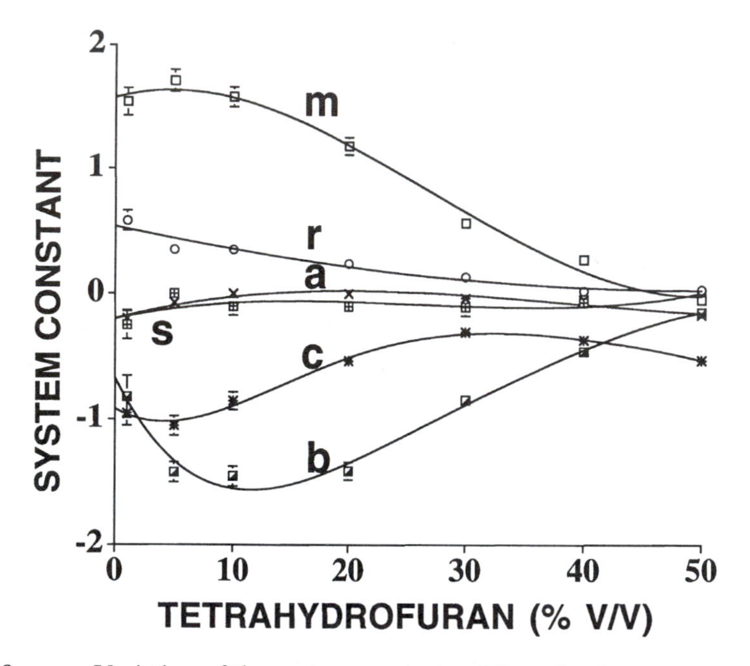

Figure 9. Variation of the system constants of the solvation parameter model with composition of the water/tetrahydrofuran mobile phase for a spacer-bonded propanediol silica sorbent for solid-phase extraction. From Seibert et al., 1996 with permission.

where R_T is the total extraction recovery, n the number of extractions, R_o the extraction efficiency, and R_{ret} the limiting extraction recovery used to account for irreversible solute retention by the sorbent. A plot of $-\ln(1 - R_T)$ against n (or the equivalent volume of elution or wash solvent) yields a straight line from which the extraction efficiency and the R_{ret} factor are calculated from the slope and intercept, respectively. Within this model favorable conditions for matrix simplification are recognized as those in which the extraction efficiency (slope of the plot) is low for the analytes and high for the matrix in the wash step and high for the analytes in the elution step. In many cases the plots prove to be non-linear, but this does not prevent a qualitative assessment of the recovery and matrix simplification procedure by inspection of the experimentally derived curves.

VI. SORBENT TRAPPING OF VOLATILE COMPOUNDS FROM AIR

Sorbent trapping using short tubes packed with a porous polymer or carbon (or occasionally other materials such as silica or alumina) is widely used in air monitoring and occupational hygiene (Namiesnik, 1988; Poole et al., 1990; Harper, 1993; Ciccioli 1993). The sampling characteristics of these devices are adequately represented by Equation 8 (Lovkvist and Jonsson, 1987) with certain reservations (Harper 1993). The cartridges are packed with a coarse particle material in short bed lengths to enable fast sampling flow rates with a low-pressure drop. Conditions of sampling result in low numbers of theoretical plates and a sampling performance that is easily affected by the packing density and dimensions of the sorbent bed (Senum, 1981; Van der Straeten, 1985; Sunesson et al., 1995). The general condition of ideal chromatography assumed for the derivation of Equation 8 must be considered doubtful for some practical applications, since adsorption isotherms at low concentration are more likely to be curved than are partition isotherms (Harper, 1993; Simon et al., 1995). This results in breakthrough volumes with a pronounced concentration dependence, even for the trace sample concentrations that are common in air sampling.

It is routine practice to employ sampling tubes with two sorbent sections. The second and shorter backup section is analyzed separately from the main section, and used to infer whether breakthrough has occurred during sampling. This provides a practical solution to the problem of identifying early breakthrough due to sorbent sampling conditions. Breakthrough volumes can be measured directly by drawing air through the sorbent tube from a standard atmosphere of the analyte and analyzing the effluent from the sorbent bed continuously with a sensitive detector such as the flame ionization detector (Harper, 1993; Simon et al., 1995; Sunesson et al., 1995). If the results are to be meaningful, then the experimental conditions must model, very closely, those used in practice. This makes the measurements more difficult and, generally, a range of likely sampling conditions must be evaluated corresponding to a range of possible breakthrough volumes in practice.

A simpler method is to use gas chromatography to determine the retention volume of the analytes on the sorbent as a function of reciprocal temperature, and then extrapolate the results to the temperature at which sampling is performed. Unfortunately, the direct and the gas chromatographic method rarely give the same values for the breakthrough volume and the differences can be quite large (Harper, 1993). Both methods may be affected differently by the conditions of non-ideal chromatography. The large difference in the number of theoretical plates between the sampling cartridge and a typical gas chromatography column calls into question the accuracy of the kinetic contribution to the breakthrough volume. Finally the gas chromatographic method is limited by assumptions of a constant adsorption enthalpy over the complete extrapolation temperature range. These factors and variations in the sampling conditions and matrix effects for real samples result in the recommendation of a safe sampling volume of about 0.7 V_B for field applications.

The prediction of breakthrough volumes by the solvation parameter model, Equation 12, should be applicable to gas phase sampling after modification to the form familiar for gas chromatography, Equation 15 (Abraham et al., 1990; Abraham et al., 1991; Poole et al., 1992; Poole and Poole 1995a). The new solute descriptor log L^{16} is introduced to account for the additional dispersion interactions that occur when a solute is transferred from the gas phase to the sorbent.

$$SP = c + rR_2 + s\pi_2^H + a\alpha_2^H + b\beta_2^H + l \log L^{16} \qquad (15)$$

For transfer between condensed phases the dispersion interactions in both phases are approximately equal and self-canceling. Assuming ideal behavior in the gas phase then dispersion interactions are negligible, and there is no possibility of cancellation. Consequently, a new term is required to account for the combination of cavity formation and solute-sorbent dispersion interactions when a solute is transferred from a gas to a sorbent. Log L^{16} is the distribution constant for the analyte between a gas and hexadecane at 298 K and can be determined directly by gas chromatography or obtained by back calculation from gas chromatographic data on other phases than hexadecane (Abraham, 1993b). The system l constant determines the contributions from cavity formation and dispersion interactions to sorbent retention and in gas/liquid chromatography is generally correlated to the free energy of transfer for a methylene group. Abraham and coworkers (Abraham and Walsh, 1992; Abraham et al., 1993; Grate et al., 1995) have applied Equation 15 to characterize the retention properties of graphitized carbon and fullerene. With SP as log V_g^o, where V_g^o is the specific retention volume at 273 K, they obtained the following equation for the sorption properties of Carbotrap™[16], a form of graphitized carbon black,

$$\log V_g^o = -4.73 - 2.27\, R_2 + 2.65 \log L^{16} \qquad (16)$$

($R = 0.97$; $SE = 0.88$; $F = 318$; $n = 38$)

Polar interactions are not important for retention on this material, which is dominated by cavity formation and dispersion interactions, combined with a significant contribution from electron lone pair/lone pair repulsion. A sample of graphite flakes (and coincidentally fullerene) showed different

[16] Carbotrap is a trademark of Supelco Inc., Bellefonte, Pennsylvania.

behavior to Carbotrap which resulted in equation (17), where SP is now log K_g, the gas/solid adsorption distribution constant at 298 K

$$\log K_g = -0.86 - 0.27R_2 + 0.86\pi_2^H + 0.94\alpha_2^H + 0.46 \log L^{16} \quad (17)$$

(R = 0.97 ; SE = 0.15 ; F = 124 ; *n* = 36)

These materials are rather weakly polarizable, behave as hydrogen-bond bases, and have a significantly lower capacity for dispersion interactions than the Carbotrap material. These two equations serve to illustrate the wide range of properties common to different forms of carbon and provide one reason why carbon from different sources often behave so differently in solid phase extraction.

The porous polymer Tenax™[17], a poly(2,6-diphenyl-p-phenylene oxide), is the most popular sorbent for trapping semi-volatile organic compounds with a molecular weight greater than heptane. Brown and Purnell (1979) have provided a collection of retention volumes extrapolated to 293 K for a number of varied solutes on a Tenax sorbent cartridge that can be fitted to Equation 18 in which SP is log V_R

$$\log V_R = -3.31 - 0.37 R_2 + 0.73\pi_2^H + 0.48\alpha_2^H + 1.41 \log L^{16} \quad (18)$$

(R = 0.99 ; SE = 0.20 ; F = 566 ; n = 51)

Dispersion interactions are responsible for retention on Tenax supported by polar interactions of a dipole-type and weak sorbent hydrogen-bond acid interactions. Electron lone pair-lone pair repulsion contributes unfavorably to retention. The dominance of dispersion interactions and the capacity of the sorbent for induction interactions resulting from its polarizability are anticipated from its structure but the weak hydrogen-bond acidity is unanticipated and presumably is an indication of the presence of impurities or structural heterogeneity in the polymer. These last few studies indicate the power of the solvation parameter model to characterize materials for solid phase extraction as well as providing a basis for the prediction of sorbent retention.

VII. SOLID-PHASE MICRO-EXTRACTION

Solid-phase micro-extraction (SPME) uses a polymer-coated fused silica fiber as the sampling device (Zhang et al., 1994 ; Boyd-Boland et al., 1994). The active sampling area is a cylindrical fiber typically 1 cm long with a radius of about 0.015 cm coated with a film 10 to 100 μm thick of immobilized polymer. The fiber is either immersed in the medium to be sampled (direct sampling) or suspended in the headspace above the sample (indirect

[17] Tenax is a trademark of Enka Research Institute, Arnhem, The Netherlands.

sampling). The direct method is most suitable for liquids with a low level of matrix interference while the indirect method is applicable to a wider range of sample types that includes solids and other solvent matrices that would interfere in the direct sampling mode.

In contrast to SPE using cartridges and particle-loaded membranes, exhaustive extraction is rarely achieved. In the general case SPME is an equilibrium sampling method yielding chromatographic profiles that in appearance may show little resemblance to those obtained by SPE. This is because the amount extracted by the fiber is controlled by a series of distribution constants between the different phases involved. For the direct sampling mode the amount of material extracted by the fiber at equilibrium, w_1, is given by Equation 19 (Louch et al., 1992; Arthur et al., 1992):

$$w_1 = C_o V_1 V_2 K_c / (K_c V_1 + V_2) \tag{19}$$

where C_o is the initial concentration of the analyte in the sample solution, V_1 the volume of the fiber coating, V_2 the volume of the sample solution and K_c the distribution constant for the analyte between the fiber coating and the sample solution. The volume of the fiber coating, typically less than a microliter, is generally much smaller than the sample volume, and Equation 19 can be simplified to $w_1 \cong C_o V_1 K_c$. The amount extracted by the fiber is linearly related to the original analyte concentration and independent of the sample volume (as long as $V_2 \approx V_1$).

Equation 19, however, only considers the mass absorbed at equilibrium and does not take into account the time required to reach equilibrium. For analytes with a large distribution constant (K_c) the time required to reach equilibrium can be considerable since the analytes must diffuse through a thin film of static solvent surrounding the fiber to reach the fiber coating. The time required to reach equilibrium can be reduced by agitating the sample solution by stirring or sonication. In the extreme case of perfect agitation the rate of absorption by the fiber coating is determined only by the diffusion of analyte in the polymer coating (directly proportional to the square of the coating thickness). In reality, perfect agitation conditions exist only in theory, and the direct sampling method is of limited application for analytes with large distribution constants between the fiber coating and the sample solution.

Extraction times can be greatly reduced by sampling analytes from the headspace above the sample solution since diffusion coefficients in the vapor phase are four orders of magnitude higher than in solution (Zhang and Pawliszyn, 1993). A rapid equilibrium between the solvent and vapor phase can be achieved by continuously stirring the solvent to generate a continuously fresh surface. By virtue of indirect sampling the method can also be applied to solids and other sample matrices not suitable for direct sampling. In this case the amount of sample absorbed by the fiber coating is given by Equation 20:

$$w_1 = C_o V_1 V_2 K_c / (K_c V_1 + K_{HS} V_3 + V_2) \quad (20)$$

where K_{HS} is the distribution constant for the analyte between the headspace and the solution (or sample) and V_3 is the volume of the headspace. For most analytes, K_{HS} is relatively small and sampling from the headspace will not effect the amount of analyte absorbed by the coating if the volume of the headspace is small compared to the sample volume. In this case the amount of analyte absorbed by the fiber will be very close to the amount that would be absorbed by direct sampling and similar method detection limits result.

The headspace method is very efficient for the analysis of volatile compounds but compounds of low volatility and high solvent solubility transfer very slowly to the headspace. The rate at which equilibrium in the system is established becomes very slow and method detection limits will be poor. Because distribution constants are temperature-dependent there is usually an optimum temperature for headspace SPME. In general terms higher temperatures favor release of analytes from the sample matrix but lower absorption by the coated fiber from the vapor phase, and a compromise must be found. The distribution constants involved in SPME could be modeled by the solvation parameter technique but no work has appeared on this topic to date.

VIII. CONCLUSIONS

Solid-phase extraction has developed as a largely empirical approach to sample preparation and has been widely regarded as an economical replacement for LLE but with a greater potential for automation. The theoretical models presented in this chapter indicate the need for a more purposeful approach to understand the fundamental aspects of the technique. Only in this way will we achieve full optimization of the sampling process with the different sampling devices now in use, and be able to develop expert systems to predict the optimum sampling conditions for any analyte without resorting to trial-and-error experiments. Theoretical understanding of the SPE process is still in its infancy, although initial results obtained with the solvation parameter model for characterizing sorbent retention properties and for the prediction of retention and recovery are very encouraging.

REFERENCES

Abraham, M. H. and McGowan J. C., (1987) The Use of Characteristic Volumes to Measure Cavity Terms in Reversed Phase Liquid Chromatography. Chromatographia 23:243.

Abraham, M. H., Whiting, G. S., Doherty, R. M. and Shuely, W. J., (1990) A New Method for Characterisation of GLC Stationary Phases - The Laffort Data Set. J.Chem.Soc.,Perkin Trans. II, p. 1451.

Abraham, M. H., Whiting, G. S., Doherty, R. M. and Shuely, W. J. (1991) A New Solute Solvation Parameter, π_2^H from Gas Chromatographic Data. J.Chromatogr. 587:213.

Abraham, M. H. and Walsh, D. P., (1992) Application of the New Solvation Equation to log Vg Values for Solutes on Carbonaceous Adsorbents. J.Chromatogr. 627:294.

Abraham, M. H., (1993) Scales of Solute Hydrogen-Bonding: Their Construction and Application to Physicochemical and Biochemical Processes. Chem.Soc.Revs. 22:73.

Abraham, M. H., (1993a) Construction of a Scale of Solute Effective or Summation Hydrogen-Bond Basicity. J.Phys.Org.Chem. 6:660.

Abraham, M. H., (1993b) Solvation Parameters for Functionally-Substituted Aromatic Compounds and Heterocyclic-Compounds from Gas-Liquid Chromatographic Data. J.Chromatogr. 644:95.

Abraham, M. H., Du, C. M., Grate, J. W., McGill, R. A. and Shuely, W. J., (1993) Fullerene as an Adsorbent for Gases and Vapours. J.Chem.Soc.Chem.Commun. 1863.

Abraham, M. H., Chadha, H. S. and Leo, A. J., (1994) Relationship Between High-Performance Liquid Chromatography Capacity Factors and Water-Octanol Partition Coefficients. J.Chromatogr. A 685:203.

Abraham, M. H., Andonian-Haftvan, J., Whiting, G. S., Leo, A., and Taft, R. S., (1994a) The Factors That Influence the Solubility of Gases and Vapors in Water at 298 K, and a New Method for its Determination. J.Chem.Soc.Perkin Trans. II, p. 1777.

Abraham, M. H. and Roses, M., (1994) Effect of Solute Structure and Mobile Phase Composition on Reversed-Phase High-performance Liquid Chromatography Capacity Factors. J.Phys.Org.Chem. 7:672.

Albert, K., Brindle, R., Martin, P., and Wilson, I. D., (1994) Characterization of C18-Bonded Silicas for Solid-Phase Extraction by Solid-State NMR Spectroscopy. J.Chromatogr. A 665:253.

Arthur, C. L., Killam, L. M., Buchholz, K. D. and Pawliszyn J., (1992) Automation and Optimization of Solid-Phase Micro-Extraction. Anal.Chem. 64:1960.

Berrueta, L. A., Gallo, B. and Vicente, F., (1995) A Review of Solid Phase Extraction: Basic Principles and New Developments. Chromatographia 40:474.

Bitteur, S. and Rosset, R., (1987) Comparison of Octadecyl-Bonded Silica and Styrene-Divinylbenzene Copolymer Sorbents for Trace Enrichment Purposes. Fundamental Aspects II. Chromatographia 23:163.

Brown, R. H. and Purnell, C. J., (1979) Collection and Analysis of Trace Organic Vapour Pollutants in Ambient Atmospheres. The Performance of a Tenax-GC Adsorbent Tube. J.Chromatogr. 178:91.

Boyd-Boland, A. A., Chai, M., Luo, Y. Z., Zhang, Z., Yang, M. J., Pawliszyn, J. B. and Gorecki, T., (1994) New Solvent-Free Sample Preparation Techniques Based on Fiber and Polymer Technologies. Environ. Sci. Technol. 28:569A.

Carr, J. W. and Harris, J. M., (1988) In Situ Fluorescence Detection of Polycyclic Aromatic Hydrocarbons Following Preconcentration on Alkylated Silica Adsorbents. Anal.Chem. 60:698.

Carr, P. W., (1993) Solvatochromism, Linear Solvation Energy Relationships, and Chromatography. Microchem.J. 48:4.

Casas, M., Berrueta, L. A., Gallo, B. and Vicente, F., (1992) Solid-Phase Extraction Conditions for the Selective Isolation of Drugs from Biological Fluids Predicted using Liquid Chromatography. Chromatographia 34:79.

Ciccioli, P., (1993) Chemistry and Analysis of Volatile Organic Compounds in the Environment (H. J. Th. Bloemen and J. Burn, Ed.) Blackie, Glasgow, p. 92.

Fernando, W. P. N., Larrivee, M. L. and Poole, C. F., (1993) Investigation of the Kinetic Properties of Particle-Loaded Membranes for Solid-Phase Extraction by Forced Flow Planar Chromatography. Anal.Chem. 65:588.

Gelencser, A., Kiss, G., Krivacsy, Z., Varga-Puchony, Z. and Hlavay, J., (1995) A Simple Method for the Determination of Capacity Factor on Solid-Phase Extraction Cartridges. I. J.Chromatogr. A 693:217.

Gelencser, A., Kiss, G., Krivacsy, Z., Varga-Puchony, Z. and Hlavay, J., (1995a) The Role of Capacity Factor in Method Development for Solid-Phase Extraction of Phenolic Compounds. II. J.Chromatogr. A 693:227.

Grate, J. W., Abraham, M. H., Du, C. M., McGill, R. A., and Shuely, W. J., (1995) Examination of Vapor Sorption by Fullerene, Fullerene-Coated Surface Acoustic Wave Sensors, Graphite, and Low-Polarity Polymers Using Linear Solvation Energy Relationships. Langmuir 11:2125.

Hagan, D. R., Markell, C. G., Schmitt, G. and Blevins, D. D., (1990) Membrane Approach to Solid-Phase Extraction. Anal.Chim.Acta 236:157

Harper, M., (1993) Evaluation of Solid Sorbent Sampling Methods by Breakthrough Volume Studies. Ann.Occup.Hyg. 37:65.

Hennion, M.-C. and Coquart., (1993) Comparison of Reverse Phase Extraction Sorbents for the On-Line Trace Enrichment of Polar Organic Compounds in Environmental Aqueous Samples. J.Chromatogr. 642:211.

Hennion, M.-C. and Pichon, V., (1994) Solid-Phase Extraction of Polar Organic Pollutants from Water. Environ.Sci.Technol. 28:576A.

Hughes, D. M. and Gunton K. E., (1995) Representing Isocratic Multicomponent Solid-Phase Extraction Data by an Extension of Liquid-Liquid Extraction Theory. Anal.Chem. 67:1191.

Jandra, P. and Kubat, J., (1990 Possibilities of Determination and Prediction of Solute Capacity Factors in Reversed Phase Systems with Pure Water as the Mobile Phase. J.Chromatogr. 500:281.

Josefson, C. M., Johnston, J. B. and Tubey, R., (1984) Adsorption of Organic Compounds from Water with Porous Poly(tetrafluoroethylene). Anal.Chem. 56:764.

Kamlet, M. J., Doherty, R. M., Abraham, M. H. and Taft, R. W., (1985) An Analysis of the Factors that Influence Adsorption of Organic Compounds on Activated Carbon. Carbon 23:549

Larrivee, M. L. and Poole, C. F., (1994) Solvation Parameter Model for the Prediction of Breakthrough Volumes in Solid-Phase Extraction with Particle-Loaded Membranes. Anal.Chem. 66:139.

Liska, I., Krupcik, J. and Leclercq, P. A., (1989) The Use of Solid Sorbents for Direct Accumulation of Organic Compounds from Water Matrices -- A Review of Solid-Phase Extraction Techniques. J.High Resolut.Chromatogr. 12:577.

Liska, I., Kuthan, A. and Krupcik, J., (1990) Comparison of Sorbents for Solid-Phase Extraction of Polar Compounds from Water. J.Chromatogr. 509:123.

I. Liska., (1993) On-line Versus Off-Line Solid-Phase Extraction in the Determination of Organic Contaminants in Water. Advantages and Limitations. J.Chromatogr. A 655:163.

Louch, D., Motlagh, S. and Pawliszyn J., (1992) Dynamics of Organic Compounds from Water Using Liquid-Coated Fused Silica Fibers. Anal.Chem. 64:1187.

Lovkvist, P. and Jonsson, J. A., (1987) Capacity of Sampling and Preconcentration Columns with a Low Number of Theoretical Plates. Anal.Chem. 59:818.

Martin, P., Taberner,A., Fairbrother, A. and Wilson, I. D., (1993) An Investigation of the Effects of Carbon Loading and Endcapping on the Solid-Phase Extraction of β-Blockers onto C-18 Bonded Silica Gel. J.Pharm.Biomed.Anal. 11:671.

Martin, P., Morgan, E. D. and Wilson, I. W., (1995) Comparison of the Properties of a Normal and Base Deactivated Bonded Silica Gel for the Solid Phase Extraction of [^{14}C]-Propranolol. Anal.Proc. 32:179.

Mayer, M. L. and Poole, C. F., (1994) Identification of the Procedural Steps that Effect Recovery of Semi-Volatile Compounds by Solid-Phase Extraction Using Cartridge and Particle-Loaded Membrane (Disc) Devices. Anal.Chim.Acta 294:113.

Mayer, M. L., Poole, C. F. and Henry, M. P., (1995) Sampling Characteristics of Octadecylsiloxane-Bonded Particle-Embedded Glass Fiber Discs for Solid-Phase Extraction. J.Chromatogr. A 695:267.

Mayer, M. L., Poole, S. K. and Poole, C. F., (1995a) Retention Characteristics of Octadecylsiloxane-Bonded Silica and Porous Polymer Particle-Loaded Membranes for Solid-Phase Extraction. J.Chromatogr. A 697:89.

McGill, R. A., Abraham, M. H. and Grate, J. W., (1994) Choosing Polymer Coatings for Chemical Sensors. Chemtech. 24:27.

Miller, K. G. and Poole, C. F., (1994) Methodological Approach for Evaluating Operational Parameters and the Characterization of a Popular Sorbent for Solid-Phase Extraction by High Pressure Liquid Chromatography. J.High Resolut.Chromatogr. 17:125.

Mol, H. G. J., Janssen, H.-G., Cramers, C. A. and Brinkman, U. A. Th., (1995) On-Line Sample Enrichment-Capillary Gas Chromatography of Aqueous Samples Using Geometrically Deformed Open-Tubular Columns. J.Microcol.Sep. 7:247.

Namiesnik, J. (1988) Preconcentration of Gaseous Organic Pollutants in the Atmosphere. Talanta 35:567.

Pankow, J. F., Isabelle, L. M. and Kristensen, T. J., (1982) Effects of Linear Flow Velocity and Residence Time on the Retention of Non Polar Aqueous Organic Analytes by Cartridges of Tenax-GC. J.Chromatogr. 245:31.

Pichon, V. and Hennion, M.-C., (1994) Determination of Pesticides in Environmental Water by On-Line Trace-Enrichment and Liquid Chromatography. J. Chromatogr. A 665:269.

Pichon, V., Chen, L., Guenu, S. and Hennion, M.-C., (1995) Comparison of Sorbents for Solid-Phase Extraction of the Highly Polar Degradation Products of Atrazine (Including Anmmeline, Ammelide and Cyanuric Acid). J. Chromatogr. A 711:257.

Poole, S. K., Dean, T. A., Oudsema, J. W. and Poole, C. F., (1990) Sample Preparation for Chromatographic Separations: an Overview. Anal.Chim.Acta 236:3.

Poole, C. F. and Poole, S. K., (1991) Chromatography Today. Elsevier, Amsterdam, The Netherlands.

Poole, C. F., Kollie, T. O and Poole, S. K., (1992) Recent Advances in Solvation Models for Stationary Phase Characterization and the Prediction of Retention in Gas Chromatography. Chromatographia 34:281.

Poole, C. F. and Fernando, W. P. N., (1993) Comparison of the Kinetic Properties of Commercially Available Precoated Silica Gel Plates. J.Planar Chromatogr. 6:357.

Poole, S. K. and Poole, C. F., (1995) Influence of Solvent Effects on the Breakthrough Volume in Solid-Phase Extraction Using Porous Polymer Particle-Loaded Membranes. Analyst 120:1733.

Poole, S. K. and Poole, C. F., (1995a) Chemometric Classification of the Solvent Properties (Selectivity) of Commonly Used Gas Chromatographic Stationary Phases. J.Chromatogr. A 697:415.

Poole, C. F., Poole, S. K., Seibert, D. S. and Miller, K. G., (1996) "New Approaches to Solid-Phase Extraction" Methodological Surveys in Bioanalysis of Drugs (E. Reid, H.M. Hill and I.D. Wilson, Eds.) Royal Society of Chemistry, Cambridge, UK, 24, p. 194.

Przyjazny, J., (1985) Evaluation of the Suitability of Selected Porous Polymers for the Preconcentration of Organosulphur Compounds from Water. J.Chromatogr. 346:61.

Schoenmakers, P. J., Billiet, H. A. and De Galan, L., (1979) Influence of Organic Modifiers on the Retention Behaviour in Reversed-Phase Liquid Chromatography and its Consequences for Gradient Elution. J.Chromatogr. 185:179.

P. J. Schoenmakers, Billiet, H. A. and De Galan, L., (1983) Description of Solute Retention Over the Full Range of Mobile Phase Composition in Reversed-Phase Liquid Chromatography. J.Chromatogr. 282:107

Seibert, D. S. and Poole, C. F., (1995) Retention Properties of a Cyanopropylsiloxane-Bonded Silica-Based Sorbent for Solid-Phase Extraction. J.High Resolut.Chromatogr. 18:226.

Seibert, D. S. and Poole, C. F., (1995a) Influence of Solvent Effects on Retention in Reversed-Phase Liquid Chromatography and Solid-Phase Extraction Using a Cyanopropylsiloxane-Bonded, Silica-Based Sorbent. Chromatographia 41:51.

Seibert, D. S., Poole, C. F. and Abraham, M. H., (1996) Retention Properties of a Spacer Bonded Propanediol Sorbent for Reversed-Phase Liquid Chromatography and Solid-Phase Extraction. Analyst 121:511.

Senum, G. I., (1981) Theoretical Collection Efficiency of Adsorbent Samplers. Environ.Sci.Technol. 15:1073.

Shoup, R. E. and Mayer, G. S., (1982) Determination of Environmental Phenols by Liquid Chromatography / Electrochemistry. Anal.Chem. 54:1164.

Simon, V., Riba, M.-L., Waldhart, A. and Torres, L., (1995) Breakthrough Volume of Monoterpenes on Tenax TA: Influence of Temperature and Concentration for α-Pinene. J.Chromatogr. A 704:465.

Subra, P., Hennion, M.-C., Rosset, R. and Frei, R. W., (1988) Recovery of Organic Compounds from Large-Volume Aqueous Samples Using On-Line Liquid Chromatographic Preconcentration Techniques. J.Chromatogr. 456:121.

Sunesson, A.-L., Nilsson, C.-A. and Andersson B., (1995) Evaluation of Adsorbents for Sampling and Quantitative Analysis of Microbial Volatiles Using Thermal Desorption Gas Chromatography. J.Chromatogr. A 699:203

Thurman, E. M., Malcolm, R. L. and Aiken, G. R., (1978) Prediction of Capacity Factors for Aqueous Organic Solutes Adsorbed on a Porous Acrylic Resin. Anal. Chem. 50:775.

Van der Straeten, D., Van Langenhove, H. and Schamp, N., (1985) Comparison Between Theoretical and Experimental Sampling Efficiences on Tenax GC. J.Chromatogr. 207:207.

Weidolf, L. O. G. and Henion, J. D., (1987) Liquid-Solid Extraction Conditions Predicted by Liquid Chromatography for Selective Isolation of Sulfoconjugated Steroids from Equine Urine. Anal.Chem. 59:1980.

Werkhoven-Goewie, C. E., Brinkman, U. A. Th. and Frei, R. W., (1981) Trace Enrichment of Polar Compounds on Chemically Bonded and Carbonaceous Sorbents and Application to Chlorophenols. Anal.Chem. 53:2072.

Zhang, Z. and Pawliszyn J., (1993) Headspace Solid-Phase Micro-extraction. Anal.Chem. 65:1843.

Zhang, Z., Yang, M. J. and Pawliszyn J. B., (1994) Solid-Phase Micro-extraction. Anal.Chem. 66:844A.

7

SECONDARY INTERACTIONS AND MIXED-MODE EXTRACTION

Brian Law, Zeneca Pharmaceuticals, Mereside, Alderley Park, Macclesfield, United Kingdom

I. INTRODUCTION

Since the inception of solid-phase extraction (SPE), a massive range of stationary phases have become available. However, most of the published work has been carried out using non-polar phases. A survey of the recent literature (e.g., Majors, 1995; Majors, 1996) showed C18 to be the most commonly used phase, which together with the other non-polar phases, consistently account for between 50 and 80% of all reported applications. The predominance of the non-polar phases probably stems from the view of both manufacturers and the early users, that the use of non-polar cartridges was merely a convenient extension of the widely used technique of reversed-phase high-performance liquid chromatography (RP-HPLC). Thus, all the general rules of HPLC were believed to apply i.e., analytes will be retained if applied in a weakly eluotropic solvents such as water, but would elute with strong solvents such as methanol. It was also assumed that the relative eluotropic strength of the common HPLC solvents, methanol, acetonitrile and tetrahydrofuran, would be translate from HPLC to SPE. It quickly became apparent, however, that SPE was different to HPLC and much more complex than had been originally believed. It also became apparent that many of the rules that applied so widely and effectively in HPLC did not apply in SPE.

These differences were attributed to so-called "secondary interactions," which although present in HPLC and often considered as detrimental to the separation process, seemed to play a more significant role in SPE — a role that, as you will see in this chapter, is potentially beneficial. These secondary effects were thought to include hydrogen bonding with unreacted silanols on the silica surface and ionic or coulombic interactions with ionized acidic residual silanols. The effects of trace metals in the silica, acting either directly or indirectly through their influence on neighboring silanols, was also believed to play a role.

The evidence for the secondary interaction when using reversed-phase cartridges is mainly empirical and is based on observations of the unusual elution behavior of basic compounds. Thus, it is frequently observed that bases are over-retained on silica-based reversed phases and are difficult to elute with simple organic or aqueous/organic solvents. Moreover, acetonitrile which is a more powerful solvent than methanol in RP-HPLC, appears to be much weaker than methanol when used for the elution of basic drugs from reversed-phase cartridges (Ruane and Wilson, 1987; Marko et al., 1990; Soltes et al., 1983; Law et al., 1992). While the evidence for secondary interactions has been mainly indirect it is frequently cited as being the cause of the observed phenomena.

The importance of these secondary interactions becomes obvious when one considers the energies involved. Reversed-phase dispersive interactions are relatively weak (< 10 kcal/mol) in contrast to coulombic (ionic) processes (100 kcal/mol) or even hydrogen bonding interactions (10 kcal/mol). To date, there have been only a few systematic studies of these effects (Ruane and Wilson, 1987; Marko et al., 1990; Law et al., 1992; Law and Weir, 1992).

II. CATION-EXCHANGE EFFECTS

The first published report implicating cation-exchange effects in the reversed-phase extraction of basic compounds came from Soltes and co-workers (Soltes et al., 1983). These workers found that the basic drug pentacaine (pK_a 8.6) could not be eluted from a Sep-Pak™[18] C18 cartridge with pure acetonitrile. They demonstrated similar findings for other compounds such as stobadin (Marko, 1988; Marko et al., 1990) and propranolol (Marko et al., 1990). The very high elution volumes for these compounds when using methanol as the eluent (up to 25 bed volumes) was further evidence for interactions other than the expected hydrophobic interaction. These findings were confirmed by Ruane et al. (1988) in the study of a several β-blocker type drugs using C18 cartridges from six different manufacturers.

In experiments carried out in the author's laboratory (Law et al., 1992), the elution characteristics of the β-blockers propranolol and atenolol (Figure 1.) were studied using Bond Elut C2 and C18 cartridges. Following application in water the cartridges were successively washed with portions of aqueous methanol (1 mL) where the concentration of methanol was increased from 0 to 100% in 20% increments. The cumulative elution over the whole series of solvents (6 x 1 mL) was never more than 7%. The test combination involving the most polar compound and the more polar cartridge, atenolol/C2, where the reversed-phase interaction should have been minimal and hence the elution more facile, actually showed the lowest overall elution — only 0.27% of the applied atenolol. The poor eluting power of acetonitrile in comparison to methanol was also demonstrated. Acetonitrile's low eluotropic strength has never been fully explained, although it may be due to the fact that acetonitrile only possesses hydrogen bond acceptor ability whereas methanol possess both donor and acceptor properties (Law et al., 1992).

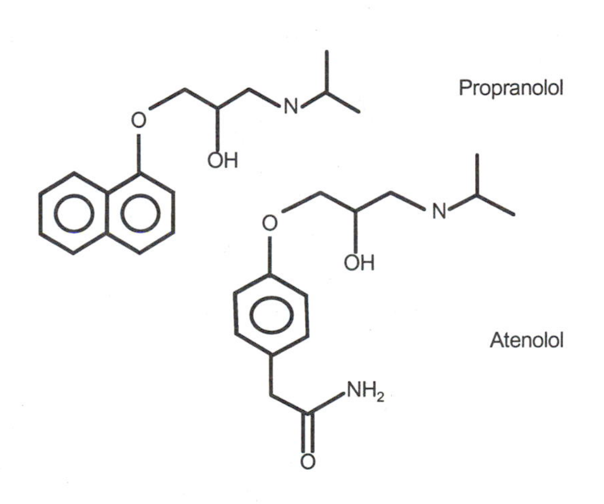

Figure 1. Structure of propranolol and atenolol.

[18] Sep-Pak is a trademark of Waters Inc., Milford, Massachusetts.

A. ELIMINATING SECONDARY CATION INTERACTIONS AT ELUTION

In a well-optimized SPE method it should be possible to elute the analyte of interest in two to three bed volumes, i.e., < 0.5 mL for a 100 mg cartridge. For the extraction of basic compounds using a reversed-phase cartridge this has been brought about by the use of methanol or acetonitrile containing an acid (Harrison et al., 1985), a base (Moors and Massart, 1991) and either an organic buffer (Ruane and Wilson, 1987) or an inorganic buffer (Musch et al., 1989). Under these circumstances the drug substance has been eluted in a relatively small volume with good efficiency. Because of the variation in the nature of the additive to the methanol or acetonitrile (the eluents contain cationic species, anions capable of ion pairing and have pHs that are far from neutral), it is unclear what mechanisms of retention were operating in each case.

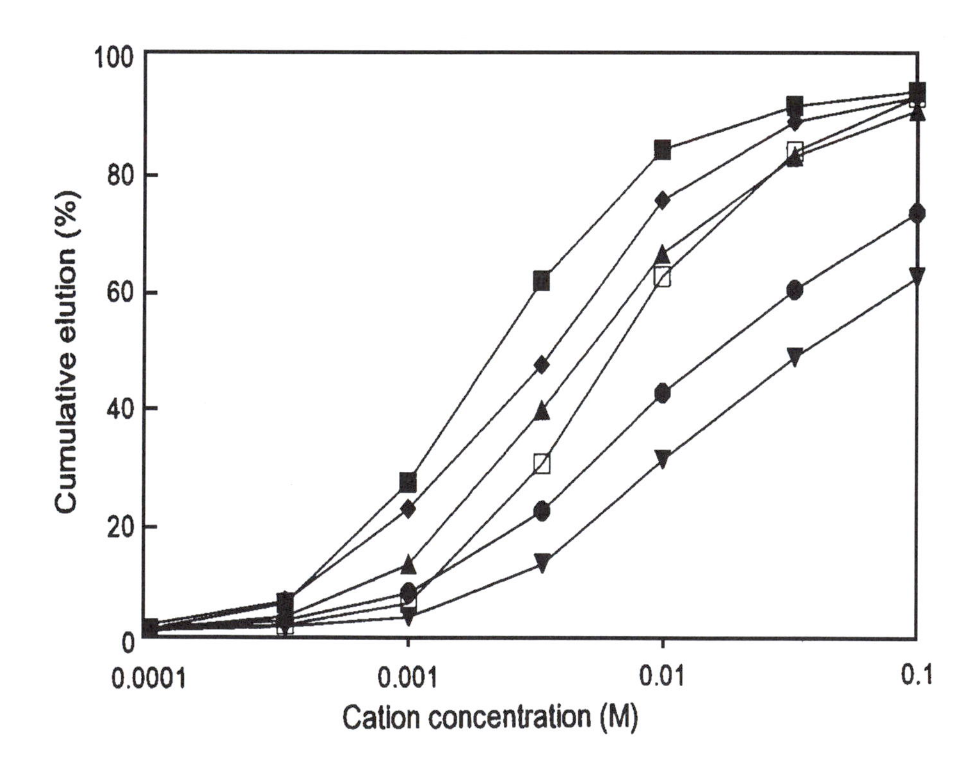

Figure 2. The cumulative elution profiles for the β-blocker propranolol eluted from a C18 cartridge with 1 mL aliquots of methanol/water (50/50) containing a series of different cations, as their acetate salts. Cations investigated are Pb^{2+} ■; Cu^{2+} ♦; K^{+} ▲; NH_4^{+} □; Na^{+} ●; Li^{+}.▼

If the basic compounds were being retained by an ionic interaction, it was postulated that they should be eluted with neutral eluents containing a displacing cation. Furthermore, the eluting power of a range of cations should correlate with their strength as predicted by classical cation-exchange chromatography.

Figure 2 shows such an experiment (Law et al., 1992) where the drug propranolol was eluted from a C18 Bond Elut cartridge with a range of cations (as acetate salts) dissolved in methanol/water (50/50, v/v). Following application in water, the propranolol was eluted from the cartridge by applying a series of solutions (1 mL) containing a single cation of increasing concentration (10^{-5} to 10^{-1} molar).

There is clearly a significant difference in the elution power of the cations, with lead and copper bringing about elution at much lower concentrations than lithium for example. With one or two exceptions, where the eluting power of the cations is ambiguous (Paterson, 1970), the relative strengths of the cations observed here agrees very well with data for the same cations in classical ion-exchange chromatography (Kraak, 1982; Simpson and Van Horne, 1993). As part of the same experiments, the organic cations tetrabutyl ammonium and triethylammonium were studied and found to show elution strength greater than or comparable to lead and copper and certainly greater than the common buffer and eluent additives such as ammonium or sodium ions.

In further studies, the use of 1:1 (v/v) methanol/water containing increasing concentrations of ammonium acetate (up to 10^{-1} molar) was found to be of insufficient strength to elute the most lipophilic compounds (e.g., tamoxifen, logP = 6.63, pK_a 8.57) even from the more polar C2 cartridge. This clearly indicated that to obtain elution it is necessary to overcome both the reversed-phase and the ionic interaction by the use of an organic modifier and a cationic displacer.

Interestingly, there is evidence that the two modes of retention do not need to be overcome simultaneously. For example, in the analysis of chlorpromazine and its metabolites using a C8 Bond Elut cartridge (Smith et al., 1987), it was possible to obtain elution with methanol by first washing the cartridge with 0.25% sodium carbonate. Washing with methanol prior to application of the sodium carbonate solution gave no elution of the analytes.

III. HYDROGEN BONDING EFFECTS

With polar phases such as cyanopropyl and aminopropyl, hydrogen bonding effects are expected and in fact this type of interaction is probably the main mode of retention. Where the stationary phase is relatively non-polar such as with C2 or C18, hydrogen bonding interactions would be considered unlikely. However, systematic experiments carried out in the author's laboratory (Law and Weir, 1992) shows that hydrogen bonding plays an important role in the retention with the shorter chain phases such as C2 and probably, though to a lesser degree, on the more lipophilic phases also.

For example, if one examines the elution properties of the two β-blockers atenolol and propranolol from C2 and C18 cartridges (Figure 3) there is clear evidence for the presence of hydrogen bonding with the polar solute atenolol on the C2 phase (Law and Weir, 1992). Propranolol, the more lipophilic compound and with poor hydrogen bonding capability, elutes far more readily (with 50% aqueous methanol containing ammonium acetate) from a C2 than a C18 cartridge, as would be expected. In contrast the highly polar atenolol elutes more readily from a C18 than a C2 cartridge. This over-retention of atenolol on C2 can be explained by the amide (CH_2CONH_2) on the ring of atenolol. This phenylacetamide moiety has strong hydrogen bond acceptor properties (Abraham et al., 1989) and on the C2 phase where the silanols are more readily accessible it is retained by both hydrogen bonding and reversed-phase interactions. Such an effect would be expected with any analyte having strong hydrogen bonding properties when extracted using a phase with a high level of readily accessible silanols.

A simple rule, then (one which we may have been able to predict from considering the nature of the bonded phase and the details of a retention process) is that over-retention of polar analytes is likely to be greater on shorter chain bonded phases and on unendcapped bonded phases.

Unlike the situation with ionic-interactions, it is not possible to overcome these secondary interactions by the addition of a competitor other than water. Since water is a much stronger hydrogen bond donor and acceptor than most organic compounds, it is an effective masking agent for the residual silanols. Thus, through the use of water rich eluents it is possible to minimize to a certain degree the hydrogen bonding effects. However, this approach would need to be balanced with the need to have a high organic concentration in the eluent to overcome the reversed-phase interactions.

There are a number of literature reports where poor elution of polar analytes have been observed when polar phases (C2 and CN) have been used (Leloux et al., 1989; Musch et al., 1989). It is probable that the effects observed in these studies were due to hydrogen bonding of the polar analytes. In addition, Mills et al. (1993) claim to have demonstrated hydrogen-bonding interactions between triazine herbicides and the sulfonic acid groups of a mixed-mode stationary phase.

In practice hydrogen bonding interactions are less of a problem for the analyst in contrast to the ionic effects discussed above. Furthermore, since the manipulation of these effects is difficult, they have to be tolerated in the knowledge that they probably lead to interesting selectivities. However, it is always a good policy to keep an open mind about the true nature of the retention mechanism and examples do exist (Bland, 1986) where it appears that hydrogen bonding secondary interactions have been used to advantage.

IV. MINIMIZATION OF SECONDARY INTERACTIONS

For many workers, secondary interactions are still considered to be a problem that should be eliminated wherever possible, and a number of strategies have been developed to overcome them. Several manufacturers have produced a range of modified phases, including partially and thoroughly endcapped materials as well as so called base-deactivated phases that are analogous to modern HPLC packings. Although they minimize the effects of residual silanols, these materials are not without their disadvantages. For example, Martin et al. (1993) compared a range of stationary phases with different degrees of carbon loading and endcapping. With the heavily endcapped materials retention of polar solutes was poor, indicating the importance of the ionic interactions to the overall retention process. Similar studies comparing standard and base-deactivated reversed phases (Martin et al., 1995) showed both good retention and recovery from the base deactivated phase which appeared to have much reduced silanol interactions.

A. THE CONDITIONING STEP AND CONTROL OF SECONDARY INTERACTIONS

The approach adopted by most workers involves the use of a silanol blocking agent in a similar manner to that used in reversed-phase HPLC. The use of buffers at the conditioning stage or co-mixed with the sample will also have a degree of silanol blocking effect, as will the endogenous components of plasma or urine. For example, Doyle et al. (1987) showed that the recovery of a basic vasodilator-β-adrenoreceptor antagonist from a Bond Elut C18 material was good from plasma and urine but relatively poor (only 27%) from water. Pretreatment of the column with an aqueous solution of di-n-butylamine (0.1%) prior to sample application allowed quantitative recovery when the compound was applied in water.

As part of a systematic study carried out in the author's laboratory (Law et al., 1992), an investigation was carried out into the effect of cartridge conditioning with a range of silanol blocking agents. Particular attention

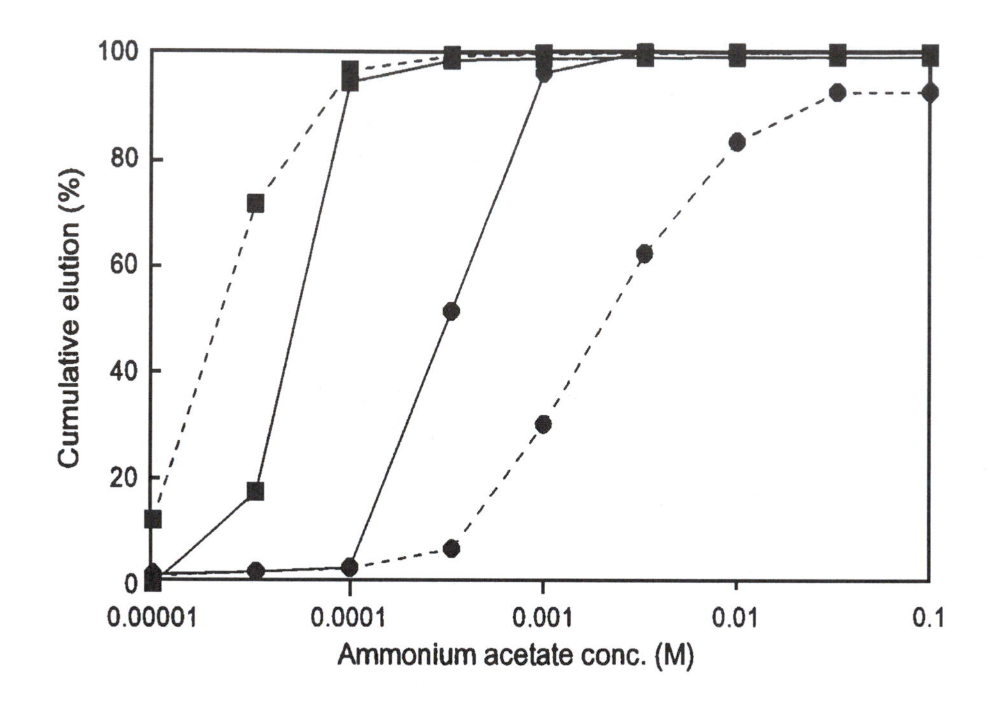

Figure 3. The cumulative elution profiles for the β-blockers atenolol and propranolol eluted from a C2 and a C18 cartridge. Elution was effected with aliquots (1 mL) of ammonium acetate of increasing concentration dissolved in methanol/water (50/50). Atenolol ■; propranolol ●; C2 solid line; C18 broken line.

was paid to those cations (e.g., sodium and potassium) that are common constituents of conditioning buffers and sample diluents.

Initial experiments involved conditioning cartridges with methanol, 50% aqueous methanol (with or without added salts, 0.1 M) and then water prior to application of the radio-labeled test compounds, either propranolol or atenolol, in water. The cartridges were then washed with water (1 mL), 50% aqueous methanol (1 mL), and finally methanol (2 x 1 mL). The application and elution volumes were separately collected and subjected to liquid scintillation counting.

The results of these experiments for propranolol on a C2 cartridge are shown in Figure 4. The use of a salt solution at the conditioning stage clearly permits facile elution of the test compound without the need to investigate complex elution systems. When the salt solution was omitted, elution of propranolol was negligible. Thus, the inclusion of the cationic species during conditioning has minimized the secondary interactions and al-

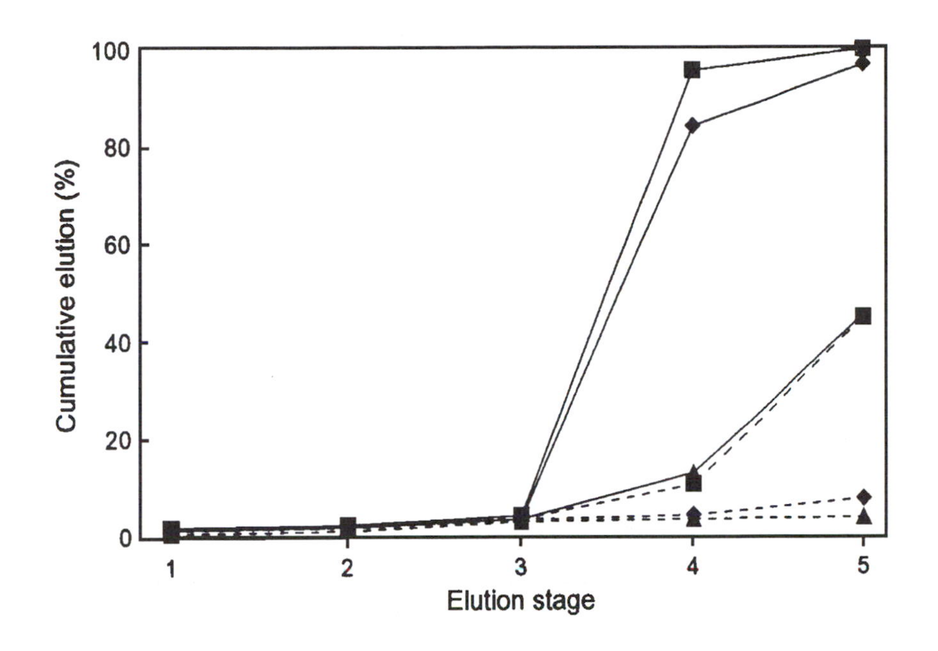

Figure 4. Cumulative elution profiles for propranolol eluted from a C2 cartridge which has been conditioned with sodium acetate (solid line) and ammonium acetate (broken line) at three different concentrations, 1.0 M ■; 0.1 M ◆; and 0.01 M ▲. Elution stage 1 is the application volume (1 mL), stages 2 to 5 refer to elution with 1 mL aliquots of: water (2), 50% methanol (3); methanol (4); methanol (5).

lowed the sorbent to function as a true reversed phase. The effectiveness of the ions at the conditioning stage was similar to that seen at the elution stage although non-parallel nature of the curves made exact comparisons difficult.

Further experiments were carried out to investigate the effect of concentration of the salt solution and the nature of the conditioning solvent. In a comparison of three concentrations (0.01, 0.1, and 1.0 M) of either sodium or ammonium acetate, the higher concentration was much more effective at eliminating secondary interactions. These results are consistent with the calculated concentrations of silanols on the stationary phase and the concentration of cations required to block them.

The influence of the solvent in which the cation was applied was also found to have a marked effect. Figure 5 shows a comparison of sodium and potassium acetate applied at a sub-optimal concentration (0.1 M) in three different solvents: water, 50% aqueous methanol, and methanol. These data clearly show that solutions of salt ions applied in methanol are far more ef-

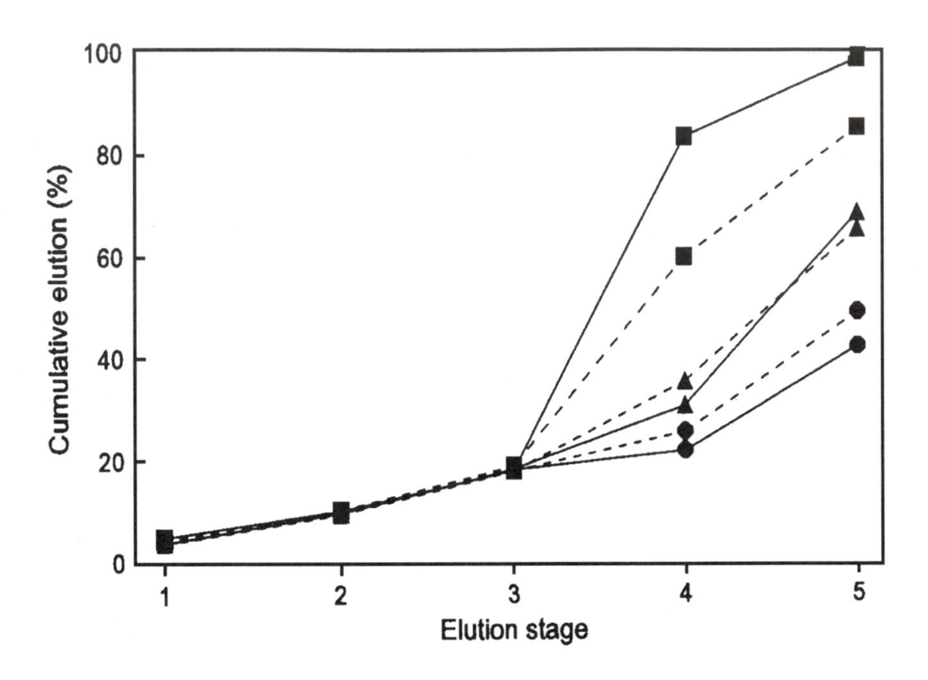

Figure 5. Cumulative elution profiles for atenolol eluted from a C2 cartridge showing the effect of conditioning with potassium acetate (solid line) and sodium acetate (broken line), both 0.1M, in one of three different solvents, methanol ■; methanol/water (50/50) ▲ and water ●.

fective conditioning agents, if our goal is to eliminate or moderate secondary interactions, than aqueous or aqueous/methanol solutions. The difference in blocking strength between the sodium and potassium ions is also clearly demonstrated.

The solvent effects are difficult to explain, but they may be related to the properties of the hydrated cations. For example, the more lipophilic hydration sphere afforded by methanol, versus that of water, may allow easier penetration of the cation through the hydrocarbon layer.

These results have important implications for the routine application of SPE and in particular the transfer of methods between laboratories or the establishment of literature methods. Many procedures involve the use of buffers, either for conditioning the cartridge, dilution of the sample prior to application, or analyte elution. Even in the recent literature, many papers still fail to report the exact make-up of the buffer used (e.g., Wells et al., 1994; Yuan et al., 1995). Bearing in mind the marked differences in the effect of sodium and potassium, this should be considered unacceptable. Furthermore, the fact that most workers employ aqueous buffers with the concentration 0.1 M or less suggests that the silanol blocking is probably incomplete in these methods, and a significant uncontrolled variable may have been introduced.

V. MIXED-MODE EXTRACTIONS

From the foregoing discussion and much of the published literature it should be obvious that the secondary interactions are actually of benefit in the development of extraction procedures. Their presence gives a greater degree and control of selectivity than would have been possible if the cartridge acted as a pure reversed phase.

The disadvantage to employing the silanols, however, is that they are actually by-products of the bonding procedure and their concentration and activity may vary from one batch of material to another (Marko et al., 1991; Law et al., 1992). Furthermore, silanols are relatively weak acids and hence their ionization is easily suppressed, such that the ion-exchange capability is lost. If a method is to be developed that deliberately utilizes the silanol interaction mechanisms, then it should be developed on sorbents that are tested by the manufacturer for normal-phase activity as well as reversed-phase activity. Some monofunctional C18 sorbents and cyano phases fall into this category.

To overcome the above problems and provide a controlled source of cation exchange sites several manufacturers have developed the so-called mixed-mode stationary phases (e.g., Clean Screen DAU™[19], Narc-2™[20] and Bond Elut Certify™[21]). These materials are silica-based with a mixed bonded phase consisting of a hydrophobic element (e.g., C8) and a strong cation exchanger, such as benzene sulfonic acid. The controlled and built-in ion exchange capability of these materials should allow more robust methods to be developed than would be possible with standard reversed-phase materials. The retention through ion exchange means that these columns can be washed with relatively strong solvents such as methanol, hence giving effective removal of anionic and neutral interferences without seriously affecting the recovery of the analyte.

Although mixed-mode materials were introduced in 1986, the number of literature reports in relation to standard reversed-phase cartridges is still relatively small. This may reflect the conservative nature of the practicing analyst, or possibly the fact that for the isolation of single basic analytes, they offer little advantage over a well-optimized procedure using a C2 or

[19] Cleen Screen is a trademark of United Chemical Technologies, Bristol, Pennsylvania.
[20] Narc-2 is a trademark of J.T. Baker Chemical Company, Phillipsburg, New Jersey.
[21] Certify is a trademark of Varian Inc., Palo Alto, California.

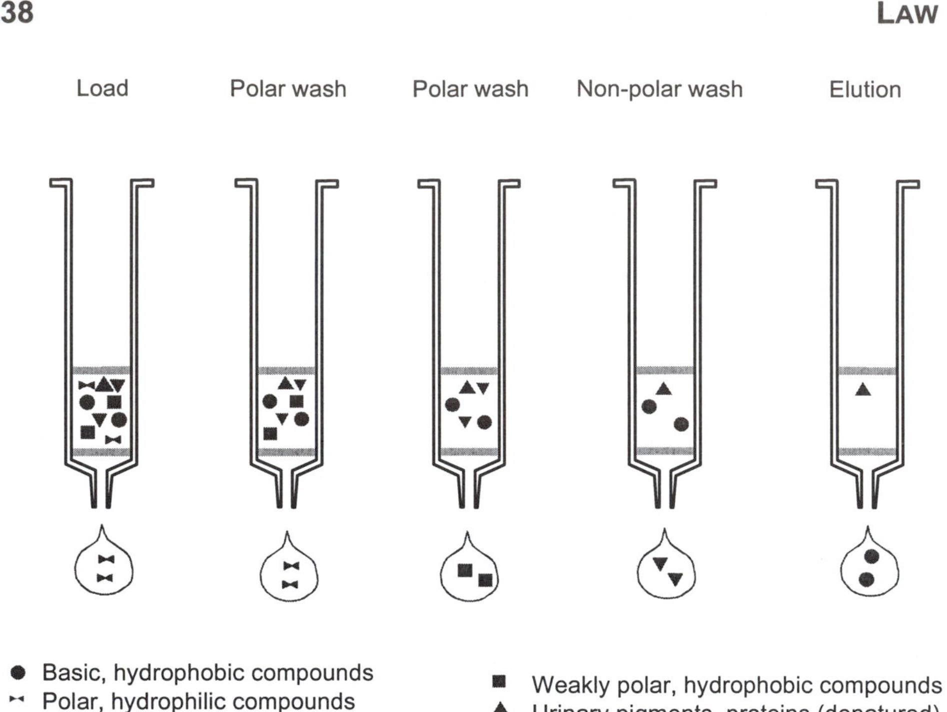

Figure 6. A schematic of a typical mixed-mode extraction of a urine sample. Sample components, grouped into chromatographic classes, as shown eluting from the cartridge (e.g., Bond Elut Certify, Chen et al., 1992a) as each wash/elution aliquot is applied.

C8 cartridge. It may also reflect the fact that these products were initially targeted by the manufacturers at the extraction of abused drugs in biological samples — a field with a narrow range of analytes and sample matrixes. There has been little discussion in the literature on the mechanisms of extraction with these cartridges, virtually all reports merely detailing the methods used, with no attempt at rationalizing the conditions used.

One area, however, where the mixed C8/SCX has been studied in some detail and where it offers an advantage over reversed-phase materials is in multi-analyte isolation procedures such as would be encountered in drug screening. By careful manipulation of the application and elution conditions, these cartridges can be used not only to extract but also to fractionate drugs into their general classes, i.e., acids, bases and neutrals.

The mixed-mode cartridges are used in a similar manner to the standard materials although the elution conditions are usually different. The following procedure, depicted in Figure 6, which is typical of that reported in the literature (Thompson et al., 1989; Chen et al., 1992a; Chen et al.,

1992b), describes how these materials can be used to separate drugs into their respective classes.

The cartridge is first conditioned in the normal manner with methanol and water, and then the analytes are applied under acidic conditions (pH approximately 6). The basic analytes ($pK_a > 6$) will be ionized and retained by ionic interactions. The acids, although partially ionized, should be retained along with the neutral drugs by hydrophobic, reversed-phase interactions, providing the application solvent is weakly eluotropic (e.g., water). The pH on the cartridge can then be adjusted by the passage of acetic acid solution (pH approx. 3) which should enhance the ionic interaction of the bases. This pH change also reduces the ionization of the acids, which along with the neutral compounds can be eluted in a non-polar solvent such as dichloromethane or acetone/chloroform. The ionization state of the stationary phase is unaffected since the sulfonic acid groups are very strong acids. Finally, by eluting with a solvent of intermediate polarity containing a small amount of ammonia, both the ionic and the hydrophobic interactions are overcome and the basic compounds are then eluted. The acetic acid solution also acts as a wash solution, improving the cleanliness of the final extracts. Depending on make of cartridge used and the exact nature of the compounds to be extracted, it is also possible to include additional washes using, for example, methanol or hexane.

Although less well known, mixed-mode materials are also available with C8/strong anion exchange (SAX) phases and mixed SCX/SAX/polar/non-polar phases. The former of these (Certify II) was specifically designed for the analysis of Δ^9-tetrahydrocannabinol-11-oic acid (Wimbish and Johnson, 1990). The latter (AccuCAT™[22]) was intended for the analysis of catecholamines (Dixit and Dixit, 1991). Neither of these have seen use as widespread as for the cation/hydrophobic phases. Some explanation as to why they are less widely used is given in Chapter 9, which deals with the application of SPE to drug screening in race horses and other animals.

VI. CONCLUSIONS

Secondary interactions between the analyte and the residual silanols on a reversed-phase extraction cartridge play an important and often dominant role in the retention and selectivity for basic solutes in SPE procedures.

The secondary cation-exchange interactions can be controlled by the use of a suitable cationic species at either the conditioning or elution stages.

[22] AccuCAT is a trademark of Varian Inc., Palo Alto, California.

It is probable also that the endogenous components of biological fluids act to minimize these interactions. The hydrogen bonding effects, although less easy to control, can lead to significant over-retention for certain polar analytes and may result in interesting selectivity differences.

A logical extension of these secondary interactions are the so-called mixed-mode phases where an ion-exchanger and a hydrophobic phase are combined to give controlled, mixed-retention mechanisms. The combined C8/SCX phase shows particular promise as a general purpose phase for the simultaneous isolation of a wide range of drug types including acids, bases, and neutrals.

REFERENCES

Abraham, M.H., Duce, P.P., Prior, D.V., Barratt, D.G., Morris, J.J. and Taylor, P.J., (1989) Hydrogen Bonding. Part 9. Solute Proton Donor and Proton Acceptor Scales for Use in Drug Design. J.Chem.Soc.Perkin Trans., 2:1355.

Bland, R., (1986) Applications of Solid Phase Extraction to the Determination of Salbutamol in Human Plasma by HPLC. Proceedings of The Third Annual International Symposium on Sample Preparation and Isolation Using Bonded Silicas, Cherry Hill, NJ.

Chen, X.-H., Wijsbeek, J., Franke, J-P. and de Zeeuw, R.A., (1992a) A Single-Column Procedure on Bond Elut Certify for Systematic Toxicological Analysis of Drugs in Plasma and Urine. J.Foren.Sci., 37:61.

Chen, X.-H, Franke, J-P., Wijsbeek, J. and de Zeeuw, R.A., (1992b) Isolation of Acidic, Neutral and Basic Drugs from Whole Blood Using a Single Mixed-Mode Solid-Phase Extraction Column. J.Anal.Toxicol., 16:351.

Dixit, V and Dixit, V.M., (1991) Sample Preparation for the Analysis of Catecholamines and their Metabolites in Human Urine. J.Liquid Chromatogr., 14:2779.

Doyle, E., Pearce, J.C., Picot, V.S. and Lee, R.M., (1987) Analysis of Drugs from Biological Fluids Using Disposable Solid Phase Columns. J.Chromatogr., 411:325.

Harrison, P.M., Tonkin, A.M. and McLean, A.J., (1985) Simple and Rapid Analysis of Atenolol and Metoprolol in Plasma using Solid-Phase Extraction and high-performance Liquid Chromatography. J.Chromatogr., 339:429.

Kraak, J.C., (1982) *Techniques in Liquid Chromatography* (Simpson, C.F. ed), Wiley, Chichester.

Law, B., Weir, S., and Ward, N.A., (1992) Fundamental Studies in Reversed-Phase Liquid-Solid Extraction of Basic Drugs; I: Ionic Interactions. J.Pharm. Biomed.Anal., 10:167.

Law, B., and Weir, S., (1992) Fundamental Studies in Reversed-Phase Liquid-Solid Extraction of Basic Drugs; II: Hydrogen Bonding Effects. J.Pharm.Biomed. Anal., 10:181.

Leloux, M.S., De Jong, E.G. and Maes, R.A.A., (1989) Improved Screening Method for Beta-Blockers in Urine Using Solid-Phase Extraction and Capillary Gas Chromatography-Mass Spectrometry. J.Chromatogr., 488:357.

Majors, R.E., (1995) Sample Preparation Perspectives: New Approaches to Sample Preparation. LC.GC, 13 (2);82-94.

Majors, R.E., (1996) Sample Preparation Perspectives: Trends in Sample Preparation. LC.GC, 14 (9);754-766.

Marko, V., (1988) Determination of Stobadine, a Novel Cardioprotective Drug, using Capillary Gas Chromatography with Nitrogen-Phosphorus Detection after its Selective Solid-Phase Extraction from Serum. J. Chromatogr., 433:269.

Marko, V., Soltes, L. and Novak, I., (1990) Selective Solid-Phase Extraction of Basic Drugs by C18 Silica. Discussion of Possible Interactions. J.Pharm.Biomed.Anal., 8:297.

Marko, V., Radova, K. and Novak, I., (1991) Variations in Solid-Phase Extraction of Basic Drugs Using Bonded Silica. II Batch-to-Batch Variations. J.Liq.Chromatogr., 14: 1659.

Martin, P., Taberner, J., Fairbrother, A. and Wilson, I.D., (1993) An Investigation of the Effects of Carbon Loading and Endcapping of the Solid-Phase Extraction of Beta-Blockers onto C_{18} Bonded Silica Gel. J.Pharm.Biomed.Anal., 11:671.

Martin, P., Morgan, E.D. and Wilson, I.D., (1995) Comparisons of the Properties of a Normal and Base Deactivated Bonded Silica Gel for Solid Phase Extraction of [^{14}C]-Propranolol. Anal.Proc., 32:179.

Mills, M.S., Thurman, E.M. and Pedersen, M.J., (1993) Application of Mixed Mode, Solid-Phase Extraction in Environmental and Clinical Chemistry. Combining Hydrogen-Bonding, Cation-Exchange and Van der Waals Interactions. J.Chromatogr., 629:11.

Moors, M. and Massart, D.L., (1991) Evaluation of Solid-Phase Extraction of Basic Drugs from Human Milk. J.Pharm.Biomed.Anal., 9:129.

Musch, G., Buelens, Y. and Massart, D.L., (1989) A Strategy for the Determination of Beta Blockers in Plasma using Solid-Phase Extraction in Combination with High-Performance Liquid Chromatography. J.Pharm.Biomed.Anal., 7:483.

Paterson, R., (1970) *An Introduction to Ion Exchange*, Heyden and Sons Ltd, London.

Ruane, R.J. and Wilson, I.D., (1987) The Use of C18 Bonded Silica in the Solid Phase Extraction of Basic Drugs - Possible Role for Ionic Interactions with Residual Silanols. J.Pharm.Biomed.Anal., 5:723.

Ruane, R.J., Wilson, I.D. and Tomkinson, G.P., (1988) *Bioanalysis of Drugs and Metabolites* (Reid, E, Robinson, J.D. and Wilson, I.D. (eds)) The Use of Secondary Interactions for the Solid-Phase Extraction of some 'β-Blocking' Drugs on C-18 Bonded Silica. p295.

Simpson, N.J.K. and van Horne, K.C., (1993) the Handbook of Sorbent Extraction Technology, 2nd Edition, Varian Sample Preparation Products, Harbor City, CA.

Smith, C.S., Morgan, S.L., Greene, S.V. and Abramson, R.K., (1987) Solid-Phase Extraction and High-Performance Liquid Chromatographic Method for Chlorpromazine and Thirteen Metabolites. J.Chromatogr., 423:207.

Soltes, L., Benes, L. and Berek, D., (1983) Selective Preseparation of Pentacaine from Biological Materials by SEP-PAK C_{18} Cartridge. Methods Findings Exp.Clin.Pharmacol., 5:461.

Thompson, B.C., Kuzmack, J.M., Law, D. W. and Winslow, J.J., (1989) Copolymeric Solid-Phase Extraction for Quantitating Drugs of Abuse in Urine by Wide-Bore Capillary Gas Chromatography. LC-GC, 7:846.

Wells, M.J.M., Riemer, D.D. and Wells-Knecht, M.C., (1994) Development and Optimization of a Solid-phase Extraction Scheme for Determination of the Pesticides Metribuzin, Atrazine, Metolachlor and Esfenvalerate in Agricultural Runoff Water. J.Chromatogr., A 659:337.

Wimbish, G.H. and Johnson, K.G., (1990) Full Spectral GC/MS Identification of Δ^9-Carboxy-Tetrahydrocannabinol in Urine with the Finnigan ITS40. J.Anal.Toxicol., 14:292.

Yuan, Z., Russlie, H.Q. and Canafax, D.M., (1995) Sensitive Assay for Measuring Amoxicillin in Human Plasma and Middle Ear Fluid using Solid-Phase Extraction and Reversed-Phase High-performance Liquid Chromatography. J.Chromatogr., B 674:93.

8

SOLID-PHASE EXTRACTION FOR BROAD-SPECTRUM DRUG SCREENING IN TOXICOLOGICAL ANALYSIS

Rokus A. de Zeeuw and Jan Piet Franke, Groningen Institute for Drug Studies, University Centre for Pharmacy, Groningen, The Netherlands

I. INTRODUCTION

There are three major tasks in the toxicological analysis of a given specimen: First, we must determine if the specimen contains any potentially harmful substance(s); second, we must identify the substance(s) involved; and third, we must determine the quantity of the substance(s) involved and interpret the result in regard to the reason for carrying out the analysis.

These three steps are mandatory in all areas of toxicological analysis, including clinical, forensic, workplace, occupational, environmental, veterinary, and food toxicology. The first two steps relate to qualitative analysis and often go hand in hand. Depending on the circumstances or the purpose, two approaches can be distinguished, namely:

- The directed search, geared to a limited number of substances such as in-workplace testing of urine samples, testing for growth promoters in animal tissue, or for alcohol in samples taken after a traffic violation or suspected drunk-driving arrest.
- The undirected search, also called systematic toxicological qualitative analysis (STA).

STA can be defined as the undirected chemical-analytical search for a potentially toxic substance (or substances) whose presence is uncertain and whose identity is unknown. We shall concentrate on the latter in this chapter, although rules will be given for deriving directed search methods and Chapter 9, dealing with forensic veterinary testing, will explore strategies for extracting specific drugs or drug classes.

Obviously, STA is required if little or no information is available as to which toxic agent is involved. However, it should be stressed that broad spectrum drug screening has a much wider area of application. Even when the toxic agent is known or where there is strong suspicion as to its nature, STA remains necessary to check for additional toxic agents hitherto not known to be present or, of equal importance, to demonstrate the absence of other toxicologically relevant agents.

It is clear that the analytical toxicologist faces a formidable task in STA. Our environment surrounds us by thousands of compounds, many of which have harmful properties, such as pesticides, herbicides, bactericides, household products, cosmetics, drugs of abuse, and doping agents. Moreover, the samples an analyst has to work with are usually based upon a very complex biomatrix, in which the substances of toxicological relevance are present in trace amounts. Against this background we can define three key steps, comprising several smaller tasks:

Sample work up/isolation/concentration: Potential techniques include hydrolysis, digestion, protein removal, liquid-liquid extraction (LLE), solid-phase extraction (SPE), supercritical fluid extraction (SFE), immuno-affinity chromatography (IAC).

Differentiation/detection: We must take into account the widest variety of substances, including possible metabolites, matrix components. The most often used techniques are immunoassays (IA) or receptor assays (RA), thin-layer chromatography (TLC) combined with color reaction, gas chromatography (GC) combined with various detector systems such as FID, ECD, NPD, MS, MS/MS, high-performance liquid chromatography (HPLC), combined with various detector systems, such as UV, DAD, ECD, MS, MS/MS.

Identification: This is commonly achieved by comparing analytical data found for the unknown substance(s), using differentiation/detection techniques with similar data for reference substances compiled in data bases.

When trying to develop suitable strategies for these tasks, the following performance goals should be kept in mind:

- Sample work-up techniques should retain all toxicologically relevant substances while, at the same time, removing all non-relevant substances and interferences.
- The techniques chosen must give maximum differentiation in a minimum amount of time and detection must be universal and sensitive.
- Adequate identification requires the presence of comprehensive and updated databases of toxicologically relevant substances. If the unknown substance under investigation is not in the database, it either cannot be identified or may be misidentified.

In this chapter, we will concentrate on sample work-up approaches for STA, particularly for organic substances such as drugs. The traditional sample work-up technique in analytical toxicology was LLE. Although LLE proved to be suitable in a substantial number of cases, the disadvantages of this technique (for example unpredictable and low recoveries or no recovery at all, matrix interferences, emulsion formation, use of large amounts of hazardous solvents) have frequently troubled the analyst. In recent years the development of suitable materials for SPE has provided a new impetus to investigation of extraction approaches. It has been demonstrated in many studies that these new media can be a very useful means for sample work-up (Abusada et al., 1993; Ellerbe et al., 1993; Harkey et al.,

1991; Kikura et al., 1992; Leyssens et al., 1991; Roy et al., 1992). Most of these publications, however, have been geared towards the isolation of one drug or a limited number of related substances (directed analysis).

In STA the undirected approach is required, in which case a sample work-up step cannot be optimized towards a given substance or a given class. Instead, it must compromise between an acceptable recovery of a great many different substances and an adequate removal of matrix compounds within a reasonably short period of time.

When we started our studies in this field in the early nineties, one of the first questions we faced was the choice of a suitable SPE material. Given the above prerequisites, and taking into account that our substances of interest can have acidic, neutral, or basic properties, we presumed that a single SPE extraction mechanism would not suffice. Therefore we opted for a so-called mixed-mode cartridge, containing a solid phase that can exert at least two types of interactions. The validity of this approach has been confirmed by subsequent research by other groups (Lin et al., 1997; Solans et al., 1995).

In view of our interest in toxicologically relevant substances, we chose a mixed-mode bonded silica cartridge in which the silanol groups were partly derivatized with medium length alkyl chains and partly with cation exchange substituents (Simpson and Van Horne, 1993). It was anticipated that at suitable pH, acidic and neutral substances would be retained by hydrophobic interactions with the alkyl chains and that the basic substances would interact with the cation exchange groups. Most of our work was performed on Bond Elut Certify®[23] cartridges, containing 130 mg of sorbent and with a reservoir of 10 ml. However, the developed procedures were found to be easily amenable to similar mixed-mode SPE cartridges from other manufacturers, such as Worldwide Monitoring Corp., Ansys Inc., and E. Merck (Platoff and Gere, 1991).

II. A GENERAL SOLID-PHASE EXTRACTION PROCEDURE

In our initial investigations towards the use of mixed-mode SPE cartridges, we concentrated on plasma and urine as matrices. For the detection and quantitation (recovery calculations), we selected gas chromatography with flame ionization detection (GC/FID). Even though FID is not the most sensitive detector for the majority of drugs containing nitrogen, sulfur, or phosphorus atoms, its universal detection properties are desirable because

---

[23] Certify is a registered trademark of Varian Inc., Palo Alto, California.

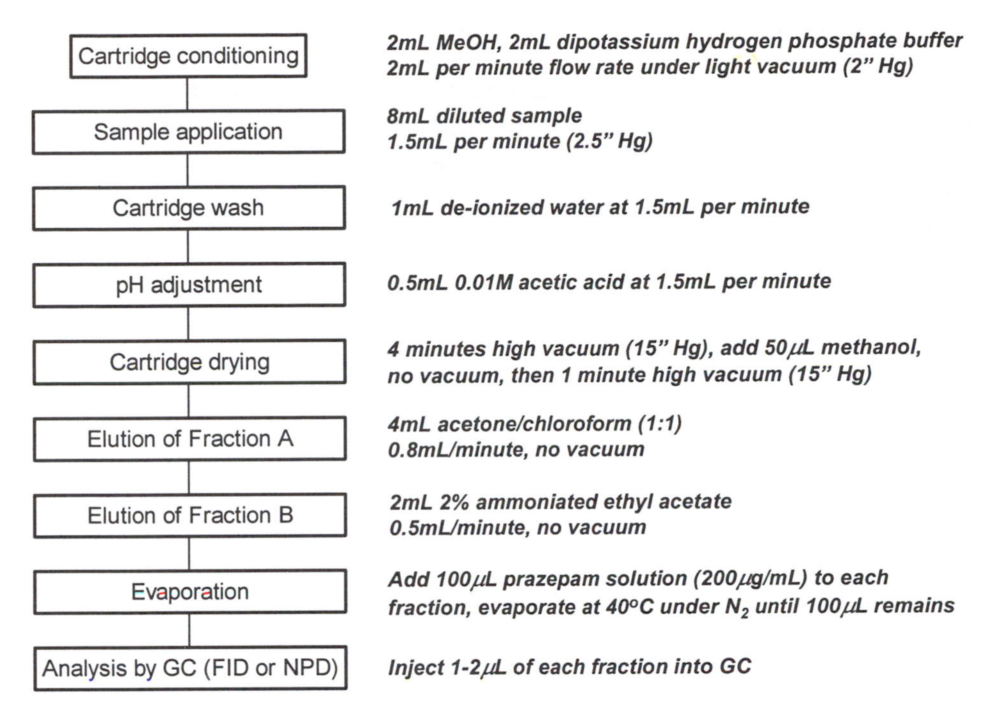

Figure 1. Overview of the solid-phase extraction procedure for plasma and urine.

this universality provides a reliable and simple way of achieving good insight into the co-extracted matrix components. This approach was taken a stage further (Wilson and Nicholson, 1987; Wilson, 1994) by employing nuclear magnetic resonance (NMR) as a detection system for ibuprofen metabolites in extracts from C18 cartridges of biological fluids. In this case, non-specific proton NMR was used to show all the co-extractants, including several that may not have appeared in a GC chromatogram. Figure 1 gives a schematic overview of the general procedures for Bond Elut Certify cartridges with a bed mass of 130 mg and a sample reservoir of 10 ml (Chen et al., 1992A). What follows, it should be remembered, is not the only way to use these mixed-mode SPE cartridges; however, it was found to yield excellent results in our laboratory. The following steps are required, using a suitable vacuum manifold:

A. CARTRIDGE CONDITIONING

The cartridge is prepared by conditioning with 2 mL of methanol, followed by 2 mL of phosphate buffer (0.1 mol/L, pH 6.0), preferrably potassium phosphate (the importance of the counter ion is demonstrated in Chapter 7). The cartridge must not become dry before sample application.

B. SAMPLE PRETREATMENT AND APPLICATION

Blood plasma or urine (2 mL) containing the analyte is diluted with 6 mL of the same phosphate buffer used to condition the cartridge, and the mixture is briefly vortexed. The diluted sample is applied to the top of the cartridge and pulled through slowly under light vacuum at a flow rate of approximately 1.5 mL/min.

C. CARTRIDGE WASH

Washing of the cartridge takes place with 1 mL of deionized water.

D. pH ADJUSTMENT

The pH of the extraction system is adjusted to a slightly acidic pH value by applying 0.5 mL of 0.01 mol/L acetic acid (pH=3.3). Many drugs have acidic or basic properties, and it can be anticipated that their cartridge retention and elution behavior will be affected by the pH of the extraction system. At pH 3.3, acidic, neutral, and some weakly basic drugs behave as relatively nonpolar compounds, and are retained on the cartridge by the hydrophobic groups of the sorbent, while other basic drugs behave as charged species, and are adsorbed by the negative ionic groups of the sorbent.

This brings up an important point: Plenty of data is available on the properties of commonly abused drugs. The pK_a values of functional groups and solubility, for example, have been measured with great accuracy. Care must be taken when applying these data during the SPE method development process. For example, quoted pK_a values pertain to very specific conditions of temperature, concentration, and environment. The effective acidity or basicity of a functional group close to a bonded silica surface may be very different. Consequently, you may use quoted values of molecular properties as a guide for selection of experimental conditions of pH, say, but it is always wise to verify the selection by experiment.

E. CARTRIDGE DRYING

The sorbent bed is dried by sucking air through the cartridge at a vacuum of at least 10″Hg, for 4 minutes, followed by the application of 50μL methanol and a second vacuum drying step for 1 minute. Cartridge drying is very important when using GC as the final method of analysis as water in the extract may damage the GC injection liner and column, or disrupt a post-extraction derivatization step.

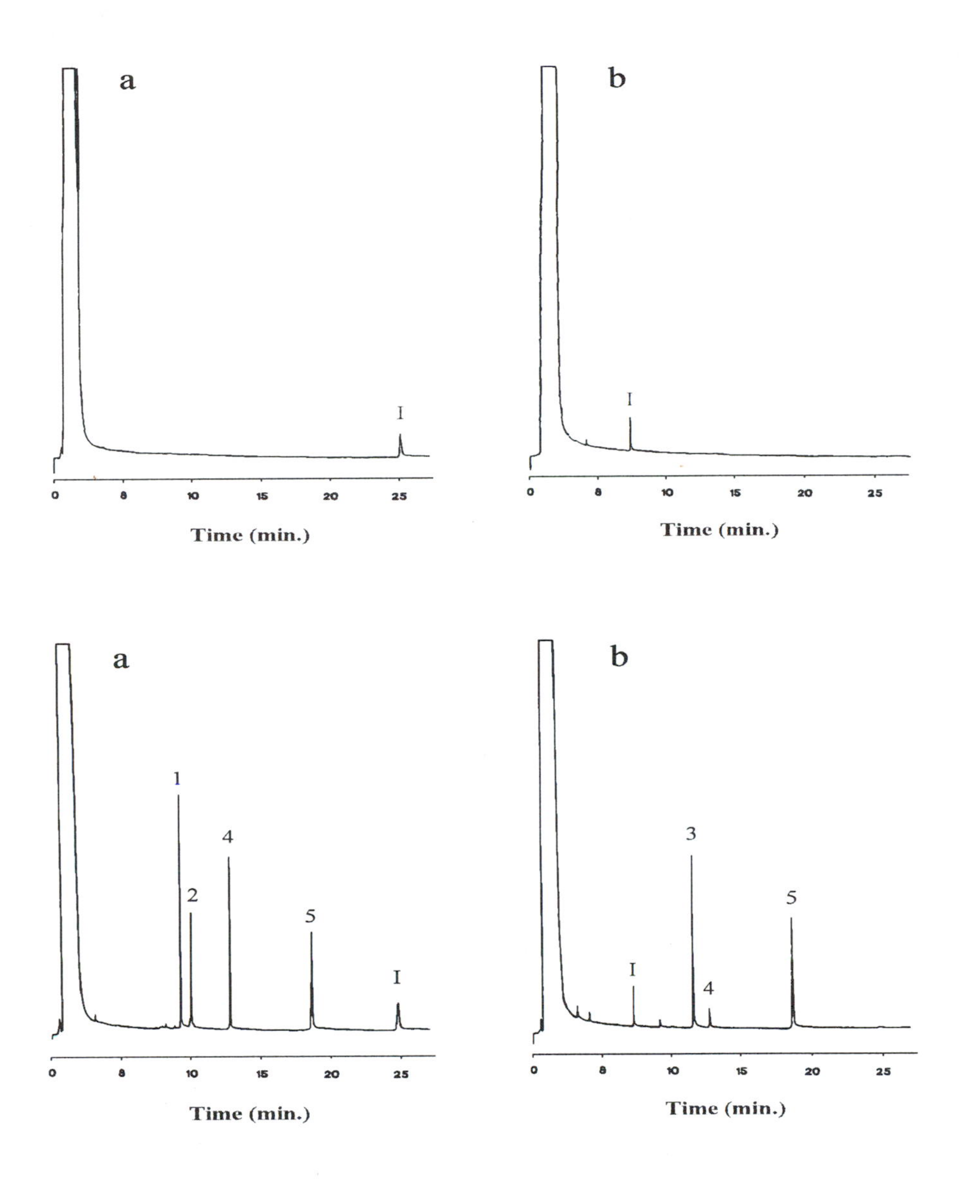

Figure 2. Chromatograms of Fractions A and B from blank (top) and spiked (bottom) plasma, using GC/FID. 1=butalbital, 2=meprobamate, 3=procaine, 4=methaqualone, 5=prazepram (standard), I=impurity.

F. ELUTION OF FRACTION A

The analytes retained by the hydrophobic groups of the sorbent are eluted using 4 mL of chloroform/acetone (1:1) at a flow rate of 0.8 mL/min.

G. ELUTION OF FRACTION B

The basic substances retained by the cation exchange groups of the sorbent in their protonated form are eluted by 2 mL ethyl acetate containing 2% of ammonia (33%) at a flow rate of 0.5 mL/min. A slow elution is important as, compared to those retained by van der Waals interactions, the displacement of species retained by ion exchange is slow.

The time required to extract ten samples simultaneously, from cartridge conditioning through to elution of fraction B, is approximately 30 minutes.

H. EVAPORATION

The two fractions are evaporated separately. Evaporation to dryness has the advantage that, upon re-dissolution, one can chose an organic solvent that is optimal for the subsequent chromatographic analyses, such as TLC, GC, and HPLC. Remaining traces of water in the extracts will also be removed. However, evaporation to dryness often results in loss of the more volatile analytes such as amphetamines or barbiturates from the extracts. Therefore, especially for further TLC or GC analysis, we prefer to add 100 μL of a prazepam internal standard solution (200 μg/ml in methanol/ethyl acetate 1:1) to each extract and then evaporate the extracts under nitrogen down to a volume of approximately 100 μL. Thus, prazepam serves as a chromatographic internal standard (Chen et al., 1992A) and not as an extraction internal standard.

An alternative approach is to use a keeper solvent, such as dimethyl formamide or toluene, or some other solvent that can azeotrope the water out of the eluent (Dixit and Dixit, 1991). One other technique that is sometimes applicable is to adjust the pH of the eluent such that the volatile species becomes a salt. This approach has been used for amphetamine extraction (Dixit et al., 1992).

III. SPECIFIC CASES

The results from this general approach, combined with different detection systems and applied to different sample matrixes, are shown in the following sections.

A. SPE OF DRUGS IN PLASMA AND URINE, FOLLOWED BY GC/FID ANALYSIS

Table 1 presents extraction yields of drugs at a concentration of 10 μg/mL from plasma, showing that acidic and neutral drugs (barbiturates, mepro-

Table 1. Recoveries of 25 drugs from calf plasma, at a spiking level of 10 μg/ml

Drug	Recovery (%), n = 5		Total recovery (%)	RSD[a] %
	Fraction A	Fraction B		
Metharbital	95.1	ND	95.1	4.5
Probarbital	82.4	ND	82.4	10.2
Pentobarbital	100.2	ND	100.2	6.3
Butalbital	93.3	ND	93.3	5.4
Secobarbital	100.6	ND	100.6	3.8
Hexobarbital	96.1	ND	96.1	5.0
Heptabarbital	100.5	ND	100.5	2.7
Oxazepam	91.1	ND	91.1	5.4
Lorazepam	93.8	ND	93.8	7.0
Diazepam	83.9	14.6	98.5	1.9
Clonazepam	87.1	ND	87.1	3.8
Nitrazepam	58.6	35.2	93.8	7.2
Methaqualone	87.0	11.8	98.8	4.9
Meprobamate	99.2	ND	99.2	5.5
Amphetamine	ND	98.9	98.9	8.9
Methamphetamine	ND	90.2	90.2	6.8
Mepivacaine	ND	103.7	103.7	4.2
Trimipramine	ND	105.5	105.5	3.5
Levallorphan	ND	101.6	101.6	4.0
Procaine	ND	98.6	98.6	5.0
Promethazine	ND	95.6	95.6	5.3
Cocaine	ND	96.2	96.2	6.2
Imipramine	ND	94.4	94.4	3.5
Codeine	ND	92.3	92.3	2.5
Morphine	ND	97.8*	97.8	5.2

[a] Relative standard deviation, %
ND = not detected.
- = eluted with two 2 mL aliquots of 2% ammoniated ethyl acetate. The first 2 mL of solvent eluted 61.9% of morphine and the second eluted 35.9% of morphine.

bamate) were eluted in Fraction A, while basic drugs were eluted in Fraction B. Weakly basic drugs with pK_a values near 3, such as methaqualone (2.5) and the benzodiazepines diazepam (3.3) and nitrazepam, eluted in both fractions.

The total recoveries obtained were, with the exception of probarbital and clonazepam, all between 90 and 100% with relative standard deviations of 10% or less. The advantage of two separate fractions is that acidic and basic substances with similar GC behavior will not coincide or overlap (e.g mepivacaine and heptabarbital). Figure 2 gives an example of the chromatograms of blank and spiked plasma. Using the same extraction scheme, urine samples gave comparable recoveries, as shown in Table 2.

Table 2. Recoveries of 7 drugs from human urine, spiking level 10μg/mL

Drug	Recovery (%), n = 4		Total recovery (%)	RSD[a] (%)
	Fraction A	Fraction B		
Pentobarbital	97.0	ND	97.0	4.5
Secobarbital	97.1	ND	97.1	2.0
Hexobarbital	99.2	ND	99.2	1.6
Diazepam	65.5	32.5	98.0	5.0
Methamphetamine	ND	97.9	97.9	5.7
Mepivacaine	ND	104.3	104.3	2.7
Imipramine	ND	101.7	101.7	1.9

[a] Relative standard deviation, %

B. SPE OF DRUGS IN PLASMA AND URINE, FOLLOWED BY GC/NPD ANALYSIS

Using GC/FID, drugs were determined at a concentration level of 10 μg per milliliter of biological fluid. These levels reflect typical therapeutic concentrations for the acidic drugs (e.g., barbiturates, meprobamate), and toxic concentrations for the basic drugs. For the deter-mination of basic drugs at therapeutic levels, the sensitivity of the GC/FID was too low. Thus, for the final analysis step CG/NPD was evaluated to determine basic drugs at lower concentrations (100-200 ng/mL) reflecting therapeutic levels (Chen et al., 1992B). It appeared that for plasma the same extraction procedure could be used for both GC/FID and GC/NPD. This resulted in clean extracts with extraction yields of 77-102% and relative standard deviations less than 7.3% (Table 3).

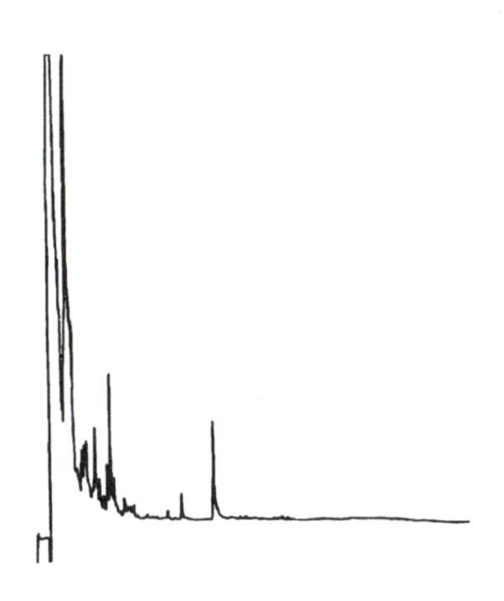

Figure 3. Chromatogram of Fraction B from blank urine without an extra wash step, using GC/NPD.

The use of this procedure for urine samples, however, resulted in dirty extracts, especially in the first part of the chromatogram (Figure 3). By introducing an extra wash step in the SPE procedure with 1.0 mL of 20% acetonitrile in water, between the cartridge wash and pH adjustment steps in the scheme of Figure 1,

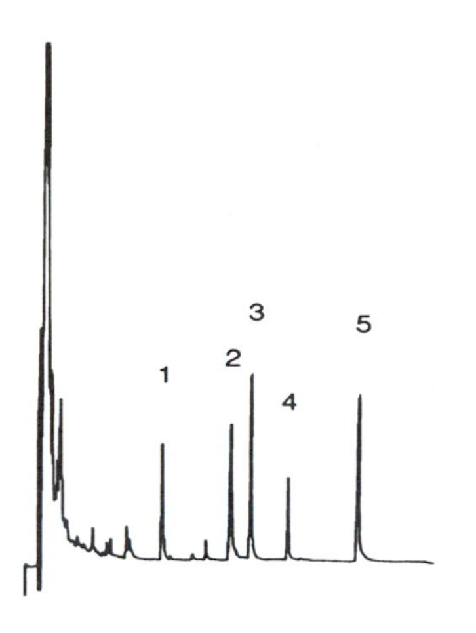

Figure 4. Chromatogram of Fraction B from spiked urine, 200 ng/ml of each drug, with an extra wash step in the SPE procedure. Analysis by GC/NPD. 1=phencyclidine, 2=methadone, 3=trimipramine, 4=codeine, 5=prazepam (chromatographic standard).

interfering compounds could be removed without influencing the re-coveries of the basic analytes under investigation (Figure 4; Table 3). In this way, basic drug recoveries of over 80% with relative standard deviations of less than 7.3% were obtained. Given the diversity of analytes extracted using this application it is tempting to propose that this 20% acetonitrile/80% water wash can be used indiscriminately when polar species are found to interfere with the analysis of a urine sample.

C. SPE OF DRUGS IN WHOLE BLOOD

Since whole blood or diluted whole blood will clog the extraction cartridge, the whole blood sample has to be pretreated before application to the extraction cartridge. Several methods have been described for the precipitation of the proteins, including blood cells. It appeared that this type of pretreatment lowered extraction yields by 50% or more (Chen et al., 1992). Based on the work of Tebbett (1987), a pretreatment procedure was developed using sonication (15 min) of the blood sample (1 mL), followed by dilution with 6 mL potassium phosphate buffer and centrifugation. The sonication process disrupts the cell membranes to the extent that no clogging occurs when the supernatant after centrifugation is applied to the conditioned SPE cartridge. The analytes are then eluted in two fractions, as described in Figure 1.

Table 4 shows that at a concentration level of 2 µg/mL good extraction yields were obtained (over 80%) with good reproducibilities (RSD less than 8.2%). Apparently, the sonication also helps to release drugs from their protein binding sites. Moreover, the chromatograms (GC/FID) are virtually free from interferences (Figure 5).

Table 3. Recoveries of basic test drugs from plasma and urine using GC/NPD at concentrations of 200 ng/mL or 100 ng/mL (N = 3).

Drug	Plasma		Urine	
	Recovery (%)	RSD[b] (%)	Recovery (%)	RSD[b] (%)
Clonazepam	90.5	6.38	84.7	4.48
Cocaine	95.6	2.88	NT[c]	-
Codeine	96.1	0.50	91.9	7.90
Cyclizine	85.5	6.01	94.2	4.49
Dibenzepin	90.9	4.58	104.5	1.20
Diphenhydramine[a]	88.5	5.98	85.1	1.21
Dipipanone	88.0	6.76	NT	-
Imipramine	90.9	2.47	NT	-
Levallorphan	95.5	2.40	NT	-
Loxapine	81.8	0.70	81.9	2.49
Mepivacaine	102.4	0.96	NT	-
Methadone	98.2	0.97	91.4	2.49
Mianserin[a]	98.6	3.05	95.4	2.73
Nomifensine	86.6	7.31	89.3	4.31
Phencyclidine	82.4	2.61	83.2	5.86
Promazine	77.7	3.22	85.3	2.72
Trimipramine	88.6	0.87	84.5	5.18
Tripelennamine[a]	99.4	2.27	89.0	0.53

[a] At a concentration of 100 ng/mL
[b] Relative standard deviation
[c] NT = not tested

D. SPE OF DRUGS IN TISSUES

The analysis of tissue samples, particularly liver and brain, is an important area in forensic toxicology, but suitable work-up and extraction methods for broad-spectrum screening purposes are extremely difficult to develop, even when using SPE. Homogenized tissue samples cannot be applied directly onto the SPE cartridges because this will result in clogging of the cartridges. An alternative SPE technique, matrix solid-phase dispersion, is described in Chapter 13. We opted to work with commercially available SPE cartridges and therefore investigated various pretreatment procedures, using liver tissue as an example (Huang et al., 1996).

Table 4. Extraction yields of fifteen drugs extracted from whole calf blood at a spiking level of 2μg/mL.

Drug	Extraction yield (%), n = 3			RSD[a] (%)
	Fraction A	Fraction B	Total	
Amitriptyline	ND[a]	98.2	98.2	7.90
Butetamate	ND	87.5	87.5	8.18
Codeine	ND	90.9	90.9	3.16
Glutethimide	90.5	ND	90.5	7.75
Hexobarbital	91.9	ND	91.9	5.10
Ketazolam	89.2	ND	89.2	2.46
Lidocaine	ND	94.4	94.4	4.35
Mepivacaine	ND	87.7	87.7	3.25
Methadone	ND	95.0	95.0	1.21
Methamphetamine	ND	91.5	91.5	6.31
Methaqualone	74.3	28.2	102.5	6.10
Pentobarbital	93.1	ND	93.1	4.31
Phencyclidine	ND	81.2	81.2	8.14
Propiomazine	ND	90.0	90.0	1.99
Trimipramine	ND	102.5	102.5	2.22

ND = not detected
[a] Relative standard deviation, %.

Sonication of the homogenized liver sample followed by centrifugation and application of the clear supernatant to the mixed-mode SPE cartridge, using the procedure of Figure 1, gave suitable recoveries for most acidic and neutral drugs, but the recoveries for basic drugs were unacceptably low. Apparently, the latter remain bound to the particulate cell material and are subsequently removed in the pellet. We therefore subjected the pellet to enzymatic digestion, using the protease Subtilisin Carlsberg (Sigma, St. Louis, MO) in a phosphate buffer at pH 10.5. Digestion was carried out for 1 hour at 60°C, the pH was then adjusted to 6 with phosphoric acid, the suspension was centrifuged at 6000 g for 10 minutes and the supernatant was applied to the SPE cartridge, from which the acidic/neutral fraction had already been eluted. The result is a suitable screening procedure for liver samples, as depicted in Figure 6. Extraction yields for a variety of different drugs from spiked tissues are given in Table 5. Chromatograms of Fractions A and B are shown in Figure 7.

It should be noted that enzymatic digestion of the homogenized liver and direct application to the SPE cartridge at step 2 of the procedure gave unsatisfactory results. The chromatogram of Fraction A thus obtained showed many interfering peaks, conceivably lipophilic compounds such as fatty acids from the liver matrix. In the procedure described in Figure 6, the fatty acids are washed away in step 8 of the SPE procedure. Another

Table 5. Extraction yields of twenty drugs extracted from liver tissue after enzymatic digestion. Spike level 2μg/100mg wet tissue.

Drug	Recovery (%)			RSD[a] (%)
	Fraction A	Fraction B	Total	
Allobarbital*	80.3	ND	80.3	1.8
Benzocaine	45.5	ND	45.5	3.6
Codeine	ND	87.3	87.3	6.2
Diazepam	49.4	35.3	84.9	3.9
Doxepin	ND	94.7	94.7	2.5
Flunitrazepam	68.8	14.6	83.4	7.4
Glutethimide*	84.2	ND	84.2	2.5
Heptabarbital	101.8	ND	101.8	7.6
Ketazolam*	51.5	18.5	70.0	7.5
Lidocaine	ND	86.6	86.6	1.4
Mepivacaine*	ND	91.0	91.0	1.4
Meprobamate	100.2	ND	100.2	2.6
Methamphetamine	ND	74.6	74.6	4.8
Methadone	ND	86.5	86.5	5.7
Methaqualone	75.1	8.5	83.6	4.3
Nortriptyline	ND	87.5	87.5	8.4
Pentobarbital	79.7	ND	79.7	3.5
Promethazine	ND	52.1	52.1	5.0
Secobarbital	86.4	ND	86.4	5.1
Trimipramine*	ND	81.4	81.4	3.3

[a] Relative standard deviation (%), based on n = 4, substances marked with * n = 3.
ND = not detected

interesting observation was that the use of an alkaline solution of Subtilisin Carlsberg in Tris-buffer [tris (hydroxymethyl) aminomethane] as recommended by Osselton (1977), gave interfering peaks in the chromatogram of Fraction B with retention times between 7-7.5 min. When Subtilisin Carlsberg was dissolved in 1 M phosphate buffer pH 10.5, these interferences were no longer present, but the efficiency of the digestion itself was not effected. Finally, the amount of liver introduced in the sample work-up procedure appeared to be critical — it should not exceed 100 mg wet weight. Higher amounts of liver resulted in lower recoveries of various drugs, indicating that the maximum binding capacity of the SPE cartridges had been reached. Yet, when 100 mg of wet tissue is used, extraction and detection of drugs at the 1 μg/100 mg tissue is possible with CG-FID. If desired, lower concentrations can be detected with CG-NPD or CG-MS.

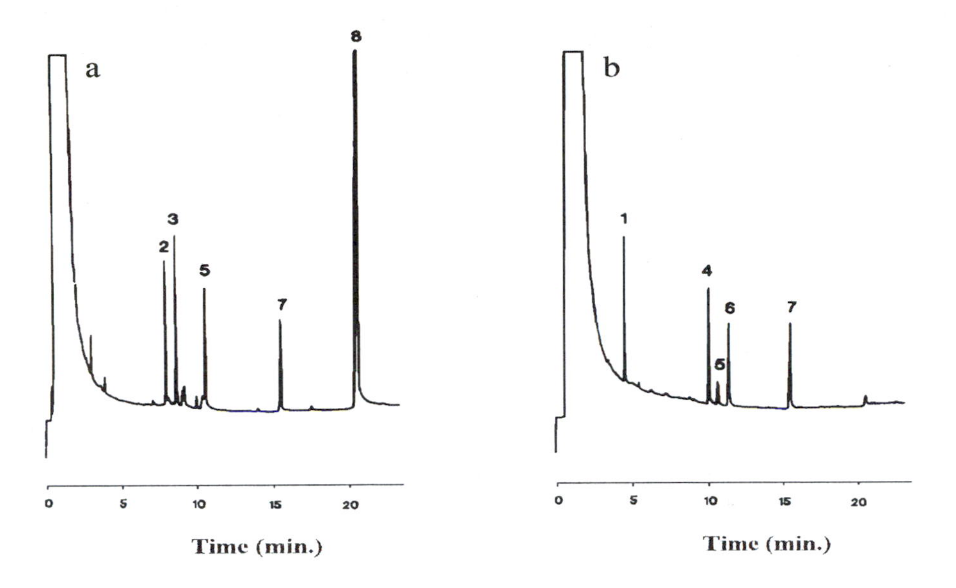

Figure 5. Chromatograms of Fractions A and B from spiked blood, using GC/FID. 1=methamphetamine, 2=pentobarbital, 3=hexobarbital, 4=mepivacaine, 5=methaqualone, 6=trimipramine, 7=prazepam (standard), 8=cholesterol.

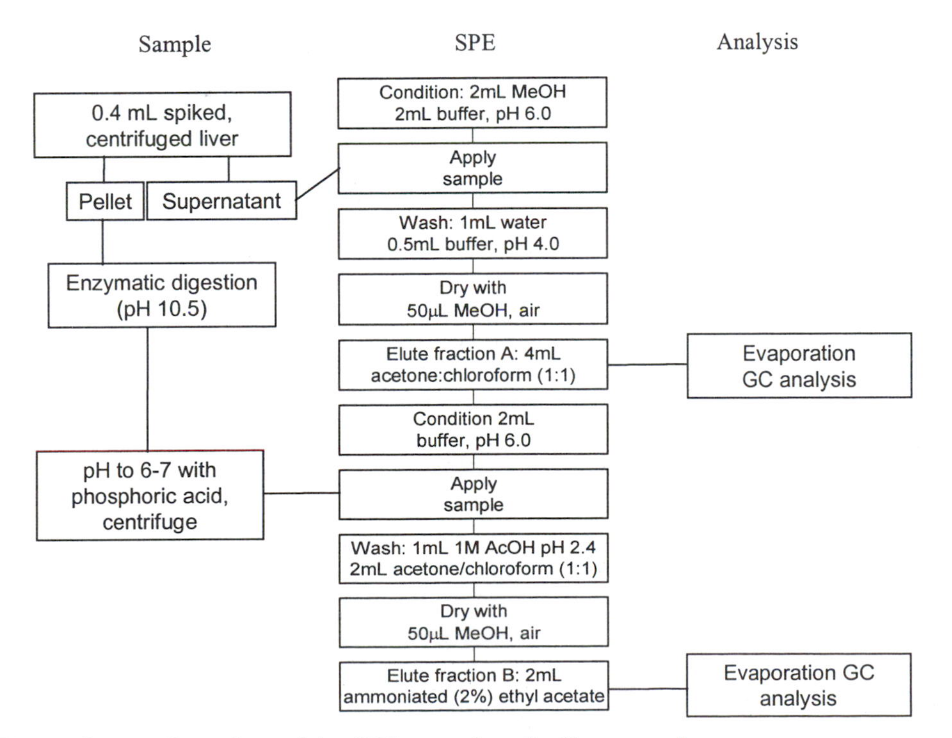

Figure 6. Overview of the SPE procedure for liver samples.

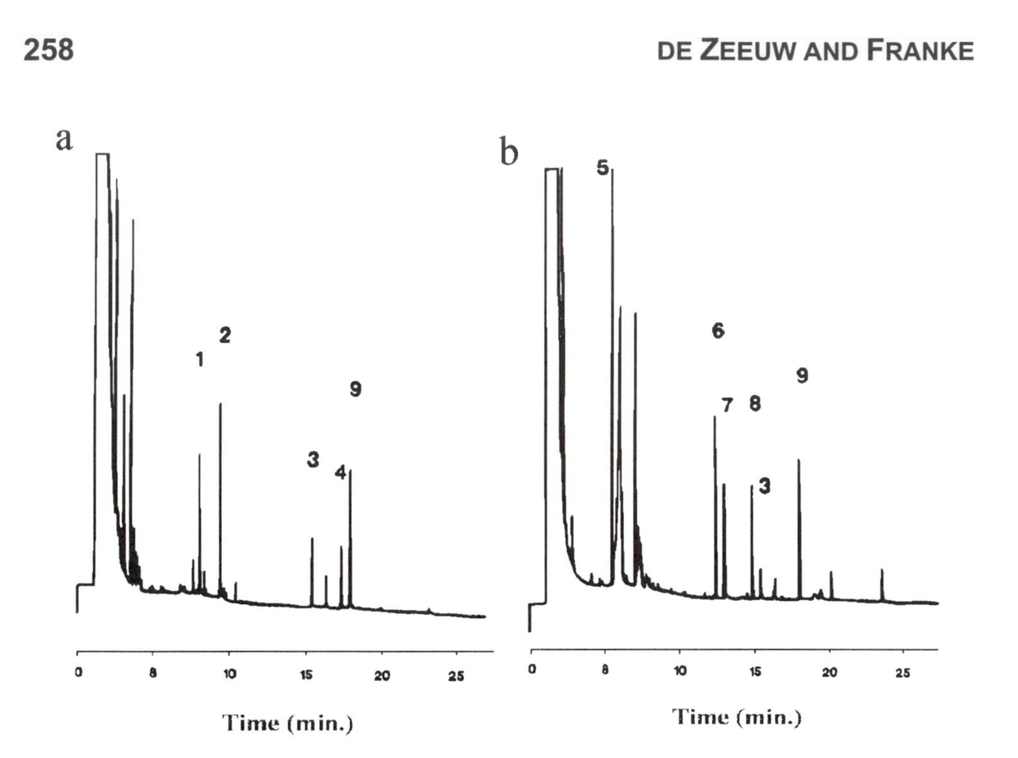

Figure 7. Chromatograms of Fractions A and B from spiked liver tissue, using GC/FID.

IV. GENERAL CONSIDERATIONS

Several other factors impinge upon the problem of systematic toxicological analysis and directed searching for drug residues. Cost of the analysis, reliability of the commercially available products, and "external" costs such as glassware usage and safety overheads are all important matters for a laboratory manager to consider. What follows addresses two of these concerns.

A. REUSABILITY OF SPE CARTRIDGES

Cost always remains an issue when looking for appropriate analytical methods. Although SPE cartridges are relatively inexpensive, we investigated the reusability of Bond Elut Certify cartridges for the extraction of toxicologically relevant drugs from plasma. The cartridges were found to be reusable after regeneration according to Figure 8 (Chen et al., 1993C). Table 6 lists recoveries of five test drugs as a function of increasing numbers of reuse. It can be seen that the extraction capacity of the cartridges decreased slightly with the number of reuses, so that a single cartridge should not be used more than 3-4 times. The reusability of cartridges for the extraction of

Table 6. Recoveries of drugs, spiked at 10μg/mL, from plasma, as a function of increasing re-use of the SPE cartridge (n = 4).

Drug	New	1st re-use	2nd re-use	3rd re-use	4th re-use
	($X^a \pm SD^b$)	($X^a \pm SD^b$)	($X^a \pm SD^b$)	($X^a \pm SD^b$)	($X^a \pm SD^b$)
Pentobarbital	97.6±6.0	94.7±9.8	85.3±3.9	80.3±11.9	74.0±5.9
Hexobarbital	94.0±9.1	90.0±8.1	86.9±2.0	83.8±4.9	78.2±1.4
Mepivacaine	100.1±2.6	96.9±4.3	93.1±4.5	88.6±7.5	79.2±3.7
Trimipramine	101.5±4.1	107.8±4.9	106.1±2.3	99.0±6.6	96.9±5.7
Clonazepam	97.4±9.0	102.4±8.1	100.4±7.3	100.4±5.3	96.6±2.3

[a] X = mean recovery (%).
[b] SD = standard deviation.

other matrices, such as whole blood or tissues, has not been investigated so far.

Labor costs, solvent usage and solvent disposal cost should always be compared with the cost of the cartridge when re-use is attempted, and great care must be exercised to ensure that there is absolutely no carry-over between samples extracted on the same cartridge.

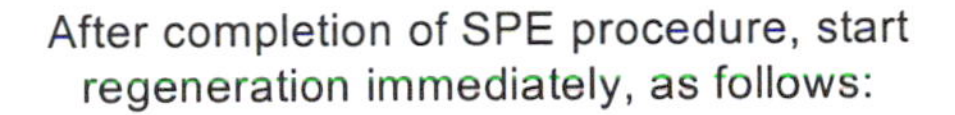

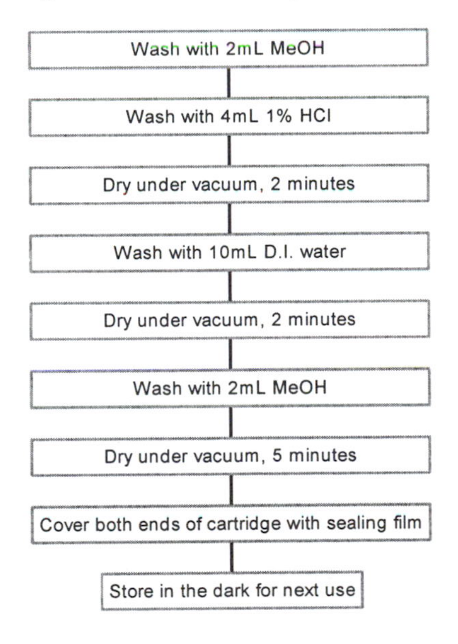

Figure 8. Regeneration program for Bond Elut Certify Cartridges.

B. LOT-TO-LOT REPRODUCIBILITY OF SPE CARTRIDGES

The extraction yields for a given experiment described in the paragraphs above were always determined with one batch of SPE cartridges. When these procedures are being used for routine work or when these procedures are adopted by other laboratories the lot-to-lot reproducibility becomes an important factor. Lot-to-lot reproducibility of the cartridge materials was studied by extracting five drugs from whole blood using 12 lots of Bond Elut Certify cartridges and six lots of CleanScreen™[24] DAU cartridges. The Bond Elut

[24] CleanScreen is a trademark of Worldwide Monitoring Corporation, Horsham, Pennsylvania

Certify cartridges were analyzed in two groups, the six lots of group II having been tested one year after the six lots of group I. Figures 9 and 10 show that good extraction yields can be obtained with excellent reproducibilities. Analyses of variance were performed on these data, again indicating a good lot-to-lot reproducibility. From the 12 lots of Bond Elut Certify cartridges, only one lot showed a somewhat higher recovery. This appeared also to be the case for the six lots of CleanScreen DAU cartridges, where one lot showed somewhat higher extraction yields (Chen et al., 1993A).

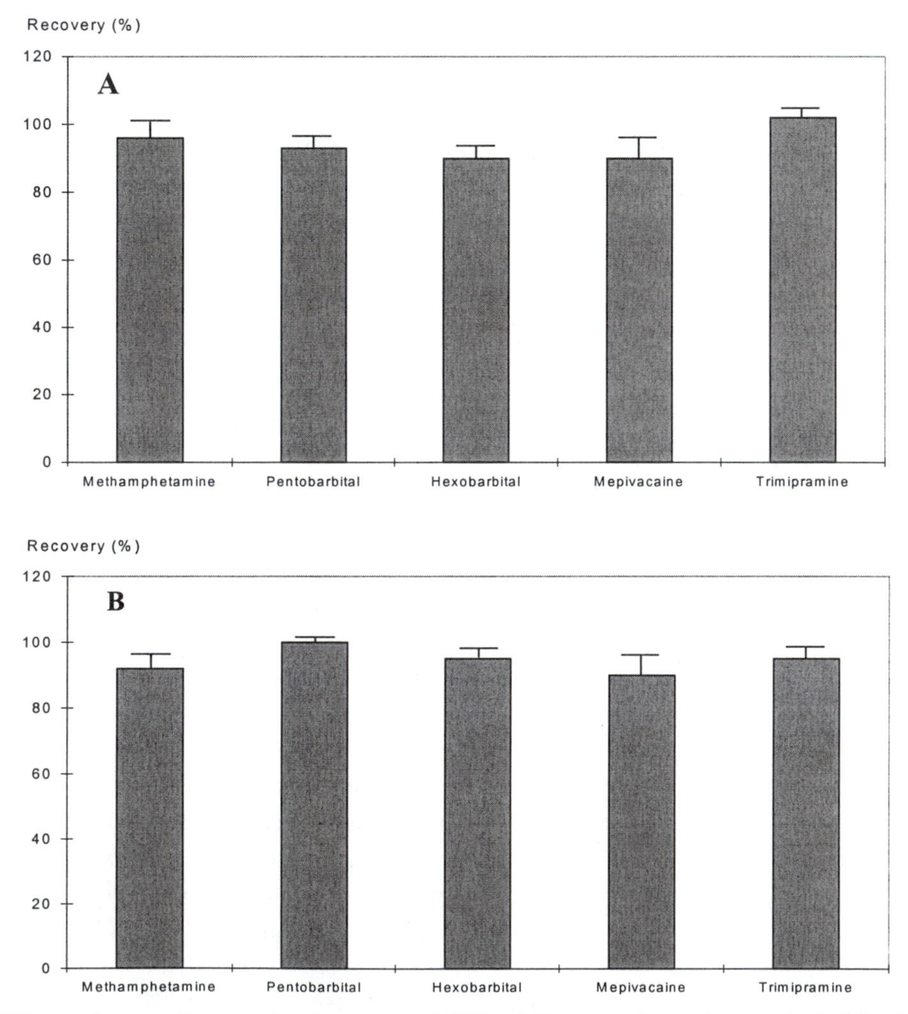

Figure 9. Recoveries (means and SD) of five test drugs from whole blood for two groups of Bond Elut Certify Cartridges. A: Group I, six different lots. B: Group II, six different lots produced one year later.

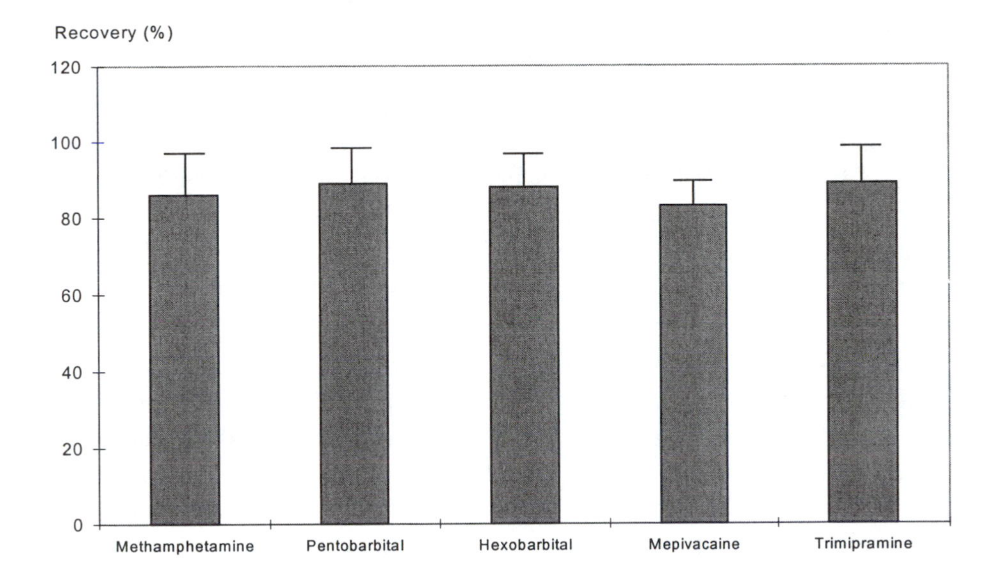

Figure 10. Recoveries (means and SD) of five different test drugs from whole blood for six lots of CleanScreen DAU.

V. AUTOMATION AND SUBSEQUENT METHOD MODIFICATIONS

The SPE procedure developed for screening is a rather complex and labor-intensive one, with cartridge conditioning, sample application, wash steps, cartridge drying, and two separate elution steps. All these steps have to be carried out with care to achieve reliable and reproducible results.

Automation may improve the experimental precision, may decrease the amount of repetitive work performed by technicians and may result in a higher throughput. To test these assertions a fully automated procedure was developed, performed on an Automated Sample Preparation with Extraction Cartridges (ASPEC) system (Chen et al., 1993B). A current list of other manufacturers or distributors of SPE automation is supplied in Appendix A of this book. For this complex extraction procedure no standard program was available. Therefore, we developed a program of our own through the Gilson Manager Software. The sample pretreatment step and the solvent evaporation step were carried out off-line.

A. AUTOMATED SPE PROCEDURE FOR DRUG SCREENING IN PLASMA AND WHOLE BLOOD

Two options for automated SPE exist: 1) sequential extraction or; 2) batch extraction (also known as parallel extraction). The ASPEC system (Gilson Medical Electronics, Villiers le Bel, France) is designed to carry out extractions in either a batch mode or in a sequential mode. In the batch mode, each step in the extraction procedure is carried out on all samples in the sample rack. In the sequential mode, the entire procedure is performed sample by sample. The sequential mode was chosen for this work — the automated system accomplishes the complete extraction process on the first sample and then continues with the next sample. The batch mode is not suitable when a large number of samples are to be prepared and large volumes of solvents, buffer, sample, and eluents are required. This results in too long a period of time between application of solvents to the first and the last cartridges, so that the cartridges may dry out before the next application of solvent. A consequence of this may be low extraction yields and poor reproducibilities.

Starting with the manual extraction procedure, the different steps in the process were optimized, especially in relation to the flow rates of the solvents. In the ASPEC system, the flow of the sample, solvents and air (for

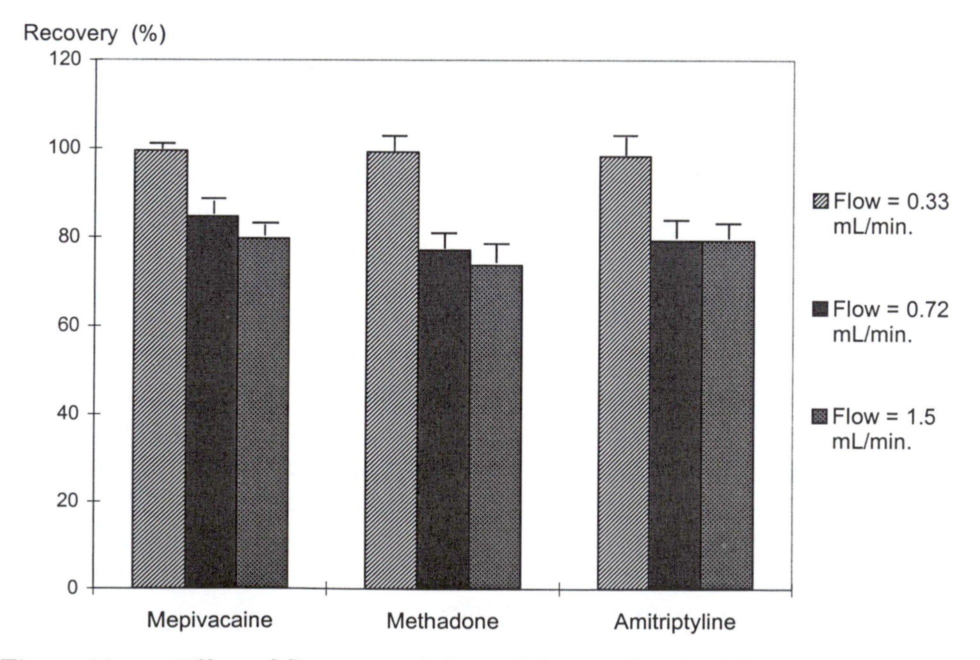

Figure 11. Effect of flow rates of eluent B (ammoniated ethyl acetate) on the recoveries of three test drugs from plasma with the fully automated SPE procedure on the ASPEC system (n=3).

drying) are not established by vacuum but by displacement: i.e., the solvents are not sucked through the cartridge but pushed through the cartridge. The flow through the cartridge (system) can be accurately established by the speed of the pistons. It appeared that a low flow rate is essential to obtain high and reproducible recoveries. Figure 11 shows this effect for the solvent used to elute Fraction B (ammoniated ethyl acetate). The extraction yield increased from about 80% to 95% when lowering the flow rate from 1.5 to 0.33 mL/minute. In a manual procedure it will be difficult to maintain such low flow rates for 6 minutes. Table 5 demonstrates that the automated assay shows excellent reproducibility. Even for whole blood samples relative standard deviations of less than 5% were obtained.

The speed of an automated assay is dependent on the purpose of the extraction, the volumes of sample and eluents, and the complexity of the procedure. In this screening procedure, where relatively large volumes of sample, washing solvents, and eluents are used, one extraction took 40 minutes. The ASPEC system can work unattended and overnight, for a total of 16 samples in sequence (11 hours). In circumstances where faster sample throughput is required either a faster, simpler procedure could be employed or multiple SPE workstations or parallel-processing workstations could be used.

Table 7. Recoveries of drugs from plasma with the comprehensive SPE procedure and HPLC with UV detection at 205 nm, spiking level 1 μg/mL.

Drug	Recovery (%)		RSD (%), n = 3
	Fraction A	Fraction B	
Allobarbital	94.0		3.8
Flufenamic acid	97.1		2.1
Paracetamol*	71.0		9.4
Pentobarbital	98.9		2.2
Salicylic acid	69.1		7.0
Secobarbital	98.9		0.3
Diazepam		94.2	6.7
Diphenhydramine		90.6	3.1
Ketazolam		98.0	4.1
Mepivacaine		98.6	1.6
Methaqualone		94.0	6.0
Papaverine		106.8	2.4
Trimipramine		92.0	1.5

* Acetaminophen

B. A COMPREHENSIVE SPE PROCEDURE FOR PLASMA AND WHOLE BLOOD FOR USE WITH HPLC

Although the above procedures worked well for the majority of toxicologically relevant substances, when it was applied in other laboratories, more polar substances as well as strongly acidic compounds were sometimes difficult to detect. It appeared that this was caused by rather low recoveries. These substances (e.g., salicylic acid, paracetamol, morphine) are also not readily analyzed by GC without prior derivatization. The recoveries of these compounds were found to be affected by the starting pH in the SPE procedure (pH 6.0). When the sample is buffered at this pH the more strongly acidic drugs are present in their dissociated, ionic forms. These ionic species retain poorly by non-polar mechanisms, compared to the neutral species. Therefore, we changed the starting pH to 2.2 and combined the SPE procedure with a gradient HPLC method, employing UV- or Diode Array Detection (DAD).

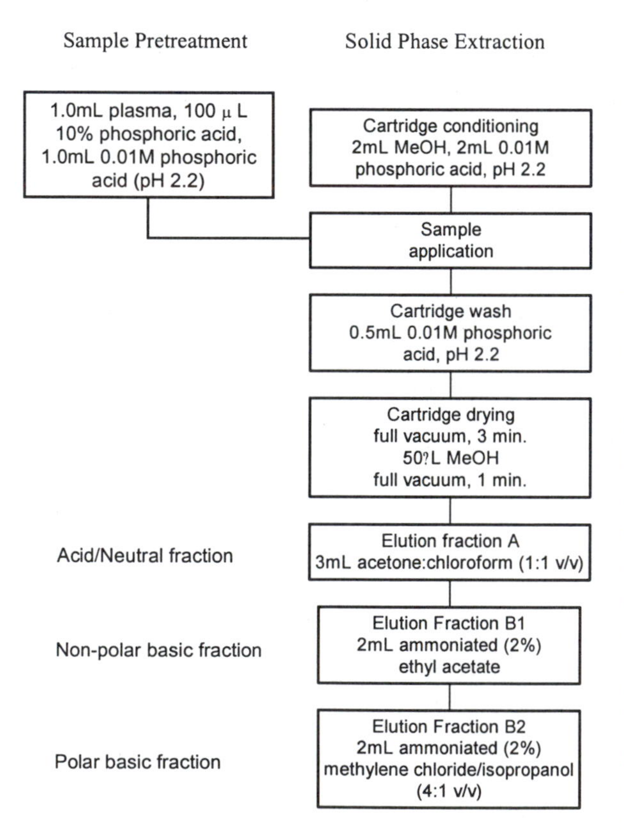

Figure 12. Comprehensive SPE procedure for broad spectrum drug screening, including more polar compounds in plasma and whole blood for use with HPLC.

The outline of the comprehensive procedure is given in Figure 12 (Huang et al., 1997). The starting pH of 2.2 results in less ionization of the acidic drugs and hence, better retention on the cartridge. It should be emphasized, however, that the amounts of water in the sample application step and in the wash step should be kept as small as possible. If not, the more polar acidic compounds will be partly washed away. Therefore, plasma samples are only diluted 1:1 with phosphoric acid pH 2.2 (compare this with a fourfold dilution in the original procedure in Figure 1) and the wash step uses only 0.5 ml of pH 2.2 phosphoric acid (compare

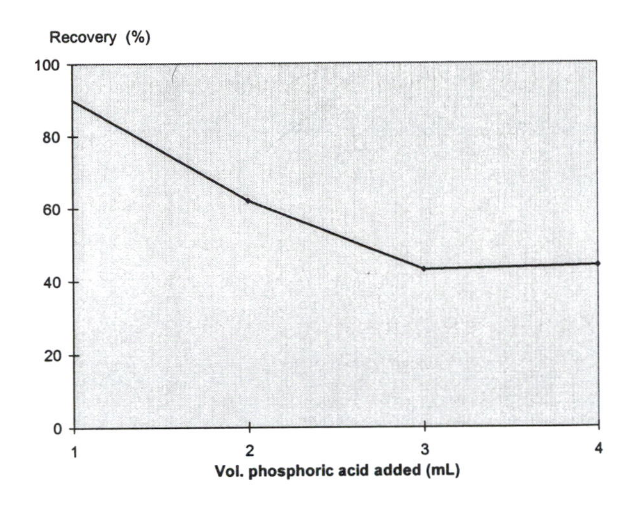

Figure 13. Influence of sample dilution on the recovery of paracetamol from plasma (plasma volume 1 ml).

this with 1.0 mL of pH 6.0 phosphate buffer, shown in Figure 1).

Figures 13 and 14 show the impact of using more diluent or more wash fluid, respectively, on the recovery of paracetamol (acetaminophen) from plasma. A consequence of the lower volume of wash fluid is an increased number of matrix peaks in the chromatograms, especially in Fraction A (See Figure 15). Figure 16 depicts chromatograms of spiked plasma samples. Typical recoveries from plasma at the 1 µg/ml level are given in Table 7.

Unfortunately, the comprehensive SPE procedure as outlined in Figure 12 did not work for whole blood. A 1:1 dilution of the blood in the sample preparation step followed by sonication, proved to be insufficient to prevent the resulting specimen from clogging the cartridge. The only suitable solution to this problem was dilution of the blood sample with 3 volumes of phosphoric acid pH 2.2, followed by sonication, centrifugation and specimen application.

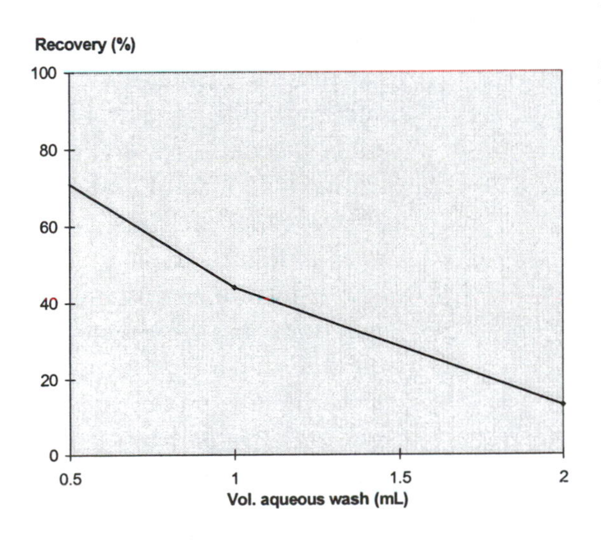

Figure 14. Influence of volume of wash fluid on the recovery of paracetamol from plasma.

As expected, this compromise resulted in lower recoveries of the more polar acidic and neutral drugs, as can be seen in Table 8. Thus, to obtain satisfactory ex-traction yields for the more polar acidic compounds, there are two prerequisites:

1) The pH of the extraction system must be kept at around 2.2 until the elution of fraction A and

2) The dilution of the sample and the volume of

Table 8. Recoveries of drugs from whole blood, spiked at 1 μg/ml. using the modified comprehensive SPE procedure and HPLC with UV detection at 205nm.

Drug	Recovery (%)		
	Fraction A	Fraction B	RSD (%), n = 3
Flufenamic acid	42.8		16.2
Paracetamol*	36.9		7.9
Salicylic acid	56.1		2.1
Diphenhydramine		89.4	3.8
Ketazolam		93.7	3.8
Mepivacaine		92.5	1.0
Papaverine		86.7	4.7
Trimipramine		84.9	2.2

* Acetaminophen

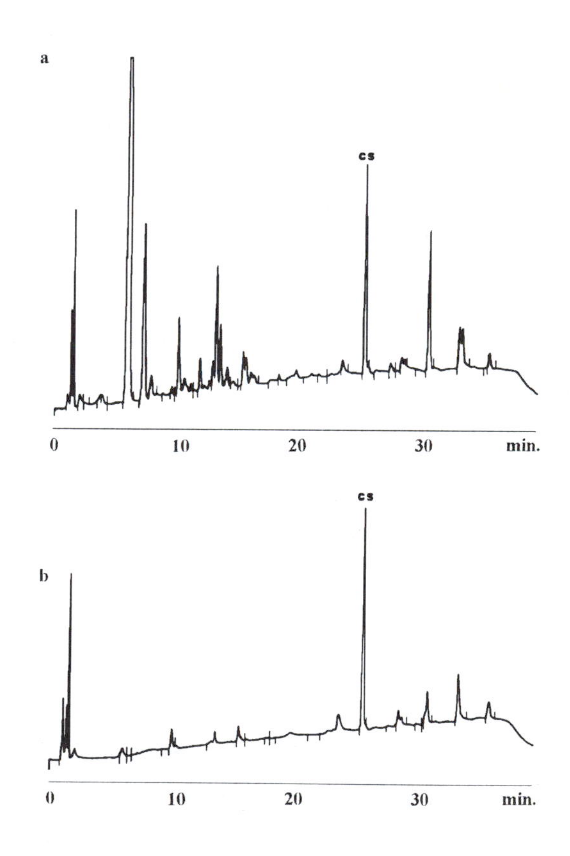

Figure 15. Chromatograms of Fractions A and B from blank plasma using the comprehensive SPE procedure and HPLC with UV detection at 205 nm. CS indicates the chromatographic standard, diclofenac.

the wash fluid must be kept small.

For the more polar basic compounds (e.g. morphine and benzoylecgonine), it was found that a single elution of fraction B with 2 ml ammoniated ethyl acetate did not always give optimum recoveries. This problem was overcome by carrying out a second elution under alkaline conditions, either with another 2 ml of ammoniated ethyl acetate or with 2 ml ammoniated dichloromethane /isopropanol 4:1, as can be seen in Figure 12.

After elution of the various fractions, the eluents were carefully evaporated to dryness under a stream of nitrogen in a water bath at 40°C. The residues were redissolved in 100 μl of a mixture of triethylammonium phosphate buffer (0.025 M, pH 3.0) -

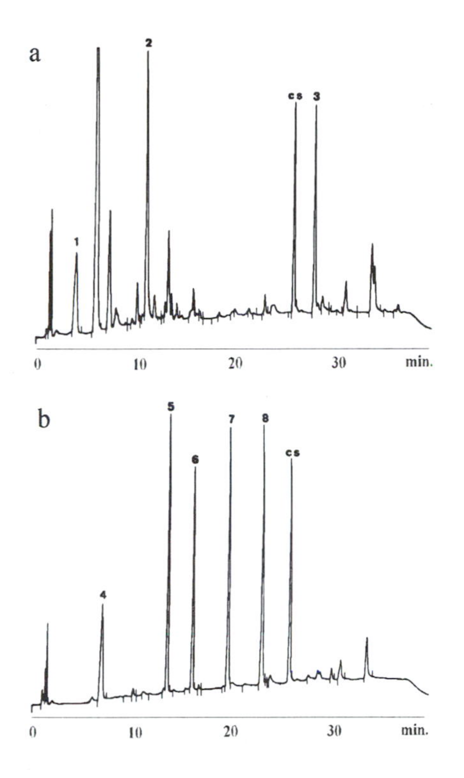

Figure 16. Chromatograms of Fractions A and B from spiked plasma (1μg/ml) using the comprehensive SPE procedure and HPLC with UV detection at 205 nm. CS=diclofenac (chromatographic standard), 1=paracetamol, 2=salicylic acid, 3=flu-fenamic acid, 4=mepivacaine, 5=papaverine, 6=diphen-hydramine, 7=trimipramine, 8= ketazolam.

acetonitrile 2:1, containing 2 μg diclofenac as a chromatographic standard. About 80 μl was subjected to HPLC analysis. The column was a Lichrocart[25] RP-Select B (5 μm) column, 100 x 4 mm, together with a RP-Select B (5 μm) guard column (Merck, Darmstadt, Germany).

Gradient elution was applied at a flow rate of 1.0 ml/min. Mobile phase A consisted of triethyl-ammonium phosphate buffer (0.025 M, pH 3.0), mobile phase B was HPLC-grade acetonitrile. The gradient program was:

90% A + 10% B to 30% A + 70% B in 30 min (linear); 30% A + 70% B for 5 min; 30% A + 70% B to 10% A + 90% B in 5 min (linear); back to a; equilibration for 10 min.

This HPLC system allowed good separation and sensitive detection of a great many drugs of interest, without the need for derivatization. It should be noted, however, that HPLC with UV-detection is less suitable for substances with low UV absorptivity (Huang et al., 1997).

VI. MODIFICATION OF BROAD-SCREEN EXTRACTIONS FOR CONFIRMATION TESTING

If a compound can be extracted as one component of a suite of drugs extracted during a screen, it follows that the compound can be extracted alone, following the same procedure. However, the analyst has the advantage that

---

[25] Lichrocart is a trademark of E. Merck, Darmstadt, Germany.

there is no need to compromise on any extraction parameter in order to achieve high recoveries of other compounds. Thus a confirmation extraction should be able to yield higher recoveries, better clean-up, a simpler, faster procedure or a combination of all of these three things. To illustrate this a basic drug extraction for the primary urinary metabolite of cocaine, benzoylecgonine, is discussed.

A. BENZOYLECGONINE FROM URINE

Extraction of basic drugs alone is usually performed according to a scheme similar to figure 1. In other words the sample is loaded at pH 6. At this pH the baseline from a urine extract is clean. This applies for several of the published applications that are referenced in this chapter, including extraction of cocaine. For these drugs acceptable elution can be achieved with a single aliquot of ammoniated methanol or isopropanol/ methylene chloride (1:4) and acceptable clean-up can be achieved by a simple methanol wash. The principles of retention at each stage are the same as for the general drug screen, but the requirements on the solvents used are less stringent and the clean-up can be more effective.

This is one example of how a directed approach to extraction of just one drug, or a reduced range of drugs, can be made. Taking this analysis further, we will focus on benzoylecgonine, which has a structure that permits non-polar retention (a seven-membered aliphatic ring and a phenyl group) and a cationic site (a tertiary amine). However, unlike most of the drugs discussed so far in this chapter, benzoylecgonine also possesses a carboxylic acid functional group. This will cause problems unless its ionic state is addressed: If it is allowed to remain as an anion when a wash is applied to the column that disrupts non-polar interactions, then it will not retain, since the ability to interact by ion exchange is compromised. In general, zwitterions (for this is what benzoylecgonine is at pH 6) will retain poorly, if at all, by ion exchange. In our general procedure for broad-spectrum drug screens given in Figure 1, the drug would elute in fraction A. Thus, we would see a basic drug metabolite appear in an otherwise acidic drug fraction.

The solution is to adjust the pH accordingly. For example, we could wash the cartridge with a 1M solution of acetic acid prior to the organic wash step. During the course of the acidic wash the benzoylecgonine zwitterion is converted into a simple cation, which can now retain just like a regular basic drug. The acid/neutral elution step can, in this case, be performed using methanol, rather than a combination of chloroform and other solvents and the benzoylecgonine elutes in ammoniated isopropanol/methylene chloride (1:4 v/v).

VII. CONCLUSIONS

From the above paragraphs one may conclude that SPE is a very useful tool for sample workup, isolation and concentration in STA. The strategy of a mixed-mode cartridge providing hydrophobic and cation exchange interactions, combined with a pH-dependent sample application and extraction, can give high recoveries of analytes from plasma, urine, whole blood, and tissues, and the resulting SPE eluates are easily amenable to subsequent GC- and HPLC-analysis. The chromatograms show almost no interference from endogenous matrix components, so that toxicologically relevant substances could be easily detected and quantitated.

Since the general STA method described in this chapter is a compromise, taking into account as many as possible substances with a large variety in structures, it is relatively long (approximately 40 min) and it must be carried out with care. However, the method lends itself to automation, which can increase the throughput and substantially reduce the amount of manual labor.

What has been attempted in this chapter is to show how an initial method evolved as the analytical parameters (drug types, analytical technique, sample matrix) change. Although a large variety of substances show good extraction with the methods described, it remains important to extend our experiences with additional drugs, especially when new ones are introduced to the market. Another area of interest is to investigate the extraction behavior of metabolites with the present SPE procedures, but this is highly dependent on obtaining suitable specimens and/or the availability of metabolites as appropriate reference standards.

Further attempts to speed up and/or miniaturize the extraction process have recently led to the introduction of SPE discs or micro-columns with extraction path lengths in the order of 0.5-2 mm, solid-phase microextraction techniques. In the latter, the sorbent is coated on a needle that is first brought in contact with the liquid sample (immersed or headspace) and subsequently introduced into a GC injection port. The potentials of these newer techniques for STA are still being investigated.

REFERENCES

Abusada, G.M., et al., (1993) Solid-Phase Extraction and GC/MS Quantitation of Cocaine, Ecgonine Methyl Ester, Benzoylecgonine, and Cocaethylene from Meconium, Whole Blood and Plasma. J.Anal.Toxicol., 17: 353-357.

Bond Elut Certify (1988) Instruction Manual. Varian Sample Preparation Products, Harbor City, CA, USA.

Chen, X.H., Wijsbeek, J., Franke, J.P. and De Zeeuw, R.A., (1992 A) A Single Column Procedure on Bond Elut Certify for Systematic Toxicological Analysis of Drugs in Plasma and Urine. J.Forensic Sci. 37: 61-72.

Chen, X.H., Franke, J.P., Wijsbeek, J. and De Zeeuw, R.A., (1992 B) Isolation of Acidic, Neutral and Basic Drugs from Whole Blood using a Single Mixed-Mode Solid Phase Extraction Column, J.Anal.Toxicol. 16: 351-355.

Chen, X.H., Franke, J.P., Wijsbeek, J. and De Zeeuw, R.A., (1993 A) A Study of Lot-to-Lot Reproducibilities of Bond Elut Certify and CleanScreen DAU Mixed-Mode SPE Columns in the Extraction of Drugs from Whole Blood. J.Chromatogr. 617: 147-151.

Chen, X.H., Franke, J.P., Wijsbeek, J. and De Zeeuw, R.A., (1993 B) Pitfalls and Solutions in the Development of a Fully Automated SPE for Drug Screening Purposes in Plasma and Whole Blood. J.Anal.Toxicol. 17: 421-426.

Chen, X.H., Franke, J.P., Wijsbeek, J. and De Zeeuw, R.A., (1993 C) Reusability of Bond Elut Certify Columns for the Extraction of Drugs from Plasma. J.Chromatogr. 619: 137-1422

Dixit, V. and Dixit V.M., (1991) Solid Phase Extraction of 11-nor-Δ-9-tetrahydrocannabinol-9-carboxylic Acid from Human Urine with Gas Chromatographic-Mass Spectrometric Confirmation. J.Chromatogr., 567: 81-91.

Dixit, V., Pocci, R. and Dixit, V.M., (1992) Solid Phase Extraction and GC/MS Confirmation of Amphetamine and Methamphetamine with Enantiomeric Resolution in Human Urine. "SPE At Work, No.9", Varian Sample Preparation Products, Harbor City, California, USA.

Ellerbe, P., et al., (1993) Determination of Amphetamine and Methamphetamine in a Lyophilized Human Urine Reference Material. J.Anal.Toxicol., 17(3); 1165-170.

Harkey, M.R., Henderson, G.L. and Zhou, C., (1991) Simultaneous Quantitation of Cocaine and its Major Metabolites in Human Hair by Gas Chromatography-Chemical Ionization Mass Spectrometry. J.Anal.Toxicol., 15(5):260-265.

Huang, Z.P., Chen, X.H., Wijsbeek, J., Franke, J.P. and De Zeeuw, R.A. (1996) An Enzymic Digestion and Solid Phase Extraction Procedure for the Screening of Acidic, Neutral and Basic Drugs in Liver Using Gas Chromatography for Analysis. J.Anal.Toxicol. 20: 248-254.

Huang, Z.P., Franke, J.P., Wijsbeek, J. and De Zeeuw, R.A., (1997) "A Single Column SPE Procedure for Screening for a Broad Range of Substances Including More Polar Compounds in Plasma an Whole Blood by HPLC". Proc. 34th Int. Meeting of TIAFT, Interlaken, Switzerland, In press.

Kikura, R., Ishigami, A. an Nakahara, Y., (1992) Studies on Comparison of Metabolites in Urine Between Deprenyl and Methamphetamine. III. Enantiomeric Composition Analysis of Metabolites in Mouse Urine by HPLC with GITC Chiral Reagent. Japan.J.Toxicol. Environ.Health, 38(2): 136-141.

Lai, C-K., Lee, T., An, K-M. and Chan, A, Y-W., (1997) Uniform Solid-phase Extraction Procedure for Toxicological Drug Screening in Serum and Urine by HPLC with Photo-diode Array Detection. Clin.Chem., 43(2): 312-325.

Leyssens, L., et al., (1991) Determination of Beta-receptor Agonists in Bovine Urine and Liver by Gas Chromatography-Tandem Mass Spectrometry. J.Chromatogr.Biomed.Applic., 102(2): 515-527.

Osselton, M.D., (1977). The Release of Basic Drugs by Enzymic Digestion of Tissues in Cases of Poisoning. J.Forensic Sci.Soc. 17: 189-104.

Platoff, G.E. and Gere, J.A., (1991) Solid Phase Extraction of Abused Drugs from Urine. Forensic Sciences Review 3(2).

Roy, I.M., Jefferies, T.M., et al., (1992) Analysis of Cocaine, Benzoylecgonine, Ecgonine Methyl Ester, Ethylcocaine, and Norcocaine in Human Urine using HPLC with Post-Column Ion-Pair Extraction and Fluorescence Detection. J.Pharm.&Biomed.Anal., 10(10-12): 943-948

Simpson, N.J.K. and van Horne, K.C., Eds. (1993) The Handbook of Sorbent Extraction Technology, 2nd Edition. Varian Sample Preparation products, Habor City, CA. Pp. 85-98.

Solans, A., Carnicero, M., de la Torre, R and Segura, J., (1995) Comprehensive Screening Procedure for the Detection of Stimulants, Narcotics, Adrenergic Drugs and their Metabolites from Human Urine. J.Anal.Toxicol., 19: 104-114.

Tebbet, I.R., (1987) Rapid Extraction of Codeine and Morphine in Whole Blood for HPLC. Chromatographia 23: 377-378.

Wilson, I.D., (1994) Sample Preparation for Biomedical and Environmental Analysis (Eds. Stevenson and Wilson), Plenum Press, New York: 37-52

Wilson, I.D. and Nicholson, J.K., (1987) Solid Phase Extraction Chromatography and Nuclear Magnetic Resonance Spectrometry for the Identification and Isolation of Drug Metabolites. Anal.Chem., 59: 2830-2832.

9

THE APPLICATION OF SPE TO VETERINARY DRUG ABUSE

Paul M. Wynne, Racing Analytical Services Limited, Flemington, Victoria, Australia

I. INTRODUCTION

The drug testing of racing animals is designed to dissuade participants from using performance-altering drugs in competition and to maintain a standard of animal health such that an animal is not raced when unfit, nor forced beyond its normal endurance. Industry research impacts aspects of veterinary therapeutics and animal nutrition.

As with most drug testing operations, the samples collected are normally blood, urine, or saliva. Testing may extend to fecal matter, stomach contents, postmortem samples, feed and other agricultural samples, pharmaceutical and other preparations, and various items of industry hardware. With the notable exceptions of bicarbonate and some polypeptides, typical target analytes are small organic molecules with a molecular weight of less than 1000 Daltons. In some cases of doping, the drug dose is sub-therapeutic, as the physiological effect of a clinically significant dose may be too great. The urinary excretion products are often not the parent drug and may be metabolites for which there is little or no recorded data, particularly in the case of large animals in which few metabolic studies have been documented. Metabolism gives rise to more polar compounds than the parent drug, which often pose additional problems in separation and analysis. Unlike most toxicological or postmortem drug testing, samples are presented with limited history and no clinical notes.

A. VETERINARY, THERAPEUTIC, AND OTHER DRUGS

The classes of drugs that may be abused in racing include legitimate veterinary therapeutics, although many jurisdictions allow the use of anti-infective agents that have no significant performance-altering effects. Commonly used veterinary therapeutics include non-steroidal anti-inflammatory drugs (NSAIDs, e.g., phenylbutazone, flunixin, ketoprofen), corticosteroids (dexamethasone, methylprednisolone, hydrocortisone), local anaesthetics (lignocaine, procaine), anabolic agents (testosterone, nandrolone, boldenone, and stanazolol), and tranquilizers (acepromazine and reserpine). These drugs find wide application in overcoming race and training injuries, and care must be applied in withholding each drug for advised periods prior to racing. While many drugs are cleared in several days, some such as the long acting anabolic steroids and the sedative reserpine, may act (and be detected) for long periods after administration. All these classes of drugs are capable of affecting performance, and their use during racing is prohibited by most jurisdictions.

Stimulants include the classical stimulants (cocaine, amphetamine, caffeine, and nicotine), narcotics (etorphine, morphine, fentanyl, and dextromoramide), sedatives (diazepam and acepromazine), and anabolic steroids. Minor tranquilizers in man and other mammals are often stimulants at low doses in the horse.

Sedatives (or stoppers) may be used to reduce an animal's performance while it appears that the animal and rider are still attempting to perform to their maximum ability. The drugs find particular application in greyhound racing where there is no driver or jockey to slow the animal. Reducing performance is invariably tied to manipulation of betting or pre-selection for future events. The group of drugs includes anesthetics (chloral hydrate and barbiturates), tranquilizers (benzodiazepines, phenothiazines, and reserpine) and β-blockers (see example A). Sedatives are also used in dressage, where a quiet horse is required, and in high-precision human sports, such as archery and shooting, where tension or a high heart rate can affect performance.

Diuretics have long been employed in the belief that by diluting the urine, the urinary concentration of other drugs will be reduced. Generally, the matrix background is also reduced and the urine appears thin and watery. Low viscosity equine urine is considered suspicious and is normally investigated for evidence of diuretics. While the diuretics are generally grouped together because of their common action, they must be subdivided according to their acid/base properties when considering their extraction.

Other drugs used in both human and veterinary medicine, as well as many preparations used by herbalists and homoeopaths from various cultural backgrounds, may also be prohibited by racing authorities.

B. COMMONLY USED METHODS OF DRUG DETECTION

The nature of the industry requires rapid testing for an almost limitless number of compounds within a tight financial budget. This goal is only achieved with extensive use of class-specific screening and the selective use of target compound analysis. Screening of samples is carried out by either immunoassay (RIA, EMIT, ELISA, FPIA) or by pre-extraction of the sample followed by analysis using a chromatographic method. Most laboratories employ gas chromatography (GC) and high-pressure liquid chromatography (HPLC) with a variety of detectors. Thin-layer chromatography (TLC) has, in many laboratories, fallen from favor as it is labor-intensive and non-specific.

Immunoassay provides some class-specific capability, but it is not a viable means of testing for structurally diverse groups of drugs. Further, the non-definitive nature of immunoassay, because of cross-reactivity, makes the method unacceptable for the confirmation of most target groups.

C. CHEMICO-LEGAL REQUIREMENTS

As positive results invariably must be defended in some form of tribunal or court, analytical data must be unequivocal. Confirmation of the presence of drugs or their metabolites is carried out almost exclusively by mass spectrometry, as no other methods can produce unequivocal identification of a compound with sufficient sensitivity at biological concentrations (Vine, 1996). The detection of drug metabolites has become more important following the realization that they allow positives to be called when no parent drug is excreted, thereby establishing that administration of the drug has occurred.

II. THE USE OF SPE IN FORENSIC RACING CHEMISTRY

In the early 1980s, it became increasingly apparent that extraction technology was not sufficiently developed for the increasing range of drugs and metabolites for which testing was required. In 1983, there were no reports published in the AORC proceedings using SPE methods, while by the late 1990s, the use of SPE in racing chemistry was commonplace. From the point of view of general screening, the timetable of method development parallels the evolution of SPE.

A. THE MATRIX BACKGROUND AND ENDOGENOUS COMPOUNDS

The most common biological matrix tested in racing is urine. Horse urine presents an analytical challenge for the extraction of all drugs because of the relatively high viscosity of the samples, particularly where the animal was dehydrated by exercise prior to sample collection. The urine is often too thick to pass through a solid-phase cartridge and sometimes remains so after enzyme hydrolysis. For the reliable automated solid-phase extraction of horse urine, it should be hydrolyzed, insoluble materials removed by centrifugation and diluted with an equal volume of a buffer suitable for the application. These and other considerations described below apply equally to urine samples collected from other species and to sample prepared from feeds, preparations and other biological samples.

Horse urine typically contains compounds that will extract into most SPE fractions. Many dietary components or their metabolites are structurally similar to target analytes and it is therefore important when preparing extracts for broad based screening that these compounds be co-extracted and resolved in a later chromatographic step. Various interfering compounds are discussed below in conjunction with the drug groups that they affect.

Dog urine, while rarely viscous, is often highly concentrated. However, dogs are occasionally affected by *Diabetes insipidus* and other similar temporary conditions called "water diabetes." These conditions result in large urine volumes and significant analyte dilution. The high protein diet of dogs may lead to the formation of nitrogen containing heterocyclic compounds that may interfere with some screens.

B. ANABOLIC STEROIDS

The major phase 2 metabolic pathway for many anabolic steroids in the horse is the formation of sulfate conjugates. Some glucuronides and free steroid are also excreted. As the enzymatic hydrolysis of alkyl sulfates is ineffective, conjugates are generally cleaved by acid-catalyzed hydrolysis to give the free steroid (or alcoholysis to give the corresponding ether). The free steroids are readily derivatized prior to analysis by GC/MS.

While the sulfate group provides a potential site for ion-exchange on an NH2 phase, the urine must be subjected to extensive desalting and pre-extraction to remove endogenous urinary acids which would otherwise compete at the ion-exchange surface. Such pre-extraction may result in the loss of the free steroids and their glucuronide conjugates.

The functionality of the anabolic steroids is quite diverse. Some, such as testosterone, are relatively non-polar while others, such as stanozolol and its metabolites, are more highly functionalized and more water-soluble. The use of selective elution solvents to separate a fraction from other urinary products (such as many of the diet derived plant phenolics) is therefore unwise when the extracted material is to be *screened* for as many anabolic agents as possible. Most methods for the separation of anabolic steroids therefore employ a C18 phase SPE cartridge for desalting and crude extrac-

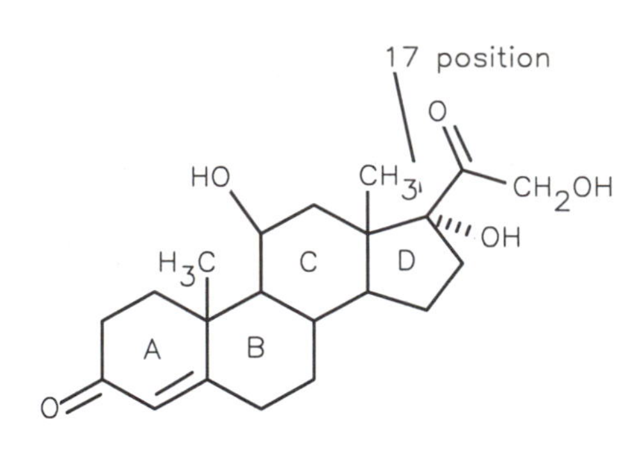

Figure 1. Hydrocortisone, a corticosteroid showing the pendant group at the 17-position. Other corticosteroids may be functionalized differently on the A, B and C rings and several contain substituents in the D ring.

tion of the steroidal fraction. Final separation is achieved by GC prior to MS analysis. The extraction of anabolic steroids from urine is detailed in example I. Careful consideration of the data is essential as many cases of interference by plant phenolics have been observed in the author's laboratory, during screening for anabolic steroids.

C. CORTICOSTEROIDS

In the horse, the corticosteroids are excreted mainly as glucuronide conjugates and are usually enzyme-hydrolyzed prior to extraction. Two different phases have found equal application for the recovery of these drugs from urine. Like anabolic steroids, the corticosteroids are well recovered by retention on a C18 phase, from which they may be eluted with dichloromethane, methanol or ethyl acetate, prior to derivatization and analysis by GC/MS or LC/MS. The major functions of SPE are desalting of the sample and concentration of the target analytes into a non-aqueous extract. Chromatography is relied upon to separate the derivatized corticosteroids from other co-extracted materials (Delahunt et al., 1997).

An alternative SPE method using a silica cartridge allows the more selective retention of the corticosteroids, which are eluted by varying the ratio of dichloromethane (a wash solvent) and ethyl acetate (an elution solvent). A multiple SPE cartridge method has been described in which corticosteroids were eluted from a Chem Elut™[26] cartridge with dichloromethane onto a sodium sulfate cartridge. The analytes were further eluted from the sodium sulfate cartridge, with ethyl acetate/dichloromethane (1:1 v/v), onto a silica cartridge from which they were desorbed with ethyl acetate/ethanol prior to analysis by HPLC (Samuels et al., 1994).

It is perhaps not surprising that immuno-affinity column chromatography has been reported as a method suitable for the recovery of corticosteroids prior to chromatographic analysis. The pendant group for hydrocortisone at C-17 is shown in Figure 1. This site is common to all corticosteroids and would therefore provide an ideal site for the formation of a class-

[26] Chem Elut is a trademark of Varian Inc., Palo Alto, California.

specific antibody. The same region of the steroid is likely to be responsible for the efficient retention of the compounds on a silica phase.

It is anticipated that similar SPE methodology may be applied to the isolation of the aglycones of a diverse group of plant derived cardio-active steroids such as digoxin. The alkaloidal steroids could also be further purified from such an extract on a SCX phase as described below for basic drugs.

D. BASIC DRUGS

Simple basic drugs are readily recovered from urine by solvent extraction. Variation of solvent polarity, from relatively non-polar solvents, such as diethyl ether or 1-chlorobutane, to highly polar solvent mixtures, such as chloroform/2-propanol (3:1 v/v), allows some selectivity in determining target groups. The co-extraction of interfering species is often minimized by the judicious selection of the extraction solvent or by back extraction of the bases into acidic solution. Low polarity solvents give clean extracts that have traditionally been screened by GC-NPD or TLC. High polarity solvent mixtures have been used to recover target groups such as the opiates which have been screened by GC/MS following derivatization, or by TLC using appropriate spray reagents. Clarke's Isolation and Identification of Drugs (Moffat, 1986) and similar monographs describe the principles by which this methodology can be adapted for forensic veterinary work.

The extraction of bases from plasma samples is subject to far greater interference from co-extracting fatty acids and neutral compounds and generally requires back extraction into acid to obtain a clean and useable extract. The utility and effectiveness of solvent extraction for the recovery of basic drugs initially resulted in slow acceptance of SPE alternatives. SPE methodology does, however, offer distinct advantages in analyte recovery over more traditional methods, as illustrated in example C.

Extraction using either a SCX or mixed-mode (SCX/non-polar) cartridge like Certify®[27] gives extracts that may be used for screening bases of both high and low polarity, with or without derivatization and without the need for back extraction or other lengthy clean-up procedures. These mixed-mode cartridges allow the rapid recovery of such diverse groups as β-blockers (see example A), β-agonists (salbutamol and clenbuterol), opiates (morphine and etorphine) and other narcotic analgesics (fentanyl and dextromoramide), alkaloidal drugs (quinine and strychnine) and some basic diuretics (amiloride, triamterene, and clopamide). Importantly, bases may be isolated from urine at neutral pH, allowing the extraction of compounds that may be otherwise unstable (see example E). The method allows the

[27] Certify is a registered trademark of Varian Inc., Palo Alto, California.

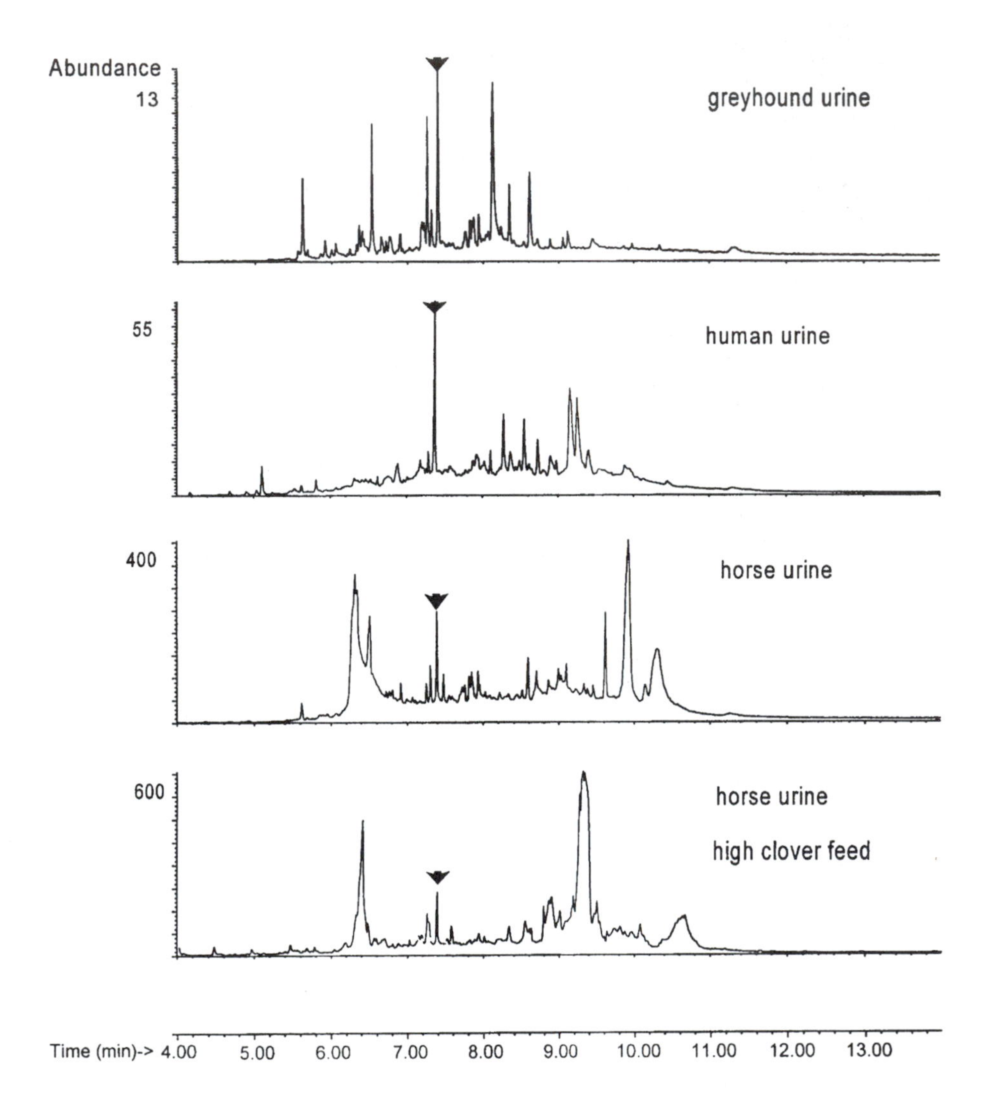

Figure 2. The GC/MS total ion chromatograms of the diazomethane-treated acid/neutral fraction obtained following the extraction of urine samples. The arrow indicates the internal standard, niflumic acid, spiked at 10μg/mL. Note that the baselines for these traces (the bottom one from a horse fed on clover-rich pasture and the one above it from a horse fed only on bagged grain and dry feed) differ.

facile extraction of highly polar compounds that might otherwise involve lengthy sample preparation. Some selectivity in elution can be achieved by varying the polarity of the ammoniacal elution solvent mixture using ethyl acetate (eluting less polar compounds), dichloromethane and 2-propanol, or methanol (eluting highly polar compounds).

One early method of solid-phase extraction of bases involved applying urine samples to a conditioned C18 phase, which was washed with water and hexane. The cartridge was eluted with ether and the eluate passed through an NH2 cartridge from which the retained bases were subsequently eluted with methanol (Leferink, et al., 1987). The NH2 cartridge in this case was used for the retention of the abundant urinary acids. The elution of bases prior to acids and neutrals (as would occur on a SAX/non-polar sorbent like Certify II) is of limited value in the extraction of horse urine. This is because the endogenous acids are capable of swamping available anion exchange sites, resulting in breakthrough of acidic interferences into the basic eluate. Further to this, as the bulk of low level target analytes are basic drugs, it is desirable for screening purposes to obtain a basic eluate that is low in co-eluting neutral substances. As many of these bases are highly functionalized, cation exchange allows efficient recovery of these compounds, while allowing acids and neutrals to be eluted with highly polar solvents.

Some basic drugs and/or their metabolites may be regarded as amphoteric and may exist in zwitterionic form. Worked examples of the extraction of several of these compounds are given in examples D, H, and J. The formation of the ammonium phenoxide inner salt of morphine is proposed as one reason for its poor extraction into various solvents when compared to its O-methyl derivative, codeine. The formation of the zwitterion is suppressed during SPE extraction on SCX and Certify cartridges by application of the sample at neutral pH, at which both the phenol and tertiary amine functions are protonated. SPE extraction of morphine and related basic drugs yields extracts that are significantly cleaner than comparable solvent extraction methods. This approach is the same as that for extraction of benzoylecgonine from human urine where the carboxylic acid, of an otherwise basic drug metabolite, must be neutralized before effective ion exchange on a SCX cartridge can occur.

E. ACIDIC DRUGS

The recovery of acidic drugs from equine urine is complicated by the presence of high concentrations of endogenous urinary acids and phenols. These species are far less abundant in human or dog urines (see Figure 2 for a comparison).

The compounds range in polarity and structure from long chain fatty acids to simple aryl and alkyl carboxylic acids. Horse urine contains particularly high concentrations of hippuric acid and N-phenyl-acetylglycine. Extraction of carboxylic acid drugs from urine is therefore difficult, and most methods rely on crude isolation of an acidic fraction followed by sepa-

Table 1. Comparative recoveries of NSAIDs spiked into horse urine at 10μg/mL, by SPE extraction and solvent extraction. (1) % recovery of SPE extract relative to solvent extraction with hexane/ethyl acetate/dichloromethane (4:3:3) or (2) chloroform/2-propanolol (3:1). Use of mixed mode SPE cartridges provided cleaner extracts than liquid/liquid extraction was able to deliver.

Compound	Recovery % (1)	Recovery % (2)
Carprofen	115	78
Cicloprofen	111	100
Clanobutin	171	287
Diclofenac	126	52
Diflunisal	468	106
Fenoprofen	126	55
Flufenamic acid	162	42
Flunixin	105	80
Flurbiprofen	161	55
Ibuprofen	123	69
Indomethacin	169	88
Ketoprofen	247	125
Meclofenamic acid	88	25
Mefenamic acid	107	63
Naproxen	133	46
Oxyphenbutazone	753	165
Phenylbutazone	105	80
Sulindac	446	226
Tolfenamic acid	139	26

ration of the target analytes by chromatography. The endogenous compounds often cause severe interference with TLC, GC and GCMS methods.

As many of these compounds are acids of relatively high water solubility and/or low molecular weight, reversed-phase HPLC has been found to be effective for the screening of target groups.

While some target analytes are simple carboxylic acids, others such as diflunisal are capable of forming a stable intramolecular hydrogen bond and may exhibit some pseudo-zwitterionic character in aqueous solutions. Others, such as clanobutin, oxyphenbutazone, and sulindac, are functionalized with sterically accessible water-solubilizing groups, which may reduce the efficiency of extraction into organic solvents, particularly those with a high non-polar component.

The SPE of acidic drugs on a neutral phase offers the opportunity for improved recovery of such compounds. Intra- and intermolecular hydrogen bonding and pseudo-zwitterionic interactions are minimized as the sorbent and, in some cases, the eluting solvent may be regarded as aprotic.

The concentration of the endogenous acids has limited the usefulness of SAX and Certify II phases in the extraction of acidic drugs. Most reported methods employ C18 or C8/Certify as a method for desalting and isolating neutral and acidic fractions, either separately or as a combined eluate.

Successful clean-up of an acidic C18 fraction on a DEA cartridge has been reported for the carboxylic acid drugs such as the NSAIDs and some diuretics by Walden et al., (1994).

The Certify method described in example C has been used to isolate a diverse range of acidic and neutral drugs (Batty et al., 1994). The acids were retained as their conjugate bases and subsequently converted to the protonated form on the SPE cartridge before elution in an aprotic organic solvent. Elution of the NSAIDs in the acid/neutral fraction was comparable in many cases to that achieved by common solvent extraction methods. Improved recoveries were noted for polar or pseudo-zwitterionic drugs using this SPE method, as shown in Table 1. It has been adapted from similar methods described in Chapter 8.

The isolation of low level carboxylic acid drugs and metabolites requires considerable sample preparation. One method that has been used successfully to eliminate the hippuric acid and N-(phenylacetyl)glycine interferences from a Certify acid/neutral eluate involved treating the extract with trifluoroacetic anhydride. This treatment cleaved the amide bonds of the glycine conjugates. Any TFA esters could be destroyed by treatment with methanol while the residual benzoic acid, trifluoroacetic acid and phenylacetic acid could be successfully removed by evaporation. The residual material was suitable for GC analysis or further purification on SAX or Certify II phases.

The problems associated with the SPE extraction of carboxylic acids also applies to other weakly acidic drugs such as phenylbutazone, isopyrin, phenytoin, and the barbiturates.

F. STRONGLY ACIDIC DRUGS

Few free sulfonic acids are used as therapeutic agents. Some examples include 3,5-dibromo-4-hydroxyphenylsulfonic acid (a topical disinfectant) and ethamsylate (a hemostatic agent). The dihydroxyphenylsulfonic acid in ethamsylate has been isolated from urine and plasma using an ion-pairing reagent (Popot et al., 1992). Typically, urine at pH 4 was treated with tetrabutylammonium hydroxide and applied to a conditioned Chem Elut cartridge. The cartridge was eluted with dichloromethane, which was collected and evaporated. The residue was dissolved in methanol/water and applied to a SAX cartridge conditioned with methanol and acetic acid. The cartridge was washed, the sulfonic acid was eluted with methanol containing sulfuric acid and was analyzed directly by HPLC.

G. WEAKLY BASIC AND OTHER NEUTRAL DRUGS

There are several weakly basic drugs such as the xanthines (caffeine, theophylline, and theobromine) and the local anaesthetic benzocaine, that are poorly retained on SCX or mixed-mode phases. During routine extraction of biogical samples, these compounds may be considered as neutral species. Neutral drugs other than the steroids, which have been discussed above, in-

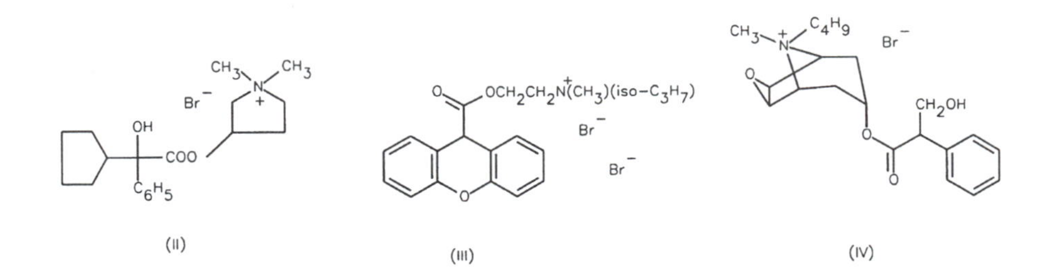

Figure 3. Structures of Glycopyrolate (II), propanthalene (III) and N-butylhyoscine (IV)

clude the aryl and aryloxy glycols and their derivatives (guaphenesin and methocarbomal), the terpenoids (camphor and menthol), amides (chloramphenicol), and many of the diuretics. The reduction metabolites of the nitrated benzodiazepines are more basic than their parents and are readily recovered using the methodology described above for basic drugs.

The weakly basic and neutral drugs in general have a high degree of polar functionality and are therefore co-extracted with plant-derived phenolics, some fat-soluble vitamins, and some acidic compounds when extracted on a neutral phase (C18 or C8) or a polar phase (silica). Improved backgrounds have been reported for many neutral drugs by further extracting a C18 eluate on a silica SPE cartridge (Foster et al., 1990). As with other drug groups, the elimination of the interfering species from the SPE eluate would result in considerable loss of many target analytes. Fortunately, many of these drugs are used at high doses and their detection, following GC, LC, or TLC separation, is relatively simple.

The detection of drugs or metabolites present at low levels may be subject to greater matrix interference than is observed by alternative solvent extraction. Of these, the xanthines and benzodiazepines are discussed in more specific terms in section IV, below.

H. QUATERNARY AMMONIUM COMPOUNDS

A small group of quaternary ammonium drugs find application in veterinary medicine. Propanthalene, glycopyrrolate, and N-butyl hysoscine bromide are typical examples of the group. Their structures are shown in Figure 3. Screening for such compounds has been carried out by acid hydrolysis of the urine to liberate the carboxylic acid portion of each drug, which was then isolated by either solvent extraction or by SPE on a C18 phase. The acids were methylated and screened by GC/MS.

Alternatively, the quaternary ammonium compounds have been isolated by SPE on a CBA cartridge. The extraction of glycopyrrolate is described in example G (Medonca et al., 1992). In some cases, the base load and salt concentration may need to be reduced prior to application of the sample to the CBA cartridge.

III. EXAMPLES OF SUCCESSFUL EXTRACTIONS

A. ACIDIC AND BASIC METABOLITES OF β-BLOCKERS

The β-adrenergic blockers (β-blockers) are used to control hypertension and other cardiovascular related conditions, and are effective "stoppers" that have been abused in human, equine, and canine sports (Wynne et al., 1996). In general, these compounds are metabolized by aryl hydroxylation and oxidative N-dealkylation, to yield either a primary amine metabolite, or lactic acid and glycol metabolites. The metabolic scheme for the β-blocker, propranolol in the greyhound, is shown in Figure 4.

The collection of both an acid-neutral fraction and a basic fraction from a Bond Elut Certify cartridge allows the effective screening of all the major metabolites. Enzyme-hydrolyzed urine from greyhounds that had been administered one of the β-blockers was diluted with an equal volume of phosphate buffer (pH 6.0) and was passed through a Bond Elut Certify cartridge, previously conditioned with methanol (2 mL), water (2 mL) and 0.1 M phosphate buffer (pH 6.0, 2 mL). The cartridge was washed with 0.1 M phosphate buffer (2 mL) and 1 M acetic acid (2 mL) for pH adjustment and then dried with air at 45 psi for 3 minutes. Acids and neutrals were eluted with chloroform/acetone (3:1, 2 mL, acidic fraction), the cartridge was washed with methanol (2 mL), and again dried with air for 3 minutes. The base fraction was eluted with ethyl acetate/dichloromethane/2-propanol (5:4:1) containing 2% concentrated aqueous ammonia (2 mL, basic fraction). Both the basic and acidic fractions were evaporated to dryness prior to derivatization and analysis by GC/MS.

The acid neutral fraction gave good recovery of the glycol and lactic acid metabolites, while the basic fraction showed excellent recovery of the parent drugs and their hydroxylated and other basic metabolites. In this case, the detection of the metabolite classes was desirable as published reports suggested significant inter-species differences in metabolism. Studies had also shown that for some β-blockers, the parent drug was excreted unchanged for only a short time following administration. The method allowed for the extraction of samples to screen for a large group of drugs and

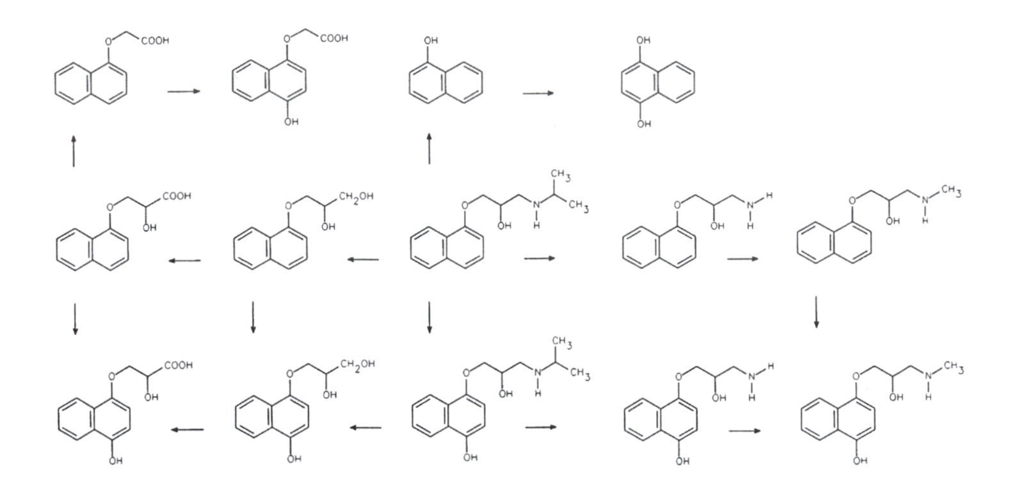

Figure 4. The metabolism of propranolol in the greyhound.

did not require modification for differences in the drug, species of animal, or period that had elapsed after administration of the drugs.

B. THE METABOLISM OF TIMOLOL AND ZWITTER-IONIC COMPOUNDS

Timolol (V), a β-adrenergic blocker that is commonly prescribed in human medicine, has had a reputation as an undetectable "slower," presumably because it is used at doses considerably lower than other β-adrenergic blockers. The drug is also highly functionalized and is subject to considerable first pass metabolism. Many of the basic metabolites are readily recovered from enzyme hydrolyzed urine using the method described above for other β-blockers, above (Vine et al., 1995; Gannon et al., 1997).

The lactic acid metabolite (Figure 5. VI) and the morpholino-cleaved acid metabolite (Figure 5. VII) have been identified in the dog (Tocco et al., 1980) and the horse (Law, 1993; Duffield et al., 1992). Both compounds are amphoteric and are difficult to recover from urine by solvent extraction. Recovery of the amphoteric and basic metabolites was achieved by passing enzyme hydrolyzed urine through a conditioned Bond Elut Certify cartridge. The cartridge was rinsed with water and the pH adjusted with pH 4 citrate buffer. Interfering neutral and acidic material was washed off the cartridge with methanol. Timolol and its metabolites may be detected when the column is eluted with isopropanol/ammonia (10:1 v/v). This fraction requires silylation prior to analysis by GC/MS.

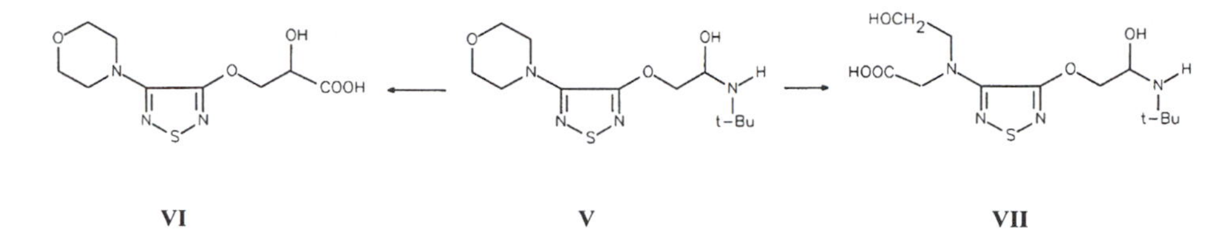

Figure 5. Structure of timolol (V) and its amphoteric primary metabolites (VI and VII).

Similar methodology may be applied to the recovery of benzoylecgonine, the major equine urinary metabolite of cocaine, and the carboxylic acid metabolite of detomidine, a veterinary tranquilizer. A modification of the method for the recovery of more polar amphoteric compounds, such as some α-amino acids, is described below in example J. The use of the Certify method described above gives cleaner extracts from equine urine than previously reported procedures using a C18 phase (Chiu et al., 1997)

C. BASIC DRUGS AND INTERFERENCES FROM HORSE URINE ON A NON-POLAR/SCX MIXED MODE SORBENT

If horse urine is extracted as described in example A, or extracted with 1-chlorobutane or chloroform/isopropanol (3:1 v/v) by liquid-liquid techniques, marked differences in recovery of analytes will be observed. The efficiency of analyte recovery in the basic eluate is dependent on several competing factors including polarity and functionality, volatility, stability to hydrolysis and the non-polar steric surface of the analyte. The influence of these factors and comparative recoveries of some drugs are listed in table 2.

Nicotine is volatile and also readily bound to glass surfaces. The superior recovery of nicotine by SPE may reflect reduced evaporative losses and minimized exposure to reactive surfaces. Fenspiride is recovered using SPE at the same improved ratio when compared with both solvent systems. When solvent extraction is carried out close to the pKa value for an individual compound, incomplete neutralization can result in reduced recoveries. Increasing the pH of the extraction system may give better recovery of more basic drugs, but other compounds may show reduced recoveries due to ionization of phenolic OH groups and other functional groups. For general screening methods, the optimum pH is the value at which the best recoveries are achieved for the maximum number of target analytes.

Table 2. Comparative recoveries of volatile bases spiked in horse urine at 10 μg/mL, by SPE extraction and solvent extraction. Percent recovery of each component in the SPE extract relative to solvent extraction with solvent a (1-chlorobutane) and solvent b (chloroform/2-propanolol (3:1)). Indicators of the compounds' properties are qualitative only. (+) = high, (-) = low.

Compound	% RECOVERY Solvent a	Solvent b	Polarity	Volatility	Hydrolytic stability	Basicity
Benzocaine	10	10	+	-	+	-
Benzydamine	104	63	-	-	+	+
Chlorpromazine	92	49	-	-	+	+
Desethyllignocaine	121	69	+	-	+	+
Dextromoramide	112	51	-	-	+	+
Diazepam	74	28	+	-	+	-
Fenspiride	153	154	+	-	+	+
Lignocaine	88	70	-	-	+	+
Hydroxylignocaine	312	78	+	-	+	+
Hydroxyprolintane	266	170	+	+	+	+
Methyl phenidate	79	264	-	+	-	+
Nicotine	>>5000	>>5000	+	++	+	+
Procaine	110	1170	-	-	-	+
Prolintane	180	2280	-	++	+	+
Propoxycaine	108	223	-	-	-	+

Retention of organic bases using the strong cation exchange sorbent, SCX, dependent on the protonation of the bases to their cationic form, and this mechanism is therefore capable of extracting compounds with a greater range of pKa values. It has particular application in the case of ring-hydroxylated metabolites of fenspiride, whose high basicity, combined with high polarity, might result in poorer recovery by solvent extraction.

Analysis of the methanol wash solution that is applied prior to the base elution shows the presence of both benzocaine and diazepam. While the compounds initially retain well on the sorbent by non-polar interactions, their acid/base properties result in poor retention at the sulfonate sites and their subsequent elution in the methanolic wash. The 4-amino group of benzocaine is only weakly basic due to the destabilization of the protonated form by the electron withdrawing ester group. The structurally related propoxycaine has a tertiary amine moiety that is capable of ion exchange on the SCX phase, and it therefore retains efficiently and elutes in the basic fraction.

SPE of a lignocaine administration sample shows not only acceptable recovery of lignocaine, but also good recoveries of hydroxylated and N-dealkylated products. It is difficult to obtain similar recoveries by solvent extraction, as these species require the use of solvents of such high polarity that unacceptably dirty backgrounds often result.

In keeping with the trends in the recovery of more polar basic compounds, we see an increase in the detection rate of loline, hydroxylated lupanine and phenylethylamine alkaloids, and water-soluble B-group vitamins (notably metabolites and artifacts of thiamine). While the extraction and detection of such compounds increases the data processing load per sample, it is desirable that they are extracted. These compounds are similar in structure and functionality to many target analytes and therefore provide a valuable indicator of the SPE performance. The failure to extract these compounds could be indicative of the failure to extract functionally similar drugs.

For example, hordenine, a common alkaloid found in green barley and various grasses, has the molecular weight of and a similar mass spectrum to the ring-hydroxylated metabolite of methamphetamine. The two are readily differentiated following derivatization. However, manipulation of the extraction methodology to eliminate hordenine from the extract would also eliminate most hydroxylated sympathomimetic amines. Notably, the basic fraction of the SPE extract from equine urine is virtually free from the abundant neutral compounds such as equol and other plant-derived phenolic compounds that are endogenous to equine urine. The method also eliminates compounds such as cholesterol, vitamin E, and fatty acids.

The extraction of greyhound urine using the method described above for horses, shows similar recoveries of the drugs described. Greyhound urine, as expected, shows no evidence of plant alkaloids or their metabolites. However, several substituted quinolines and other polycyclic nitrogen-containing compounds associated with a diet high in protein (derived from cooked meat) are readily observed. Some of these compounds are highly polar species and are not observed following solvent extraction. Again, the presence of such compounds, while increasing the data processing load per sample, are an indicator of the broader extractive power of the mixed-mode SPE method compared to solvent extraction. The extract is virtually free of cholesterol and fatty acids, which are prominent in fractions recovered from dog urine by solvent extraction.

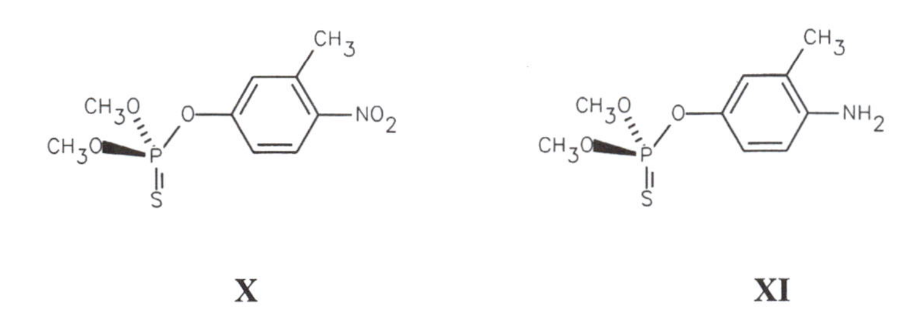

Figure 7. Fenitrothion (X) structure and that of its reduction metabolite (XI).

D. A HIGHLY WATER-SOLUBLE DRUG AND METABOLITE

2-Amino-4-picoline (VIII) is an active compound that was investigated for use as an analgesic in the 1950s. The drug has been demonstrated to have significant potency, primarily as a stimulant of β-endorphin release. It has also been found to be active as a respiratory and cardiac stimulant by stimulating the release of norepinephrine. Its non-specific activity has resulted in its withdrawal from the pharmaceutical market.

The drug, which may be considered performance-enhancing, is extensively metabolized in the horse to a ring-hydroxylated metabolite (Figure 6), which is recovered in low yield from urine by solvent extraction. Illicit use of the compound was thought to be virtually undetectable by doping chemists because of the compound's high water solubility, high polarity, and low molecular weight.

The SPE procedure described in Example A gave excellent recovery of both the metabolite and parent drug (Wynne et al., 1998). The sample was evaporated in a cool tube, to prevent evaporative loss, and acetylated with acetic anhydride prior to analysis by GC/MS. The result was an extract clean enough to allow the isolated metabolite to be studied by NMR spectroscopy, thereby permitting the assignment of the structure 6-amino-4-methyl-2-pyridone (IX).

E. COMPOUNDS SENSITIVE TO EXTREMES OF pH

The solvent extraction of basic drugs usually requires neutralization of an amine moiety between pH 9.5-11.5. Compounds containing a carboxylate group are typically extracted between pH 2 and 4. Some drugs and artifacts, such as the reduction metabolite of fenitrothion (Figure 7), are hydrolyzed by such pH adjustment. Fenitrothion is used as a grain fumigant and its amino metabolite is observed in many horse urine samples.

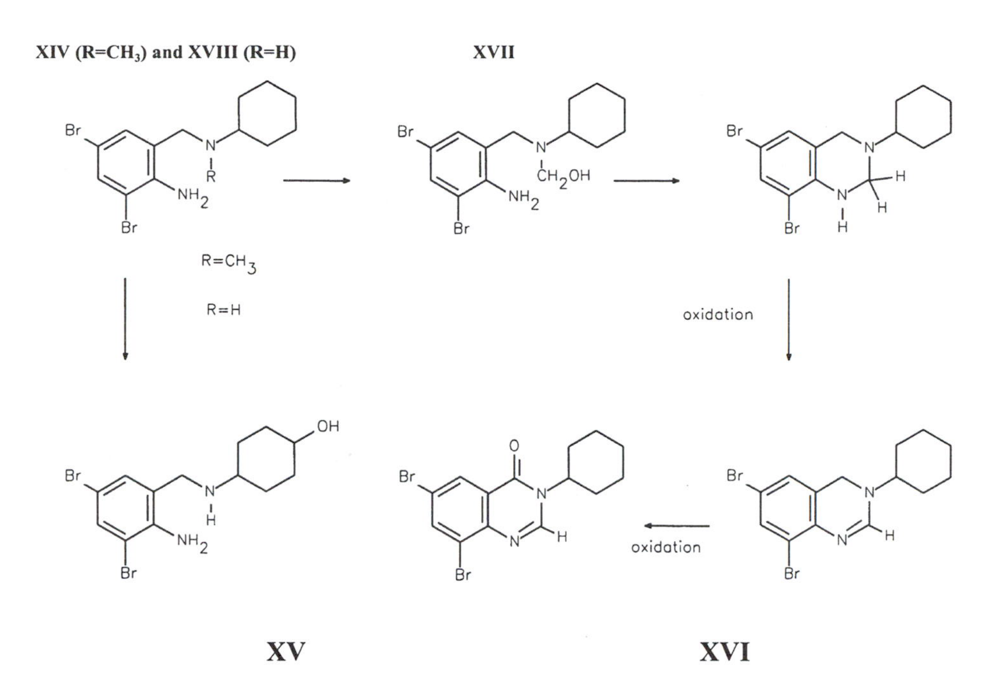

Figure 8. Bromhexine (XIV) metabolism pathways and reactions.

The extraction of urine samples, using the procedure described in Example A, provided good recovery of the metabolite (Wynne et al., 1994). The compound was not well recovered from urine samples subjected to extremes of pH.

F. ARTIFACTS FROM COMPOUNDS SENSITIVE TO EXTREMES OF pH

The recovery of unstable drug metabolites and artifacts poses interesting problems in analytical toxicology. In the case of aminofenitrothion, the use of SPE allowed the isolation of a compound that might otherwise go undetected. Here we explore a different issue, arising from the ability of SPE to pull many compounds out of a urine sample under gentle pH conditions (Wynne, 1997).

Bromhexine (Figure 8, XIV) is a mucolytic agent used in veterinary medicine and is converted in vivo to its active metabolite ambroxol (XV). Solid-phase extraction of enzyme-hydrolyzed horse urine, collected following administration of bromhexine, showed a compound by GC/MS that was detectable for a longer period of time than either XIV or XV. The compound was shown to be 6,8-dibromo-3-(4'-oxo-cyclohexyl)-4-(3H)-quinazolinone (XVI), an artifact of an unstable hemi-aminal metabolite (XVII).

Solid-phase extraction at pH 6.3 provides conditions suitable for the isolation of the transient hemi-aminal species (which subsequently undergoes cyclization and oxidation to XVI). The compound was not recovered by solvent extraction, nor was the compound recovered by SPE from samples that were subjected to extremes of pH. The results of this study suggested that the major urinary metabolite of bromhexine in the horse is 2,4-dibromo-6-(N-cyclohexyl-N-(hydroxymethyl)-amino) methylaniline (XVII) which is subsequently demethylated to desmethylbromhexine (XVIII), before being further metabolized to ambroxol.

It is significant that SPE permitted the identification of an important metabolite that had previously not been observed in urine extracts. In the absence of parent drug it was initially regarded as an unknown because the mass spectrum of the artifact was significantly different from both the parent and the decomposition products observed following solvent extraction.

A change in methodology from solvent extraction to SPE must therefore be accompanied by a change in criteria for the further investigation of unusual compounds detected during testing.

G. ISOLATION OF GLYCOPYRROLATE FROM EQUINE URINE

Glycopyrrolate is an anticholinergic drug that has been abused in racing events. Weak cation exchange is the preferred method of extraction as the quaternary ammonium moiety makes the compound difficult to isolate by other means. Typically, a urine sample is adjusted to pH 7 and applied to a CBA cartridge that has been conditioned with methanol and phosphate buffer at pH 7. The cartridge is washed with buffer, methanol, chloroform, and methanol again. Glycopyrrolate is eluted with methanol containing 5% acetic acid. The eluate is further purified by passage through a conditioned C18 cartridge, which is washed with ammonium acetate buffer and water before being dried with air and washed with methanol. The glycopyrrolate is eluted with ammonium acetate buffer/acetic acid/methanol (1:2:47 v/v/v). Following concentration of the eluate, the sample is analyzed by LC/MS (Medonca et al., 1992).

This method may be applied successfully to the extraction of other quaternary ammonium compounds such as propanthalene, N-butylhyoscine bromide, their polar metabolites, and other similar drugs from urine.

H. ISOLATION OF GABAPENTIN FROM SERUM

Isolation of basic drugs from serum and plasma may be achieved using similar extraction techniques to those used for urine. The fatty acid component of blood, particularly in samples collected from greyhounds, results in

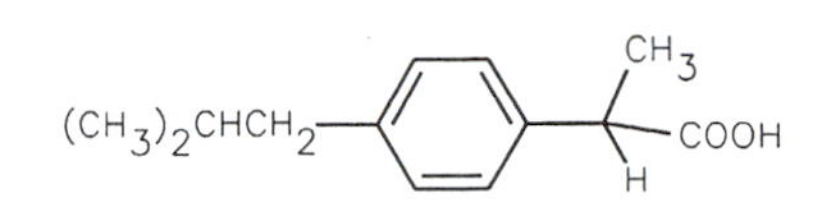

Figure 9. Gabapentin (XX) and ibuprofen (XXI) structures.

a matrix interference in the fraction that would be expected to contain acidic drugs, making the isolation of acidic analytes more difficult.

Wolf et al., (1996) have described the isolation of gabapentin (Figure 9, XX) from serum. An aliquot of the serum (500 μL) was combined with 20% acetic acid (500 μL) and applied to a C18 SPE cartridge that had been conditioned with methanol, water, and dilute HCl. The cartridge was washed with water, ethyl acetate and hexane, dried, and the target analyte eluted with methanol containing 2% ammonia. The sample was derivatized prior to analysis by GC/MS.

Ethyl acetate and hexane were employed as washing solvents as the protonated form of gabapentin was known to have little solubility in them. Selective elution of analytes from a C18 SPE cartridge by variation of the elution solvent polarity is employed routinely for anabolic steroid analysis. In the case of gabapentin, SCX or Certify approaches may also have proved effective.

This method also deals with the problem of extracting species that contain a carboxylic acid group from blood. Isolation of simple carboxylic acid drugs which contain no other "active" functional groups, such as ibuprofen (Figure 9, XXI), is more difficult. While some degree of clean-up on a C18 phase may be achieved with selective elution mixtures, it is not possible to remove all fatty acid material. Further, selective elution is of limited value for screening purposes. The diverse structures of various non-steroidal anti-inflammatory drugs (NSAIDs) requires that they all be eluted into the same fraction with a general elution mixture. Fortunately, many acidic drugs such as the NSAIDs are used at relatively high doses and the target analytes may be separated from other material by chromatography.

I. SEPARATION OF ANABOLIC STEROIDS

There are few significant variations in SPE methods employed for the extraction of anabolic steroids. Typically, the aglycone and sulfate conjugated steroids are isolated from enzymatically hydrolyzed urine by applying the sample to a C18 SPE cartridge conditioned with methanol and water. The cartridge is washed with water to remove salts and other soluble material and with hexane to both remove some non-polar material and aid in drying the cartridge. The steroid fraction may be eluted directly with methanol for subsequent methanolysis in the presence of methanolic HCl. The fraction may also be further partitioned using Sephadex™[28] LH-20 chromatography. Alternatively, the steroids may be eluted by a methanolysis solvent of ethyl acetate/methanol (9:1 v/v) containing sulfuric acid. Following methanolysis, the sample is neutralized and washed with water to remove excess acid reagents and derivatized prior to analysis by GC/MS.

The steroid fraction may be partitioned by eluting free steroids and glucuronides using ether. The sulfate conjugates may then be eluted using methanol. Use of these various isolates to establish the administration of anabolic steroids has been well described by Houghton (1992).

J. THE RECOVERY OF α-AMINO ACIDS FROM URINE

Example B described the modification of a mixed-mode non-polar/SCX method for the recovery of bases that allowed recovery of amphoteric compounds. The method is unsuitable for some highly polar compounds such as the α-amino acid, tyrosine. Presumably the zwitterionic form applied to the Certify cartridge at pH 6 has sufficiently high polarity that it is not retained by the C8 phase, nor is their sufficiently distinct cationic character in the protonated amino group for retention at the sulfonate exchange sites. Failure of the ion-exchange mechanism is due primarily to the proximity of the anionic carboxylate group which repels the zwitterion from the sulfonate site. Elimination of the anionic characteristics of the analyte prior to extraction of the sample leads to successful retention and elution.

Typically, enzyme-hydrolyzed urine was adjusted to pH 1 and passed through a preconditioned Bond Elut Certify cartridge. The cartridge was washed with 0.01M HCl (pH1-2), dried, and then washed with methanol and the analytes eluted with methanol/aqueous ammonia (10:1 v/v). This fraction was evaporated to dryness and derivatized with bis(trimethylsilyl) trifluoroacetaminde (BSTFA) prior to analysis by GC/MS.

K. THE RECOVERY OF CATECHOLAMINES AND RELATED COMPOUNDS FROM PRESERVED URINE

Catecholamines and related compounds have been successfully extracted from acid-preserved urine by SPE using alumina. The phase must be activated and is therefore packed into cartridges immediately before use. In a typical extraction, acidified urine is treated with an anti-oxidant solution containing mercaptoethanol, sodium metabisulfite, and ascorbic acid, adjusted to pH 8.5 and extracted onto the activated alumina phase. The alumina phase is washed with phosphate buffer and water, then dried. Cate-

---

[28] Sephadex is a trademark of the Pharmacia Company, Uppsala, Sweden.

chols are eluted with methanol/acetic acid (0.4 M) and the eluate is evaporated to dryness prior to derivatization and analysis. This method has been found to give good recovery of dihydroxyphenylacetic acid (DOPAC) but variable recoveries of dopamine and DOPA. Variations of the alumina adsorption method have been described by many authors (Anton et al., 1962; Letellier et al., 1997).

The use of PBA cartridges has been found to give more consistent recoveries of DOPA and dopamine. Care must still be taken when using either extraction mechanism, to eliminate potential interference from diet-derived catechols.

IV. LIMITATIONS OF SPE FOR VETERINARY SAMPLES

Generally, methods that fail for particular drugs or groups of drugs are not reported. Often poorer results for one compound are the result of a compromise between effective recovery of that compound and its inclusion in a general rather than targeted screen. Such compromises are required in terms of speed and efficiency.

Caffeine is an excellent example of such a compromise. It is a weakly basic to neutral molecule (its acid salts are dissociated rapidly in aqueous solution) that is well recovered from urine by solvent extraction with chlorobutane at pH 9 to 10. Typically, solvent extraction under such conditions co-extracts few interfering species. Extraction of hydrolyzed urine on a C18, C8 or mixed-mode cartridge results in recovery of caffeine (and the other neutral xanthines) in the neutral or acid/neutral fraction. This fraction is rich in the endogenous urinary acids and plant phenolics found in horse urine. These co-extracting species may interfere with the chromatography and detection of caffeine and increase the number of "dirty" samples run on an instrument. It is usually not cost effective to introduce a separate SPE extract for the xanthines in such cases. In the author's laboratory, where SPE provides the basis of almost all drug screening, caffeine is still covered separately by solvent extraction, followed by analysis using GC/NPD.

Like the xanthines, the benzodiazepines have traditionally been recovered from aqueous solutions by solvent extraction at neutral or basic pH. These drugs are somewhat more polar than the xanthines, necessitating the use of more polar elution solvents if high recoveries are desired. Yet this requirement for more polar extraction solvents also results in increased levels of co-extracted matrix components. Many analysts have a bias towards including such structurally and clinically similar analytes in a single screening method. While such a screen is highly desirable when possible, it

Table 3. pKa data for a selection of benzodiazepines. The general structure is shown in the protonated form. Data is taken from Clarke (1989) and the Merck Index, 12th edition. The general structure is shown in the protonated form.

DRUG	R_1	R_2	R_3	R_4	R_5	pKa_1	PKa_2
Clonazepam	H	=O	H	NO_2	Cl	1.5	10.5
Clorazepate	H	$-(OH)_2$	COOH	Cl	H	3.5	12.5
Diazepam	CH3	=O	H	Cl	H	3.3	-
Flurazepam	$CH_2CH_2N(Et)_2$	=O	H	Cl	F	1.9	8.2
Lorazepam	H	=O	OH	Cl	Cl	1.3	11.5
Medazepam	CH_3	H_2	H	Cl	H	6.2	-
Midazolam		(A)	H	Cl	F	6.2	-
Nitrazepam	H	=O	H	NO_2	H	3.2	10.8
Nordazepam	H	=O	H	Cl	H	3.5	12.0
Oxazepam	H	=O	OH	Cl	H	1.7	11.6
Temazepam	CH_3	=O	OH	Cl	H	1.6	-
Triazolam		(B)					
Alprazolam		(B)				2.4	

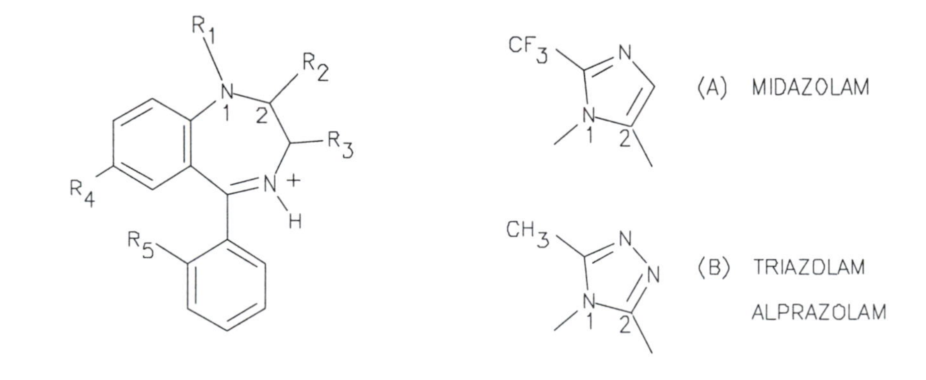

is not simple to develop a SPE method which covers all members of the group. Some common benzodiazepines and their pKa values are listed in Table 3. The value pKa_1 refers to the deprotonation of the N-4 position in the cationic form of the parent, while pKa_2 generally refers to the deprotonation of the N-1 position to yield an anionic species. Exceptions are clorazepate where the first deprotonation is of the carboxylic acid function, and flurazepam where the pKa_2 is related to the N,N-diethylamino group.

When developing an extraction regime for the benzodiazepines, it is important to remember that many are subject to considerable metabolic change. Often little of the parent drug is excreted — a case typified by diazepam, which is excreted in the human mainly as the metabolites oxazepam, temazepam, and nordazepam. Most benzodiazepines are eluted in the acidic-neutral fraction when a general extraction method such as that described in example A is applied. Some, such as the nitro-reduced metabolites of nitrazepam and clonazepam and the basic drugs such as flurazepam, are sufficiently well retained by the SCX phase to be eluted in the basic fraction. Acidic drugs such as clorazepate may be poorly eluted in the acidic-neutral fraction if the pH adjustment step is not sufficiently acidic to neutralize the carboxylate group. While some clorazepate would be eluted in the basic fraction, without derivatization it is unlikely to show sufficiently good GC characteristics for analysis. The acidic-neutral fraction from horse urine is generally too high in endogenous acids and neutrals to be suitable for mass-screening by GC methods, though it may be amenable to further analysis by HPLC.

As might be anticipated the neutral forms of many benzodiazepines have been extracted from human urine at a variety of pH values on both neutral and mixed-mode phases. One potentially useful approach for targeted extraction and clean-up of the group from a complex matrix has been described by Berrueta et al. (1993). Bond Elut C2 columns were conditioned with methanol and a pH 6 buffer. The urine was adjusted to pH 6 and passed through the cartridge, which was then washed with various mixtures of methanol and water. Target analytes were eluted from the column with a different mixture of methanol and water.

This method makes use of the ability of the benzodiazepines to form a hydrogen bond with uncapped silanol groups while still having sufficient hydrophobic surface character to be retained by the C2 phase. It is likely that many of the endogenous acids and phenolic compounds would also be retained by the phase, but careful selection of the wash and elution solvents allows a degree of clean-up to be achieved. Not surprisingly, the same authors have reported a similar method using a silica phase for the recovery of benzodiazepines from both blood and urine (Casas et al., 1993). Optimization of such a method is described in Chapter 10.

The need to obtain high recoveries of as many benzodiazepines as possible is driven, in part, by the limitations of alternative screening methods. Structural variations in the group have resulted in poor coverage of some analytes by immunoassay. Such methods are generally designed for maximum cross-reactivity with the bicyclic species such as diazepam. Unfortunately, the methods often show poor cross-reactivity towards the tricyclic analogues such as midazolam, alprozolam, and their metabolites.

A further example of a method that initially showed little promise by SPE was described by Chui et al. (1994). Chromoglycate, a highly functionalized diacid, was found to be poorly recovered using a Bond Elut Certify II cartridge because of the high endogenous acid content of equine urine. Better recovery of the compound was achieved by extraction of the urine at pH 1.5 with ethyl acetate, after which the organic-soluble material was esterified to form the bis-tert-butyl ester. The derivatized extract was then purified by retaining it on a conditioned C18 SPE cartridge and eluting the target analyte with methanol/water (9:1) prior to LC/MS analysis. While an ion-exchange phase (Certify II) proved to be inappropriate in this case, SPE remained an effective choice for the recovery of the derivatized analyte.

The examples described in this section all illustrate an important point: While direct adaptations of analyses developed around solvent extraction may have failed, the SPE portion of each adaptation still resulted in the extraction of the target analyte or analytes. Where the SPE adaptations were unsuccessful it was because the extracts were unsuitable for further analysis using methodology that had been developed for the analysis of solvent-extracted samples. In considering the development of SPE methods for drug screening, it is important to remember that the target analytes may not always be recovered in the same fraction in which they were recovered by alternative methods. For example, the acidic-neutral eluate described in Example A may be used for the screening of the xanthines and a wide variety of benzodiazepines. Many analysts may consider that these compounds are suitable analytes for GC analysis. Because of unacceptably high matrix backgrounds, the SPE eluate is not suitable for GC analysis. However, the eluate may, with careful planning, be screened for these and many other target groups by LC or LCMS.

V. THE CRYSTAL BALL

A. NUMBER OF TARGET ANALYTES

One commonly held belief amongst industry observers is that the drugs abused in racing are becoming more sophisticated. However, a review of the drugs commonly identified in animal urines shows that the older drugs do not appear to be on the way out. They are used because they are effective, easily administered, stable without special storage requirements and, most significantly, inexpensive and readily available. Conversely, many new drugs and synthetic polypeptides are unstable and require careful preparation and administration. More recently developed pharmaceuticals are often expensive and less commonly available. Despite this, it is likely that in the medium to long term, the number of target analytes and categories will increase both because of the introduction of new therapeutics and a continued use of older drugs.

B. INSTRUMENTATION

Extraction technology has yet to evolve to the point where the industry can consider moving its focus from chromatography as the final step in separation prior to analysis. The requirement for analytical results that can be defended under both scientific and legal scrutiny demands the use of a technique that provides unequivocal results. The "fingerprint" of a digitized mass spectrum provides more definitive (and therefore more legally defensible) identification of a compound at biological concentrations than other techniques. The number of bits of informative power in a mass spectrum is only exceeded by NMR spectroscopy, a technique that is limited in sensitivity (it is inherently insensitive to carbon) and not usually interfaced with a chromatographic system. Recent advances in stop flow LC/NMR suggest that tandem instruments may become available in the near future however the capital cost of such instruments would make their routine use unlikely.

Mass spectrometry is currently undergoing a rapid expansion in the area of LC/MS. The LC interface allows the introduction of aqueous samples into the mass spectrometer and may reduce the need for derivatization of some compounds. The method is limited, in comparison to GC, by poorer peak shape and resolution, and use of the technique will demand effective sample extraction and concentration. However, the method has found significant application in the analysis of water-soluble compounds (sulfonates, phosphates, and quaternary ammonium compounds), sulfonamides, and carboxylates for which HPLC is the preferred chromatographic method. The use of high-resolution mass spectrometry is not yet routine in the field of racing chemistry. Its application in human sports testing has largely centered on the detection of low levels of anabolic steroids.

Capillary electrophoresis/mass spectrometry (CE/MS) has not found regular use in racing chemistry although its ability to separate amphoteric compounds may benefit some applications. The method would require sample preparation that was at least as extensive as that currently used for LC or GC applications.

In the context of SPE extraction of biological samples, the most significant developments to laboratory instrumentation have been in the area of automated (robotic) sample extraction. Although other chapters have considered SPE robotics, there are some unique considerations in the automated sample handling of equine urine.

The most limiting factor in handling equine (and other herbivore) urine is the viscosity. Animals that have become dehydrated during racing may pass particularly viscous urine which, when refrigerated, can produce a gelatinous semi-solid that is difficult to break up. Either enzymatic or acid hydrolysis, followed by centrifugation and dilution of the sample is necessary to prevent the clogging of cartridges or fluid handling systems. Clogging is particularly problematic when SPE robots are being used during unsupervised periods of operation. The delivery system components and the means of applying liquids and the sample to the SPE cartridge (diameter of fluid lines, syringe or vacuum pumps, or gas pressure, for example) are particularly important. If a pressure overload device is tripped too easily during sample application, a large number of samples may fail to elute correctly or automated runs may unnecessarily be aborted.

Three commercially available systems were made available by their manufacturers and evaluated in the author's laboratory, for their suitability in the extraction of equine and canine urine and other racing samples. Instruments were evaluated against each other and against a manually operated vacuum manifold. The comparison was carried out using the extraction method described in example A for a series of acidic and basic drugs of varying polarity using Bond Elut Certify and HCX Confirm™[29] cartridges.

The Spe-ed Wiz™[30] SPE was promoted as being capable of processing 30 samples in 30 minutes. On evaluation, the instrument was found to dispense all reagents in series, rather than parallel, even though the instrument has a 30 port manifold. Sample loading was only possible under vacuum and it was this step (along with the positioning of appropriate collection tubes) that required operator intervention. The gas sealing system (the means of achieving positive pressure elution) required modification in the laboratory to achieve a satisfactory seal. This instrument was found to be inappropriate for unattended operation and with vacuum drawing of the sample, there was danger that the cartridges could be deconditioned or run dry inadvertently.

The ASPEC™[31] has an X-Y-Z tracking probe which delivers all reagents and samples to the cartridge. The cartridges are contained in a block moving in the X-axis over the instrument bed on which racks of sample and collection tubes rest.

The Benchmate™[32] II uses a robotic arm to move the sample and elution tubes from collection racks to a weigh station that is able to sense sample loading and analyte elution (by change in mass of the tubes). The cartridge is transferred to an SPE station that operates by driving a fluid-handling plunger into the SPE cartridge, thereby compressing the sorbent bed and allowing uniform application of all samples and reagents to the bed. This instrument is easy to program but limited in the selection of cartridge and tube sizes that it accepts. It also delivered a preprogrammed air push behind each reagent or sample delivered to the cartridge. For some

---

[29] Confirm is a trademark of International Sorbent Technology, Hengoed, Wales.
[30] Spe-ed Wiz is a trademark of Applied Separations, Lehigh Valley, PA.
[31] ASPEC is a trademark of the Gilson Medical Electronics Company, Villiers le Bel, France.
[32] Benchmate is a trademark of Zymark Corporation, Hopkinton, Massachusetts.

applications, slow flow rates had the potential to increase the run time to unacceptable lengths.

The Benchmate II appeared to provide much more efficient drying of the column than the ASPEC, although both used compressed gas. This variable was found to be particularly important for good recovery of bases using the optimized elution scheme described in example A. Sorbent bed compression by the plunger in high-quality cartridges with sorbents of narrow particle size ranges results in less channeling of the fluids through the sorbent bed and therefore more reliable stationary phase-mobile phase interactions. Further, superior drying of the sorbent prior to elution of both the acid-neutral and basic fraction leads to more effective elution of analytes and less carry over into subsequent eluates.

The ASPEC and the Benchmate II both gave acceptable recoveries for acidic and neutral drugs (the first eluate), compared to a manual SPE method. However, the Benchmate II gave superior recoveries for basic drugs (the second eluate) than those obtained using either the ASPEC or vacuum manifold. Fewer cartridges blocked on the Benchmate II than on the other systems investigated. The Zymark Rapid Trace, which is equipped with a similar plunger system to the Benchmate II, proved to be significantly less capable of loading equine urine samples that had been enzymatically hydrolyzed and diluted. This is because the Benchmate II has more powerful syringe pump motors combined with a well-sealed sample application system (the plunger mechanism). As a result, the instrument was less likely to abort an extraction run when samples of higher than normal viscosity were encountered.

C. SPE, IMMUNO-AFFINITY CHROMATOGRAPHY, AND OTHER EXTRACTION TECHNIQUES

The concentration of drugs in biological samples such as urine, and the nature of the matrix limit the application of micro-SPE methods in racing chemistry drug testing. Extraction of sufficient analyte to exceed the detection limits on most of the screening instruments employed, usually requires the extraction of a relatively large volume of urine (> 2 mL) and also requires a high degree of clean-up.

In the immediate future we are likely to see continued growth in the use of SAX, CBA, SCX, and NH2 phases. Such growth will reflect the increased availability of LC/MS and the ability to better extract highly polar or water-soluble drugs. Use of other specialty phases with aqueous elution solvents will doubtless be explored and they will become of greater importance in the recovery of low level residues, highly functionalized species and the investigation of previously unextracted target groups.

The nature of the urine matrix has meant that until now analysts have sacrificed clean extracts to ensure that as many target analytes as possible will be found in each extract. Historically, analysts have attempted to replace solvent extraction methods with SPE methods that mirror the recovery of target analytes. Increased use of specialty phases may lead to a redistribution of target analytes amongst the fractions for analysis with simultaneous reduction of the matrix background.

Perhaps the most specific of specialty phases are those employed for immuno-affinity chromatography (IAC). Two techniques have been described, one employing a covalently bound (immobilized) antibody cartridge and the second using an immobilized protein cartridge to which the antibody is bound. The latter method requires elution of the antibody-antigen pair for recovery of the analyte and may be regarded as somewhat similar to the use of ion-pairing reagents to effect extraction. Use of IAC has been limited by the expense of the cartridges and the need for other clean-up procedures such as SPE or solvent extraction to be used to further purify the sample before application to the IAC column. Provided care is taken to avoid denaturing of the antibody or the prosthetic groups by heat, solvents or extremes of pH while achieving complete dissociation of the antigen antibody complex during the elution step, these cartridges can be reused. The cartridges may also be subject to microbiological degradation and must be protected accordingly. IAC is unlikely to be widely used in the short to medium term unless the range, availability and stability of the phases are increased and the unit cost reduced. IAC will continue to be used in conjunction with other methods, for the further investigation of unusual screening results.

The use of covalently-bound antibody cartridges and their application to racing chemistry has recently been reviewed by Beumier et al., (1996) and Delahunt et al., (1997). The method has found limited application for the isolation of both β-agonists and corticosteroids, which are often difficult to isolate from other urinary excretion products because they are used at relatively low doses. IAC has also been successfully applied to the recovery of atrazine and its metabolites (Rollag et al., 1996). Each family of these compounds has a common structure suitable for the generation of a class-specific antibody.

IAC techniques may offer some potential in the extraction of reserpine and other low dose drugs for which current extraction methodology may prove of limited effectiveness. The benzodiazepines which remain a difficult class of compounds to extract effectively because of variations on the basic structure, as previously described, may also be better recovered by IAC. Nedved et al. (1996) have used immuno-affinity extraction with non-covalently bound antibody on a protein C cartridge for the recovery of selected benzodiazepines from pure solutions and crude mixtures. It might be anticipated that while a generic antibody may be raised to many bicyclic benzodiazepines, the simultaneous extraction of the tricyclic and further substituted species with such an antibody may be less efficient. In such circumstances, multiple antibody cartridges may be viable.

The use of natural products and hormones as therapeutic agents may expand the number and types of compounds targeted by drug screening laboratories. Further increases in the determination of levels of endogenous substances should also be expected. A move away from the readily detected NSAIDs and corticosteroids may necessitate the monitoring of inflammatory response by determining the levels of prostaglandins and leukotrienes. Specialty phases or modified phases such as argentation media may be adapted to SPE preparation of samples but these phases are unlikely to find routine application in the medium term. Rather, such sorbents are likely to be used for the further separation of compounds recovered by more conventional SPE methods.

VI. CONCLUSIONS

It is important to realize that SPE is a clean-up method and that the final separation of analytes requires the use of a separation technique. However, use of the technology has permitted more flexible use of chromatographic or spectroscopic instruments and has given us new methods to enhance analyte recovery and reduce the matrix background.

SPE can be highly reproducible if a rugged method is developed, and provides clean extracts for analysis by GC/NPD, HPLC, GC/MS, LC/MS and many other analytical techniques. As the acceptance of specific applications grows, we shall see a shift in the way we classify drugs based on their liquid-liquid extraction properties.

We have shown in this chapter how, with rare exceptions, when chosen correctly SPE phases can not only match but can beat traditional extraction techniques. In addition, SPE is safe and readily automated. Therefore, exposure to large quantities of harmful solvents is minimized and labor requirements are significantly reduced. For all these reasons we should expect to see a significant increase in the use of SPE throughout the racing industry over the next few years.

REFERENCES

Anton, A.H. and Sayre, D.F., (1962) Factors affecting the aluminium oxide-trihydroxyindole procedure for the analysis of catechol amines. J.Pharmacol.Exp.Ther., 138: 360-375.

Batty, D.C., Wynne, P.M. and Vine, J.H., (1994) Extraction of acidic and basic drugs from equine and canine blood and urine using a single solid phase extraction cartridge, Proc. 12th ANZFS Symposium, Wellington, New Zealand, 12-19 Nov., 1994.

Batty, D.C., Wynne, P.M. and Vine, J.H., (1996) Comprehensive drug screening using a single solid-phase extraction cartridge, Proc. 11th Int.Conf.Rac.Analyst.Vet., Ed. E. Houghton and D. Auer, R&W Publications (Newmarket, UK), 197-202.

Beaumier, P., Sarkar, P. and Leavitt, R., (1996) Applications of immunoaffinity chromatography: a review, Proc. 11th Int.Conf.Rac.Analyst.Vet., Ed. E. Houghton and D. Auer, R&W Publications (Newmarket, UK), 123-132.

Berrueta, L.A., Gallo, B. and Vicente, F., (1993) Analysis of oxazepam in urine using solid-phase extraction and high performance liquid chromatography with fluorescenecedetection by post column derivatisation, J.Chrom.Biomed.Appl., 616: 344-348.

Budavari, S., (1989) The Merck Index 11th edition. An Encyclopedia of Chemicals, Drugs and Biologicals. Merck and Company, Rahway, NJ.

Casas, M., Berrueta, L.A., Gallo, B. and Vicente, F., (1993), J.Pharmaceut.Biomed. Anal., 11: 277.

Chen, X.H., Wijsbeek, J., Franke, J.P. and De Zeeuw, R.A., (1992) A single column procedure on Bond Elut Certify for systematic toxicological analysis of drugs in plasma and urine, J.Forensic Sci., 37: 61-72.

Chiu, F.C.K., Damani, L.A., Li, R.C. and Tomlinson, B., (1997) Efficient high performance liquid chromatographic assay for the simultaneous determination of metoprolol and two main metabolites in human urine by SPE and fluorescence detection., J.Chromatogr. B., 696: 69-74.

Chui, Y.C., Esaw, B., Laviolette, B. and Yu, N., (1994) Detection of sodium chromoglycate in horse urine and serum following administration of cromovet, Proc. 10th Int.Conf.Rac.Analyst.Vet., Ed. P. Kallings, U. Bondesson and E. Houghton., R&W Publications (Newmarket), 233-238.

Delahunt, Ph., Jacquemin, P., Colemonts, Y., Dubois, M., DeGraeve, J. and Deluyker, H., (1997) Quantitative determination of several synthetic corticosteroids by gas chromatography - mass spectrometry after purification by immunoaffinity chromatography, J.Chrom. B, 696: 203-215.

Duffield, A.M., Reilly, P.J., Matyr, S., Wise, S. and Suann, C.J., (1992) An equine urinary metabolite of oral timolol and the detection of etorphine in urine by NICI GC/MS, Proc. 9th Int.Conf.Rac.Analyst.Vet., Ed. C.E. Short, ICRAV and Louisiana State University (Louisiana), 159-164.

Foster, S.J., Chalmers, P. and Dunnett, N., (1990) An automated solid phase extraction procedure for some diuretics and similar drugs in horse urine with HPLC detection, Proc. 8th Int.Conf.Rac.Analysts Vet.; 3-10.

Gannon, J.R., Wynne, P.M. and Vine, J.H., (1997) Timolol slow release devices in the racing greyhound, Proc. 11th Int.Conf.Rac.Analyst.Vet., Ed. E. Houghton and D. Auer, R&W Publications (Newmarket, UK), 388-394.

Houghton, E., (1992) Anabolic steroids in the horse - a review of current knowledge, Proc. 9th Int.Conf.Rac.Analyst.Vet., Ed. C.E. Short, ICRAV and Louisiana State University (Louisiana), 3-16.

Ishimitsu, S., Fujimoto, S. and Ohara, A., (1989) High performance liquid chromatographic determination of m-tyrosine and o-tyrosine in rat urine, J.Chrom.,Biomed.Applic., 489: 377-383.

Law, W.C., Leung, K.K. and Crone, D.L., (1993) Timolol metabolites, Proc. 9th Int. Conf. Rac. Analyst. Vet., Ed. C.E. Short, ICRAV and Louisiana State University (Louisiana), 255-261.

Lettelier, S., Garnier, J.P., Spy, J. and Bousquet, B., (1997) Determination of the L-DOPA/L-tyrosine ratio in human plasma by high performance liquid chromatography. Usefulness as a marker in metastatic malignant melanoma. J.Chrom. B, 696: 9-17.

Leferink, J.G., Dankers, J. and Schotman, (1987) Doping analysis using solid-phase extraction and GC/MS with automated data handling, Proc. 6th Int.Conf.Rac.Analyst.Vet., Ed. D.L. Crone, Macmillan Publishers (Hong Kong), 171-176.

Medonca, M., Ryan, M. and Todi, F., (1992) Glycopyrrolate: Detection and elimination in the horse, Proc. 9th Int.Conf.Rac.Analyst.Vet., Ed. C.E. Short, ICRAV and Louisiana State University (Louisiana), 327-336.

Moffatt, A.C. (Editor) (1986) Clarke's Isolation and Identification of Drugs in Pharmaceuticals, Body Fluids and Post-mortem Materials, 2nd Edition, Pharmaceutical Press (London), 87-263.

Muskiet, F.A.J., Thomasson, C.G., Gerding, A.M., Fremouw-Ottevangers, D.C., Nagel, G.T. and Wolthers, B.G., (1979) Determination of catecholamines and their 3-O-methylated metabolites in urine by mass fragmentography with use of deuterated internal standards, Clin.Chem., 25(3): 453-460.

Nedved, M.L., Habibi-Goudarzi, S., Ganem, B. and Henion J.D., (1996) Characterization of benzodiazepine "combinatorial" chemical libraries by on-line immunoaffinity extraction, coupled column HPLC-Ion Spray Mass Spectrometry-Tandem Mass Spectrometry, Anal.Chem., 68: 4288-4236.

Popot, M.A., Bonnaire, Y. and Plou, P., (1992) Screening by HPLC/fluorimetric detection and confirmation by MS/MS of ethamsylate in horse urine and plasma samples, Proc. 9th Int.Conf.Rac.Analyst.Vet., Ed. C.E. Short, ICRAV and Louisiana State University (Louisiana), 111-120.

Rollag, J.G., Beck-Westermeyer, M. and Hage, D.S., (1996) Analysis of Pesticide Degradation Products by Tandem High-Performance Immunoaffinity Chromatography and Reversed-Phase Liquid Chromatography, Anal.Chem., 68: 3631-3637.

Samuels, T., Teale, P. and Houghton, E., (1994) Applications of bench-top LC/MS to drug analysis in the horse: 1. Development of a quantitative method for urinary hydrocortisone, Proc. 10th Int.Conf.Rac.Analyst.Vet., Ed. P. Kallings, U. Bondesson and E. Houghton., R&W Publications (Newmarket), 115-118.

Stanley, S.M.R. and Rodgers, J.P., (1994) Use of immunoaffinity chromatography with HPLC-particle beam MS and GC/MS for the confirmation and analysis of corticosteroids in equine urine, Proc. 10th Int.Conf.Rac.Analyst.Vet., Ed. P.

Kallings, U. Bondesson and E. Houghton., R&W Publications (Newmarket), 226-232.

Tocco, D.J., Duncan, A.E.W., DeLuna, F.A., Smith, J.L., Walker, R.W. and Vandenheuvel, W.J.A., (1980) Timolol metabolism in man and laboratory animals, Drug Metabol.Dispos., 8(4): 236-240.

Vine, J.H., (1996) The central role of mass spectrometry, Proc. 11th Int.Conf.Rac.Analyst.Vet., Ed. E. Houghton and D. Auer, R&W Publications (Newmarket, UK), 151-157.

Vine, J.H., Wynne, P.M., Jenkins, M.A., Lind, K.L. and Dyke, T.M., (1995) Metabolism and excretion of timolol in the racing greyhound, 12th Aust. N.Z. Int.Conf. on Forensic Sci., Auckland.

Walden, M., Dunnett, N. and Chalmers, P., (1990) A solid phase extraction procedure for acidic drugs with HPLC separation and UV fluorescence detection, Proc. 8th Int.Conf.Rac.Analyst.Vet., 11-15.

Wolf, C.E., Saady, J.J. and Poklis, A., (1996) Determination of gabapentin in serum using solid-phase extraction and gas-liquid chromatography, J.Analyt.Toxicol., 20: 498-501.

Wynne, P.M., Vine, J.H. and Amiet, R.G., (1998) The metabolism and urinary excretion of 2-amino-4-methylpyridine in the horse. Proc. 12th Int.Conf.Rac.Analyst.Vet, (in press) R&W Publications (Newmarket, UK).

Wynne, P.M., (1996) Several artefacts from equine urine, Proc. 11th Int. Conf. Rac. Analyst. Vet., Ed. E. Houghton and D. Auer, R&W Publications (Newmarket, UK), 481-487.

Wynne, P.M., Batty, D.C. and Vine, J.H., (1994) Urinary artefacts from feed additives, Proc. 10th Int.Conf.Rac.Analyst.Vet., Ed. P. Kallings, U. Bondesson and E. Houghton., R&W Publications (Newmarket), 249-252.

Wynne, P.M., Dyke, T.M., Vine, J.H., Lind, K.L., Jenkins, M.A. and Wickramasinghe, I., (1996) Beta-blockers in greyhounds II: Beta-blocker metabolism and analytical detection, Proc. 11th Int.Conf.Rac.Analyst.Vet., Ed. E. Houghton and D. Auer, R&W Publications (Newmarket, UK), 470-477.

10

SOLID-PHASE EXTRACTION OF BIOLOGICAL SAMPLES

Steen H. Ingwersen, Novo Nordisk A/S, Maaloev, Denmark

I. INTRODUCTION

Solid-phase extraction is widely used for the preparation of biological samples for further analysis in areas as diverse as clinical chemistry, forensic science, and biomedical and pharmaceutical research. The popularity of

SPE is due in part to its utility in achieving high selectivities and recoveries, its ease of use and minimal consumption of hazardous extraction solvents. This, and the following two chapters address the specific challenges and requirements of the most commonly encountered biological matrices and clinical/toxicological applications for which SPE has proved effective.

This chapter does not offer a comprehensive literature review; rather its intent is to present the author's experience from developing a number of SPE procedures for new drug candidates. A review of SPE applications is provided by Scheurer and Moore (1992). A summary of properties and components of common matrix types is provided in most clinical text books. This is important because identifying and eliminating interferences and troubleshooting low recoveries requires good knowledge of the matrix composition. The procedures used for preparation of biological samples have been reviewed by McDowall et al. (1989), Tippins (1987), and Watson (1994), and recommendations for development of SPE procedures have been presented by Simmonds et al., (1994). However, a somewhat different approach is suggested in this chapter.

The examples presented below have been developed for pharmacokinetic studies during drug development. Modern drug candidates are often very potent substances, for which reason the doses administered in preclinical and clinical studies are relatively low. Thus, assay sensitivity is a major goal of pharmacokinetic studies; sensitivity must be high enough to allow estimation of the terminal plasma half-life in vivo. In order to obtain such sensitivities, clean plasma extracts are essential.

The emphasis in this chapter has been laid on extraction of blood plasma for HPLC assay. This should not imply serious limitations because many of the basic principles for extracting plasma are generally applicable to other biological matrices and to different analytical techniques. Section IV deals with special problems encountered when extracting other biological sample types such as urine or tissue samples.

II. GENERAL CONCERNS WITH BIOLOGICAL SAMPLES

Extraction of biological samples before injection into an HPLC system serves a number of objectives. Two of these, concentration and clean-up, were introduced in Chapter 1, and indeed apply to almost every SPE extraction. In addition, we can add three additional and important objectives:

- Prevention of clogging of analytical columns
- Elimination of protein binding

- Elimination of enzymatic degradation of the analyte

These objectives may be met merely by removal of the bulk of protein constituents from the sample. This is usually achieved by default when using SPE, so the emphasis during method development may be placed on optimization of the analyte recovery and removal of residual interfering compounds.

A further opportunity to concentrate the sample emerges if the final extract is evaporated and the residue is dissolved in a smaller volume than that of the extracted sample. When possible, this may be exploited for improving the assay sensitivity. For a 0.5 mL sample the limit of quantitation (LOQ) can typically be reduced three to fivefold using this approach, provided that the purity of the extracts are high. Giese (1994) has shown that some SPE methods introduce buffer salts into the final extract. When evaporating extracts to dryness prior to reconstitution, consideration should be given to these salts, as they can mask analytes and result in poor resolubilization.

III. A STRATEGY FOR METHOD DEVELOPMENT FOR PLASMA SAMPLES

A. RATIONALE

The following outline proposes a strategy for performing the experiments necessary to develop and optimize a solid-phase extraction, although in practice, requirements for additional experiments will inevitably emerge during the process of optimization. For instance, based on results of the optimization experiments, it may become necessary to repeat previous steps under different conditions, or an unexpected finding may suggest an approach which was not originally anticipated. In principle, however, the optimization steps described below will always have to be considered when developing a new SPE procedure.

As described elsewhere in this book, the physicochemical properties of the analyte and of the sorbent are of paramount importance to the retention of the analyte on a particular sorbent material. We have also seen how modeling and theory can help us anticipate the best sorbent and choice of conditions for extraction of an analyte. Nevertheless, the strategy proposed here takes a more empirical approach to method optimization. This has been done for three reasons.

First, secondary interactions may operate under a range of conditions for many analytes; such interactions should be exploited whenever possible because they are potentially beneficial (Bland, 1986; Law, 1992). Theoretical modeling of the secondary interaction contribution to retention/elution may be difficult as the magnitude of these effects depends on many factors, some of which will be unknown for a particular combination of analyte and sorbent. Second, most resources spent during method development are likely to be spent on efforts to remove interfering substances from the extracts rather than on optimizing the analyte recovery. As the nature of the interfering substances in general will be unknown, we shall have to rely on experimental data to determine the success of each approach. Third, by using the empirical approach suggested here, a large range of sorbent materials, pH values, washing solutions etc., will be tested, thereby avoiding a premature rejection of possible extraction conditions.

B. PRACTICAL CONSIDERATIONS

Before attempting SPE optimization, some practical questions need to be answered. One of these is the choice of elution procedure. Four options for eluting samples and solvents through SPE cartridges are available: Vacuum, gravity flow, positive pressure (by a single syringe or pump or using a positive pressure manifold), and centrifugation. The choice of method will depend on the requirements for batch size, available equipment, time available per extraction and often, to a surprisingly large degree, on personal preferences.

For vacuum-mediated elution, commercial manifolds are available with capacities in the range of 8 to 30 cartridges. Where larger capacities are required, manifolds may be easily constructed using polycarbonate sheets and similar materials. Irrespective of the manifold capacity, it is important for the reproducibility of the procedure that the vacuum is continuously regulated during the elution in order to obtain a near constant flow through the cartridges. If a regulator is not supplied with the manifold, vacuum control may be achieved by placing a vacuum distillation controller in the vacuum line. The main advantage of the vacuum-mediated elution is the ease of handling because the SPE cartridges can remain on the manifold lid throughout the extraction process. One drawback is a potential carry-over from previous extractions due to adsorption of analytes or impurities onto the surface of manifold needles if these needles are re-used. If not encountered or investigated during method development, this concern will need to be addressed during method validation.

Elution by centrifugation offers a convenient way of reproducibly eluting SPE cartridges, the capacity being dependent only on the size of the available centrifuge. In order to obtain well-defined elution conditions the centrifuge should be controlled by a thermostat, although this is not an absolute requirement. In practice, the elution is performed with the SPE cartridge flange resting on the lip of a centrifuge tube of appropriate dimensions (e.g. 13 x 100 mm for 3 mL reservoir cartridges). The tubes need not be emptied between each elution step unless the cartridge tip dips into the eluate at any step of the extraction. However, clean tubes should be mounted before desorption of the analyte from the cartridge. For this purpose conical glass tubes are suitable as they allow for handling small volumes of reconstituted extracts. Conditioning and elution steps are performed at relatively low centrifugal force (100-200 x g, depending on the actual conditions) to prevent the sorbent from drying out. This contrasts with the sample loading step, which may use high centrifugal force (if the sample is viscous, for example) in order to achieve an appropriate flow rate. It may be advantageous to perform the last washing of the cartridges by means of a stronger centrifugation in order to obtain cleaner extracts. The cartridge tips should be wiped in order to remove residues of sample constituents before desorption of the analyte.

Positive pressure is becoming more popular as a means of passing liquids through the SPE cartridge, both because of the introduction of commercially available positive pressure manifolds for manual sample processing and because of the growth in automated systems that use positive pressure or physical displacement rather than vacuum. The relative benefits of vacuum and positive pressure have not been fully evaluated at the time of writing (Wynne, 1997). There are some obvious specific cases where positive pressure is superior to vacuum processing (for example, use of nitrogen compressed gas, rather than laboratory air drawn through under vacuum, will benefit recoveries of analytes that are oxygen-sensitive). Other considerations, such as the lower cost of using vacuum or the greater control permitted by positive pressure systems, will probably influence the choice made by an analyst.

C. PRETREATMENT OF SAMPLES AND SPE CARTRIDGES

It is important to centrifuge plasma or urine samples (1500 x g, 10 min) before transfer to the SPE cartridge in order to prevent clogging. Other samples, such as tissue, will need different pretreatments before sample application (section IV). Even with properly pretreated samples, some blocking of the SPE device can occur. This may be overcome by increasing the centrifugation speed, the vacuum, or the pressure applied to the cartridge. Some kinds of sorbent materials such as extraction discs are particularly susceptible to clogging. Manufacturers have addressed this problem by placing a filter above the disc. This solution is proven to be effective although, inevitably, it results in a higher effective bed volume and risks adding product-related interferences to the extract.

Optimization experiments may begin with either aqueous solution or the matrix of interest. The latter is highly recommended as the retention properties of the analyte(s) in water may differ from those in other matrices such as plasma. As will be shown in section III, matrix components can have a dramatic influence on sorbent/analyte and analyte/sample interactions. Moreover, by introducing the biological matrix from the beginning, an early impression can be obtained of the purity of the extracts without the need to repeat some of the initial experiments.

For each SPE condition tested, two to four samples including blanks and spikes should be extracted and assayed. For the initial pH profiling experiments, in which retention versus sample pH is investigated, single determinations are preferred in order to cut down on the number of samples, whereas in subsequent experiments duplicate determinations may be used. Time may be saved by using the duplicate extracts injected sequentially onto the LC system to identify any late eluting peaks from one extract appearing in the next analytical run. The alternative is to run a clean mobile phase (blank) sample after the extracted samples from each SPE condition, especially during the exploratory extractions where very dirty sample extracts can be expected. Later on in the method development process when washing steps and elution steps have been optimized, assays of 5-6 replicates of the spiked sample for each condition may be used. This will allow for optimization of the precision of the extraction.

The concentration of the spiked samples should be sufficiently high to allow for estimation of recoveries even if interfering peaks are present. For methods using UV detection, concentrations in the range 1to10 μmol/l (0.3-3 μg/mL for a substance with a molecular weight of 300 Daltons) will usually suffice. On the other hand, by using concentrations that are too high, the capacity of the sorbent may be exceeded, thus resulting in low recoveries.

SPE columns need to be conditioned before sample application. This is usually accomplished by passing 0.5 to 2 mL of a suitable solvent (most commonly methanol for non-polar and ion exchange sorbents) followed by 0.5 to 2 mL of the liquid (usually an aqueous buffer) for diluting the samples. In some instances, however, it may prove advantageous to use distilled water instead of buffer for conditioning. The chapter on secondary interactions gives clear examples of where the often-overlooked conditioning step can have an enormous impact on analyte recovery. It is important to pay special attention to factors such as the buffer counter ions at this stage and to use the cartridge as soon after conditioning as possible, to prevent the sorbent drying out.

D. SCREENING SORBENTS AND pH VALUES

Since we must be able to elute a very high percentage of our compound from the sorbent in order to know how effective retention is, a quick test should be run before examining retention on various sorbent types and under various loading conditions. One approach is to load a blank sample onto a cartridge and then spike the analyte, in a small volume of stock standard solution, such that none of the standard solution passes through the cartridge. The analyte can then be eluted with a strong elution solvent such as methanol acidified with 1% glacial acetic acid or basified with 2% concentrated ammonium hydroxide solution. If a check for recovery indicates quantitative elution we can proceed to investigate retention.

The first experiment should be a combined screening of sorbent types and pH values of the dilution buffer. For each sorbent tested, buffers with pH values of 2, 3, 4, 5, 6, 7, 8, and 9 are used for diluting the sample before application to the SPE cartridge. For this purpose, 0.1 to 0.2 M buffer solutions are used. The buffers used by the author for these experiments are: phosphoric acid (pH 2.1); potassium phosphate (pH 6 - 8); sodium formate/triethylamine (pH 2.5 - 4); sodium acetate (pH 4.5 - 5.5); and tris (pH 9). Remember to specify the counter-ion again. For an initial screening experiment, it is sufficient to use steps of one pH unit whereas for subsequent optimization experiments, smaller pH steps may be called for. The number of sorbents to be included in the initial screening experiments is limited by the need to keep the experiments within reasonable size. On the other hand, 6 different sorbents can easily be tested within two days of work. This will allow for the testing of a range of reversed-phase sorbents including at least one type of extraction disc.

Mixed-mode sorbents should be tested as well at this stage, as these will often yield extracts of higher purity than conventional reversed-phase cartridges. It can be relevant to test identical sorbent types (e.g. C18) from different manufacturers as these will often possess different characteristics with respect to secondary interactions, as well as showing small differences in their primary retention mechanism.

A typical protocol for an initial pH profiling experiment would be to mix 1.5-2 mL of various buffer solutions with 0.5 mL of plasma before application to the cartridge. The cartridge should be washed twice with 2 mL of water and eluted with 2 mL of methanol. The extract is then evaporated, redissolved in mobile phase and injected into the HPLC.

The results of a pH profiling experiment are conveniently evaluated by plotting the recovery against pH for each sorbent tested or by overlaying chromatograms of blank samples. This will often immediately yield an impression of which pH value(s) will give optimal combination(s) of recovery and selectivity for each individual sorbent type.

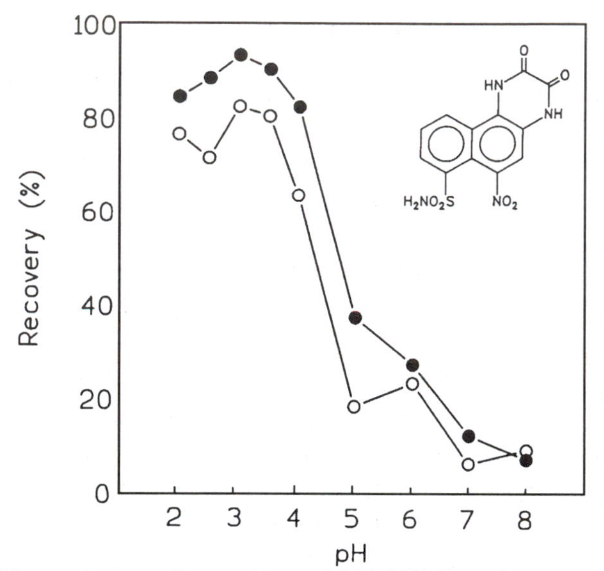

Figure 1. Recoveries of NBQX from human plasma extracted on Bond Elut Certify II (o) and Bond Elut C8 (•) at different pH values of the buffer used to dilute the sample (Ingwersen, 1993, with permission).

The following example studies the effect of pH on extraction of a drug for two different sorbent types. Aliquots of plasma spiked with NBQX (0.5 mL) were diluted with buffers at various pHs (1.5 mL) and 0.1 mL of internal standard solution. Each sample was applied to a preconditioned Bond Elut Certify® II[33] or a preconditioned Bond Elut C8 cartridge. The cartridge was washed with 2 mL of water and 2 mL of 1 mM solutions of the diluent buffer. The analytes were desorbed with 1 mL methanol/water (1:1 v/v), evaporated to dryness and the residue reconstituted in HPLC mobile phase prior to analysis. The buffers used for diluting plasma were: 0.1 M phosphoric acid, pH 2.1, 0.1 M formic acid adjusted to pH 2.5, 3.0, 3.5 or 4.0 with triethylamine, 0.1 M sodium acetate buffer, pH 5.0 and 0.1 M phosphate buffer, pH 6.0, 7.0 and 8.0.

An example of recoveries obtained in one pH profiling experiment using two different sorbent types are seen in Figure 1 for the experimental drug NBQX. Acceptable recoveries were obtained for this acidic analyte at pH values between 2 and 4 using both Bond Elut C8 and Certify II (containing a mixture of non-polar and anion exchange functionalities). Optimal recoveries were obtained at pH 3.0 on both sorbents, but inspection of the blank chromatograms (not shown) revealed that the cleanest chromatograms were obtained using the Certify II cartridges at pH 3.5.

The analyte studied in this example is a relatively hydrophilic weak acid with an octanol-water partition coefficient (log P_{oct}) value of 0 and a pK_a value of 6.4. These characteristics account for the almost complete lack of retention on non-polar sorbents at pH values above 6, but the optimal recoveries at pH 3.0 and the optimal purity at pH 3.5 for Certify II car-

[33] Bond Elut is a trademark and Certify is a registered trademark of Varian Inc, Palo Alto, California.

tridges could not be accounted for using the available physicochemical properties or other theoretical considerations.

When evaluating the initial pH profiling results and designing follow-up experiments, it is important not to reject sorbents based on lack of extract purity as this can be improved upon by optimizing the washing and elution steps. Choose instead the two to five combinations of sorbents and pH values with recoveries above 70% which yield the purest extracts.

E. OPTIMIZING THE WASHING PROCEDURE

In the pH profiling experiment described above, the sorbents were washed by water only. Having chosen the combinations of sorbents and pH values for further investigation, the next step to optimize is the cartridge wash. The parameters to optimize during this step are the composition of the washing solvent, the amount of washing solvent and the procedure used to eliminate the residual washing solvent before elution. The composition of the washing solvent should be investigated at the initial stage whereas the other parameters are conveniently investigated during the final optimization and simplification of the procedure (see Section III.G).

For a simple non-polar extraction, the composition of the washing solvent should first be varied using increasing concentrations of methanol or acetonitrile in water. Note that several other water-miscible solvents may be used but in practice it is seldom necessary to use more than these two common solvents. Typically the wash solvent methanol composition is increased in steps of 10% from 100% water through to 100% methanol. The extraction protocol would then require that the sample (0.5 mL) diluted with buffer (1.5 – 2.0 mL) be applied to the preconditioned cartridge. The cartridge would then be washed with water and the methanol/water combination to be tested. The remaining analyte should then be eluted and assayed for recovery by an appropriate technique.

An example of recoveries obtained by washing with increasing concentrations of methanol is shown in Figure 2. In this experiment drug X was extracted using SPEC® PLUS[34] C8 extraction discs. For each sample 0.5 mL of plasma was mixed with 1.15 mL of water, 0.25 mL of phosphoric acid (0.1%) and 0.1 mL of internal standard solution before application to the preconditioned extraction disc. The sorbent was washed twice with 2 mL of water followed by 0.5 mL of one of the methanol/water mixes. The analyte and internal standard were desorbed by eluting twice with methanol/ ammonia 1:1 (v/v). Evaporated extracts were dissolved in 150 µl of mobile phase/ water 1:1 (v/v) and 100 µl injected into the HPLC.

---

[34] SPEC is a trademark of Ansys Corporation, Irvine, California.

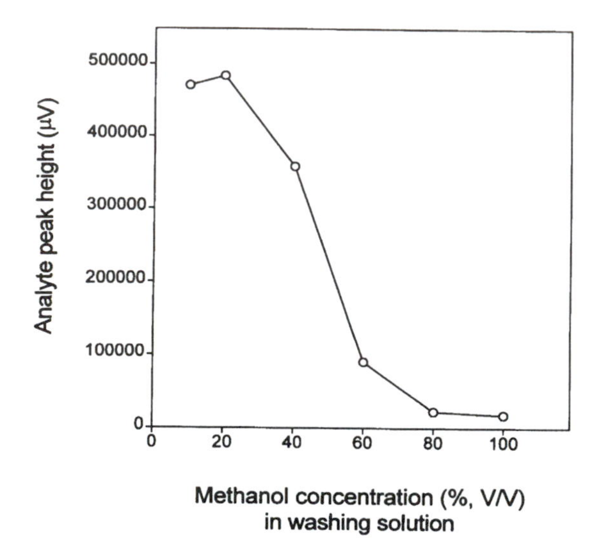

Figure 2. Effect of washing the sorbent with increasing concentrations of methanol on the recovery of drug X from human plasma (Ingwersen, unpublished results).

The analyte has a log P_{oct} value of 1.5 at pH 7.4 and contained carboxylic acid and tertiary amine functionalities with pK_a values of 3.4 and 9.4 respectively. As seen, the discs could be washed with up to 20% methanol without affecting the recovery. However, inspection of the blank chromatograms (not shown) revealed that methanol concentrations of at least 60% were required to achieve adequate clean-up and thus, this was not a feasible approach. The results shown in Figure 2 indicate that the methanol content of the elution solvent could be reduced from 100% to 80% without affecting recovery. While the method was not developed further in this case, the information that a weaker elution solvent than pure methanol could be used to desorb the analyte was obtained from this experiment. Such information can be useful in developing alternative methods, particularly if sample purity remains a problem during the elution step.

Disappointingly, as in the above experiment, the use of mixtures of methanol and water instead of water for the washing step rarely improves the extract purity dramatically for plasma samples (this is in contrast to extractions on other biological samples such as urine). A much more efficient washing solution would be pure acetonitrile. Surprisingly, this solvent can often be used for the washing of non-polar sorbents without substantial loss of recovery for biologically active compounds, an indication that secondary interactions are likely to be responsible for the retention of the analyte (refer to Chapter 7 for a full discussion). The difference in activity of methanol compared with acetonitrile may be explained by the ability of methanol to disrupt sorbent/analyte hydrogen bonding, leading to elution of the analyte in methanol. Acetonitrile is aprotic and is therefore less able to disrupt hydrogen bonding.

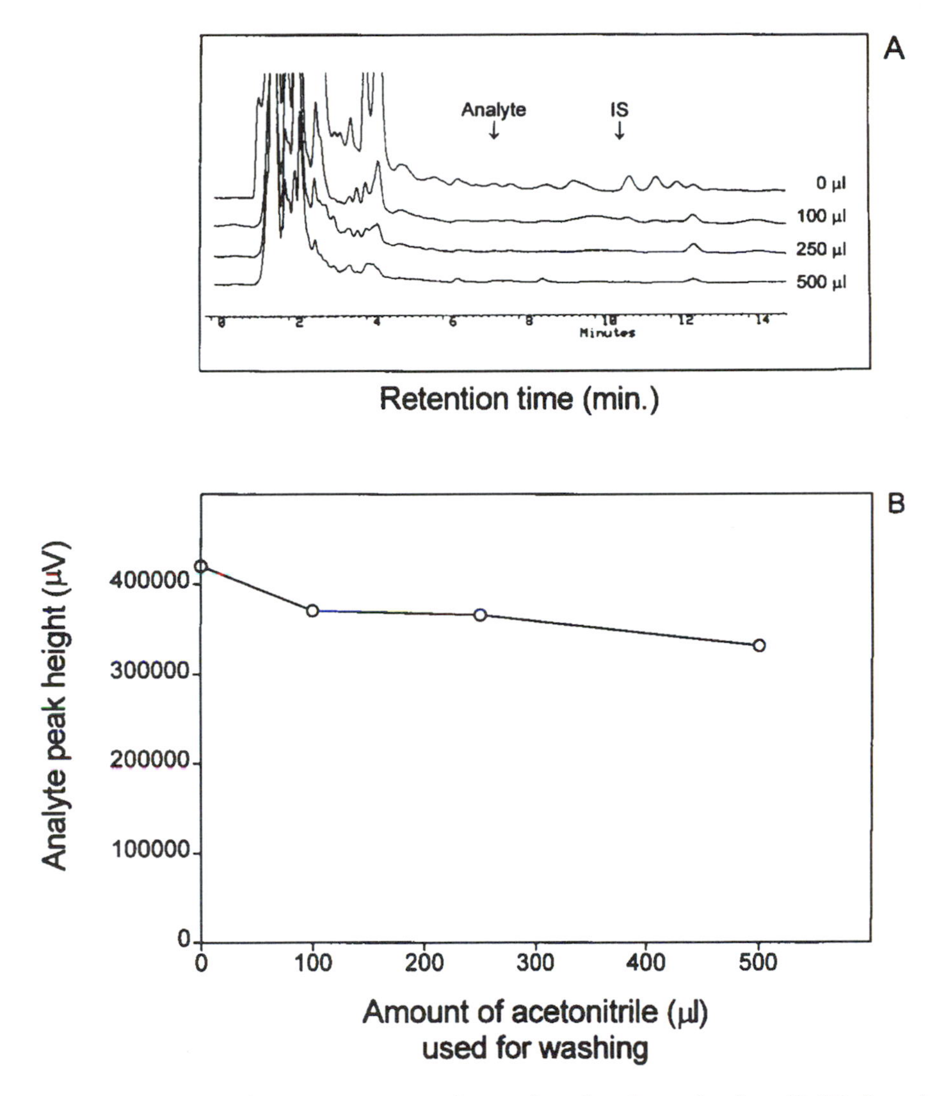

Figure 3. Purity/recovery versus volume of wash solvent for drug X (B) from human plasma using SPEC PLUS C8 extraction discs. The extraction conditions were as described in the text, except that the sorbents were washed twice by 2mL of water followed by 0, 100, 250, or 500 µl of acetonitrile before the elution of analyte and internal standard from the discs (Ingwersen, unpublished results).

Whenever possible, the option of washing with pure acetonitrile or another non-H-bonding solvent (acetone or even hexane have been used, Bland, 1986) should be exploited as it is a simple step to add to a procedure and can be easily optimized, as shown in Figure 3. If the analyte clearly H-bonds with its surroundings (for example a glucuronide-bound metabolite or a sulfated metabolite) it may be worth testing an acetonitrile wash during the pH profiling step. A positive outcome can lead to a very quick subsequent method development, and may free the analyst from selecting an inconvenient pH simply because the background is poor at other values of pH. An example of the use of pure acetonitrile as washing solution is shown in Figure 3 for drug X from the above experiment. The extraction discs in this case could be washed by 0.5 mL acetonitrile with only minor loss of extraction recovery (Figure 3.B), resulting in somewhat cleaner extracts (Figure 3.A).

A large number of washing solutions may be investigated for the removal of persistent interfering peaks. Solvents in which the analyte are insoluble are good choices for testing as a wash solvent as there is little risk of desorbing the analyte and thereby reducing recovery. Occasionally, complicated washing regimes will improve purity. However, exotic or complex washing schemes should be avoided where possible as they invariably lengthen the extraction and require more testing for ruggedness than might otherwise be necessary. Strategies for removing persistant interferences are dealt with in section III.H.

F. OPTIMIZING THE ELUTION PROCEDURE

The parameters to optimize in the elution step are:

- The elution solvent
- The volume of the elution solvent
- The number of elution steps

Optimization of the elution solvent is described in this section whereas optimization of the volume of the elution solvent and the number of elution steps are dealt with in section III.G.

If either the recovery is low or interfering impurities persist, substantial improvements may be obtained by changing the elution solvent. By far the most frequently used elution solvent is methanol and if at this stage, acceptable recoveries and purity have been obtained using methanol there is no need to proceed further, especially in those cases where the extracts need to be evaporated and redissolved before analysis. It should be remembered that methanol is not a perfect solvent, however. One example of this is its tendency to form UV-active oxidation products over time. Acetonitrile avoids this problem but is not as effective in all cases.

The ultimate goal is to desorb the analyte from the sorbent using the weakest possible eluting solvent. As mentioned above (section III.E), one possibility would be to use the information obtained from the washing experiments to investigate the use of mixtures of water and methanol or acetonitrile as elution solvents. Further experiments may require application of the sample, mixed with a suitable diluent, to a second cartridge. The cartridge should then be washed following the scheme in section III.E, and the analyte desorbed with an elution solvent prior to estimation of recovery by an appropriate technique.

The results of an experiment using two different sorbent types and mixtures of methanol and water for the elution of NBQX are seen in Figure 4. The solvent composition for obtaining optimal recoveries in this case was 50% methanol/water (v/v) for both sorbent types. Thus, it seemed that the hydrophilic nature of this analyte required that water be present in the eluting solvent for optimal recovery.

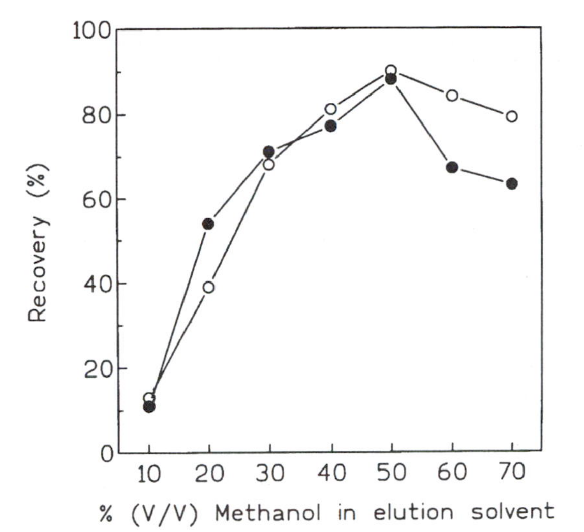

Figure 4. Effect of increasing methanol concentrations on the elution of NBQX from human plasma extracted on Bond Elut Certify II (o) and Bond Elut C8 (•) cartridges. All extraction conditions except elution were as described in the legend to Fig. 1 using the procedure that employs pH 3.5, 1 mL elution solvent (From Ingwersen, 1993, with permission).

If concentrating the sample before the final assay is unnecessary for a particular application, the mobile phase used for the HPLC procedure, or a mobile phase-compatible system is an elegant choice as eluting solvent, assuming it is effective.

An alternative approach is shown in Figure 5. In this case a gradient LC pump pushes a 0 to 100% methanol mobile phase through the SPE cartridge which has been connected to a UV detector by low-pressure fittings. Injection of the analytes yields a series of elution curves, from which it is easy to tell the solvent composition that completely elutes the most strongly retained analytes. At the same time the composition of the strongest wash solvent that does not elute the analytes can also be identified. Caution is necessary in interpreting this data, however, for it assumes that retention/elution is identical for purely aqueous samples and real-life samples such as plasma.

Another variable is the pH of the elution solvent. For acidic analytes a base may be included and for basic analytes an acid may be included in the elution solvent in order to improve the recovery of strongly retained species. Much information can be obtained from the results of the initial pH profiling experiments (section III.D) when considering this approach. Remember to use volatile acids (e.g. acetic acid or formic acid) and bases (e.g. ammonia or ethylamine) for inclusion in the elution solvent if evaporation

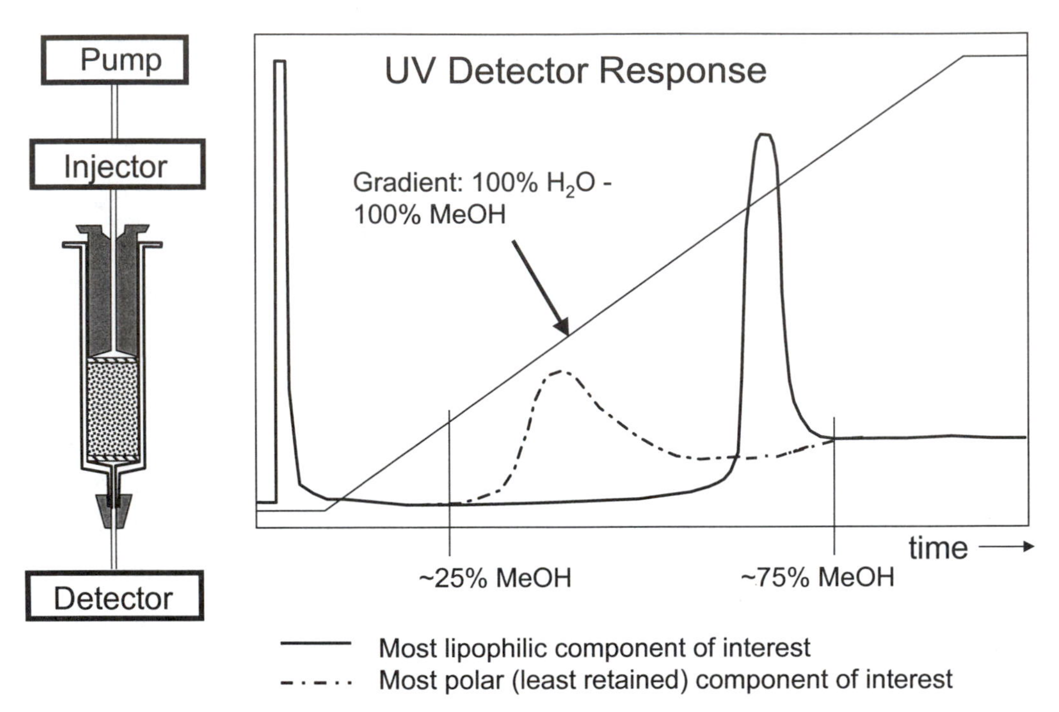

Figure 5. A novel way for optimizing elution and wash solvent strength. Analyte is injected onto a SPE cartridge through which a gradient mobile phase (100% aqueous at t = 0 to 100% methanol at the end of the run) is pumped. The elution profiles of the analytes are easily observed using this technique.

of the extracts is required. Usually, 0.05 to 0.2 N solutions of acids or bases will suffice, but the concentration needs to be optimized for each application. These modifiers may also be thought of as performing like HPLC mobile-phase modifiers since one aspect of a typical non-polar retention is the secondary interaction between sorbent and analyte. Such interactions may be weakened or strengthened during wash and elution steps by changing the effective pH of the applied solution.

G. FINAL OPTIMIZATION AND SIMPLIFICATION

Having obtained a procedure with acceptable recovery (at least 50%) and purity, it is time to work on final optimization and simplification of the procedure. This step is most important if the assay is going to be used for large numbers of samples, where incremental savings in time and solvent will have a substantial payback in cost and efficiency. Some of the experiments described in this section aim to improve the precision of the procedure, for which reason replicate assays (e.g., 4 - 6) of spiked samples should be per-

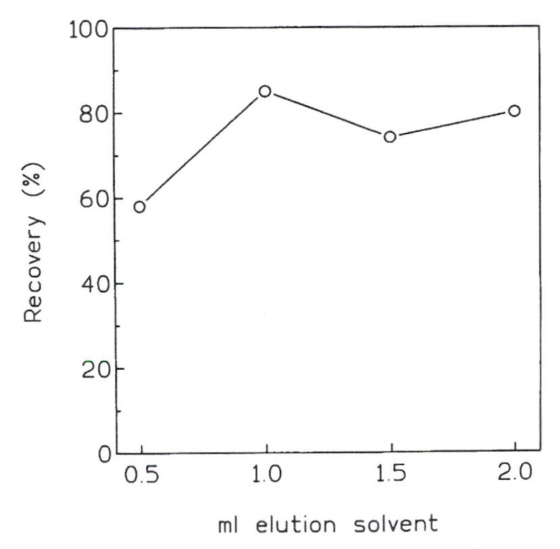

Figure 6. Effect of the volume of elution solvent used to desorb NBQX from Bond Elut Certify II cartridges. The extraction conditions were as described in the legend to Fig. 1 (procedure at pH 3.5) except that the analyte and the internal standard were eluted from the sorbent by 0.5 (A), 1 (B), 1.5 (C), or 2.0 mL (D) of methanol-water, 50:50 (v/v) (From Ingwersen, 1993, with permission).

formed at each extraction condition investigated. At a minimum, the following experiments should be considered:

- Simplifying the sample application
- Reducing the number of washing steps and the amount of washing solvent
- Optimizing the elution procedure for the washing solvent
- Optimizing the volume and number of steps used for eluting the analyte

In most published methods the sample is diluted with buffer before application to the SPE cartridge. However, the transfer of the mixture from vials to cartridges either by pipetting or by decanting, constitutes an operation which may be unnecessarily time consuming, in particular for large batches of samples. Thus, the alternative of mixing the sample with buffer directly on top of the SPE cartridge should be investigated, taking into consideration the effect on both recovery and precision. The prerequisite for this simplified procedure is that the sample does not spontaneously pass through the cartridge by gravity, in which case a uniform treatment of the samples in an assay run can not be obtained. Care must also be taken to obtain sufficient mixing of sample and buffer on the cartridge.

If not investigated during the initial experiments, the possibility of cutting down on the number of washing steps and on the volume of the washing solution should be considered. Thus, the initial water wash can often be omitted without effect on the purity, and the volume of the washing solution may be reduced, in particular when using extraction discs. This will cut down on the required volume of the reagent and it may even obviate the need for an additional emptying of the vials used to hold the cartridges when using centrifugation as a sample-processing mechanism.

When performing the final wash before desorbing the analyte, air drying should be tested. Air drying is accomplished by applying prolonged vacuum or by using high speed centrifugation (e.g., 1000 x g) during elution. Some investigators even include an additional washing step using a small volume of an organic solvent (e.g., 25 μl of acetone) in order to remove water residues on the cartridge. If this is the method chosen, care must be taken to ensure this volume does not begin to elute analytes.

Drying of the sorbent may improve the purity and even increase the recovery but again, care must be taken. Sufficient drying can reduce the injection peak tail considerably for reversed-phase HPLC of non-polar extracts of plasma and other biofluids. Too much drying, however, can lower recoveries (Nau and Pocci, 1997), though whether this is caused by evaporative loss, oxidation from the air or change in retention mechanism as water and conditioning solvent are removed from the bonded phase remains unclear.

The volume of the solvent used to elute the analyte should be minimized, as this will reduce the time required to evaporate the extract and will minimize background. As seen in Figure 6, the volume of solvent used to elute NBQX from Bond Elut Certify II cartridges could be reduced from 2 to 1 mL without affecting the recovery. For extraction discs, a volume of 0.5 mL or even less of the elution solvent will often result in complete desorption of the analyte.

The number of applications of the elution solvent may be important for recovery and precision. Thus, we have repeatedly observed that eluting twice with 1 mL of solvent will increase the recovery compared with the use of a single elution step of 2mL of the same solvent. Even if the improved recovery obtained by a two-step elution appears marginal or insignificant, this procedure may result in an improved precision. Thus, although a two-step elution procedure is more cumbersome, it might be of interest to consider at this stage.

One final consideration, and one that is nearly always overlooked, is optimization of the sorbent bed mass. Wilson and Nicholson (1994) used the technique of proton nuclear magnetic resonance (NMR) to investigate the retention of ibuprofen metabolites from urine in a semi-quantitative way. Their results showed that while the initial application on 500 mg C18 Bond Elut cartridges gave good recovery, much less sorbent was actually needed for "good" retention. The assessment of "good" was based on a subjective inspection of the NMR spectrum signal-to-noise. Indeed, the ibuprofen metabolites were appreciably retained on only 25 mg of sorbent. Meanwhile co-retained interferences (for example, citrate, hippurate and creatinine) were vastly reduced. While not all these species will appear as interferences in a GC or LC chromatogram, they will adversely affect the instrument operation, possibly resulting in more regular maintenance of the GC injector and detector or more regular replacement of the LC guard column.

H. STRATEGIES FOR REMOVING PERSISTENT INTERFERENCES

Persistent interferences often constitute the main problem during method development and months can be spent on attempting to remove impurities with little progress. In the author's experience, a strategy for removing interfering impurities should consider the following steps, which are discussed in more detail below:

- Test pure acetonitrile as a washing solution.
- Search for a weaker eluting solvent for desorbing the analyte.
- Test sorbents from different manufacturers or with different functionalities.
- Change the detector wavelength or consider an alternative detection technique.
- Change the chromatographic procedure (stationary or mobile phase).

As pointed out above (Section III.E), the use of complicated washing regimens rarely results in dramatic purity improvements, the main exception being the use of pure acetonitrile or even pure methanol as washing solutions. Remember: If the analyte is completely insoluble in a solvent, then that solvent should be tested as a wash solvent if improved clean-up is required.

If pure methanol or a mixture of methanol and an acid or a base is used for elution during method development, alternative solvents should be investigated. Thus acetonitrile in some cases can be substituted for methanol as elution solvent, and if acidic or basic methanolic solutions are used, the effect of reducing the concentration of methanol while increasing the water content should be tested.

Sometimes dilute aqueous acids or bases with little or no organic content can be used for desorbing relatively hydrophilic and potentially ionizable analytes from reversed phase SPE cartridges. This will almost always produce extracts of higher purity than if methanol were included in the elution solvent.

Alternative sorbents (either identical sorbent types from different manufacturers or completely different sorbent types) should be considered. This is where the benefit is realized of not narrowing down the sorbents under examination too early in the method development process. If a procedure has been optimized on one brand or type of reversed phase sorbent, the method may generally be applied to another sorbent without further development or with minor modifications. Should one of these sorbents show promise it may be worthwhile to re-optimize the procedure for that sorbent. When carrying out such investigations, the inclusion of mixed-mode sorbents and extraction discs should be considered. Both types of device have been shown to give superior results for some applications. In the case of extraction discs the improvement has been linked to lower bed mass (Wilson and Nicholson, 1994).

An elegant way of dealing with interference problems is to modify the detection procedure. Fluorescence detection is substantially more selective than UV detection and advantage can be taken of this for fast and efficient method development since less stringent requirements are placed on the SPE clean-up. If fluorescence is not applicable for a given analyte, the use of higher wavelengths should be considered. Usually, the absorption peak is chosen as the wavelength used for UV detection. However, if this peak λ_{max} is well below 300 nm, interferences from plasma are likely to occur and the use of higher UV wavelengths should be considered. This principle was exploited for the assay of NBQX which has an absorption maximum of 294 nm. As seen in Figure 7, the chromatograms of blank plasma were considerably cleaner when the wavelength was increased from 294 nm to 380 nm, and although this was far from the peak wavelength, a limit of quantitation of 2 ng/mL (corresponding to 5.9 nmol/l) was obtained. Further, Figure 7 shows that a mixed-mode sorbent (Bond Elut Certify II) yielded a superior purity than a conventional C8 sorbent for this analyte.

Finally, if all other efforts have failed, the use of a modified chromatographic system with different selectivity should be considered. Such a change may be particularly beneficial if the problem consists of a single interfering peak. For example, when developing a method for the assay of NBQX in urine, the analytical column used for plasma samples (Lichrospher™[35] RP-18) had to be replaced with a Chromspher™ B C18 column in order to obtain sufficiently clean extracts (Ingwersen, 1993). The change from a reversed-phase to a normal-phase system is required in some of the toughest cases. Although normal phase HPLC procedures are considered more cumbersome and time consuming, such procedures often perform well in combination with reversed-phase SPE yielding very clean blank chromatograms. This highlights a point familiar to those who use column-switching techniques in HPLC. A change in retention modes between extraction and analysis is good practice. Even when it is not possible to use a different retention mode, the use of SPE and LC sorbents with widely differing capacities and selectivities (e.g., C2 SPE cartridge and C18 HPLC column) is often advantageous.

---

[35] Lichrospher and Chromspher are trademarks of E. Merck, Darmstadt, Germany.

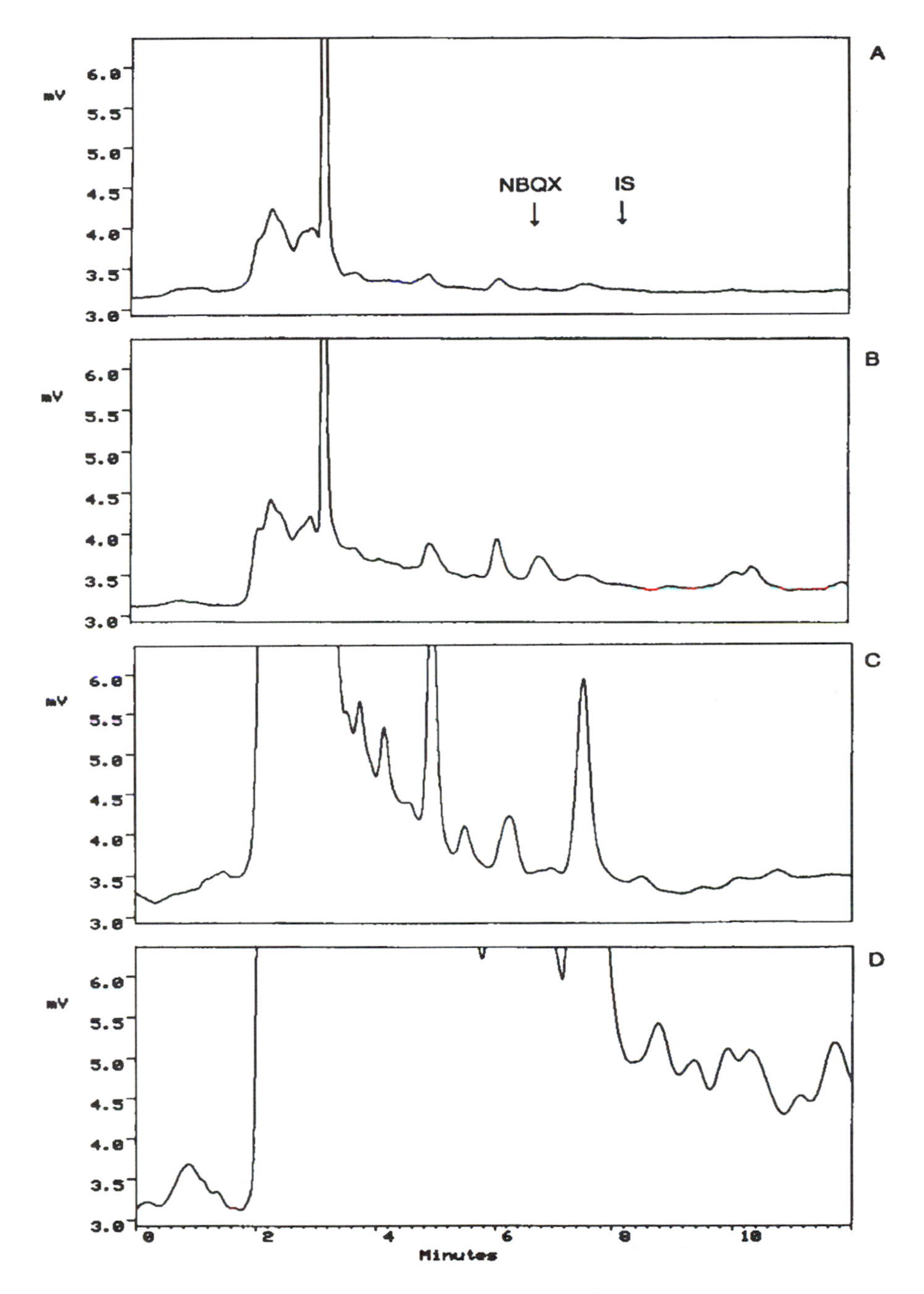

Figure 7. Chromatograms at 380 nm of blank human plasma extracted on Bond Elut Certify II (A) and Bond Elut C8 (B). For comparison, chromatograms obtained at the UV absorption maximum of NBQX (294 nm) are shown for Certify II (C) and C8 cartridges (D). The extraction procedure was as described in the legend to Fig. 1, using the procedure at pH 3.5 (from Ingwersen, 1993, with permission).

I. IMPROVING LOW RECOVERIES

If, during the initial screening of sorbents and pH values, low recoveries (i.e. below 50%) are obtained throughout, the first issue to be resolved is whether the low recovery is due to incomplete retention, loss during washing or incomplete desorption. The sample should be appropriately diluted and applied to a preconditioned SPE cartridge, which should then be washed and eluted using optimal conditions (as determined during method development). All fractions, including the sample waste should be collected and assayed for their content of the target analyte. A mass balance calculation will indicate the cause (for example, see Bouvier, 1995).

The problem of incomplete retention is rarely encountered on reversed-phase sorbents if careful screening of sorbent types and pH values has been performed. However, very hydrophilic analytes can be difficult to retain on reversed-phase silica sorbents, and in such cases the use of an ion pairing agent in the dilution buffer is an option. This principle was applied to retention of the experimental drug Y on SPEC C8 extraction discs. 0.5mL aliquots of spiked plasma were mixed with 1 mL each of buffers at various pH values before application to preconditioned extraction discs. The discs were then washed by 2 mL of water and eluted by 0.5 mL of methanol. The buffers used for diluting plasma were 0.1 M phosphoric acid, pH 2.1, 0.1 M formic acid adjusted to pH 2.5, 3.0, 3.5, or 4.0 with triethylamine, 0.1 M potassium acetate buffer, pH 5.0 and 0.1 M phosphate buffer, pH 6.0, 7.0, and 8.0. The results at pH 4.0 are not shown due to clogging of the discs at this pH value. Drug Y contains a phosphonic acid function ($pK_a < 2$) and has log P_{oct} values of -0.27 and -2.26 at pH 2 and pH 7, respectively. As seen in Figure 8, addition of 10 mM tetrabutyl ammonium hydroxide to the buffers used for diluting the sample substantially increased the recovery of this analyte over the pH interval studied. Thus, almost any analyte can be retained on non-polar sor-

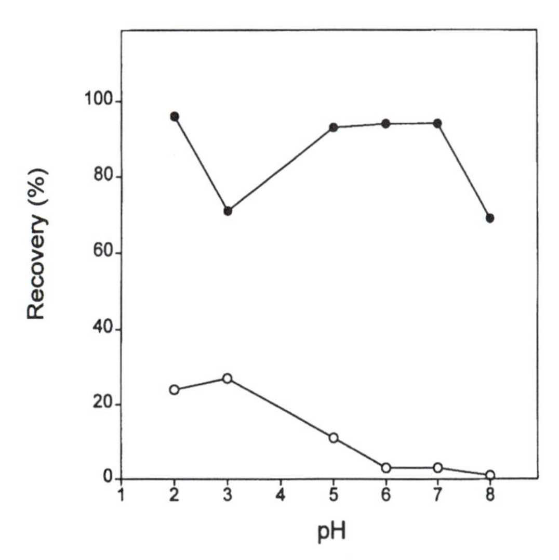

Figure 8. Recoveries of drug Y from human plasma obtained by SPE extraction at various pH values with (•) and without (o) addition of an ion pairing agent (10mM tetrabutyl ammonium hydroxide) to the buffer used for diluting the samples (Ingwersen, unpublished results).

bents, irrespective of the lipophilicity of the analyte.

In theory, incomplete retention can also be due to tight protein binding of the analyte. However, this is rarely encountered as a problem because the sorbent, if chosen correctly, will usually strip the drug from the binding proteins when the sample is passed through the cartridge. Furthermore, a correct choice of pH value will often eliminate protein binding in the sample. However, if protein binding needs to be eliminated at other pH values, this can be done by adding a protein denaturing agent such as urea (8 M) or an organic solvent like acetone or acetonitrile to the buffer used for diluting the sample. It should be remembered that the addition of these solvents may have an adverse effect on non-polar retention of the analyte on the SPE phase. Alternatively, the proteins may be precipitated prior to sample application, although care must be taken to avoid coprecipitation of the analyte(s).

Low recovery of standards spiked into a biosample containing proteins has traditionally been attributed to protein binding. Theory held that when a drug binds to a protein of greater than about 15,000 daltons, the protein, being too large to penetrate into the sorbent pore structure (assuming a 60 Ångström mean pore diameter), would pass through the sorbent unretained and would drag the bound analyte with it. Interestingly, though protein binding has long been invoked as the reason for low recovery of drugs in biosamples, there is very little unequivocal evidence that this theory correctly describes the phenomenon.

Contrary evidence has been gathered (Ingwersen, 1994) that suggests that, in at least one apparently classic case of “protein binding,” it is the presence of other compounds in the sample, notably amine-containing species, that are responsible for disrupting secondary interactions between analyte and sorbent. This matrix-related effect consequently reduces retention or hindering elution of the analyte. In one specific case the author attempted disruption of protein binding by addition of a concentrated urea solution. This apparently broke up protein binding but it also reduced the strength of secondary interactions, permitting more facile elution by pure methanol or acetonitrile. Addition of a "modifier", such as ammonium hydroxide, to the methanol has been shown to increase recovery from an untreated sample to the same level as achieved by loading a sample that had been treated with 8 M urea. This would not have been possible if protein binding had prevented retention of analyte.

The problem of incomplete desorption is most likely due to strong secondary interactions between the analyte and the sorbent. As mentioned above (section III.A), secondary interactions can often be exploited to the benefit of an SPE procedure. However, as pointed out by Law et al. (1992), certain lipophilic basic analytes are so strongly retained on some reversed-phase sorbents that incomplete desorption is encountered. The solution to such problems would be to substitute a less hydrophobic sorbent (e.g., C2 in place of C18) and to include a silanol blocking agent such as ammonium acetate in the elution solvent. Alternatively, a silanol blocking agent such as 0.1 M potassium acetate may be included in the methanol used for conditioning the SPE cartridges in order to reduce ionic interactions (Law, 1992).

IV. EXTRACTING URINE AND SOLID TISSUE SAMPLES

Urine is characterized by a low protein content and by a less well defined sample matrix than plasma. For instance, urine pH varies between 4.5 and 8 in normal subjects and the content of electrolytes also varies considerably, depending on the diet and the rate of urine production. Furthermore, due to the occasional presence of bacteria in urine samples, the analyte stability in this matrix is sometimes compromised. Procedures for the extraction of plasma can often be applied to urine directly or with minor adjustments (Chen et al., 1992). Occasionally, adjusting a method for urine can be a difficult task, particularly where the extraction requires isolation of a hydrophilic analyte in the presence of high levels of the endogenous hydrophilic compounds typically found in urine. This was discussed above (Section III.H) and illustrated by the assay of NBXQ in urine, which the author derived from a procedure developed for plasma samples. If it is necessary to develop a method for urine from scratch, we can proceed along the lines indicated for plasma.

A common feature of urine extraction is a hydrolysis step, since many drugs are excreted as sulfates, glucuronides or other conjugates forms. Use of acid hydrolysis to cleave off these groups is common but it cannot be used in all cases. For example, 6-monoacetyl morphine, a heroin metabolite, would be hydrolyzed under the conditions used to release the other conjugated opiates from their glucuronide forms. In such cases more subtle approaches, such as the use of a specific enzyme like a glucuronidase, is preferable.

Solid tissue samples can also be extracted by SPE but a pretreatment is required before application to the cartridge unless one is to use the MSPD technique (refer to Chapter 13). The pretreatment aims at liberating the analyte from solid tissue constituents and removing particulate tissue residues that would otherwise block the cartridge. The pretreatment usually consists of homogenizing the sample in the presence of a suitable buffer (e.g. 0.1 M perchloric acid), or an organic solvent. Alternatively, the tissue can be disrupted by ultrasonication and hydrolyzed by incubating with proteolytic enzymes such as Subtilisin Carlsberg. The homogenate is centrifuged and adjusted to a suitable pH before application to the SPE cartridge. If organic solvents have been used in the pretreatment step, evaporation to dryness and reconstitution may be required before transfer to the SPE cartridge. From there on, the method development can proceed as described above for plasma with the exception that extraction discs are usually unsuitable for solid tissue samples due to possible clogging. Fortunately, procedures for the extraction of plasma are often directly applicable to properly pretreated tissue samples. Some procedures used for solid-phase extraction of tissue samples have been reviewed by Scheurer and Moore (1992) and by Simpson and Van Horne (1993).

V. CONCLUSIONS

Solid-phase extraction is a powerful technique for biological sample preparation, but method development may be time consuming, in particular if high sensitivities are required. Because a large number of combinations of cartridge types, dilution buffers, washing solutions, and elution solutions are available, it may seem almost impossible to identify the optimal extraction conditions. However, the guidelines presented in this chapter should help in designing experiments for obtaining suitable, if not optimal, conditions of extraction of plasma and serum samples, without the danger of arriving in a dead end at an early stage of method development.

Chapters 8 and 9 cover related ideas to those introduced in this chapter but apply them to other matrix types, notably human urine and animal urine and plasma, and for different purposes. While not explicitly stated in terms of method development, the actual methods described will all have been developed following similar rigorous method development practice. Chapter 14 will address the aspect of automated SPE method development — a discipline that is currently of great importance due to the introduction of high-throughput robotic systems and SPE formats, and the consequent need to ensure that the SPE step does not become the slowest part of the overall analysis.

REFERENCES

Bland, R., (1986) Applications of Solid Phase Extraction to the Determination of Salbutamol in Human Plasma by HPLC. Procedings of the Third International

Symposium on Sample Preparation and Isolation Using Bonded Silicas, Cherry Hill, NJ.

Bouvier, E.S.P., (1995) Sample Preparation Perspectives: SPE Method Development and Troubleshooting. LC.GC 13(11): 852-856.

Chen, X.H., Wijsbeek, J., Franke, J.P. and De Zeeuw, R.A., (1992) A Single Column Procedure on Bond Elut Certify for Systematic Toxicological Analysis of Drugs in Plasma and Urine. J.Forensic Sci. 37: 61-72.

Crawley, Wang, X., Murugaiah, V., Naim, A. and Giese, R.W., (1994) Masking as a Mechanism for Evaporative Loss of Trace Analyte, Especially After Solid Phase Extraction. J.Chromatogr. A., 699:395-402.

Ingwersen, S.H., (1993) Combined Reversed Phase and Anion Exchange Solid Phase Extraction for the Assay of the Neuroprotectant NBQX in Human Plasma And Urine. Biomed.Chromatog. 7:166.

Ingwersen, S.H., (1994) A Note on Solid Phase Extraction of the Highly Protein-bound Dopamine Uptake Inhibitor GBR 12909, Sample Preparation for Biomedical and Environmental Analysis (D. Stevenson & I.D. Wilson, Ed.) Plenum Press, New York, p. 139.

Law, B. and Weir, S., (1992) Fundamental Studies in Reversed-Phase Liquid-Solid Extraction of Basic Drugs; I: Ionic Interactions. J.Pharmaceut.Biomed.Anal. 10:167.

McDowall, R.D. Doyle, E. Murkitt, G.S. and Picot, V.S., (1989) Sample preparation for the HPLC analysis of drugs in biological fluids. J.Pharmaceut.Biomed.Anal. 9:1087.

Nau, D. and Pocci, R., (1997) Studies of Recovery Against Drying Time for the Primary Urinary Metabolite of nor-Δ^9 THC Using Bond Elut Certify II (personal communication).

Scheurer, J and Moore, C.M., (1992) Solid-Phase Extraction of Drugs from Biological Tissues - a Review. J.Anal.Toxicol. 16:264.

Simmonds, R.J. James, C. and Wood, S., (1994) A Rational Approach to the Development of Solid Phase Extraction Methods for Drugs in Biological Matrices, Sample Preparation for Biomedical and Environmental Analysis (D. Stevenson & I.D. Wilson, Ed.) Plenum Press, New York p. 79.

Simpson, N.J.K., and Van Horne, K.C., (1993) The Handbook of Sorbent Extraction Technology, 2nd. Edition Varian Associates, Appendices F, G, H, p. 124.

Tippins, B., (1987) Selective Sample Preparation of Endogenous Biological Compounds Using Solid Phase Extraction. Am.Biotech.Lab., Jan/Feb 1987.

Wilson, I.D., (1994) Clinical Analysis from Biological Matrices, Sample Preparation for Biomedical and Environmental Analysis (D. Stevenson & I.D. Wilson, Ed.) Plenum Press, New York, p. 71.

Wilson, I.D. and Nicholson, J.K., (1994) Proton Nuclear Magnetic Resonance Spectroscopy: A Novel Method for the Study of A Solid Phase Extraction, Sample Preparation for Biomedical and Environmental Analysis (D. Stevenson & I.D. Wilson, Ed.) Plenum Press, New York, p. 37.

Wynne, P.M., (1997) Racing Analytical Services Limited, Flemington, Victoria, Australia. Personal communication.

11

SOLID-PHASE EXTRACTION MEDIATED BY COVALENT BONDING: APPLICATIONS OF IMMOBILIZED PHENYLBORONIC ACID

I.D. Wilson and P. Martin, Zeneca Pharmaceuticals, Mereside, Alderley Park, Macclesfield, Cheshire, United Kingdom

I. INTRODUCTION

As described elsewhere in this volume the bulk of SPE methodologies are based on either non-polar (Van der Waals), polar, or ionic interactions, which are relatively non-specific. In contrast, covalent bond formation, based on the presence of particular functional groups in the analyte, allows for the development of extraction methods of much greater specificity. Boronic acids, immobilized in one way or another have been widely used in affinity chromatography for the isolation of proteins, and have also been used in analytical chemistry for the isolation or chromatography of lower molecular mass compounds with suitable functional groups. Following on from the initial development of boronic acids immobilized on materials such as sepharose gels, has been the development of applications using phenylboronic acid covalently linked to silica gel. Indeed this material is probably the most widely available material for covalent SPE based on the use of immobilized phenylboronic acid (PBA). As will be reviewed below, this covalently-bonded sorbent, that is able to extract compounds by among other mechanisms, covalent bond formation, affords some useful opportunities for building selectivity into the sample preparation step. Here the relative advantages and disadvantages of methods based on covalent extraction using immobilized boronic acids will be described, with the main emphasis on the silica-based material. However, applications which do not employ silica-based materials will also be considered where these are either unique, having no silica-based counterpart, or where we consider them to contain useful lessons or methodologies that could be used equally well with the modern SPE phase.

II. MECHANISMS OF INTERACTION OF ANALYTES WITH IMMOBILIZED PBA

A very good account of the mechanism of extraction onto PBA was produced over a decade ago by Stolowitz (1985). The mechanism by which analytes can be induced to interact with boronic acids to form cyclic esters is illustrated in Figure 1. As shown in this figure, in order for the reaction to proceed the boronate must be in the reactive $-B(OH)_3^-$ form, which is obtained by conditioning the cartridge with an alkaline buffer. In this form the bond angles between hydroxyl groups attached to the boron central atom are approximately 109°. The sample is then applied and the formation of the covalent bond takes place only with those components containing the appropriate functional groups. Typically these would be vicinal diols, present in sugars or as catechols, but suitable functional groups also include α-hydroxy acids, aromatic O-hydroxyacids and amides, 1,3-dihydroxy-, diketo-, triketo-, and aminoalcohol-containing compounds. Once the covalent bond has been formed the cartridge may be washed with solvents to

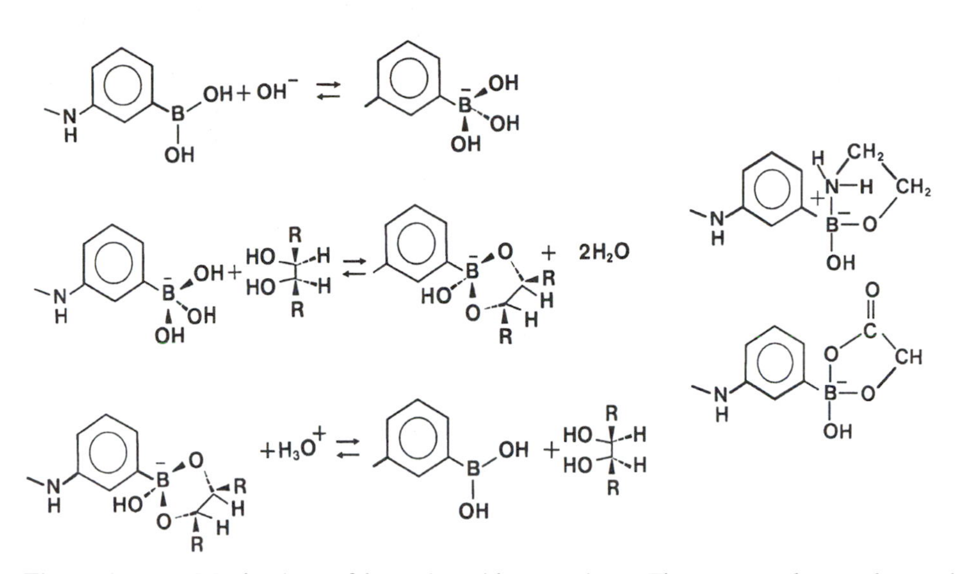

Figure 1. Mechanism of boronic acid extraction. The upper scheme shows the activation of the boronic acid in the presence of alkali, the middle scheme the formation of a covalent bond with a vicinal diol, and the lower scheme shows the use of acid to break the covalent bond and regenerate the boronic acid and the free diol. The type of covalent bonding seen with aminoalcohols and hydroxyacids is shown at the side of the main reaction scheme.

remove any non-covalently-bound contaminants providing that an alkaline pH is maintained. Elution of the compound(s) of interest is then achieved by washing with an acidic buffer/solvent which hydrolyzes the covalent bonds to give the free analyte and the PBA in the $-B(OH)_2$ form, as mentioned above.

The geometry of the boron atom is important because the five-membered ring that forms during retention is stable with an O-B-O bond angle of 109°. The ring becomes strained when the bond angle increases to 120°, as occurs when the boronate sheds a hydroxyl group and returns to the $-B(OH)_2$ form, which is trigonal planar. We may expect to see a pH dependence of retention that is the reverse of this when we attempt to extract a compound that has OH groups separated by three carbon atoms. In this case the resulting ring would be six-membered; a bond angle of 109° induces strain on a six-membered ring.

In addition to the specific interaction with the boronic acid described above, a number of non-specific interactions are also possible on the PBA phase. For example, as the PBA is immobilized on silica, interactions are possible with residual silanols. The PBA group is attached onto the silica gel substrate via an aminopropyl linker, providing sites for polar ionic and non-polar hydrophobic van der Waals interactions, while the phenyl ring affords the opportunity for pi-pi interactions. The boronic acid moiety itself offers an interesting array of potential interactions in addition to the covalent bonding of suitable compounds, including the ability to act as a hydrogen bond donor and thus to retain compounds via a normal-phase type of interaction. Further, cations can bind to the boronic acid group, and the potential for the formation of charge-transfer complexes with unprotonated amines also exists (Stolowitz, 1985). If an amine-boronic acid charge-transfer complex can form then the result is a pK_a that is lower than the uncomplexed organoboron moiety, potentially increasing the capacity of the phase. This increased capacity has important consequences for the choice of buffers (see below).

All of these interactions may come into play when a sample is loaded onto a cartridge, and some may be of assistance in the overall retention of the analyte on the phase (for example, see the discussion of glucuronide retention below). Those compounds co-extracted with the analyte, but retained solely by these non-specific interactions may be readily removed by washing with an appropriate solvent, providing that an alkaline pH is maintained. Clearly these "secondary" interactions will predominate at pHs below 7. Thus, if there is any doubt as to whether or not the extraction is being mediated by secondary interactions or covalent bond formation, it may be possible to isolate which is the dominate mechanism by comparing the results at a range of different pHs.

A. BUFFER SELECTION

The selection of the buffer for SPE on PBA phases has been discussed in some detail by Stolowitz (1985). Clearly, to ensure a good extraction efficiency the first requirement is that an alkaline pH is used for conditioning and sample application. Accordingly, the equilibration step can be accomplished with 0.1 to 1.5 M alkaline buffer at pH 10 to 12, followed by 0.01 to 0.05 M buffer at pH 8.0 to 8.5. The use of zwitterionic buffers has a number of advantages and HEPES, glycine, diglycine, and morpholine have all been employed.

The use of buffers which are themselves capable of forming covalent adducts with the PBA is clearly to be avoided, as this will effectively remove the possibility of performing SPE based on covalent bond formation with the analyte. Buffers which should generally be avoided are thus bicine, tricine, tris, and 1',2',3'-ethanolamine, though this does not mean that they can never be used (Benedict and Risk, 1984).

Elution can usually be carried out simply by reducing the pH of the eluent to 5 or below (pHs below 3 may be required to elute amines involved in charge transfer complexes). Suitable acidic modifiers include acetic, trifluoroacetic and phosphoric acids, with the addition of an organic modifier to help overcome non-polar or silanophilic secondary interactions. In some cases, where the covalent adduct is particularly stable, elution with buffers such as lactic acid or salicylic may be necessary (an example of this is given in the case of ecdysteroids extraction, described below). Competitive elution with borate buffers can provide an alternative to the use of acidic eluents when the extraction of acid labile compounds is to be performed (Stolowitz, 1985).

III. SPE APPLICATIONS OF IMMOBILIZED PBA

Some of the published applications of SPE with PBA are described below. For simplicity these are considered as classes of compound in alphabetical order.

A. ADENOSINE

The use of PBA-gels for the extraction of nucleosides from biological samples has been the subject of a number of studies, such as that of Pfadenhauer and Tong (1979), who used such a gel for the isolation of inosine and adenosine from human plasma.

More recently Echizen et al. (1989) have developed a method based on the use of silica gel-immobilized PBA for the simultaneous extraction of adenosine and dopamine from human urine. The extraction was performed on 100 mg silica-PBA cartridges which were activated by washing the phase sequentially with 0.1 M formic acid (5 mL) and then 0.25 M ammonium acetate (pH 8.8). The urine sample (0.5 mL) was applied to the cartridge following the addition of (±)-isoproterenol and 2-chloroadenosine, the internal standards, in 50 µl of ammonium acetate (2.5 M, pH 8.8). The sample was allowed to flow through the sorbent bed until the liquid meniscus just reached the top of the layer, and was then followed by a wash of 1mL of 0.25 M ammonium acetate (pH 8.8). The analytes were eluted with 1 mL of 0.1 M HCl-methanol (4:1 v/v). Following filtration 40 µl of the eluate was injected directly onto the HPLC system where, following separation on a C18 cartridge, the analytes were detected using either UV (adenosine) or electrochemical detection (dopamine).

The recoveries of the analytes through this procedure were 88 to 104% with CVs of less than 5%. The methodology was found to give results as good as those obtained with "conventional" PBA-gels.

B. ALIZARIN

In a short report, Schmid and Kupferschmidt (1986) described studies using the *cis* diol-containing tricyclic anthraquinone dye alizarin as a model compound with which to study SPE on silica-bonded PBA. The highly colored nature of the analyte meant that it was possible to assess the efficiency of extraction visually. The phase was conditioned by sequential washing with methanol and 0.1 M HEPES buffer at pH 8.6. Following application of a 0.1% solution of the dye in water the dye was observed to be quantitatively retained as a sharp band on the cartridge. Elution was achieved using methanol/0.1 M HCl (3:1, v/v). The authors then conducted a variety of experiments to show that retention was only complete between pH 7 to 10 and that, provided that subsequent wash solvents were kept alkaline, organic solvents such as methanol, ethanol or acetonitrile could be used without elution of the analyte. In addition, they found that good extractions were possible even from solvents containing up to 70% of an organic solvent as long as they were alkaline. Attempts at extraction of alizarin from blood plasma were not particularly successful, possibly due to protein binding. If the dye was first extracted from the plasma on to a C18 cartridge, from which it was then eluted with a methanol-alkaline/buffer solution, extraction on to the PBA phase was possible with high efficiency.

The authors also proposed that by monitoring UV spectral changes between potential candidates for SPE on PBA and phenyl boronic acid itself, it might be possible to predict which compounds would be effectively extracted.

C. β-BLOCKERS

The use of PBA SPE cartridges for the extraction of aminoalcohol-containing drugs such as β-blockers has been investigated by Martin et al. (1993). In this study, the extraction of a range of β-blockers, including propranolol, epanolol, ICI 118551, and practolol (structures in inset to Figure 2) from aqueous buffer and rat and human plasma was investigated and compared with extraction on a number of other SPE phases. The SPE protocol used in this work involved extraction from 0.1 M glycine buffer at pH 8.2 on to 100 mg PBA cartridges. Prior to sample application the cartridges had been conditioned with 1 mL of methanol followed by 5 mL of glycine buffer. Once the sample had been applied the cartridges were eluted with 1 mL of deionized water and then 3 mL of methanol-water (40:60 v/v) in order to remove non-specifically extracted sample components. Elution of the analytes was effected using 3 mL of methanol-water TFA (50:50:1 v/v).

The extraction of the various analytes from buffer was clearly pH dependent, with the optimum seen at pH 8 (Figure 2). Structure was also im-

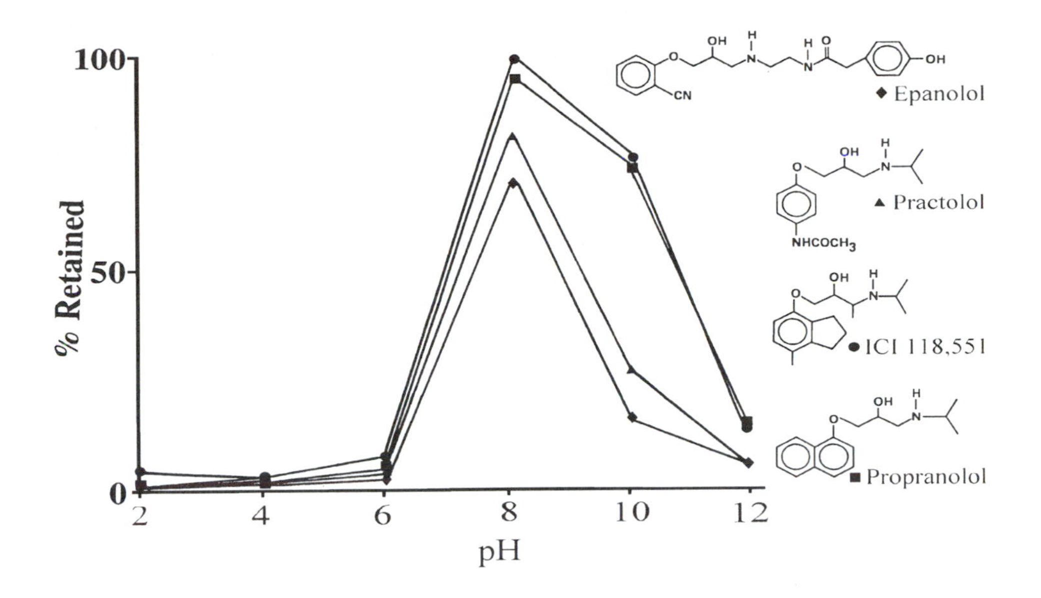

Figure 2. The pH extraction profiles obtained for four β-blockers on PBA. Structures and the key to the symbols used for the graph are shown as insets.

portant for good extraction, retention, and elution. Thus, both propranolol and ICI 118551 were efficiently extracted (greater than 90%) using this protocol with small losses at the application and wash steps. In contrast, losses of practolol at the application step amounted to over 7%, while both practolol and epanolol suffered significant losses at the wash step (8.8 and 16.5% respectively). An improvement in the losses seen at the wash step was obtained by reducing the amount of methanol in the eluting solvent.

While the results for the extraction of propranolol, practolol, and ICI 118551 from rat plasma were similar to those seen for buffer, very much more significant losses on the application and wash steps were noted for epanolol. This matrix effect on the extraction of epanolol could be minimized by the dilution of the sample with glycine buffer before application, and was probably the result of plasma protein binding.

Clearly, while quite efficient for certain of the drugs tested, the utility of immobilized PBA as a general method for the selective extraction of β-blockers, as a class, is less apparent. More work would clearly be of benefit in this area to help define those structural features that result in high extraction efficiency.

D. CATECHOL AMINES, DOPA, AND RELATED MATERIALS

The bulk of the applications that use SPE with immobilized boronates of one sort or another as the sorbent are to be found in the area of the determination of various catecholamines, DOPA (l-dihydroxyphenyl alanine), and related materials in biological fluids. Indeed, the first application of silica-bonded PBA was to catecholamines from urine. However, as a result of the importance of this area of analysis there are many examples of procedures developed on non-silica-based boric acid gels, etc. For the sake of both clarity and brevity, a comprehensive review of the entire field has not been attempted, and instead we have concentrated on selecting the more recent procedures, generally involving the use of silica-based materials.

1. DOPA

Benedict and Risk (1984) developed a two stage SPE method for the isolation of DOPA from plasma and urine fluids, with [^{14}C]-DOPA added to the sample prior to extraction as an internal standard to correct for any losses that might occur during the clean up.

For urine their methodology involved removing interfering urochromes by first extracting the sample onto a cartridge made up of two layers of sorbent; thus, an upper layer of SCX and a lower layer of C1 silica were packed into a syringe and conditioned with methanol (5 mL) and 0.2 M HCl, after which the 2.8 mL of urine and two bed volumes of 0.2 M HCl were applied. The eluate from this cartridge was then brought to pH 7.5-7.7 with 1.5 mL of 2 M Tris buffer, and one fourth of the total was then applied to 200 mg of PBA. The PBA cartridge had previously been conditioned with methanol (1 mL) and 0.2M Tris (1 mL). The cartridge was then washed with methanol (2 mL) and 0.1 M Tris (1 mL) in order to remove unwanted contaminants, with the DOPA recovered with 0.1 M HCl (0.3 mL).

For plasma samples (1 mL), following precipitation and removal of plasma proteins with ice cold perchloric acid and centrifugation, the supernatant (at pH 7.5-7.7 with 2 M Tris) was applied to the PBA cartridge and eluted as described for urine. The overall recovery of DOPA from the samples using this methodology was good (80% for urine and 84% for plasma) with good precision (SD = 2 to 3%), and detection limits by HPLC-electrochemical detection of 10 to 15 pg.

2. 5-S-L-Cysteinyl-L-DOPA

There have been several reports of the use of PBA for the extraction of urinary 5-S-L-cysteinyl-L-DOPA. In the first of these, Kagedal and Pettersson (1983) used a dual extraction involving a cation exchanger followed by a phenylboronate affinity gel. The 5-S-D-diastereoisomer was used as an internal standard.

More recently Huang et al. (1988) described a method which, like that of Kagedal and Petersson, was based on a combination of extraction first on strong cation exchange and then onto silica-PBA. Following elution from the PBA the analytes, which included epinephrine and dopamine in addition to 5-S-L-cysteinyl-L-DOPA, were analyzed by ion-pair reversed-phase HPLC. Urine samples plus internal standard (5-S-D-cysteinyl-L-DOPA) were first passed through an SCX cartridge, previously activated with successive washes of 1 mL of methanol and 1.0 mL of 0.1 M HCl. The cartridges were then washed with 0.1 M HCl and then placed on top of the PBA cartridges onto which the analytes were eluted using two column volumes of 1 M dipotassium hydrogenphosphate. The silica PBA cartridges had been pre-treated with one column volume of methanol followed by one column volume of 1 M dipotassium phosphate buffer. After removal of the SCX cartridge each PBA cartridge was washed with water and eluted with 0.5 mL of 0.1 M HCl (containing 10 mg/l of ascorbic acid). A 20 μl aliquot of this sample was then used for HPLC with detection via an amperometric detector, to give a reproducible method with a limit of detection of 5 μg/mL. The recovery of the analyte through the extraction procedure was essentially quantitative with good within day and day to day CVs.

3. Dopamine, Norepinephrine, and Epinephrine

The use of boric acid gels for the extraction of catecholamines from biological samples has been the subject of a number of reports (e.g., Oka et al., 1982; Maruta et al., 1984; Imai et al., 1988). In addition Wu and Gornet (1985) have investigated the use of silica-PBA for the extraction and HPLC analysis of norepinephrine and epinephrine in urine, comparing the method with extraction schemes based on acid-washed alumina, weak cation exchange resin and weak cation exchange resin followed by alumina. They found that neither the weak cation exchange nor the alumina alone was suitable for routine work because of co-extracted interferences.

The PBA-based method gave better results than either of these two phases alone; the combination of the two phases gave the best results in terms of sensitivity and specificity but was very time consuming. Thus, although not giving as good results as the combination of cation exchange and alumina, the PBA method was sufficiently sensitive and specific for the analytes, and allowed the samples to be analyzed more quickly than a double cartridge extraction method.

This method involved the use of two SPE cartridges connected in series with a PSA (Primary and Secondary Amine) cartridge preceding the PBA cartridge. The PBA cartridge was first conditioned by washing with methanol (1 mL) and then 0.1 M HCl (1 mL) and then the PSA cartridge was inserted. The combination was then further conditioned with 2 mL of methanol followed by 4 mL of aqueous ammonia solution and finally 4 mL of 5 mM phosphate buffer (pH 8.5). The urine sample, containing 1 mg/l of dihydroxybenzylamine as internal standard, was then applied (1 mL at pH 5 with ammonia), followed by 4 mL of pH 8.5 phosphate buffer. After a further rinse with 2 mL of the alkaline phosphate buffer, the PSA cartridge was discarded and the PBA cartridge was washed successively with 1 mL of methanol and 1 mL of acetonitrile-phosphate buffer 1:1 (v/v). The analytes were recovered with 1 mL of 0.1 M HCl. Aliquots of the eluates (20 μl) were then analyzed by HPLC with electrochemical detection. The sample preparation time was approximately 45 min for 10 samples and the assay gave a limit of detection of 50 pg for norepinephrine and 100 pg for epinephrine.

The work of Huang et al. (1988) described above for 5-S-L-cysteinyl-DOPA also included methodology for dopamine [for dopamine see also Echizen (1989)], epinephrine, and norepinephrine. According to this method, urine samples (1 mL) to which an internal standard (3,4-dihydroxybenzylamine) had been added, were diluted to 5 mL with water. The pH of the sample was then adjusted to 6.5-7.0 with sodium hydroxide prior to loading on to the SCX and PBA cartridges. The SCX and PBA extraction cartridges were prepared by washing with three column volumes of 1.0 M HCl followed by two column volumes of methanol and 0.01 M ammonium acetate buffer (pH 7.3). The sample was then applied to the SCX cartridge, and then washed with two column volumes of methanol and two column volumes of 0.01 M ammonium acetate prior to elution with 3 x 500 μl of perchloric acid. Having neutralized the eluate with a saturated solution of sodium carbonate (400 μl) the partially purified sample was loaded onto the PBA cartridge. This was washed with two column volumes of methanol and two of water followed by elution with 2 x 500 μl of 0.1 M perchloric acid. An aliquot of between 40 and 80 μl was then analyzed by HPLC with electrochemical detection. This application gave limits of detection of 1 μg/l for norepinephrine and 2 μg/l for dopamine with essentially quantitative recoveries and good between-day and within-day reproducibilities.

E. ECDYSTEROIDS

The ecdysteroids are a diverse group of relatively polar polyhydroxy steroids (see structures in Figure 3) that form an interesting class of natural products. To date, in excess of 250 ecdysteroids have been isolated from various sources. They are important in insects and crustaceans as developmental hormones, but have also been found in many species of plant where it seems probable that they function as part of these organisms' chemical defense mechanisms against phytophagous insects and other threats. A wide variety of sample preparation procedures has been developed for these compounds in the course of research on their occurrence and biological ef-

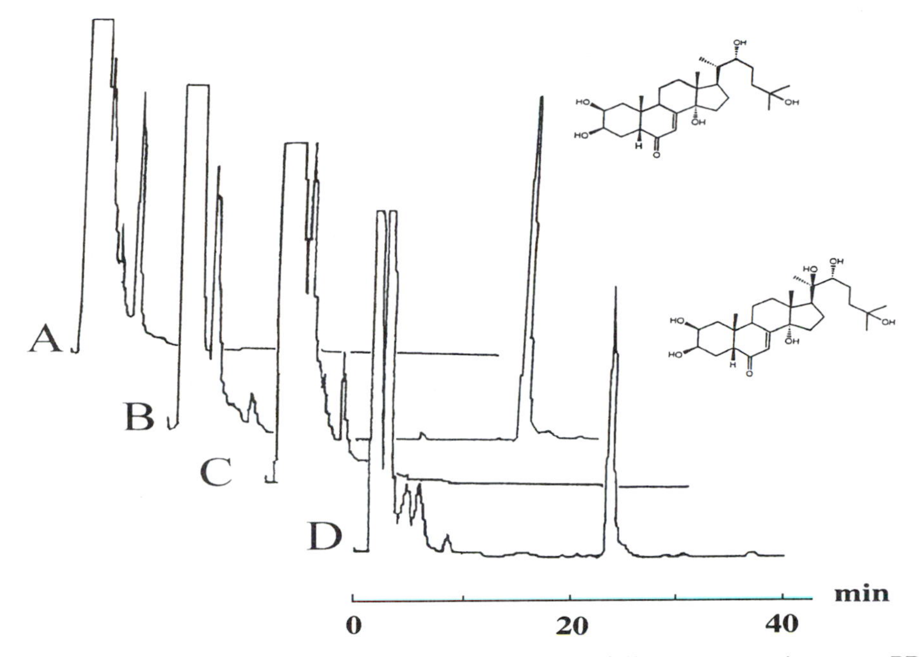

Figure 3. Reversed-phase HPLC of ecdysteroids following extraction on to PBA cartridges and subsequent elution with A) buffer, B) 70% methanol/water, C) 70% methanol in acidic buffer and D) 70% methanol with 3% lactic acid. Ecdysone, which is non-specifically retained on the PBA cartridge elutes in fraction B, while 20-hydroxyecdysone, which is specifically retained, elutes in fraction D. Structures of ecdysone and 20-hydroxyecdysone are given as insets.

fects and these methods include liquid-liquid, and solid-phase extraction methods as reviewed by Wilson et al. (1990). A notable feature of the structures of many of the ecdysteroids is the presence of one or more vicinal diols, usually in the A ring of the steroid nucleus (typically at C2 and C3), and the steroid side chain (C20 and C22). The use of SPE with PBA cartridges has been investigated for these compounds and found to give selective extraction of ecdysteroids with a C20,22 diol, such as 20-hydroxyecdysone (structure inset to Figure 3), but not those containing only the C2,3 diol, for example ecdysone, also shown inset to Figure 3 (Wilson et al., 1990; Murphy et al., 1990).

The extraction protocol developed for C20,22 diol-containing ecdysteroids involves activation of the PBA cartridges with 5 mL of ethanol followed by 5 mL of alkaline buffer at ca. pH 8. The buffers investigated included borate (100 mM pH 8.0), phosphate (100 mM, pH 8.0) and borate (100 mM, pH 8.2). Recovery was not observed to be affected by switching between these buffers, however. Ecdysteroids such as ecdysone and 2-deoxyecdysone, which do not contain a 20,22-diol group, were retained on

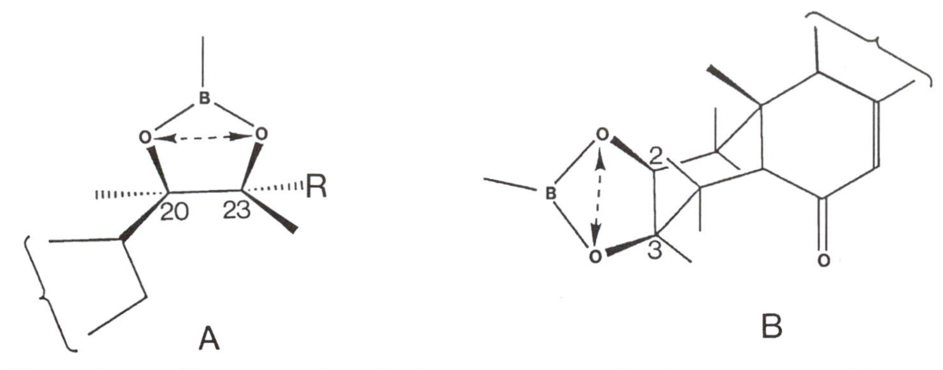

Figure 4. Formation of cyclic boronate esters of ecdysteroids containing (A) the C-20,22 diol function, where the O-O interatomic distance for the vicinal diol group is 2.52X and (B) the C-2,3 diol, where the equivalent distance is 2.80X, and the closest approach of the hydrogen nuclei of the parent OH groups is 1.20X and 1.68X in (A) and (B) respectively. These differences explain why the formation of covalent bonds with the C-20,22 diol grouping is favored.

the phase but could be eluted using 70% methanol/30% alkaline buffer (v/v). Elution with an acidic eluent was expected to enable the recovery of those ecdysteroids retained via covalent bond formation. Surprisingly, however, such buffers were without effect and even 90% methanol containing 1% TFA failed to elute more than 20% of the adsorbed analyte. In order to obtain quantitative recovery of the adsorbed ecdysteroids the inclusion of either salicylic acid (25 mM) or lactic acid (3% w/v) in 50-70% (v/v) methanol was necessary. Interestingly, PBA bound to agarose was ineffective at extracting ecdysteroids using conditions that gave good extraction efficiencies onto the silica based-PBA. A chromatogram is shown in Figure 3.

The mechanism giving rise to the selectivity of extraction of 20,22-diol-containing ecdysteroids compared to the 2,3-diol-containing compounds probably arises from the difference in the O-O atomic distances in the two types of compound. In the case of the ring-dihydroxylated ecdysteroids this distance is 2.8X, whilst for the side chain diols this distance is 2.52X (see Figure 4). It would seem therefore that a rigid 2,3-diol would be unable to enter into the cyclic bound state, whereas the unstrained cyclic boronate can form the requisite stable cyclic structure.

F. GLUCURONIDES

The conjugation of glucuronic acid to hydroxyl or carboxylic acid groups represents a common route for the metabolism of xenobiotics such as therapeutic or abused drugs. Often the first step in the analysis of such metabolites is conversion back to the aglycone using either enzymatic hydrolysis (phenolic and carboxy-glucuronides) or alkaline hydrolysis (glucuronides of carboxylic acids) followed by extraction. However, the glucuronic acid moiety contains all of the structural features needed for SPE on the phenyl-boronic acid phase, with the potential, therefore, for developing glucuronide-specific assays. We have briefly investigated the potential of such systems with a range of model phenolic glucuronides, spiked into urine, and compared the results with the extraction of the same samples extracted onto a more conventional C18 bonded phase (Tugnait et al., 1992; Tugnait et al., 1994).

The model compounds investigated were phenolphthalein glucuronide, p-nitrophenylglucuronide, α-naphthylglucuronide and 6-bromo-2-naphthylglucuronide, all of which were spiked into human urine at a final concentration of 5 mM. The extraction protocol adopted for these model glucuronides was as follows: an aliquot of the sample (500 μl) was mixed with pH 8.5 glycine buffer (1.5 mL, 100 mM); the sample was then loaded onto a previously activated phenyl boronic acid cartridge (5 mL of 100 mM pH 10 glycine buffer then 5 mL of pH 8.5 glycine buffer); the retained glucuronides were recovered from the cartridges using methanol/1% HCl (90:10 v/v, 5 mL). For extraction onto C18 bonded SPE phases the sample was applied at pH 2 (HCl) onto either 200 or 500 mg cartridges previously conditioned with methanol (5 mL) and acidified water (pH 2, HCl). Stepwise gradient was then performed with acidified methanol/water mixtures. Finally, all eluates were then reduced to dryness, redissolved in D_2O and analyzed using ^{1}H-NMR spectroscopy (500 MHz).

The results of these experiments showed that it was indeed possible to selectively retain the model glucuronides on the PBA phase to some extent. Thus, when phenolphthalein glucuronide was extracted on cartridges containing 100, 200, 300, and 400 mg of sorbent respectively, good extraction was obtained on the 300 mg cartridge and the compound was completely extracted on the 400 mg SPE cartridges. Essentially complete recovery was obtained with the methanol/HCl elution step. Using the PBA SPE material it was possible to separate phenolphthalein glucuronide and phenolphthalein sulfate, spiked into urine, from each other based on the inability of the disulphate to form covalent bonds with the sorbent. Similarly, the quantitative extraction of 6-bromo-2-napthyl-B-D-glucuronide could be performed with 600 mg cartridges. However, although it was possible to retain these two model compounds efficiently in the case of both p-nitrophenol-glucuronide α-naphthylglucuronide, extraction efficiency was not as good. Thus, for p-nitrophenol glucuronide, SPE using 500 mg PBA cartridges resulted in less than 20% of the analyte being extracted; for α-napthylglucuronide the result was approximately 50% extraction.

All of the analytes were extracted efficiently onto the C18 phase, and could be recovered essentially free of endogenous contaminants by careful selection of the elution conditions.

Clearly, therefore, whilst it is possible in favorable instances to extract phenolic glucuronides onto PBA and to selectively fractionate sulfates and glucuronides of the same parent compound, the use of PBA is not essential to obtain clean extracts of glucuronides. The possession of a glucuronic acid moiety appears insufficient to ensure efficient extraction (glucuronic acid is not retained under the conditions used here), and the structure of the aglycone itself is also important. The structural features required of the aglycone to ensure good extraction remain to be elucidated, but from the small selection of compounds studied, pi-pi interactions may be important. In addition it should be noted that although glucuronides possess vicinal diol groups which could participate in the formation of a cyclic boronate ester, the possibility that the carboxylic acid and its adjacent hydroxyl group are involved cannot be discounted without further work.

It may also be worth noting that the requirement to use alkaline conditions for extraction of glucuronides onto PBA may well preclude the use of this as a sample preparation strategy for ester glucuronides which are unstable under such conditions.

G. GLYCOSYLATED AMINO ACIDS

Diabetes is characterized by, *inter alia*, the aminoacid lysine present in proteins reacting with glucose to form the glycosylated amino acid glucitollysine. Schmid and Pollack (1985) reported the development of an online extraction method for this substance in protein hydrolysates which also incorporated an "on-column" derivatization with o-phthalic dialdehyde (OPA) to give a fluorescent product for subsequent analysis by HPLC. In this method dihydroyboryl-Si100 (Serva, Heidelberg) with a particle size of 5 μm was packed into suitable precolumn and connected to a standard 6-port injection valve.

The extraction of glucitollysine was performed by conditioning the silica-bonded PBA with 0.1 M HCl (200 μl) to desorb any unwanted contaminants, followed by 250 μl of 0.1 M sodium hydroxide, followed by 200 μl of 0.1 M phosphate buffer (pH 8.5). The sample, at pH 8.5, was then loaded on to the phase. In the case of protein hydrolysates, co-extracted amino acids could be removed by washing with water or methanol. An attempt was made to use glucosaminic acid as an internal standard but, due to its weak binding, it was lost when large volume samples were applied or when the cartridge was washed with water. Elution of the analyte could be achieved by simply reducing the pH of the mobile phase. The methodology for use with protein hydrolysates involved application of 100 μl of sample (corresponding to approximately 1 mg of hydrolyzed collagen) at pH5, followed by a wash step of 500 µl of water. The analyte was then derivatized in situ with OPA, followed by washing with water (500 µl) methanol (500 µl) and water (500 µl). The extraction cartridge was then switched into the HPLC stream and the derivatized analyte eluted for analysis on a Nucleosil 50 C18 column.

H. REDUCED OLIGOSACCHARIDES

A final example of the use of covalent SPE is the application of the PBA sorbent to extraction of a range of reduced oligosaccharides by Stoll and Hounsell (1988). This work describes conditions for the separation of oligosaccharides from their alditols, and alditols from interfering amino acids and glycopeptides. The purification of an oligosaccharide-lipid conjugate (neoglycolipid) formed by the reductive amination of the sugar lactose with phosphatidylethanolamine dipalmitoyl (PPEADP) was also demonstrated.

The cartridges were activated by washing with methanol (2 x 1 mL), HCl 1 mL), water (2 x 1 mL) and finally sodium hydroxide (0.2 M, 4 x 1 mL). Residual base was removed by washing with water (2 x 1 mL) and the samples were then applied in 200 µl of water. Elution was performed by washing sequentially with 3 x 200 µl of water, 5 x 200 µl of 0.1 M acetic acid and 1 mL 0.1 M HCl. Individual fractions elution from the column were analyzed using TLC on silica. The authors found that, under these conditions, oligosaccharides with glucose at the reducing end were not retained by PBA, but the corresponding alditols were and could be eluted by buffers of decreasing pH. Similar behavior was observed for glycoprotein-derived octasaccharides and the corresponding alditols. Modification of these conditions by application of samples under alkaline conditions enabled the separation of the analytes from amino acids, peptides and glycoproteins. The modified protocol was also perceived to be better-suited to the resolution of siallactose from siallactol.

Conditions were also given for the purification of oligosaccharide derivatives formed by reductive amination, which opens the ring structure at the reducing end of the oligosaccharides to give acylic vicinal hydroxyl groups. For this extraction the PBA cartridge was conditioned with water (2 x 1 mL), methanol (2 x 1 mL) and a 1:1 (v/v/v) mixture of methanol and chloroform (2 x 1 mL). The reaction mixture obtained from the reductive amination of lactose and PPEADP was applied in 350 µl of the methanol/ chloroform mixture. Elution was accomplished by applying, sequentially, the methanol/ chloroform mixture (5 x 200 µl), chloroform/ methanol/ water (3:7:3 v/v, 4 x 200 µl) and finally by chloroform/ methanol/ 0.1 M acetic acid (3:7:3 v/v, 6 x 200 µl).

IV. CONCLUSIONS

For a limited range of structural types the possibility of easily reversible covalent bond formation with immobilized phenylboronic acid has clear attractions. With such a sorbent the potential for highly selective extraction exists and, given the usefulness of such an approach, it is perhaps surprising that there are not more applications in the literature. However, the list of compound classes is growing — for example, Tsuchiya (1998) has recently applied the technique to polyhydroxyflavones. However, it is clear that the overall structure of the analyte may be more important than the mere possession of an appropriate group such as a vicinal cis-diol, in governing whether or not a compound is readily extracted on to silica-bonded PBA. More work could usefully be done in this area.

REFERENCES

Benedict C.R. and Risk M., (1984). Determination of Urinary and Plasma Dihydroxyphenylalanine by Coupled-Column High-Performance Liquid Chromatography with C8 and C18 Stationary Phases, J.Chromatogr., 317: 27-34.

Echizen H., Itoh R. and Ishizaki T., (1989). Adenosine and Dopamine Simultaneously Determined in Urine by Reversed-Phase HPLC, with On-line Measurement of Ultraviolet Absorbance and Electrochemical Detection. Clin.Chem., 35: 64-68.

Huang T., Wall J. and Kabra P., (1988). Improved Solid-Phase Extraction and Liquid Chromatography with Electrochemical Detection of Urinary Catecholamines and 5-S-L-Cysteinyl-L-dopa, J.Chromatogr., 452: 409-418.

Imai Y., Ito S., Maruta K. and Fujita K., (1988). Simultaneous Determination of Catecholamines and Serotonin by Liquid Chromatography, after Treatment with Boric Acid Gel, Clin.Chem., 34: 528-530.

Kagedal B. and Pettersson A., (1983). Liquid-Chromatographic Determination of 5-S-L-Cyseinyl-L-dopa with Electrochemical Detection in Urine Prepurified with a Phenylboronate Affinity Gel, Clin.Chem., 29: 2031-2034.

Kupferschmidt R. and Schmid R., (1986). Organic Dye Compounds as an Evaluation Tool for Sample Extraction Using Bonded Silicas. Proceedings of the Third Annual Symposium on "Sample Preparation and Isolation Using Bonded Silicas", Proceedings of the 3rd Annual International Symposium, Analytichem International, pp 1-22.

Martin P., Leadbetter B. and Wilson I.D., (1993). Immobilized Phenylboronic Acids for the Selective Extraction of B-Blocking Drugs From Aqueous Solution and Plasma. J.Pharm.Biomed.Anal., 11: 307-312.

Maruta K., Fujita K., Ito S. and Nagatsu T., (1984). Liquid Chromatography of Plasma Catecholamines, with Electrochemical Detection after Treatment with Boric Acid Gel, Clin.Chem., 30: 1271-1273.

Murphy S.J., Morgan E.D. and Wilson I.D., (1990). Selective Separation of 20,22-Dihydroxyecdysteroids From Insect and Plant Material with Immobilized Phenylboronic Acid in McCaffery and Wilson (eds), Chromatography and Isolation of Insect Hormones and Pheromones, Plenum, New York. pp 131-136.

Oka K., Sekiya M., Osada H., Fujita K., Kato T. and Nagatsu T., (1982). Simultaneous Fluorimetry of Urinary Dopamine, Norepinephrine, and Epinephrine Compared with Liquid Chromatography with Electrochemical Detection. Clin.Chem., 28: 646-649.

Pfadenhauer E.H. and Tong S-D., (1979). Determination of Inosine and Adenosine in Human Plasma Using High-Performance Liquid Chromatography and a Boronate Affinity Gel, J.Chromatogr., 162: 585-590.

Schmid R. and Pollak A., (1985). Specific Extraction of a Glycosylated Amino Acid from Protein Hydrolysates Using Boronic Acid Derivatised Silica Gel in "Sample Preparation and Isolation Using Bonded Silicas", Proceedings of the 2nd International Symposium, Analytichem International, 15-20.

Stoll, M.S. and Hounsell, E.F., (1988). Selective Purification of Reduced Oligosaccharides Using a Phenyl Boronic Acid Bond Elut Column: Potential Application in HPLC/Mass Spectrometry, Reductive Amination Procedures and Antigenic/Serum Analysis. Biomed.Chromatogr., 2(6): 249-253.

Stolowitz M.L., (1985). Covalent Chromatography: Immobilized Phenylboronic Acid for Sample Preparation in "Sample Preparation and Isolation using Bonded Silicas", Proceedings of the 2nd International Symposium, Analytichem International, 41-44.

Tsuchiya, H., (1998). High-performance Liquid Chromatographic Analysis of Polyhydroxyflavones Using Solid-phase Borate-complex Extraction. J.Chromatogr. B. 720: 225-230.

Tugnait M., Ghauri F.Y.K., Wilson I.D. and Nicholson J.K., (1992). NMR Monitored Solid-Phase Extraction of Phenolphthalein Glucuronide on Phenylboronic Acid and C18 Bonded Phases. J.Pharm.Biomed.Anal., 9: 895-899.

Tugnait M., Wilson F.Y.K., Wilson I.D. and Nicholson J.K., (1994). High Resolution 1H NMR Spectroscopic Monitoring of Extraction of Model Glucuronides on Phenylboronic Acid and C18 Bonded Phases, in Stevenson D and Wilson ID (eds) Sample Preparation for Biomedical and environmental Analysis, Plenum, New York, pp 127-138.

Wilson I.D., Morgan E.D. and Murphy S.J., (1990). Sample Preparation for the chromatographic Determination of Ecdysteroids Using Solid-Phase Extraction Methods. Analytica Chimica Acta. 236: 145-155.

Wu A. and Gornet T.G., (1985). Preparation of Urine Samples for Liquid-Chromatographic determination of Catecholamines: Bonded-Phase Phenylboronic Acid, Cation-Exchange Resin, and Alumina Adsorbents Compared. Clin.Chem., 31: 298-302.

12

IMMUNO-AFFINITY EXTRACTION

Derek Stevenson, Badrul Amini Abdul Rashid and Seyyed Jamaleddin Shahtaheri, University of Surrey, Guildford, Surrey, United Kingdom

I. INTRODUCTION

The need to measure trace organic compounds such as drugs, pesticides, and environmental contaminants at low concentrations in a variety of complex matrices is a challenging but most important task. Many such measurements use sophisticated instrumental techniques, particularly high-performance liquid chromatography (HPLC), gas chromatography (GC) and more recently capillary electrophoresis. Despite the advances in instrumentation most methods involve sample extraction and preconcentration such as liquid-liquid extraction (LLE) or solid-phase extraction (Pawliszyn, 1995; Barcelo, 1991). This treatment of samples before analysis is usually the most time consuming phase of the method, and the source of the largest errors. Until relatively recently this aspect of analytical science has not received the attention its importance should merit.

With the growth of SPE technology as an attractive alternative to liquid-liquid extraction, a wide range of phases has become available. These phases use different interactions to obtain the desired retention and elution characteristics. Solid-phase methods use much less solvent and can be easily automated using commercially available equipment. Trace enrichment from large volumes is also common with this type of procedure. These methods are very suitable for broad range screening but most phases do not give the extreme selectivity of enzyme/substrate or antibody/antigen interaction. There has recently been growing interest in using antibodies covalently bonded onto a suitable support as a highly selective solid-phase extraction system (Van Ginkel et al., 1992).

II. TYPES OF AFFINITY COLUMNS

Affinity-mediated extraction offers the exciting possibility of combining the specificity of antibody-antigen interactions with the separation and confirmation possible with instrumental chromatographic methods. Simple protocols for retention and elution in small volumes of aqueous based solvents could conceivably be possible, if such an interaction could be harnessed for use in a SPE-like cartridge or column. This would be expected to permit trace enrichment from large volumes of sample.

These "designer columns" use highly specific antibody-antigen interactions to isolate the analyte of interest from sample flowing through the column. The analyte is then eluted from the column and subjected to chromatographic or other analysis. The main attraction of this approach is the potential for selectivity if extractions based solely on analyte/antibody binding can be developed.

A. ANTIBODY COLUMNS

Immuno-based methods have long been established in clinical laboratories for the measurement of drugs, proteins and other endogenous compounds. There has also been growing use of such methods in the environmental field, for example, when monitoring pesticides in water, soil, or in foodstuffs (de Frutos, 1995). Several different formats and a range of labels are available. Immunoassays are very sensitive, require only low volumes of sample, and are easily automated and economical to run. However they are often characterized by higher relative standard deviation values than chromatographic methods, and suffer from matrix effects. One very important advantage of using this same interaction as a separation mechanism is that the eluent can be introduced to spectroscopic detectors, thereby allowing absolute confirmation of identity.

The goal with antibody-mediated extraction is to use the specificity of the antibodies for extraction, and to couple this with the separation, detection and confirmation offered by chromatographic or spectroscopic methods. This chapter describes the progress that the author and other workers have made in this expanding field.

B. PRODUCTION OF ANTIBODY COLUMNS

Several different support materials and bonding techniques have been used to produce immuno-extraction columns. Support materials have included dextran, agarose, polyacrylamide, controlled pore glass beads, various proteins coated on glass beads, and silica. The technique of affinity chromatography (including immuno-affinity) is well established. However, the use of immobilized antibodies to aid sample clean-up and pre-concentration prior to instrumental chromatography is still a relatively new area. Coupling techniques to immobilize antibodies include hydrophobic adsorption and covalent bonding using reactive side chains such as epoxy, thiol, aminopropyl, alkylamine, carbonyl diimidazole, and aldehydes, particularly glutaraldehyde (Walt and Agayn, 1994; Phillips, 1989).

Much of the work published in current literature uses soft gels, which cannot withstand even moderate pressure. Non-specific binding is also common. This leads to the need for large elution volumes, necessitating subsequent preconcentration of samples. All the work in our laboratory has used rigid materials, mostly based on silica but controlled porosity glass has also been used. This has the advantage that the phases developed could be used in higher pressure systems than soft gels can withstand. Antibodies are more easily raised to large molecular weight compounds so there are more applications of immuno-extraction to these than to low molecular weight compounds.

The antibodies used for immuno-extraction have been raised in a variety of species particularly sheep, rabbits, and mice. Larger animals are favored as these produce larger quantities of antibodies. The majority of work uses polyclonal antibodies but as more monoclonal antibodies become available these will find more widespread use. Once raised, antibodies are stable for several years, provided suitable storage conditions are adhered to.

One interesting approach to developing an affinity sorbent is described by Howells et al. (1994). The authors noted in this case that the natural pigment melanin was reported to have high affinities for several basic drugs, and they surmised that if they could immobilize melanin on a sorbent the result would behave like a non-specific affinity medium. They synthe-

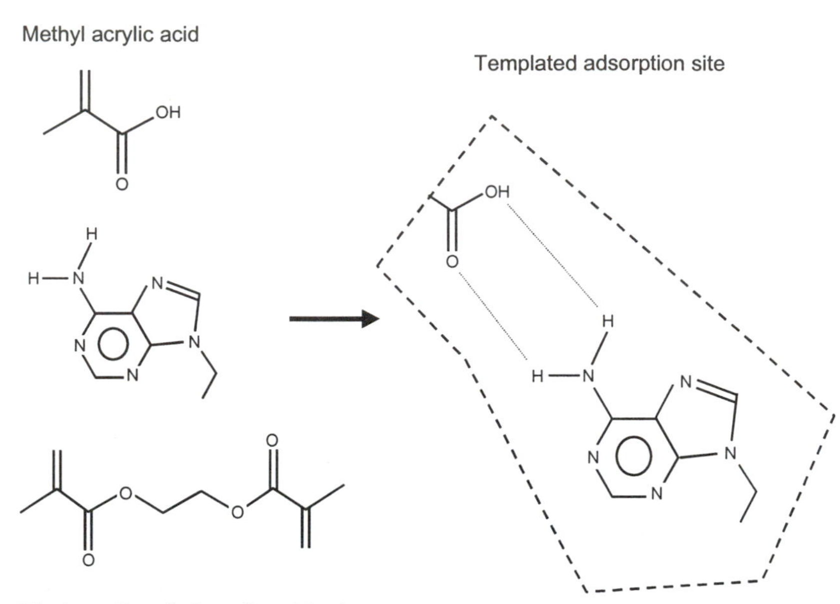

Figure 1. An illustration of how a molecule, 9-ethyl adenine, is used as a template around which polymerization occurs to yield an imprint (affinity site) of the molecule.

sized the melanin-immobilized sorbent by subjecting silica-based diol to a series of reactions. The resulting sorbent was characterized by investigation of its retention of representative drugs such as β-blockers, steroids and analgesics displaying a broad range of molecular structures.

C. MOLECULARLY IMPRINTED POLYMERS

An alternative to the use of antibodies for selective extraction is the application of molecularly imprinted polymers that mimic biologically derived antibodies. As these are based on chemical synthesis they are much cheaper and quicker to obtain than biologically derived antibodies. This may be done by taking the target molecule and building a "pocket" of monomers around it, followed by polymerization. The concept is shown in Figure 1. Subsequent removal of the target species leaves a molecular template that can behave like an affinity binding site when a sample containing the target analyte is applied to the sorbent. The molecularly imprinted polymer sorbent could then be packed into a disposable cartridge and used in much the same way as a naturally induced antibody column.

Since methacrylic acid is a common monomer used in the templating process the resulting polymer can also behave like a weak cation exchanger, so secondary interactions can be used, in addition to the affinity binding mechanism. This aspect aside, the imprinted polymers have been shown to provide similar selectivities to antibody-based sorbents in several cases. A review of molecularly imprinted polymers, comparing and contrasting these media with immunoaffinity sorbents, is provided by Sellergren (1997).

Preliminary work on their use for drug extraction has indicated that selective isolation of analyte is possible for Propranolol, Sameridine, Atrazine, Simazine and Tamoxifen (Martin et al., 1997; Andersson et al., 1997; Muldoon and Stanker, 1997; Matsui et al., 1997; Rashid et al., 1997). We are also beginning to see the use of such sorbents in the analysis of consumer products such as chewing gum formulations (Zander et al., 1998).

III. OPTIMIZATION OF ANTIBODY-MEDIATED EXTRACTION

In our laboratory polyclonal antibodies were available for the herbicides Chlortoluron, Isoproturon and 2,4-dichlorophenoxyacetic acid (2,4-D). These had been developed using conventional techniques in sheep and so several liters of antisera were available. The aims of the work were to:

- Immobilize the antibodies onto silica without loss of activity.
- Develop extractions based entirely on antibody-antigen interactions.
- Obtain quantitative recovery in a single low volume fraction.
- Develop a protocol using low volumes of non-toxic solvents.

Table 1. Parameters to be optimized for effective use of antibody-based SPE.

Column Considerations	Applications Considerations
Type of support to use	Activation of the support
Particle and pore size of support	Column pre-washing solvent
Column dimensions	pH of the sample to be loaded
How much antibody to immobilize	Volume of sample to be loaded
Quality and purity of antibodies,	Sample loading flow rate
Immobilization chemistry	Washing solvent used to elute sample components other than analyte
Column regeneration	Elution solvent composition
	Elution solvent pH

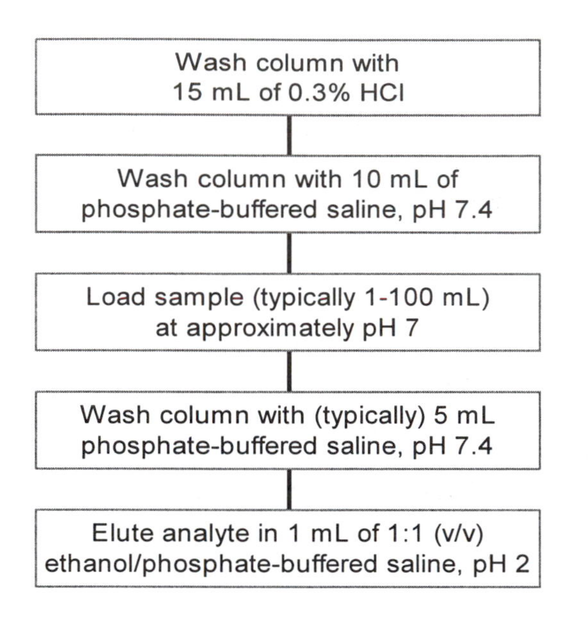

Figure 2. A basic immuno-affinity SPE protocol.

The immuno-extraction columns were prepared by adding the 2,4-D, Isoproturon or Chlortoluron antisera to aldehyde-activated porous silica (Clifmar Associates, Guildford, UK). Any unbound aldehyde groups were deactivated using glycine buffer. Different amounts of solid phase and antisera have been used but a typical column used for analytical purposes contained 1 g of dry weight solid phase and 200 μl of antisera.

Many factors have to be considered when developing and optimizing an antibody based extraction procedure. These are listed in Table 1. Optimization of all these parameters would prove time consuming so our work has concentrated on those considered most important. Once columns had been prepared initial efforts have concentrated on developing a protocol that could be used, with only minor modification, for a wide range of analytes to which antibodies were available. The basic protocol developed is shown in Figure 2. This protocol proved successful for all three compounds tried.

Results showed that loading the samples within the pH range 5-9 was necessary or the analytes could not be recovered in one fraction. The elution solvent was also critical. As the aim was to re-use antibody columns, elution solvents that would destroy the antibodies could not be used and, in any event, aqueous based systems were preferred on environmental, safety and cost considerations. Aqueous buffers with up to 50% ethanol at a pH of 2-3 gave optimum recovery. At higher pH the analytes eluted in the loading and washing fractions as well as in the elution fraction.

Although several elution solvents (ethanol, methanol, acetic acid, acetonitrile, acetone) have been tried at various ratios and pH values care has to be taken not to denature the antibodies if re-use of the column is intended. If very small quantities of antibody were coated on each column then reusability would not be an important consideration. Our preliminary work with these columns using mostly aqueous samples (tap water, river water and lake water) has indicated that they are very stable and can be reused at least 60 times. Other workers have reported using an immuno-column up to 100 times for the extraction of Zeranol and its metabolite from urine (Bagnati et al., 1991). In that application 95% acetone was used as the elution solvent. Our own preliminary work on a morphine antibody column has shown that the column can be reused about 20 times before the loss of capacity becomes significant (see Table 2). This will, of course, be affected by the type of sample put through the column and by any other sample pretreatment.

Alternative approaches to desorption have also been studied (Farjam et al., 1991). These included immuno-selective desorption with cross-reacting compounds, non-selective desorption with methanol, the use of chaotropic solutions, the use of microwaves to raise temperature and the use of surfactants and buffers to alter pH. Choice of desorption reagents needs to take into account the next step in the analytical method, particularly when this is direct injection into HPLC or GC. For example elution with HPLC mobile phase would be ideal if no further pre-concentration is needed but the mobile phase may not have the required immuno-desorption properties.

IV. APPLICATIONS

Studies on the capacity of the chlortoluron and isoproturon columns (see Table 3) showed that the columns are limited in the amount of analyte that they can retain before breakthrough occurs, and analyte is seen in the wash fractions. The maximum amount retained, 100 ng for isoproturon and 500 ng for chlortoluron on the columns coated with 100 μl of antisera, is very easily detected by modern chromatographic detectors after direct injection of the immuno-column eluent. Indeed the results described here suggest that smaller volumes of antisera and silica could be used, enhancing the attractiveness of this approach. A similar capacity (100 ng) has been reported for ciprostene extracted from plasma (Komatsu et al., 1991), though such

Table 2. Reusability of morphine affinity column.

Number of extractions	Capacity of column 1	Number of extractions	Capacity of column 2
2	150 ng	2	150 ng
14	126 ng	14	117 ng
26	73 ng	26	57 ng

comparisons are difficult to interpret without details of antisera characterization, purification, titer, avidity, etc.

Using tap water as an example, both the chlortoluron and isoproturon columns were capable of isolating analyte from 1 L of water and could be eluted in 1 ml of elution buffer, provided the weight capacity of the column for analyte was not exceeded. This preconcentration could be carried out at a sample flow rate of up to 5 ml/min. Other workers have demonstrated the possibility of using immuno-extraction on-line, emphasizing the possibilities of this approach for trace-enrichment, provided suitable immuno-columns can be produced. Automation has been shown using column switching (Farjam et al., 1991; Rule et al., 1994), using the Gilson ASTED™[36] (Sharman et al., 1992) and the Spark Holland Prospekt™[37] (Pichon et al., 1996). In the latter example, the on-line SPE system was used to concentrate the eluent from an affinity column by passing the eluent through a standard non-polar SPE sorbent before elution from this cartridge into the HPLC system.

A. NON-SPECIFIC BINDING AND CROSS-REACTIVITY

Although non-specific binding is not commonly reported, we have attempted to demonstrate that there is very little non-specific binding with our immuno-columns as this could lead to the larger elution volumes commonly reported for methods that use immuno-extraction. We have thus compared the elution profile of the chlortoluron and isoproturon immuno-columns with an activated silica column and a column bonded with antibodies from the same sheep before it had been injected with chlortoluron or isoproturon. These are shown in the full report of each method and clearly show that the retention is due principally to the antibody (Shahteheri et al., 1995; Shahtaheri et al., 1997). This may account for the small elution volume (1 ml) in which we are able to achieve quantitative recovery.

One important consideration with antibodies is the likelihood of cross-reactivity with closely related compounds. When antibodies are used in the conventional immunoassay modes (RIA, ELISA etc.), cross-reactivity is normally a disadvantage as this leads to errors in quantitation. With an antibody-mediated solid-phase extraction followed by a separation method such as GC or HPLC, cross-reactivity could be an advantage. In principle the antibody could be used to trap a closely related series of compounds which would then be eluted and separated by the instrumental chromatographic method. In practice, antibody cross-reactivity tends to occur by serendipity rather than design, but nonetheless this remains a very attractive possibility as preparation techniques and the theory of antibody production

[36] ASTED is a trademark of the Gilson Medical Electronics, SA, Villiers-le-Bel, France.
[37] Prospekt is a trademark of Spark Holland, Emmen, The Netherlands.

Table 3. Capacity of chlortoluron and isoproturon antibody columns.

Analyte	Weight of phase (g)	Volume of antibody (μl)	Column capacity (ng)
Chlortoluron	0.5	100	500
	0.5	500	2000
Isoproturon	1	100	100
	1	200	200

improve. The advantages of this cross-reactivity have been shown with phenylureas and triazines, though it did not apply to the whole group (Pichon et al., 1995a; Pichon et al., 1995b). Alternatively columns containing several antibodies capable of trapping a range of compounds can be produced (Chiabrando, 1989; Martin-Esteban, 1997).

Clearly the approach taken will depend on the purpose of the assay. An analyst interested in measuring a drug in biological fluid may wish to isolate only that compound or the drug and a small number of closely related metabolites. A water analyst might, on the other hand, favor methods capable of isolating a class of pesticides or even a wider range of organics, which can subsequently be separated and analyzed by a chromatographic method. The immuno-extraction approach can be used alone, as we have shown for water samples, or used in combination for example with a different type of solid phase. This has been demonstrated for the determination of clenbuterol in urine (van Rhijn et al., 1993) or for the determination of chloramphenicol residues in tissue homogenate (van de Water and Haagsma, 1987). Antibody extraction is particularly attractive for those hydrophilic compounds which are difficult to extract by conventional methods, such as Albuterol (Ong et al., 1989).

Some applications have been published, that initially appear to function as though affinity extraction is occurring. For example, 2-amino-1-cyclopentane-1-dithiocarboxylic acid (ACTA)-loaded silica with Platinum (IV) has been used to retain anilines from water (Goewie et al., 1984) and Silver (I) oxime sorbent has been used to extract Buturon in water (Nielen et al., 1987). However, these are more like a solid-phase extraction equivalent of immobilized metal affinity chromatography (IMAC) than a strict immuno-affininty extraction.

IV. CONCLUSIONS

Antibodies to a range of analytes can be successfully immobilized onto rigid particles such as silica. These phases are capable of extracting trace organic compounds such as drugs and pesticides from complex matrices. The custom-designed phases can be used in an identical manner to conventional, commercially available phases such as silica and bonded silicas, in both an off-line and on-line mode. They offer greater selectivity than conventional phases and could be used as either the sole sample preparation step, or in combination with other methods, to obtain clean extracts when very low detection levels are required. They also offer an alternative when analytes are difficult to extract due to hydrophilicity or stability problems in non-aqueous solutions.

The major drawback to this approach is the lack of antibodies, particularly to low molecular weight compounds. However as antibodies to more compounds become available and methods to produce them improve this aspect will become less important. Extraction of new products generated by biotechnology is one area where use of immuno-affinity SPE may prove most effective. Already, affinity gels in disposable cartridges have become commercially available which remove phenol or surfactants from reaction mixtures resulting from biotechnology syntheses.

REFERENCES

Andersson, L.I., Paprika, A. and Arvidsson, T., (1997) A Highly selective Solid Phase Extraction Sorbent for Pre-Concentration of Sameridine Made by Molecular Imprinting. Chromatographia, 46: 57-62.

Bagnati, R., Oriundi, M.P., Russo, V., Danese, M., Berti, F and Fanelli., (1991) Determination of Zeranol and B-Zeranol in Calf Urine by Immunoaffinity extraction and Gas Chromatography-Mass Spectrometry after Repeated Oral Administration of Zeranol. J.Chromatogr., 564, 493-502.

Barcelo, D., (1991) Occurrence, Handling and Chromatographic Determination of Pesticides in the Aquatic Environment - a Review. Analyst, 116: 681-688.

Chiabrando, C., Pinciroli, V., Campoleoleoni, A., Benigni, A., Piccinelli, A. and Fanelli, R., (1989) Quantitative profiling of 6-Ketoprostaglandin $F_{1\alpha}$, 2,3-Dinor-6-ketoprostaglandin $F_{1\alpha}$, Thromboxane B_2 and 2,3-Dinor-Thromboxane B_2 in Human and Rat Urine by Immunoaffinity Extraction with Gas Chromatography - Mass Spectrometry. J.Chromatogr., 495: 1-11.

de Frutos, M., (1995) Chromatography-immunology coupling, A Powerful tool for Environmental Analysis. TrAC 14, 133-139.

Farjam, A., Brungman, E.A., Henk, L. and Brinkman, U.A.Th., (1991a) On-line Immunoaffinity Sample Pre-treatment for Column Liquid Chromatography:

Evaluation of Desorption Techniques and Operating Conditions using an Anti-estrogen immuno-precolumn as a Model System. Analyst, 116, 891-896.

Farjam, A., Brugman, E.A., Soldaat, A., Timmerman, P., Lingeman, H., Jong, G.J., Frei, R.W. and Brinkman, U.A.Th., (1991b) Immunoaffinity Precolumn for Selective Sample Pretreatment in Column Liquid Chromatography: Immunoselective Desorption. Chromatographia, 31, 469-477.

Goewie, C.E., Kwakman, P., Frei, R.W., Brinkman, U.A.Th., Maasfield, W., Seshadri, T. and Kettrup, A., (1984) Pre-column Technology in HPLC for the Determination of Phenylurea Herbicides in the Presence of their Anilines. J.Chromatogr. 284:73.

Howells, L., Godfrey, M and Sauer, M.J., (1994) Melanin as an Adsorbent for Drug Residues. Analyst, 119:2691-2693.

Komatsu, S., Murata, S., Aoyama, A.T., Zenki, T., Ozawa, N., Tateishi, M. and Vrbanac, J.J., (1991) Micro-quantitative Determination of Cisprostene in Plasma by Gas Chromatography-Mass Spectrometry Coupled with an Antibody Extraction. J.Chromatogr., 568, 460-466.

Martin, P., Wilson, I.D., Morgan, D.E., Jones, G.R. and Jones, K., (1997) Evaluation of a Molecular-Imprinted Polymer for use in the Solid Phase-Extraction of Propranolol From Biological Fluids. Anal.Commun., 34: 45-47.

Martin-Esteban, A., Kwasowski, P. and Stevenson, D., (1997) Immunoaffinity-Based Extraction of Phenylurea Herbicides Using Mixed Antibodies against Isoproturon and Chlortoluron. Chromatographia 45: 364-368.

Matsui, J., Okada, M., Tsuruoka, M., and Takeuchi, T., (1997) Solid-Phase Extraction of a Triazine Herbicide using a Molecularly Imprinted Synthetic Receptor. Anal.Commun., 34: 85-87.

Muldoon, M.T., and Stanker, L.H., (1997) Molecular Imprinted Solid Phase Extraction of Atrazine from Beef Liver Extracts. Anal.Chem., 69: 803-808.

Nielen, M.W.F., Van Inger, H.E., Valk, A.J., Frei, R.W. and Brinkman, U.A.Th., (1987) Metal-loaded Sorbents for Selective Online Sample Handling and Trace Enrichment in Liquid-chromatography. J.Chromatogr. 10:617.

Ong, H., Adam, A., Perreault, S., Marleau, S., Bellemare, M., Du Souich, P. and Beaulieu, N., (1989) Analysis of Albuterol in Human Plasma based on Immunoaffinity Chromatographic Clean-up Combined with HPLC with Fluorimetric Detection. J.Chromatogr., 497: 213-221.

Pawliszyn, J., (1995) New Directions in Sample Preparation for Analysis of Organic Compounds. TrAC, 14: 113-122.

Phillips, T. M., (1989) High Performance Immunoaffinity Chromatography. Adv.in Chromatogr., 29, 133-173.

Pichon, V., Chen, L., Durand, N., Le Goffic, F. and Hennion, M-C., (1996) Selective Trace Enrichment of Immunosorbents for the Multiresidue Analysis for Phenylurea and Triazine Pesticides. J.Chromatogr., A. 725:107-119.

Pichon, V., Chen, L. and Hennion, M-C., (1995a) On-line Preconcentration and Liquid Chromatographic Analysis of Phenylurea Pesticides in Environmental Water Using a Silica-Based Immunosorbent. Anal.Chimica.Acta 311, 429-436.

Pichon, V., Chen, L., Hennion, M-C., Danial, R., Martel, A., Le Goffic, F., Abian, J. and Barcelo, D., (1995b) Preparation and Evaluation of Immunosorbents for Selective Trace Enrichment of Phenylurea and Triazine Herbicides in Environmental Waters. Anal.Chem., 67, 2451-2460.

Rashid, B.A., Briggs, R.J., Hay, J.N., Stevenson, D., (1997) Preliminary Evaluation of a Molecular Imprinted Polymer for Solid-Phase Extraction of Tamoxifen. Anal.Commun., 34, 303-305.

Rule, G.S., Mordehal, A.V. and Hennion, J., (1994) Determination of Carbofuran by On-line Immunoaffinity Chromatography with Coupled-column Liquid Chromatography Mass- Spectrometry. Anal.Chem., 66: 230-235.

Sellergen, B., (1997) Imprinted Polymers: Stable, Reusable Antibody Mimics for Highly Selective Separations. American Laboratory, June: 14-20.

Shahtaheri, J.S., Katmeh, M.F., Kwasowski, P. and Stevenson, D., (1995) Development and Optimisation of an Immunoaffinity based solid-phase Extraction for Chlortoluron. J.Chromatogr.A 697, 131-136.

Shahtaheri, J.S., Kwasowski, P. and Stevenson, D., (1998) Highly Selective Antibody-Mediated Extraction of Isoproturon from Complex Matrices. Chromatographia, 47, 453-456.

Sharman, M., MacDonald, S. and Gilbert, J., (1992) Automated Liquid Chromatographic Determination of Ochratoxin A in Cereals and Animal Products Using Immunoaffinity Column Clean-up. J.Chromatogr., 603, 285-289. J.Chromatogr., 603: 285-289.

van de Water, C., and Haagsma, N., (1987) Determination of Chloramphenicol in Swine Muscle Tissue using a Monoclonal Antibody-mediated Clean-up Procedure. J. Chromatogr., 411, 415-421. J.Chromatogr., 411: 415-421.

van Ginkel, L.A., Stephany, R.W., Van Rossum H.J. and Zoontjes, P.W., (1992) Perspectives in Residue Analysis: the Use of Immobilised Antibodies in (multi) Residue Analysis. TrAC., 11, 294-297.

van Rhijn, J.A., Traag, W.A. and Heskamp, H.H., (1993) Confirmatory Analysis of Clenbuterol Using 2 Different Derivatives Simultaneously. J.Chromatogr., 619: 243-249.

Walt D.R., and Agayn, V.I., (1994) The Chemistry of Enzyme and Protein Immobilization with Glutaraldehyde. TrAC, 13: 425-223.

Zander, A., Findlay, P., Renner, T., Sellergren, B. and Swietlow, A., (1998) Analysis of Nicotine and its Oxidation Products in Chewing Gum by a Molecularly Imprinted Solid Phase Extraction. Anal.Chem. 70: 3304-3314.

13

MATRIX SOLID-PHASE DISPERSION (MSPD)

Steven A. Barker, School of Veterinary Medicine, Louisiana State University, Baton Rouge, Louisiana

I. INTRODUCTION

From the proceeding chapters, it is clear that solid supports which have been surface-modified with bonded organic phases have revolutionized extraction science. The use of solid-phase extraction (SPE) columns, cartridges, fibers and discs has offered a mechanism for reducing the large solvent volumes often required and the complications encountered in many of the classical solvent-solvent extraction approaches. It is especially ironic that many of the older, officially sanctioned methods for conducting pollution or residue analysis generate more hazardous waste and pollution than they actually detect, monitor, or ameliorate. Many such methods have subsequently been replaced, or are being replaced, by approaches that involve the use of solid-phase extraction in one form or another. The previous chapter described, for example, the use of SPE discs and cartridges for the extraction of water samples in conducting analyses for regulatory or other monitoring purposes. Nearly all the examples so far have dealt with liquid samples or samples that have been converted to liquid form.

Standard SPE materials are primarily designed to accept aqueous or, at least, solubilized, fluid samples. Samples having too great a viscosity or that contain a moderate to heavy particulate content cannot be extracted by SPE without a dilution step, filtration, centrifugation, or some other manipulation. Solid samples cannot be directly applied to SPE materials and must, therefore, be put into a liquid, solubilized form prior to SPE extraction.

For example, the extraction of solid or semi-solid biological samples, such as tissues or rendered fats, usually begins with the disruption of the sample architecture. Classic techniques involve mincing and maceration using scissors and/or mechanical blenders, grinding of the sample with abrasives, sonication of the sample or the application of a French press. All of these techniques supply shearing forces to the sample. Some of these processes may also involve the simultaneous incorporation of organic solvents, added to and removed from the sample repeatedly. The organic extracts are then combined to obtain effective extraction of the target analyte. Most of these processes disrupt the structure of the sample as well as its cellular components, but they are often not very efficient.

In order to achieve lysis of cells and complete dissolution or disruption of the cell membrane or wall (particularly bacterial cell walls and some of the integral cellular structures), we may need to apply chemical, in addition to the physical, disruption. This may take the form of enzymatic digestion, hydrolysis with acid or base, treatment with ionic detergents or polyethers, aggressive organic solvents or hypertonic solutions. While each of these treatments may isolate the analyte effectively, each also carries with it the potential to add interferences or artefacts to the extract. Further, these agents often need to be removed after use — something which may greatly complicate the analytical process and add to solvent usage, disposal, worker exposure, and handling time.

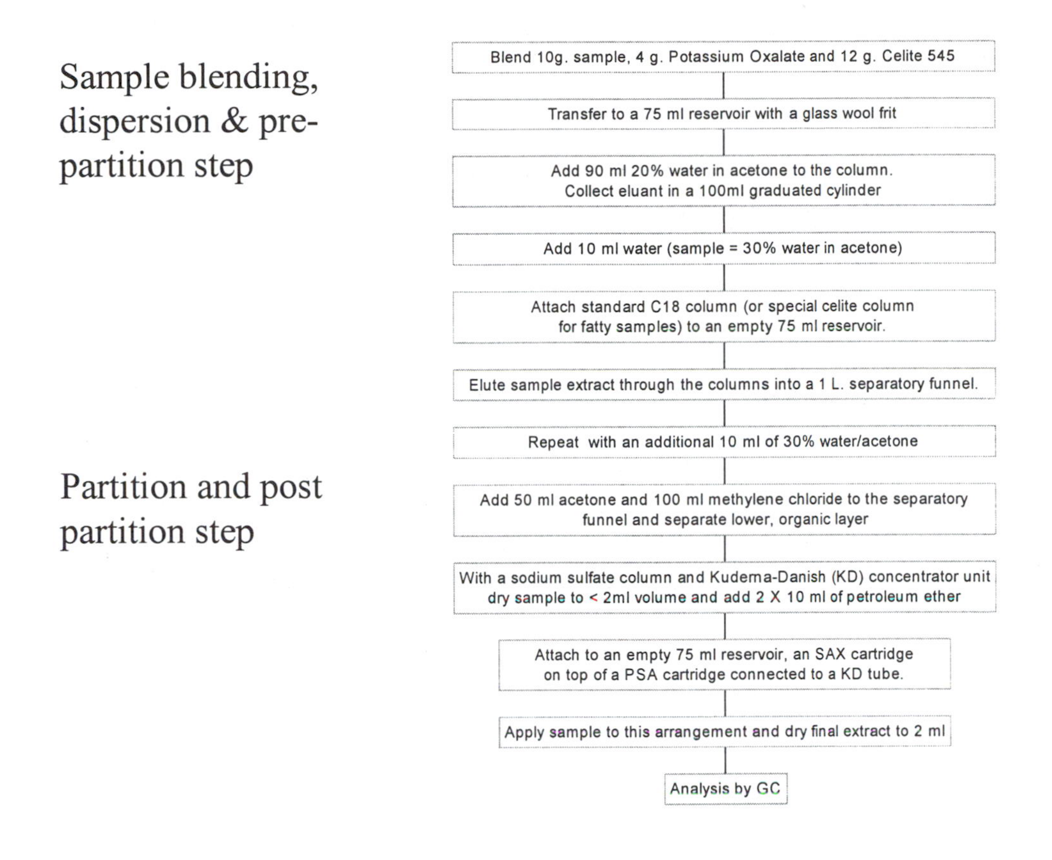

Figure 1. The"Luke" method for residue analysis in plant and fruit matrixes.

Despite all the sample manipulation implied by the above discussion, many classic SPE procedures have been developed for residue analysis in plant and animal tissues. This is attributable to the potential for SPE to clean up and fractionate complex samples. An example is given in Figure 1, of a commonly used technique for residues in fruits and crops, called the Luke method (Luke, 1995).

To address the need for faster, simpler and more efficient sample preparation, a technique called Matrix Solid Phase Dispersion (MSPD) was developed (Barker, Long and Short, 1989). MSPD is a low-tech process that can be applied as easily in the field as at the laboratory bench. It achieves disruption and extraction of solid or viscous samples, permitting the analyst to perform immunoassay or other simple tests rapidly and inexpensively. The Center for Veterinary Medicine of the Food and Drug Administration funded this research, which was designed to address the need to reduce analysis turn-around time, solvent use and disposal and to investigate and introduce new techniques for both screening and determinative analytical testing.

II. THE COMPONENTS OF AN MSPD EXTRACTION

A. THE DEVELOPMENT OF MSPD

Scientists have used granular solids to blend moist or semi-dry samples for several decades before the introduction of MSPD. The granular material was often used solely to dehydrate the sample; hence anhydrous salts, such as sodium sulfate, hygroscopic materials, such as silica gel and celite, or kieselguhr appear in articles that describe blending or grinding of samples prior to extraction. The groundbreaking realization — that bonded silicas could not only macerate or dehydrate a sample but could act as a special tool to achieve multiple sample manipulation in a single and simple step —

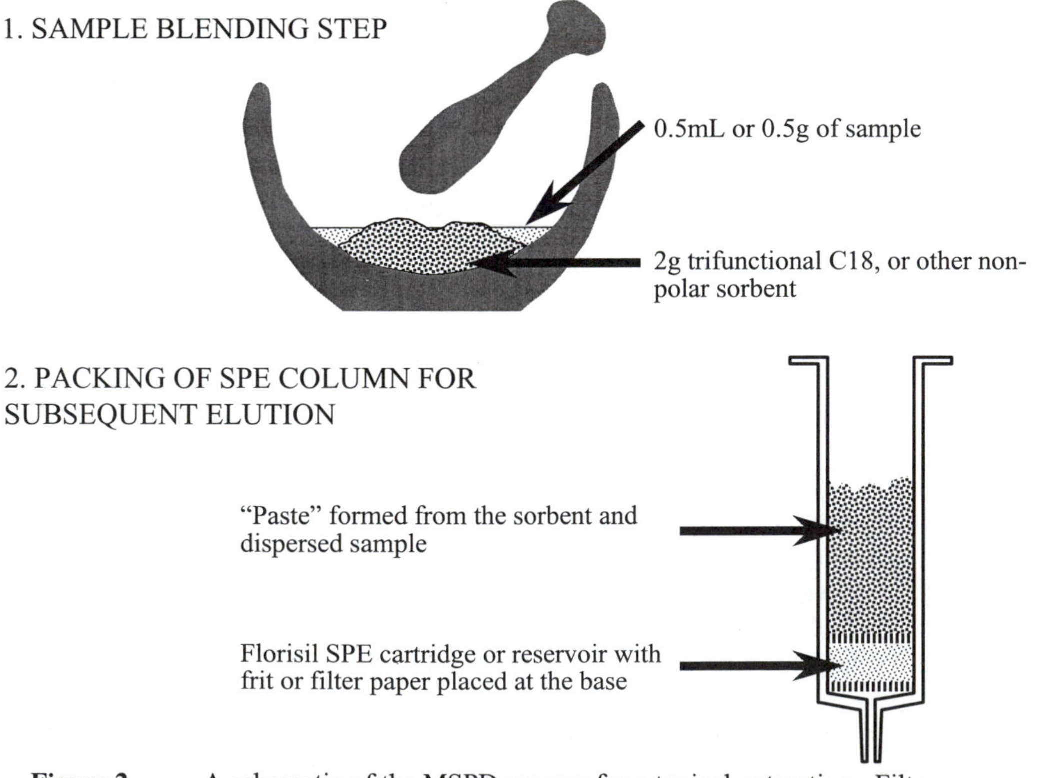

Figure 2. A schematic of the MSPD process for a typical extraction. Filter paper or a regular polyethylene frit may be used as the MSPD sorbent support. A top frit is often used to compact the MSPD blend and eliminate voids or channeling.

occurred in the late 1980s. Barker et el. introduced this new process for the disruption and extraction of target analytes from mammalian tissue samples in 1989 (Barker, Long and Short, 1989). A schematic is shown in Figure 2.

The procedure involves the blending of solid phase extraction sorbents directly with solid or viscous samples. When used in this way the solid support, most often a silica particle but possibly a more exotic material, serves as an abrasive. It destroys the sample architecture through applied shearing forces. These are provided by grinding or blending with a mortar and pestle. Glass or agate (or, more generally, ones made from non-porous materials) mortars have been most commonly used. The solid phase, for example C18, serves to assist in the sample disruption process by solubilizing integral lipids, seemingly unfolding and disrupting the cell membranes and internal substructures themselves, thus acting like a detergent. Remember, however, that the material possessing the lipophilic or detergent-like properties is a bonded phase and does not have the potential to complicate the isolation or analysis of the target analyte as occurs with the addition of solvents, detergents, or dispersants. This is because it is still, at the end of the extraction, in a different phase to the eluting solvent.

The blending process ensures that the tissue is completely disrupted and distributed over the surface of the solid support. During this blending the interactions of the solid support and bonded phase with the individual cellular components are maximized and the cellular components' interactions with one another are destroyed. The result is a dispersed sample that is, as determined by scanning electron microscopy, quite homogeneous and distributed in a layer approximately 100 Ångströms thick for applications using a common ratio of sample mass to sorbent mass (4:1 w/w). In other words our sample has become our solid phase. We postulate that this new phase exists as a bi-layer consisting of an internal silica particle and its bonded phase as an inner "membrane," for example, and an outer "membrane" of cellular components distributed so as to maximize lipophilic interactions. Polar materials and functional groups of such components would be expected to be oriented to the outside of the structure and to arrange themselves so as to minimize lipid interaction and to maximize hydrophilic interaction. This hypothetical model is represented in Figure 3.

The consistency of the resulting blended material is that of a semi-dry powder and it can thus be packed into a column or cartridge which can be eluted with solvents as if it were a standard SPE device. However, this process provides a unique form of chromatographic separation that differs from standard SPE. For example, drugs spiked into milk samples which are then extracted by MSPD and by classic SPE techniques show very different elution profiles (Barker and Fong, 1994). We must avoid making assumptions

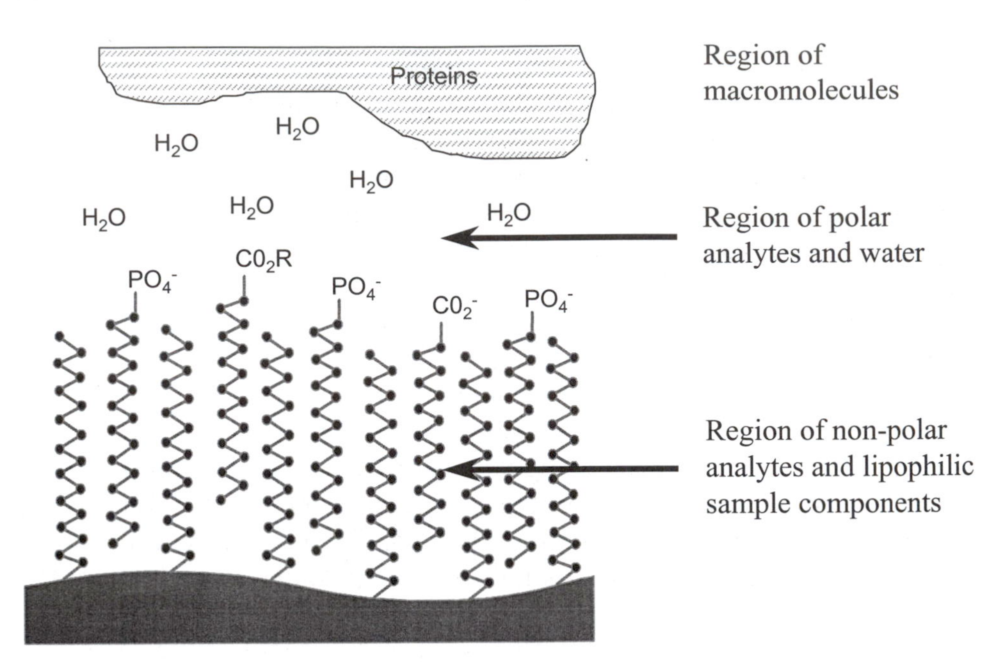

Figure 3. A hypothetical model, identifying regions of polarity and non-polarity, of a dispersed sample on the surface of an octadecyl silica sorbent.

about the nature of retention, capacity and elution based on regular SPE experiments.

The following discussion outlines the proposed mechanisms involved in the MSPD process and the factors that have been identified as influencing its performance. These results confirm the uniqueness of the process and imply that new uses and forms of bonded phase chemistries for sample extraction and preparation can be designed and applied.

B. A UNIQUE SAMPLE/SORBENT INTERACTION

MSPD differs from classical SPE in two fundamental ways: 1) The SPE samples must be in a liquid form before being added to the column, whereas the MSPD method is designed to handle solid or viscous liquid sample matrices directly; 2) the components of a sample interact very differently with the bonded-phase sorbent and with the bulk of the sample matrix when subjected to MSPD and to classic SPE. Direct blending of the sample by MSPD bypasses several steps — liquefaction, solubilization, or homogenization of the components of the sample, followed by centrifugation or filtering to remove debris. At each step we risk losing components of the sample and increasing the sample volume. MSPD removes these

Table 1. An alphabetical listing, by analyte, of references using MSPD to isolate specified drugs or substances from various types of matrices.

Reference	Analytes	Matrix
Rosen, Hellenaes, Toernquist, Shearan (1994)	acetylgestens	bovine kidney fat
McLaughlin, Henion, Kijak (1994)	aminoglycoside	bovine kidney
Yang, Fu (1994b)	amitraz, pirimicarb	fruits and vegetables
Barker, Long, Short (1989)	benzimidazoles, cephapirin, organophosphates	bovine muscle
Nau, Pocci, Nguyen (1996)	chlorinated pesticides, endrin, aldrin, endosulfan	muscle and fat tissue, fruits, prepared foods, baby foods
Long, Hsieh, Malbrough, Short, Barker (1989f)	5 benzimidazoles	bovine liver
Long, Hsieh, Malbrough, Short, Barker (1989e)	5 benzimidazoles	swine muscle
Long, Hsieh, Malbrough, Short, Barker (1989d)	7 benzimidazoles	milk
Yang, Fu (1994a)	carbofuran	corn
Boyd, Shearan, Hopkins, O'Keefe, Smyth (1991, 1993)	clenbuterol	bovine liver
Boyd, O'Keefe, Smyth (1994, 1995)		
Long, Hsieh, Bello, Malbrough, Short, Barker (1990)	chloramphenicol	milk
Long, Hsieh, Short, Barker (1991b)	9 chlorinated pesticides	bovine fat
Long, Hsieh, Short, Barker (1991a)	9 chlorinated pesticides	catfish muscle
Lott, Barker (1993a)	14 chlorinated pesticides	oysters
Lott, Barker (1993b)	14 chlorinated pesticides	crayfish hepatopancreas
Nau, Pocci, Nguyen (1996)	chlorinated herbicides, triazines, ureas	fruits and vegetables
Schenck, Barker, Long (1989)	chlorsulon	milk
Schenck, Wagner, Bargo (1993)		
Long, Hsieh, Malbrough, Short, Barker (199b)	furazolidone	milk
Long, Hsieh, Malbrough, Short, Barker (199d)	furazolidone	swine muscle
Soliman, Long, Barker (1991)	furazolidone	chicken muscle
Iosifidou, Shearan, O'Keefe (1994)	ivermectin	fish muscle

difficulties and allows the entire sample to be exposed to the extraction process. The interactions involved may be classified as:

1) Target analyte with the solid support.
2) Target analyte with the bonded phase.
3) Target analyte with the dispersed sample components.

Table 1. Continued.

Reference	Analytes	Matrix
Schenck (1995) Schenck, Barker, Long (1990)	ivermectin	bovine liver
Schenck, Barker, Long, Matusik (1989)	nicarbazin	chicken muscle
Schenck, Wagner (1995)	5 organochlorine and 5 organophosphorus pesticides	milk
Jarboe, Kleinow (1992)	oxolinic acid	catfish muscle
Long, Hsieh, Malbrough, Short, Barker (1990g)	oxytetracycline	catfish tissue
Long, Hsieh, Malbrough, Short, Barker (1990c)	oxy-, chlor-, tetracyclines	milk
Ling, Chang, Huang (1994)	polychlorinated biphenyls (PCBs)	fish
Ling, Huang (1995b)	16 organochlorine pesticides and PCBs	fish
Ling, Huang (1995a)	6 synthetic pyrethroids	vegetables
Fukushima, Taguchi, Nishumune, Suehi (1993)	5 sulfa drugs, furazolidone, clopidol	chicken tissues
Long, Hsieh, Malbrough, Short, Barker (1990b)	sulfadimethoxine	catfish muscle
Walker and Barker (1994a) Walker and Barker (1994b) Walker, Thune and Barker (1995)	sulfadimethoxine, N-acetyl metabolites	fish
Shearan, O'Keefe (1994) Shearan, O'Keefe, Smyth (1994)	sulfamethazine	swine muscle
Long, Hsieh, Malbrough, Short, Barker (1990a)	sulfonamides	swine muscle
Long, Hsieh, Malbrough, Short, Barker (1989b)	7 sulfonamides	infant formula
Long, Short, Barker (1990)	8 sulfonamides	milk
Aan Poucke, Depourcq Van Peterghem (1991)	sulfonamides	milk and muscle
Renson, Degand, Maghuin, Rogister, Delahaut (1993)	sulfonamides	livestock products
Tamura, Yotoriyama, Kurosahi, Shinohara (1994)	sulfonamides	animal tissues
Riemer, Suarez (1991) Riemer, Suarez (1992)	5 sulfonamides	salmon muscle

4) Interaction of the other sample components with the solid support.
5) Interactions of the other sample components with the bonded phase.
6) All of the above interacting with the elution solvent(s).

Finally, we should remember that these six classes of interaction do not occur in isolation, as they do when we develop a classic SPE procedure (for example, during the measurement of retention by spiking known quantities of analyte into a buffer during the method development process); rather, they all occur simultaneously.

C. THE MSPD STEP

Although MSPD has been applied to many different sample types and an equally broad range of target analytes, the developed methods have proven to be surprisingly similar. A list of references illustrating the drugs or substances isolated from a given matrix is given in Table 1. For a review of applications and individual references see Barker and Long (1992); Barker and Haley (1992); Barker (1993); Barker, Long and Hines (1993); Walker, Lott and Barker (1993); Barker and Long (1994).

A typical process involves the use of a glass, non-porous alumina or agate mortar and pestle into which the solid support/bonded phase material is placed. The sample is then placed on top of the solid support. Most studies applying MSPD have used 2.0 g of solid support to 0.5 g of sample matrix, or a similar ratio of masses. A surprisingly large number of applications have been published which yield satisfactory detection limits on just half a gram of sample, however. At this point in the process standards may be spiked onto or into the sample itself. If one wishes to modify the chemistry of the sample, antioxidants, chelating or de-chelating agents, acids, bases, etc., may also be added at this point for co-blending with the sample and solid support. These additives will affect the elution sequence or retention of the target analyte. The material is gently blended with the pestle — little force is required, even for tough samples containing a high degree of connective tissue such as kidney or fibrous plants. The analyst will note the immediate disruption of the sample architecture and, after approximately 30 to 45 seconds, almost complete dispersion of the sample. Frozen samples may take a little longer because thawing is a necessary part of dispersion of the sample.

Some combinations of sorbent and sample impart "stickiness" to the resulting blend, requiring scraping of the sides of the mortar or tip of the pestle, followed by extra blending time, to ensure complete dispersion. This is most evident with samples containing a significant quantity of connective tissue, such as kidney, and with the use of more lipophilic bonded phases, such as C18 (octadecylsilane). This problem is not as evident with shorter chain supports like C8 and C3, and with easily dispersed samples like milk or liver. While manual blending has been most common, it is possible to use a hand-held electric or battery-operated mixer fitted with a Teflon pestle. This tool blends quite well and avoids the possible repetitive motion problems that could become associated with the sequential application of the method to a large number of samples.

We have also observed that samples that exist as suspensions or that are moderately viscous can be blended onto the solid support without having to use a mortar or pestle. Such samples can be added to the solid support material contained in a snap-cap vial, tube, etc., and mixed to homogeneity by use of a stirring rod or by vortexing. This approach has been applied to whole blood, plasma, serum, and especially raw and homogenized milk samples.

D. ALTERNATIVE SAMPLE DISPERSION TECHNIQUES

In most applications a relatively small sample size, usually 0.5 mL of a liquid sample, such as milk, or 0.5 g of a solid sample, such as muscle, has been utilized to conduct an MSPD extraction. However, Van Poucke, et al. (1991), have found that they could use 5.0 to 10.0 mL of milk blended with 2.0 g of C18 material for the isolation of sulfonamide residues and still achieve approximately 95% recovery. Poorer recoveries were obtained for smaller quantities of C18, regardless of the sample size. Further, the sample was prepared in a glass syringe barrel, mixing the milk and solid support with a sealed pipette. This step eliminated the need to blend the materials in a mortar and subsequently to transfer the sample to a column — a process that can lead to some sample loss and which would require additional analyst time.

A similar approach has been applied by Schenck (1995) and Schenck and Wagner (1995) for the isolation of ivermectin and of organochlorine and organophosphorus residues in milk, respectively. In the latter case 2.0 g of the solid support is blended in the column with 1.5 mL of acetonitrile and 5.0 mL of milk. When the sample is mixed well and the aqueous phase is removed by aspiration the MSPD extraction process can then be applied. This permits the use of a much larger milk sample (5.0 mL versus the common blending of 0.5 mL) per analysis and greatly enhances the limits of detection of the overall method. Such an approach may prove applicable to other such sample matrices where increased sensitivity is an issue.

E. THE ELUTION STEP

Now that we have dispersed our sample, the resulting sorbent/sample blend can be packed into a cartridge. It is now ready for sequential washing/elution. Most studies have used a polypropylene disposable syringe barrel of approximately 10 mL fluid volume as the column, with paper or polyethylene frits to retain the blended materials. While some applications require the analyst to pack the dispersed sample/solid support into a column containing just a frit, in order to further assist the extraction process. These additional sorbents, positioned so the elution solvent must pass through them, help to remove co-eluting interferences. This can also be accomplished by stacking small columns or extraction discs under the MSPD column.

Recently the use of packed Florisil™[38] disposable cartridges has been advocated (Nau et al., 1996) as a means of retaining interferences that would otherwise be co-eluted with the analytes. Sorbents like dry Florisil also serve to remove water from the eluent as it passes through the packed bed. The sample is usually tamped down into the bottom of this column with a modified syringe plunger to approximately 4.0 mL volume and placed over a receptacle for collecting the eluate(s). Most reported applications state that 8 mL of each elution or wash solvent are allowed to flow by gravity through the column. When a higher sample throughput is desired, however, vacuum boxes have also been used to regulate the flow.

The elution sequence applied to the MSPD columns varies with the analytical needs. We may begin with a nonpolar solvent, such as hexane, and proceed with a sequence of solvent additions of increasing polarity (dichloromethane, ethyl acetate, acetonitrile, methanol, water), at each stage collecting a fraction of the sample. Alternatively, we may add a solvent to "clean" the column followed by a specific solvent to remove the target analyte while retaining undesired and potential interferences on the MSPD column.

III. THE PARAMETERS OF AN MSPD EXTRACTION

A. THE EFFECT OF THE SORBENT ON MSPD

Despite this apparent complexity, the principles of SPE and chromatography are still operable during the MSPD process. Thus, the nature of the solid support/bonded phase will affect the retention/elution of the target analytes and the corresponding sample components. To date only silica-based solid supports have been used. Studies have shown that the pore size of these materials is of little importance in the MSPD process (Barker and Floyd, 1996), but this may vary with the nature of the sample and should therefore be considered. Similarly, the particle size has been examined. As one might expect, use of very small particle sizes such as 3-20 µm, lead to extended solvent elution times and, in some cases, plugging of the MSPD column altogether. However, 40-100 µm particle size materials with a 60

[38] Florisil is a trademark of the Floridin Company, Quincy, Florida.

Ångström pore size have been used extensively and successfully. Fortunately, considering the mass of sorbent used per extraction, these materials are far less expensive.

The bonded phase will also affect the results and a range of available SPE phases should be examined for each application. It has been observed that for applications requiring a lipophilic bonded phase, in most cases, C18 and C8 may be used interchangeably. Isolation of more polar target analytes are assisted by the use of more polar solid supports and less polar analytes by less polar phases. Octadecyl sorbents having a range of carbon content (10-20 % of the total sorbent weight) have been used. These incude endcapped and non-endcapped materials. To the extent that particle size, pore size, carbon loading and endcapping status may affect an extraction, these factors should be considered for obtaining the best extraction efficiency and the cleanest sample for the analytical purpose intended.

B. THE EFFECT OF THE SAMPLE MATRIX

The biggest difference between MSPD and standard SPE lies in the influence of the matrix itself on the extraction. Unlike SPE, the entire sample is dispersed from the top to the bottom of the column in MSPD. The matrix components cover much of the sorbent, creating a new phase. Retention/elution properties no longer depend on interactions with the sorbent but with the dispersed sample and only secondarily with the solid support and the bonded phase. In other words we have created a new phase and the target analyte's distribution and interactions with this new phase are perhaps the most controlling factors in MSPD.

Indeed, this "matrix effect" has been seen by anyone who conducts chromatography. Repeated sample introduction leads to the build-up of sample components on the head of a GC or LC column and introduces a new "phase" in the column that affects the elution and retention character of target analytes that come in contact with it. The discontinuity of this phase with that of the analytical column can produce peak tailing and the formation of shoulders or multiple peaks for a single analyte. It can also lead overtime to loss of a particular analyte. In MSPD, however, this matrix effect, or deposition of sample components as an additional phase of the column, is incorporated throughout the column, providing an opportunity to establish a new level of, and consistency in, equilibria that cannot be obtained by a limited and discontinuous phase.

An example of this was perhaps observed by Reimer and Suarez in their report on the use of MSPD for isolating and screening five sulfonamides in salmon muscle tissue using thin layer chromatography (Reimer and Suarez, 1991). These researchers noted that the recovery from the MSPD process was analyst dependent: One analyst (A) used a larger, heavier pestle, and was physically larger and stronger than the second analyst (B); analyst A consistently obtained higher yields of the five sulfonamides than analyst B (70% versus 50%) and analyst A's extracts contained less fat/oil than analyst B. Regardless, each analyst obtained linear standard curves and good precision and accuracy. It was proposed that analyst A applied a larger force and obtained a more complete distribution of the sample over the solid support/C18 material used. The duration of grinding apparently had little or no effect. Recoveries from the individual analyst's columns were not improved by use of a more polar solvent. Further, it was noted that lower recoveries were obtained when dry versus methanol-conditioned C18 was used for sample preparation.

A similar phenomenon was observed by Schenck and Wagner (1995). This is consistent with observations in classical SPE where pretreatment of the solid support to "fluff-up" the bonded phase enhances recoveries and reproducibilty. In addition, conditioning of the sorbent also serves to overcome surface tension differences between the non-polar bonded phases and the more polar, water based samples. This, too, assists in the blending process and assures a more homogeneous dispersion of the sample components.

There is yet another aspect to the presence of the sample components throughout the column. It appears that the target analytes from an MSPD column elute in fractions that are not totally consistent with expected solubility behavior. This may be due to a further unique character of MSPD; the elution of a sample removes target analytes but also simultaneously fractionates the sample matrix components as well. It has been shown by mass balance experiments that the entire sample, minus a few percent of denatured macromolecules and connective tissues, can be eluted from an MSPD column (Hines, Long, Snider and Barker, 1991). Thus, an MSPD column elution of liver tissue blended with C18 removes 98% of the triglycerides in hexane, 98% of the phospholipids and steroids in dichloromethane, sugars and polyols in acetonitrile and phosphorylated sugars in water. For proteins, a sequence of elution solvents of decreasing strength can be drawn up: methanol > water > acetonitrile > ethyl acetate. Interestingly it does not follow a pattern of decreasing polarity, indicating that solubility of proteins in the elution solvent is more important than classic chromatographic retention/eluotropic strengths. Some 7% of the sample components remained on the column and consisted of connective tissues and denatured macromolecules, including DNA and related nucleotide polymers (Barker, Long and Hines, 1993).

Based on these results, the MSPD process has been used to perform lysis of rather rugged and difficult to lyse bacteria and to fractionate the sample so as to identify unique cellular components (Hines, Long, Snider and Barker, 1991; Hines, Jaynes, Barker et al., 1993; Hines and Frazier 1993).

However, in the present context, it also points to the fact that many of the unique elution properties of MSPD are due not only to the direct interactions of target analytes with the dispersed matrix, but also its association with the target analyte as the matrix is eluted. Thus, co-elution of the target analyte in association with a particular matrix component, which is also interacting with the materials remaining on the column, seems to be an important factor in the overall chromatographic character of the process.

C. ADJUSTMENT OF THE MATRIX

Since the matrix becomes the sorbent during an MSPD method, modification of the matrix results in modification of the resulting MSPD column. Several extraction studies have shown that addition of chelating agents, acids, bases, etc., to the sample before MSPD blending affects the dispersion and elution of the sample. We may be able to predict this behavior from our knowledge of how the analyte interacts with its surroundings. The intentional ionization or suppression of ionization of analytes and matrix components can greatly affect the nature of interactions of specific target analytes with the matrix and the eluting solvent(s) and should be considered, as in SPE, as a variable for attaining reproducible and efficient extractions.

D. FACTORS INFLUENCING ELUTION OF A DISPERSED SAMPLE

Similarly, the solvents we choose and the sequence we apply them in will determine the success of the MSPD experiment. As discussed above, elution profiles can be varied to obtain the optimal analytical result — balancing quantitative isolation of the target analyte with purity of the final eluent. Due to the nature of the process, one can actually isolate a range of different polarity target analytes or class of compounds in a single solvent or in different solvents passed through the same sample, making the process readily amenable to multi-residue isolation and analysis.

Elution of a MSPD column with a true gradient of solvent polarities has not been reported to date but should be applicable to the complete fraction of a sample. It has been observed in our laboratories that in an 8 mL elution most of the target analyte elutes in the first 4 mL. A similar result was obtained by Boyd, et al. (1991), wherein the drug clenbuterol was isolated from liver samples and detected by radioimmunoassay. Using the standard 8 mL elution volume, the MSPD column was eluted with hexane, followed by water; the clenbuterol was isolated in methanol and was found entirely within the first 3.5 to 4.0 mL of the methanol eluate. The elution profile will be different, of course, from sample type to sample type and from one solvent to another.

Since the tightness of the column packing could influence recovery by modifying elution rates or retention of a previously applied wash solvent in the interstitial packing, no reports so far prove that it plays a significant part in recovery. Best elution behavior is always obtained, however, if principles of chromatographic science are applied by eliminating voids and channeling.

IV. REQUIREMENTS FOR MSPD-DEDICATED SORBENTS

From the above review of applications and the study of the technique, we can propose new surface phase chemistries and devices that optimize the parameters influencing recovery and clean-up in a MSPD experiment. For example, phases that more directly accommodate cellular disruption and are more "detergent" like may prove useful for a variety of applications. Such materials could be related to the more classical materials used for such purposes, such as the Tritons and the organosulfate detergents. The latter are already available as the mixed-bed sorbents that contain C8 or other alkyl chains and sulfonic acid functional groups, for example.

Alternatives to the blending of sample and sorbent have been demonstrated: Schenck et al. (1990) and Schenck, Wagner and Bargo (1993), for example, have shown that one can slurry-mix solid-phase materials with milk, pour into a SPE column and obtain high efficiency recovery and analysis of drugs from this matrix. Such alternative procedures may similarly be used so as to increase the exposed sample size above what is feasible in a typical MSPD experiment while still avoiding the need to prepare the sample for a classic SPE extraction.

V. CONCLUSIONS

The MSPD process is designed to effect a simplification of the extraction processes commonly used for solid samples, and to assist in the evolution of such processes toward a more solventless approach. In so doing, it is predicted that a new level of efficiency may be obtained and that a significant reduction in solvent use, analyst exposure and solvent pollution and disposal will result.

The speed with which new applications are appearing in journals makes it difficult to keep up with the development of MSPD. A comprehensive list of references has been appended to this chapter to give the reader a chance to pursue a literature search. The journals listed in this section will allow you to determine on which journals to perform abstract or key word searches. Lastly, a humorous overview of MSPD can be found in "Seminars in Food Analysis" (Yago, 1996) which is an excellent introduction for scientists who are new to SPE.

REFERENCES

Barker, S.A., (1997) Analysis of Toxic Wastes in Tissues from Aquatic Species: Applications of Matrix Solid Phase Dispersion. J.Chromatogr. A. 774: 287-109

Barker, S. A., (1993) How to analyze those messy biological samples. Chemtech, 23: 42-45.

Barker, S.A. and Floyd, Z.E., (1996) Chemically Modified Surfaces: Recent Developments. Eds. J. Pesek, M. Matyzka and R. Abuelafiya. The Proceedings of the 6th International Symposium on Chemically Modified Surfaces, San Jose, June 19-21, 1995. Royal Society of Chemistry, Cambridge, UK

Barker, S. A. and Haley, R., (1992) Efficient Biological Analysis Using MSPD. International Laboratory. 46: 16-17.

Barker, S. A. and Long, A. R., (1992) Tissue Drug Residue Extraction and Monitoring by Matrix Solid Phase Dispersion (MSPD) - HPLC Analysis. J.Liq.Chromatogr. 15: 2071-2089.

Barker, S. A. and Long, A. R., (1994) Preparation of Milk Samples for Immunoassay and HPLC Screening Using Matrix Solid Phase Dispersion. J.Assoc.Off.Analyt. Chemists Int.77: 848-854.

Barker, S. A., Long, A. R. and Hines, M. E., (1993) The Disruption and Fractionation of Biological Materials by Matrix Solid Phase Dispersion. J.Chromatogr. 629: 23-34.

Barker, S.A., Long, A.R. and Short, C.R., (1989) Isolation of Drug Residues from Tissues by Solid Phase Dispersion. J.Chromatogr., 475: 363-361.

Barker, S.A. McDowell, T., Charkhian, B., Hsieh, L.C., (1990) Methodology for the Analysis of Benzimidazole Anthelmintics as Drug Residues in Animal Tissues. J.Assoc.Off.Analyt.Chemists. 73(1): 22-25.

Barker, S. A. and Walker, C. C., (1992) Chromatographic Methods for Tetracycline Analysis in Foods. J.Chromatogr. 624: 195-209.

Boyd, D., O'Keefe, M. and Smyth, M.R., (1995) Matrix Solid Phase Dispersion Linked to Solid Phase Extraction for Beta-Agonists in Liver Samples: an Update. Anal.Proc. 32: 301-303.

Boyd, D. O'Keefe, M. and Smyth, M.R., (1994) Matrix Solid Phase Dispersion as a Multiresidue Extraction Technique for Beta-Agonists in Bovine Liver Tissue. Analyst, 119: 1467-1470.

Boyd, D., Shearan, P., Hopkins, J.P., O'Keefe, M. and Smyth, M.R.., (1991) Application of Matrix Solid Phase Dispersion for the Determination of Clenbuterol in Liver Samples. Analytica Chimica Acta, 275: 221-226.

Boyd, D., Shearan, P., Hopkins, J.P., O'Keefe, M. and Smyth, M.R., (1993) Matrix Solid Phase Dispersion Linked to Immunoassay Techniques for the Determination of Clenbuterol in Bovine Liver Samples. Anal.Proc., 30: 156-157.

Fukushima, S., Taguchi, S., Nishumune, T. and Sueki, K., (1993) Determination of Five Sulfa Drugs, Furazolidone and Clopidol in Chicken Tissues by Matrix Solid Phase Dispersion Isolation Method. Osaka-furitsu Koshu Eisei Kenkyusho Kenkyu Hokoku, Shokuhin Eisei, 12: 59:52.

Hines, M.E., Long, A.R., Snider, T.G. and Barker, S.A., (1991) Lysis and Fractionation of *Mycobacterium Paratuberculosis* and *E. coli* by Matrix Solid Phase Dispersion (MSPD). Anal.Biochem. 195: 197-206.

Hines, M.E. and Frazier, K.S., (1993) Differentiation of Mycobacteria on the Basis of Chemotype Profiles by Using Matrix Solid Phase Dispersion and Thin-Layer Chromatography. J.Clin.Microbiol., 31: 610-614.

Hines, M. E., Jaynes, J. M., Barker, S. A., Newton, J. C., Enright, F. M. and Snider, T. G., (1993) Isolation and Partial Characterization of Glycolipid Fractions from *Mycobacterium Avium Serovar 2* That Inhibit Activated Macrophages. Infection and Immunity, 61: 1-7.

Iosifidou, E., Shearan, P. and O'Keefe, M., (1994) Application of Matrix Solid Phase Dispersion Technique for the Determination of Ivermectin in Fish Muscle Tissue. Analyst, 119: 2227-2229.

Jarboe, H.H. and Kleinow, K.M., (1992) Matrix Solid Phase Dispersion Isolation and Liquid Chromatographic Determination of Oxolinic Acid in Channel Catfish (*Ictalurus punctatus*) Muscle Tissue. J.Assoc.Off.Analyt.Chem.Int. 75: 428-432.

Ling, Y.-C., Chang, M.-Y. and Huang, I.-P., (1994) Matrix Solid Phase Dispersion Extraction and Gas Chromatographic Screening of Polychlorinated Biphenyls in Fish. J.Chromatogr., 669: 119-124.

Ling, Y.-C. and Huang, I.-P., (1995a) Multiresidue Matrix Solid Phase Dispersion Method for the Determination of Six Synthetic Pyrethroids inVegetables Followed by Gas Chromatography with Electron Capture Detection. J.Chromatogr. 695, 75-82.

Ling, Y.-C. and Huang, I.-P., (1995b) Multiresidue Matrix Solid Phase Dispersion method for Determining 16 Organochlorine Pesticides and Polychlorinated Biphenyls in Fish. Chromatographia, 40: 259-266.

Long, A.R., Hsieh, L.C., Bello, A.C., Malbrough, M.S., Short, C.R. and Barker, S.A., (1990) Method for the Isolation and Liquid Chromatographic Determination of Chloramphenicol in Milk. J.Agric. and Food Chem. 38(2): 427-429.

Long, A.R., Hsieh, L.C., Malbrough, M.S., Short, C.R. and Barker, S.A., (1989a) Method for the Isolation and Liquid Chromatographic Determination of Chlorsulfuron in Milk. Laboratory Information Bulletin, FDA 5(8): 3364.

Long, A.R., Hsieh, L.C., Malbrough, M.S., Short, C.R. and Barker, S.A., (1989b) A Multi-residue Method for the Isolation and Liquid Chromatographic Determination of Seven Sulfonamides in Infant Formula. J.Liquid Chromatogr., 12(9): 1601-1612.

Long, A.R., Hsieh, L.C., Malbrough, M.S., Short, C.R. and Barker, S.A., (1989c) Method for the Isolation and Gas Chromatographic Determination of Chlorsulfuron in Milk. J.Assoc.Off.Analyt.Chem. 72(5): 813-815.

Long, A.R., Hsieh, L.C., Malbrough, M.S., Short, C.R. and Barker, S.A., (1989d) A Multiresidue Method for the Isolation and Liquid Chromatographic Determination of Seven Benzimidazole Anthelmintics in Milk. J.Assoc.Off.Analyt.Chem., 72(5): 739-741.

Long, A.R., Hsieh, L.C., Malbrough, M.S., Short, C.R. and Barker, S.A., (1990a) Multiresidue Method for the Determination of Sulfonamides in Pork Tissue. J.Agric.Food Chemistry. 38(2): 423-426.

Long, A.R., Hsieh, L.C., Malbrough, M.S., Short, C.R. and Barker, S.A., (1990b) Method for the Isolation and Liquid Chromatographic Determination of Furazolidone in Milk. J.Agric.Food Chemistry. 38(2): 430-432.

Long, A.R., Hsieh, L.C., Malbrough, M.S., Short, C.R. and Barker, S.A., (1990c) Matrix Solid Phase Dispersion (MSPD) Isolation and Liquid Chromatographic Determination of Oxytetracycline, Tetracycline and Chlortetracycline in Milk. J.Assoc.Off.Analyt.Chem., 73(3): 379-384.

Long, A.R., Hsieh, L.C., Malbrough, M.S., Short, C.R. and Barker, S.A., (1990d) Matrix Solid Phase Dispersion (MSPD) Isolation and Liquid Chromatographic Determination of Furazolidone in Pork Muscle Tissue. J.Assoc.Off.Analyt. Chem., 73(2): 93-98.

Long, A.R., Hsieh, L.C., Malbrough, M.S., Short, C.R. and Barker, S.A., (1990e) Matrix Solid Phase Dispersion (MSPD) Isolation and Liquid Chromatographic Determination of Five Benzimidazole Anthelmintics in Pork Muscle Tissue. J.Food Comp.Analysis, 3:20-26.

Long, A.R., Hsieh, L.C., Malbrough, M.S., Short, C.R. and Barker, S.A., (1990f) Matrix Solid Phase Dispersion (MSPD) Isolation and Liquid Chromatographic Determination of Five Benzimidazole Anthelmintics in Beef Liver Tissue. J.Assoc.Off.Analyt. Chem. 73(6): 860-863.

Long, A.R., Hsieh, L.C., Malbrough, M.S., Short, C.R. and Barker, S.A., (1990g) Matrix Solid Phase Dispersion (MSPD) Isolation and Liquid Chromatographic Determination of Oxytetracycline in Catfish (*Ictalurus Ppunctatus*) Muscle Tissue. J.Assoc.Off.Analyt.Chem., 73(6): 864-867.

Long, A.R., Hsieh, L.C., Malbrough, M.S., Short, C.R. and Barker, S.A., (1990h) Matrix Solid Phase Dispersion (MSPD) Isolation and Liquid Chromatographic Determination of Sulfadimethoxine in Catfish (*Ictalurus Punctatus*) Muscle Tissue. J.Assoc.Off.Analyt.Chem., 73(6): 868-871.

Long, A.R. Hsieh, L.C., Short, C.R. and Barker, S.A., (1989) Isocratic Separation of Seven Benzimidazole Anthelmintics by High Pressure Liquid Chromatography with Photodiode Array Characterization. J.Chromatogr. 475: 404-411, 1989.

Long, A.R., Hsieh, L., Short, C.R. and Barker, S.A., (1991a) Multiresidue Matrix Solid Phase Dispersion (MSPD) Extraction and Gas Chromatographic Screening of Nine Chlorinated Pesticides in Catfish (*Ictalurus Punctatus*) Muscle Tissue. J.Assoc.Off.Analyt.Chem., 74: 667-670.

Long, A.R., Hsieh, L., Short, C.R. and Barker, S.A., (1991b) Matrix Solid Phase Dispersion (MSPD) Extraction and Gas Chromatographic Screening of Nine Chlorinated Pesticides in Beef Fat. J.Assoc.Off.Analyt.Chem., 74(3): 493-496.

Long, A.R., Short, C.R. and Barker, S.A., (1990) Multiresidue Method for the Isolation and Liquid Chromatographic Determination of Eight Sulfonamides in Milk. J.Chromatogr., 502(1): 87-94.

Lott, H. M. and Barker, S. A., (1993a) Matrix Solid Phase Dispersion (MSPD) Extraction and Gas Chromatographic Screening of 14 Chlorinated Pesticides in Oysters (*Crassostrea Virginica*). J.Assoc.Off.Analyt.Chem.Int., 76: 67-72.

Lott, H. M. and Barker, S. A., (1993b) Extraction and Gas Chromatographic Screening of 14 Chlorinated Pesticides in Crayfish (*Procambarus Clarkii*) Hepatopancreas. J.Assoc.Off.Analyt.Chem.Int., 76: 663-668.

Lott, H. M. and Barker, S. A., (1993c) Comparison of a Matrix Solid Phase Dispersion and a Classical Method for the Determination of Chlorinated Pesticides in Fish Muscle. Environ.Monit.Assessment, 28: 109-116.

Luke, M.A., (1995) The Evolution of a Multiresidue Pesticide Method, ACS Conference Proceedings Series, 8th International Congress of Pesticide Chemistry: Options 2000. Eds. Ragsdale, Kearney and Plummer, American Chemical Society.

McLaughlin, L.G., Henion, J.D. and Kijak, P.J., (1994) Multiresidue Confirmation of Aminoglycoside Antibiotics in Bovine Kidney by Ion Spray High-Performance Liquid Chromatography/Tandem Mass Spectrometry. Biol.Mass Spectrom., 23: 417-429.

Nau, D., Nguyen, H. and Pocci, R., (1996) Analysis of Pesticide and Herbicide Residues in Food. European Food and Drink Review. The Quarterly Review of Food and Drink Technology, Winter 1996: 63-69.

Reimer, G.J. and Suarez, A., (1992) Liquid Chromatographic Confirmatory Method for Five Sulfonamides in Salmon Muscle Tissue by Matrix Solid Phase Dispersion. J.Assoc.Off.Analyt.Chem. Int., 75: 979-981.

Renson, C., Degand, G., Maghuin-Rogister, G. and Delahaut, P., (1993) Determination of Sulfamethazine in Animal Tissues by Enzyme Immunoassay. Anal.Chim.Acta, 275: 323-328.

Reimer, G.J. and Suarez, A., (1991) Development of a Screening Method for Five Sulfonamides in Salmon Muscle Tissue Using Thin Layer Chromatography. J.Chromatogr., 555: 315-320.

Rosen, J., Hellenaes, K.-E., Toernquist, P. and Shearan, P., (1994) Automated Extraction of Acetylgestagens from Kidney Fat by Matrix Solid Phase Dispersion. Analyst, 119: 2635-2637.

Schenck, F.J., (1995) Isolation and Quantification of Ivermectin in Bovine Milk by Matrix Solid Phase Dispersion (MSPD) Extraction and Liquid Chromatographic Determination. J.Liq.Chromatogr., 18: 349-362.

Schenck, F.J., Barker, S.A. and Long, A.R., (1989) Rapid Determination of Chlorsulon in Milk Extracted Using Matrix Solid Phase Dispersion. Laboratory Information Bulletin, Dept. of Health and Human Services, Public Health Service, Food and Drug Administration. 5(10): 3381 (1-9).

Schenck, F.J., Barker, S.A. and Long, A.R., (1990) Rapid Isolation of Ivermectin from Liver by Matrix Solid Phase Dispersion. Laboratory Information Bulletin, FDA, 6(1): 3425 (1-7).

Schenck, F.J., Barker, S.A., Long, A.R. and Matusik, J., (1989) Extraction and Determination of Nicarbazin drug Residues from Tissue by Solid Phase Dispersion. Laboratory Information Bulletin, FDA. 5(8): 3354.

Schenck, F.J. and Wagner, R., (1995) Screening Procedure for Organochlorine and Organophosphorus Pesticide Residues in Milk Using Matrix Solid Phase Dispersion Extraction and Gas Chromatographic Determination. Food Add.Contam., 12: 535-541.

Schenck, F.J., Wagner, R. and Bargo, W., (1993) Determination of Chlorsulon Residues in Milk Using a Solid Phase Extraction Cleanup and Liquid Chromatographic Determination. J.Liq.Chromatogr., 16: 513-520.

Shearan, P. and O'Keefe, M., (1994) Novel Approach to the 'On-Site' Testing for Sulfamethazine in Pork Carcasses. Analyst, 119: 2761-2764.

Shearan, P., O'Keefe, M. and Smyth, M.R., (1994) Comparison of Matrix Solid Phase Dispersion with a Standard Solvent Extraction Method for Sulphamethazine in Pork Muscle Using High Performance Liquid and Thin Layer Chromatography. Food Add.Contam., 11: 7-15.

Soliman, M.M., Long, A.R. and Barker, S.A., (1991) Method for the Isolation and Liquid Chromatographic Determination of Furazolidone in Chicken Muscle Tissue. J.Liq.Chromatogr. 13(16): 3327-3337.

Stafford, S.C. and Lin, W., (1992) Determination of Oxamyl and Methomyl by High-Performance Liquid Chromatography Using a Single Stage Postcolumn Derivatization Reaction and Fluoresence Detection. J.Agric.Food Chem., 40: 1026-1029.

Tamura, H., Yotoriyama, M., Kurosaki, K. and Shinohara, N., (1994) High Performance Liquid Chromatographic Analysis of Sulfonamides in Livestock Products Using Matrix Solid Phase Dispersion (MSPD) Method with Silica Gel. Shokuhin Eiseigaku Zasshi, 35: 271-275.

Van Poucke, L.S.G., Depourcq, G.C.I. and Van Peteghem, C.H., (1991) A Quantitative Method for the Detection of Sulfonamide Residues in Meat and Milk Samples with a High Performance Thin Layer Chromatographic Method. J.Chromatogr.Sci., 29: 423-427.

Walker, C. C. and Barker, S. A., (1994a) Extraction and Enzyme Immunoassay of Sulfadimethoxine Residues in Channel Catfish (*Ictalurus Punctatus*) Muscle. J.Assoc.Off.Analyt.Chem., Int. 77: 908-916.

Walker, C. C. and Barker, S. A., (1994b) Matrix Solid Phase Dispersion Extraction and HPLC Analysis of Sulfadimethoxine and 4-N-acetyl Sulfadimethoxine Residues in Channel Catfish (*Ictalurus Punctatus*) Muscle and Plasma. J.Assoc.Off.Analyt.Chem.Int., 77: 1460-1466.

Walker, C. C., Lott, H. M. and Barker, S. A., (1993) Matrix Solid Phase Dispersion Extraction and the Analysis of Drugs and Environmental Pollutants in Aquatic Species. J.Chromatogr. 642: 225-242.

Walker, C.C., Thune, R.L. and Barker, S.A., (1995) Plasma/Muscle Ratios of Sulfadimethoxine Residues in Channel Catfish (*Ictalurus Punctatus*). J.Vet.Pharmacol.Therap. 18: 1-5.

Yago, L.S., (1996) Matrix Solid Phase Dispersion: The Next Step in Solid Phase Extraction for Food Samples? An Overview of the Technology. Seminars in Food Analysis 1: 45-54. (Chapman Hall)

Yang, R. and Fu, C., (1994a) Matrix Solid Phase Dispersion and High Performance Liquid Chromatographic Determination of Trace Carbofuran in Corn. Fenxi Ceshi Xuebao, 13: 72-75.

Yang, R. and Fu, C., (1994b) Matrix Solid Phase Dispersion and High Performance Liquid Chromatographic Determination of Trace Pirimicarb and Amitraz in Fruits and Vegetables. Hebei Daxue Xuebao, Ziran Kexueban, 14: 29-33.

14

AUTOMATION OF SOLID-PHASE EXTRACTION

Lynn Jordan, Zymark Corp., Hopkinton, Massachusetts

I. INTRODUCTION

The purpose of this chapter is to explore automated solid-phase extraction (SPE), and most importantly, put it into a realistic perspective. To accomplish this, the chapter is broken down into three sections:

1) Understanding Automated SPE
2) Workstations for Automated SPE
3) Automation of SPE Applications

The first section examines what automated SPE can achieve and will help set expectations before automation is introduced to the laboratory. The second section discusses the alternative techniques available to the chemist who wishes to automate SPE procedures and provides a framework for determining specific needs and evaluating automated systems. The third section addresses the commonly encountered pitfalls of automated SPE. Critical elements in the automation of an SPE method are discussed from the positions of automating an existing manual SPE method, optimizing a method, or using a workstation to perform automated SPE method development.

Throughout this chapter the phrase "workstation" will be used to describe the devices that are used to automate SPE applications. The term workstation was chosen because these devices are used to accomplish a specific task, or "work" in a specific location, or "station."

II. UNDERSTANDING AUTOMATED SPE

Automation is seldom considered by analysts who have only a few samples to analyze. Rather, it is the preserve for laboratories with large numbers of samples that need to be prepared for analysis quickly without increasing laboratory personnel. To ensure successful automation of SPE methods, it is essential to understand what automated SPE can realistically achieve, and even more importantly, what it can not.

A. WHAT AUTOMATION CAN ACHIEVE

Automated SPE sample preparation eliminates many of those variables associated with manual SPE. The result can be improved precision, accuracy and recovery, and this is achieved because automated equipment performs an identical sequence of steps on each sample. Therefore, with fewer manual steps in automated SPE sample preparation when compared with fully manual methods, there is less chance for error. With automation of SPE, most laboratories experience a reduction in the number of samples that must be rerun because of the enhanced consistency automated SPE can provide. In the world of ever-increasing regulatory pressures, automated SPE can provide formal documentation of how sample preparation is done, recording in electronic form, precise details of every step of every extraction.

Sample preparation is the rate-limiting step for many analyses. For example, in laboratories where liquid chromatography/mass spectrometry (LC/MS) and multiple mass spectrometry analysis (e.g., LC/MS/MS) have been introduced, more data is being produced than by traditional LC or LC/MS analysis, in a fraction of the time. The challenge becomes to keep these costly instruments operating at full capacity. Current LC/MS/MS operating practice employs sample run times of 2 to 3 minutes. The result is a backlog in sample preparation. Automation of SPE has been a vital part of these laboratories' strategies to keep these costly instruments operating at full capacity.

Operator safety is crucial; in laboratories working with biological fluids, the threat of AIDS and Hepatitis is very real. Environmental laboratories routinely handle samples that could contain unknown and lethal pollutants, and pharmaceutical companies handle potential drug candidates that have unknown risks. Reducing operator exposure to samples and reagents is of great importance in all of these situations.

The financial savings from automating SPE sample preparation can come from many areas. In removing the dull and labor intensive manual portion of SPE, operators are able to perform more challenging and value-added tasks, such as data analysis or report generation. Again, because of the consistency with which automated SPE is performed, better analytical data is achieved, and fewer sample reruns are necessary. The sample turnaround, or the time from which a sample enters the laboratory to the time in which an analytical result is available, may be greatly reduced, thus increasing laboratory throughput.

B. WHAT AUTOMATION CANNOT ACHIEVE

Even more important than the understanding of what automated SPE will achieve, is a thorough understanding of what it will not achieve. Many automation projects progress slowly or totally fail because of unrealistic goals. One of the biggest mistakes is not allowing for time to implement the automation; a workstation will not, three days after delivery, be running multiple validated methods. Likewise the operators will require time to become fully knowledgeable about the workstation. Management should plan to dedicate staff to the project, and allow them the time necessary to fully implement it.

A common concern in the laboratory is that automation will result in the reduction of laboratory staff. This can be a critical error — an automated SPE workstation is a tool to allow staff to perform their jobs more effectively. The automated workstation will require set-up prior to running samples, and manual steps must often still be performed on the samples before or after the SPE step. After the SPE step, the analytical results must be obtained, and reports generated. Routine cleaning and maintenance are critical to assuring long-term reliability for automated SPE workstations, as with any laboratory equipment.

A final important aspect of what automated SPE will not do involves the samples and SPE methods. Like manual SPE, automated SPE workstations require a clean sample, not one full of precipitates or clots. Samples will often require centrifugation, and a clean aliquot from the centrifuged sample used for the automated SPE. In addition, it is very important to recognize that if the chemistry of a manual SPE method is not optimized, automating that method will not correct it. In section IV, approaches to automating a working manual SPE method, ruggedizing the method, and performing automated SPE method development will be given.

C. CRITERIA FOR AUTOMATING SPE

As with any project, the key to success lies in choosing the correct tools. When automating SPE sample preparation, selecting the correct tools means evaluating your needs and identifying the workstation that best suits your project(s). No single workstation is correct for all uses. To correctly assess your automation needs, consider the following questions:

How many samples do you need to prepare with the automated workstation? Projects with a limited number of samples should be approached differently than projects with many thousands of samples. For laboratories with small sample loads, throughput is less of an issue. For high-volume labs it can be critical.

What is the expected project duration? How long do you expect this project to last? For example, is this work on a new chemical entity (NCE) from a pharmaceutical company that may have a short life before the next NCE appears, or is this drugs of abuse (DOA) contract work that could go on indefinitely?

What might future projects require? Whether the current project is expected to be a short- or long-term project, future use of the automated SPE workstation should always be considered. Might future projects have similar requirements to the current project, or will they be completely different, and if so, how? If your current project is long-term, would you purchase another automated SPE workstation for future projects, or use the same workstation?

How quickly do you need to provide sample results? Are sample results required immediately, or is there a reasonable period of time before the results are needed? If the workstation must be run 24 hours a day, are there three shifts of operators, or only one shift that works during the day, and who will set it up to run overnight?

How many people will use the automated SPE workstation? The larger the number of people using the automated SPE workstation, the simpler the operation should be.

How many methods will need to be run on the automated SPE workstation? The larger the number of methods to be run on the workstation, the easier it should be to switch between those methods.

What steps are required before SPE? Think about all of the steps that are performed before SPE, such as pH adjustment, centrifugation, sample aliquoting, sample dilution, hydrolysis, or the addition of internal standard. Is it necessary or desirable to automate these steps? In addition, what is the stability of the sample prior to SPE clean-up? Is it stable at room temperature and for how long? Is refrigeration necessary?

What happens to the sample after SPE? Is evaporation and reconstitution, or derivatization required? Will the final eluted sample remain stable under its current conditions? How does evaporation effect the extracted samples? Can the sample be eluted in the mobile phase for analysis? If so, the sample could potentially be either eluted on-line, or be eluted into a vial and then injected? This aspect of performance is one which can have enormous impact on efficiency. For example, on-line, high-pressure elution or low-pressure elution directly into the sample loop of an HPLC switching valve can eliminate operator involvement in handling extracts, reduce concerns about analyte stability and cut an autosampler requirement from the list of required instrumentation.

Will the workstation be used for methods development? If the workstation will be used to do automated SPE methods development, look for the ability to develop and edit a base method, trun multiple simultaneously, screen multiple sorbents, to use a variety of reagents and be able to collect multiple fractions.

How much money is budgeted? In general, the more complex the workstation the higher the cost.

III. WORKSTATIONS FOR AUTOMATED SPE

Once the needs for automation are assessed, it is time to review the available automated SPE workstation. There are numerous workstations that perform automated SPE. The key to choosing the correct workstation is applying the knowledge from the automation needs assessment to the workstation search. The approach taken here to reviewing automated SPE workstations is to look at the key areas of overall design, workstation characteristics and workstation operational control.

In the workstation design section we look at the overall picture of how the workstation functions, detailing the mode of processing samples, and all aspects of moving samples and reagents through the SPE cartridge. The section on workstation characteristics reviews operational aspects such as setting up the workstation, flexibility of the workstation towards cartridge sizes and configurations, sample containers, etc. "Workstation Control" reviews options for controlling the workstation, whether from software, a keypad, or a controller.

A listing of vendors supplying automated SPE workstations at the time of publication is supplied in Appendix 4.

A. WORKSTATION DESIGN

We focus in this section on philosophy rather than the specifics of currently available workstations. This is because while hardware designs can change in a relatively short period of time, the strategies adopted for resolving the problems of design and control of automated SPE stay current for longer periods of time. A good review of the specific approaches taken by manufacturers is given by Majors (1993). We will address the following areas:

- Mode of processing samples
- Moving samples and reagents through the SPE cartridge
- Fluid path
- Waste handling
- Laboratory requirements for the automated workstation

1. Mode of Processing Samples

There are two general modes in which automated SPE workstations process samples, **batch** and **serial**.

Batch refers to processing the entire "batch" of samples one step at a time before moving on to the next step of the method. For example, all SPE cartridges are conditioned with first conditioning reagent, all SPE cartridges are then conditioned with the second conditioning reagent. This step is followed by load, rinse and elution steps and any other steps in the method.

In the **serial** mode of processing samples, the entire SPE method is performed on an individual sample before the next sample is started. In this mode the first SPE cartridge is conditioned with the first conditioning reagent, followed by the second conditioning reagent. Load, rinse and elution steps follow. Only after the first sample is fully extracted is the second extraction started.

Many automated SPE workstations can process samples in either the batch or serial mode, and some even can process samples in both the batch and serial mode simultaneously. One way this may be achieved is through modular design, where each module operates in serial mode but several modules, receiving instructions from a single controller, operate in parallel with each other.

To determine what is best for the needs of the project, consider the following:

- The time in which the sample must sit on the workstation both before and after the SPE method
- How stable are the samples at these times?
- How long does it take to process all samples you would want to set up on the workstation?
- Is the mode of processing samples (batch or serial) acceptable for your samples?

2. Moving Samples Through the SPE Cartridge

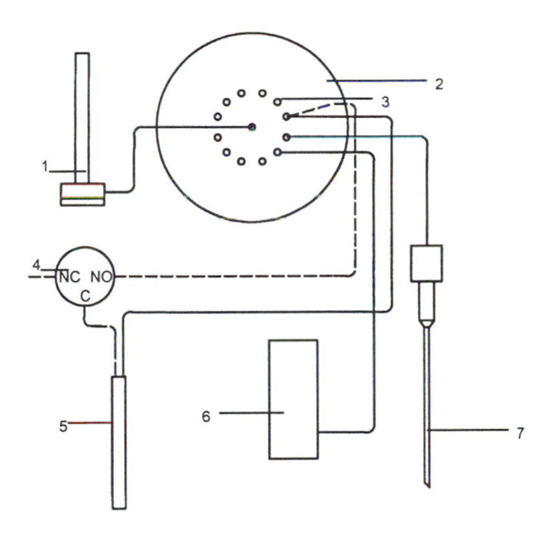

Figure 1. A schematic of a typical workstation fluidic system, with positive pressure operation provided by a syringe driven by a stepper motor.
1. Stepper Motor Driven Syringe
2. 12-Port Valve
3. Vent on 12-Port Valve
4. Gas Valve
5. 1 or 3 mL Plunger for syringe barrel SPE Cartridges
6. Mixing Vessel for methods development work
7. Cannula to load sample

Two important points — how liquids are moved through the cartridge, and how fast they are moved — are considered.

When samples are run manually using a vacuum manifold, a vacuum draws liquids through the SPE cartridge bed by negative pressure. This is the most common technique for manual sample processing because of its simplicity and convenience, and because of the availability of many different manifold types and sizes. When samples or liquids are pushed through the SPE cartridge bed using positive pressure, performance of the extraction may vary from what was observed for vacuum extractions. This variability may be attributable to sample flow characteristics under positive pressure and differences in drying effectiveness of air or a clean pressurized gas.

SPE workstations have used both vacuum and positive pressure to drive fluid flow. Positive pressure operation may be provided by a piston or syringe activated by a stepper motor, instead of by an external gas source. An illustration of such a system is shown in Figure 1.

Once you have learned how liquids are propelled through the workstation's fluid lines, investigate the rates at which the liquids move through the SPE cartridge bed. Some SPE applications may be very sensitive to flow rates. Symptoms may include lower recoveries at higher flow rates or a decrease in reproducibility. As a general rule, ion exchange extractions are more sensitive to flow rates than polar or non-polar extractions.

The steps of an extraction that are most sensitive to flow rate variations are usually the load or elute steps. If a vacuum source is used to generate the negative pressure to pull samples and reagents through the SPE cartridge bed, a specific vacuum pressure will be set, and not a flow rate. The resulting flow rate will be determined by the vacuum pressure set, the viscosity of the sample or reagent, and the packing material and the format of the SPE cartridge bed. With negative pressure operation, the pressure remains constant, while the flow rate through the SPE cartridge bed may vary. If a gas source is used to generate the positive pressure, again the pressure from the source would remain constant, allowing the flow rates through the SPE cartridge bed to vary. Some workstations that use air to generate the pressure to move liquids through the SPE cartridge bed now use a monitor to determine the flow rate of liquid through the SPE cartridge, and can vary the pressure as necessary to maintain a constant flow rate. Despite this enhancement, an automated SPE workstation that uses a pump or syringe to provide positive pressure displacement of a liquid volume (i.e. low compressibility) through the SPE cartridge bed will provide the most stable flow. Therefore you should ask the question: Are you setting pressures or a flow rate? If the answer is "I am setting a flow rate" (in other words, you are explicitly setting flow rate) this allows us to ask and find answers to the following questions:

- How sensitive is the method to flow rate variations?
- Can the system supply different flow rates for each step of an extraction?
- And if so, can it supply different flow rates for, say, the first wash step, and a different one for the second wash step?
- Or can it only apply a single rate for all steps (conditioning, loading, washing and elution)?

In many SPE methods, the recovery of the sample and reproducibility between samples may be directly related to the flow rates used in the method. If variations in flow rate during a manual method appear to have a large effect on the extraction performance, consider a workstation that uses positive liquid displacement and delivers consistent flow rates.

3. Fluid Path

One of the most significant differences between manual and automated SPE concerns the fluid path. In a vacuum box or positive pressure manifold the fluid path consists of only the SPE cartridge and possibly delivery needles. Reagents are added directly to the SPE cartridge from a dispenser, and leave the cartridge through the luer tip on the end of the cartridge. With an automated SPE workstation, there often is a common fluid path that may be used for reagents and samples. The fluid path may include valves, reagent lines, a way to introduce the sample on to the SPE cartridge bed, and a way to remove the waste generated. Because of this, there are a few more things to consider when an SPE method is automated.

First, care must be taken with reagents used. With a common fluid path, compatibility with the automated SPE workstations fluid path must be considered, as well as the compatibility with other reagents utilized in the method. To determine the compatibility of reagents with the fluid path, ask what wetted materials are in the fluid path. Next, check to see if any of the reagents used in the method are incompatible with the fluid path materials. If any incompatibilities exist, examine whether another reagent could be substituted that would accomplish the same effect in the SPE extraction which is also compatible with the wetted materials in the fluid path. To illustrate, hydrochloric acid is a common SPE reagent, used to change the pH of a sample or to modify secondary interactions on the sorbent. If the workstation has a stainless steel valve in the fluid path, halides in the hydrochloric acid will attack the stainless steel. To prevent this and reduce the stress of use on the workstation, another acid (phosphoric, nitric or sulfuric acid) may often be substituted without adversely affecting the SPE chemistry.

In addition to ensuring the compatibility of the reagents with the fluid path, the order in which the reagents are used must also be considered. In a vacuum box, it may be acceptable to follow a methanol conditioning step with a buffered reagent, because there is essentially no common fluid path. However, when a fluid path is involved, using a buffered reagent directly after an organic reagent may cause buffer salts to precipitate. In this case, we should rinse the methanol from the fluid path with water, between the methanol and buffer conditioning steps.

Second, carryover must be considered with a common fluid path. Carryover may occur when an automated SPE workstation possesses a common fluid path for both samples and reagents; some of the first sample remains in the fluid path and can "carryover" into the second sample. The method to determine carryover is to first run a blank sample, one that contains no analyte of interest. The next sample should contain a very high concentration of the analyte of interest. Follow this with two additional blank samples, and measure the amount of analyte of interest found in the final two blank samples to detect the carryover. Remember that the carryover measured is the carryover from the entire process of preparing the sample, not just from the automated SPE part — the measured carryover could come from any manual sample preparation steps, the automated SPE method, any final sample evaporation or derivatization, and the sample analysis. Determine what each portion of the sample preparation contributes to the total method carryover.

If carryover occurs, we must examine the exact fluid path the sample takes when an automated SPE workstation performs an SPE method. It is imperative to clean the automated SPE instrument's fluid path between samples in order to minimize carryover. In determining the most effective way to do this consider the following:

- What is the sample matrix?
- What solubilizes the analyte of interest?
- What steps most effectively clean the fluid path?

First, the reagents chosen to clean the fluid path must be compatible with the sample matrix. This is especially important with biological samples such as plasma and serum because these sample contain proteins; if an incompatible reagent comes into contact with any residual plasma sample, the proteins may clot, potentially causing blockage of the fluid path. If working with protein-rich samples, the first cleaning step of the fluid path must be performed with an aqueous reagent to remove the proteins. A serum sample typically has half the protein concentration of a plasma sample, but similar care should be taken. If you are working with a urine sample the choice of reagent is less critical because most urine samples contain only very low levels of proteins.

What solubilizes the analyte of interest? The best way to prevent or eliminate carryover is to remove any of the remaining analyte of interest from the fluid path. The most effective way to accomplish this is to use a reagent that solubilizes the analyte well, in order to wash it away. In many methods, the elution solvent may accomplish this task well, but there are some important issues. Miscibility of the elution reagent with the sample matrix, or the wash solvent applied directly before elution, is critical. If the elution reagent is methylene chloride (dichloromethane) and the sample matrix is plasma, the two are not miscible and this would not be a good choice for a reagent to prevent carryover. Remember with a plasma sample, the first wash must be aqueous to remove the plasma proteins; in the case above, an aqueous cleaning step should be followed by a cleaning step using a reagent that will solubilize the analyte of interest. If this reagent is organic and incompatible with the sample matrix, a post-elution cleaning step should be performed to remove the elution solvent along with any traces of the analyte and to restore fluid path compatibility for the next sample.

What actions most effectively clean the fluid path? Review the documentation provided with the automated SPE workstation, or contact a technical representative from the manufacturer to determine the best method for cleaning the fluid path. Some "tricks" have been learned, such as the use of bleach (though remember the need for compatibility with the fluid path), or the use of a non-ionic surfactant to remove proteins adhering to the fluid path walls. Keep in mind that cleaning protocols may differ depending on the sample matrix. When keeping these guidelines in mind and working to reduce carryover, the task should not be difficult.

We must also seek to reduce the occurrence of blocked fluid paths. Biofluids or environmental samples commonly contain clots or precipitates. It is good practice to provide the workstation a sample free of clots or precipitates, to ensure the reliability of the automated SPE workstations. This is most commonly done by centrifugation. Beware, however, that long delays between sample preparation and extraction increase the likelihood of clots or precipitates forming even after the sample has been centrifuged. Other preparation techniques include filtration (but be careful not to filter out particles to which the sample is bound) and sonication/dilution to kill any active clotting agent in the sample and reduce the number of particles per unit volume in the sample.

To determine what is best for the needs of the project, learn the fluid path for samples and reagents in the automated SPE workstation. If a clog develops in the fluid path, how does the workstation react? Does the workstation shut down? How does the operator learn of the clog? Is an error message generated? What happened to the other samples in the workstation that have not been prepared yet? How is the clot removed from the fluid path? What can be done to prevent this condition from occurring?

4. Waste Handling

In a vacuum manifold, the waste from conditioning the cartridges, loading sample onto the cartridge, and rinsing the cartridge goes directly into the bottom of the vacuum manifold reservoir. With an automated SPE workstation, the waste may be handled in many different ways. The three typical ways are: 1) common open area; 2) common waste system; 3) closed waste system. In a common open area, all SPE cartridge waste goes into an open flat pan, from which it is drained into a waste reservoir for disposal. The advantage of this system is its simplicity, and if any clots or precipitates form in the waste there are little, if any, adverse effects. The disadvantages are that the waste is not contained, and methods requiring harmful solvents may need to be run in a hood, where space can be limited and in high demand for other uses. In addition, this waste system may also result in more expensive waste disposal costs, because all waste (biohazard, aqueous, organic/non-chlorinated, and organic chlorinated, for example) is mixed together.

In the common waste system, all samples and reagents designated as waste go to one common location and are then moved into a waste reservoir for disposal. The advantage is that this is a more contained system and may be less subject to issues with fumes. The disadvantages of this system are that all wastes are in a common location, proteins from plasma samples may mix with other reagents and cause the formation of clots potentially forming a block in the fluid path of the waste disposal system. The common waste disposal system can also result in more expensive waste disposal costs due to mixing all forms of waste. This can be alleviated by segregation of the fluid streams, where waste may be directed to multiple different reservoirs. This is especially useful if you want to segregate out different types of wastes, for example, to segregate chlorinated wastes from the other wastes used in your SPE method. Disposal may be significantly easier and less expensive, and clotting or precipitation should be eliminated with the segregation of waste.

The third type of waste system is a closed waste system. In this system, flow through the SPE cartridge moves the waste into a holding station and then to a reservoir for disposal. A system of this type may also have the ability to segregate the waste into different streams, reducing the risk of clot formation and lowering the waste disposal costs. The major benefit of this system is that it is least likely to require operation in a fume hood. The disadvantage is that there is an additional waste disposal component in the system.

5. Laboratory Requirements for Automated Workstations

Size is always important. Not only should the length and width of the workstation be considered, but the height as well. If the workstation needs to be in the hood, make careful measurement to assure fit, and ensure the accessibility to all areas of the workstation. In determining the location for the workstation in the laboratory, consider if the workstation needs to be directly connected to a personal computer for opperation. What is the maximum distance the PC could be located from the workstation? Where is it convenient to locate the PC? Will it be more difficult if the PC is not next to the workstation?

Consider, too, what plumbing is required for reagents, and consider convenience of access to all necessary waste reservoirs and gas sources, and where the reagents, waste and gas source need to be located in relation to the workstation. Make sure the reagent bottles will be easily accessible and visible to facilitate filling and to determine reagent levels.

B. WORKSTATION CHARACTERISTICS

This section examines six aspects of automated SPE that define the suitability of a system for a given task or application. The following topics are covered:

- Set-up of the automated SPE workstation
- Number of reagents
- SPE cartridge requirements
- Sample capacity
- Sample volume
- Other functions the automated SPE workstation can perform

1. Set-up of the Automated SPE Workstation

What is required to set up the automated SPE workstation before it runs samples on a daily basis? You must consider how to get fresh reagents to the workstation daily and how to ensure that there is adequate volume in the waste reservoir for what the method will generate. Look at how the samples are placed on the workstation and in what type of a test tube or container they are supplied. If you have limited control over the type of container, make sure the workstation is flexible enough to handle the range of containers you maybe given. Elution may be into a test tube, autosampler vial or directly on-line to an analytical workstation for analysis. Again, ask the question: What and how many kinds of collection vials do I want to use? Where are the SPE cartridges placed, and do they require any special caps or seals before they are used? Do any plumbing changes need to be made to the workstation when changing between different SPE methods? The requirement for controlling the workstation will be covered in the section titled Workstation Control.

2. Number of Reagents

How many different reagents can be used on the workstation? How easily is it to changes reagents? Are there any special requirements for the reagent reservoirs; for example, type of bottle, volume, cap, or placement of the reservoir relative to the workstation? Consider the number of methods to be automated. What is the total number of different reagents in the methods? How many reagents are common to all or multiple methods? Can another reagent be made by blending two or more of the available reagents?

3. SPE Cartridge Requirements

Most automated SPE workstations work with the syringe barrel cartridge format, for a limited range of barrel sizes. Some workstations require their own special SPE housings. This is commonly the case if the workstation elutes on-line to an analytical instrument, because of the higher pressures that the cartridge may need to withstand or the low total elution volume the instrument can handle. If a special cartridge housing is required, arrangements can be made to custom pack the housings with the necessary bonded phase required for the method, so this is seldom a constraint on choice instrument. Disc cartridges are also compatible with most workstations and some of the soft gel resins can be used, though care must be taken not to crush the resin.

The manufacturers of automated SPE workstations should be contacted about restrictions on SPE cartridge types, bed masses, and tube sizes that can be used in their workstations. SPE manufacturers have been willing to pack various weights of sorbents in a single cartridge or tube size to increase the flexibility of use for workstations that can handle only one tube size or a limited range of tube sizes. Look at the current automation project, and the SPE cartridges used. If that cartridge is not compatible with the automated SPE workstation under consideration, contact the cartridge manufacturer and workstation vendor to determine if the required bonded phase can be put into the correct type of housing for that workstation.

4. Sample Capacity

The most important aspect of sample capacity is determining how you will use the workstation. Start by considering the number of samples. How do your samples arrive in the laboratory, in batches of 1000 or in small groups all day long? How quickly do analytical results need to be provided? If samples arrive in small quantities, can they be held and run in one larger batch? What is the initial manual preparation required for the samples? How long are samples stable before and after they are extracted? Must they be kept at certain temperature to remain stable? If the samples are stable for only a short period of time, consider how quickly the automated SPE workstation processes the samples. It may be more effective for those samples to be run in small batches. Once the extraction is performed, what is the next step? Are the samples to be evaporated to dryness in a batch mode, or are they able to be injected into an analytical workstation with out any additional steps? How stable are the extracted samples? A final consideration often overlooked concerns the laboratory's analytical capacity. It is not uncommon to invest in an automated SPE workstation, and be able to extract large quantities of samples only to discover the laboratory does not have the analytical workstation capacity to process these samples in a timely fashion.

5. Sample Volume

Since every workstation has a sample volume range it can accommodate, it is important to consider this in your workstation selection. Learn the minimum and maximum sample load volumes the workstation can run. Some workstations work well with a wide range of sample volumes, find out how the workstation does this. Is a larger sample container used, or must the large sample be split into many smaller containers that must be accessed. If this is the case then set-up of the workstation is probably going to be more time consuming and complex.

6. Other Functions the Automated SPE Workstation Can Perform

Learn about the overall capabilities, in addition to the SPE functions it can perform, of the automated SPE workstations under consideration. These sample preparation techniques or sample handling steps may include dilution of samples, addition of internal standard, filtration, mixing, evaporation, derivatization and injection of samples onto an analytical workstation, loading of autosampler vials, and so on. Remember more functions do not translate to better performance. Assess if these additional functions will be used for the preparation of samples to be extracted by SPE or if they could be used elsewhere in the laboratory. In addition, by having more functions, make sure you are not compromising the primary job for the workstation — to perform automated SPE.

C. WORKSTATION CONTROL

An attempt is made in this section to cover the general aspects of control of a workstation. Seven areas have been identified:

- Control of the automated SPE workstation
- Writing and editing methods
- Ability to run multiple methods
- Error conditions
- Security options
- Running samples

- Documentation

Much of this section will be concerned with the software that instructs the workstation and records the actions that workstation carries out. This field is has been in a state of rapid change for many years, so no attempt is made to identify specific characteristics of software packages. Instead focus is given to the general requirements of the controlling software, rather than the features currently available.

1. What Controls the Automated SPE Workstation

Depending on the size of the workstation and the complexity of its task, control may be provided by anything from a small, single-tasking device having a single or multi line display to a sizable computer system. Many of the workstations available at time of press utilize computers with control software supplied by the workstation manufacture. If the latter applies to the automation equipment you purchase, find out if the computer is dedicated to the workstation and if it can be networked. Determine if you can use an existing computer, or if you are required to purchase a new computer, and if a new computer is needed, find out what the specific requirements are — for example, what operating system is required.

Single or multi-line workstation control devices typically are inexpensive and simple to use, but they may lack flexibility and the ability to display the entire method at once. These devices may work well for laboratories running only a few methods on a routine basis. Many of the manufacturers offering the small single or multi line workstation control devices may also offer control software. For laboratories running numerous different methods, writing and developing new methods on a regular basis, computer control of the workstation will offer a greater flexibility and ease of use. Additionally, workstations controlled by a computer provide electronic documentation of methods, making documentation of methods and method transfer easier. Again, consider the use of the workstation in your laboratory, and even more important, the skills of the people who will use it.

2. Writing and Editing Methods

Make sure you understand what is required to write a method on an automated SPE workstation, and determine if this way of writing and editing methods will work in your laboratory.

Some workstations have software that facilitate writing a method, often using a Windows®[39] operating system with dialog boxes or drop-down menus that are preprogrammed by the software developers. For nearly

[39] Windows is a registered trademark of the Microsoft Corporation, Richmond, Washington.

every SPE method, this type of control software can produce excellent results. However you may not have the ability to control absolutely every variable of the separation. More complex software packages, based on programming languages, require more training and method writing is slower. However, they do provide control over every variable and offer an unlimited degree of flexibility. For projects with a potentially short duration, an automated SPE workstation that requires complex programming may end up sitting unused, whereas an automated SPE workstation that permits quick method development and implementation and easy operation for multiple methods will be more successful.

A rational decision on choice of software control must be based on an understanding of what variables must be controlled for your applications. An even more important question is: Who is going to be writing the methods, what are their skills, and what is their comfort level with the software? A short period of training should be anticipated and budgeted for.

3. Ability to Run Multiple Methods

Can the workstation be used to run only one SPE method at a time, or multiple methods? Again, it is important to know how you want to use the workstation. For routine work, running one method at a time may suffice. For methods development work, the ability to run multiple methods in one run is absolutely necessary.

4. Error Conditions

With any workstation, errors are inevitable. It is important to know how the workstation responds to the errors, and when and how the operator learns of the error. The most desirable circumstance is for the operator to be notified of any impending errors either before, or immediately after the run is started, before the operator walks away to start another task. These types of error might include missing containers for fractions, missing SPE cartridges or inadequate gas supply. By informing the operator of these errors, they can be easily corrected and the run restarted. Others errors, for instance running out of a reagent or having a blocked SPE cartridge usually can not be predicted in advance by the workstation. When examining error checking capability of a workstation, learn what errors can be checked for, when they are checked for, how the operator is notified of the error and what happens once the error occurs. For example, does the error cause the workstation to stop or does the error cause the workstation to abort processing of that sample, flag the error, and continue on with the next sample? And how are these errors corrected?

5. Security Options

The term security means different things to different people. To some, it can mean the ability to keep others from accessing the computer or the device which controls the workstation. For others, security may imply the ability to allow limited use of the software, but not allow any changes to be made to sample information or sample preparation method. For others it means making a method fool-proof.

Manufacturers have addressed the security issue in several ways. An example of one of the questions you may ask of the security system is: "Can I code different methods to racks, so when operators wish to run a method, they simply load samples into the correct rack, make sure the correct reagents are on the workstation and run samples all without accessing the computer?" It is most important to determine what will work best for the situation at hand, and look for security features as necessary for the laboratory and operators. Additionally, a commercially available software package could be installed on the computer for computer password security.

6. Running Samples

It is very helpful, especially in labs where many workstations operate simultaneously but on different applications, to know what method is running and how much of that method the workstation has performed. Some workstations show limited information, while others provide copious printed and real-time computer monitor reports. While the workstation is in an active mode, processing a sample information should, at a minimum, be available on the status of the sample being run, the method used and the current stage of the application the workstation is carrying out. Time reporting — knowing how far along the run is, or when the run will be completed — is found to be very helpful to users with large sample batches. When assessing the information available to the user during a run, think mostly about how it can be used. For example the time the samples will be completed is important to allow for efficient planning of instrument use and data handling.

7. Documentation

Another way in which automated SPE workstations can improve upon manual SPE is in the area of documentation. In manual SPE, there is no real way to document what has been done to the sample other than writing the manual SPE method into the laboratory notebook. With automated SPE, and a computer system for control of the workstation, documentation can be printed. Often the printouts report sample identification lists, detailing which samples and methods were run and can provide time and date stamps. This material may be printed and directly incorporated into the laboratory notebook, as positive identification of what preparation was done to each sample.

IV. AUTOMATION OF SPE APPLICATIONS

A. MANUAL-TO-AUTOMATED SPE METHOD CONVERSION

This section provides guidelines for automating a well-developed existing manual SPE method. Our goal is transfer to an automated method and then optimization of the automated SPE method. We shall assume that the manual method is rugged and well developed and that no changes are made to the SPE chemistry. The section titled “Making the Automated SPE Method Rugged” explains how an existing SPE method may be made rugged during the method development process.

Some key points to follow when automating a manual SPE method:

- Start with a *validated* manual method. Method development may be done with automated SPE workstations, but the objectives differ. This is the subject matter of a later section.
- Change only one variable at a time.
- Use a mid-level spiked sample to evaluate the effect of method changes, without losing the analyte signal.
- Make up spiked samples in the sample matrix: the primary purpose of SPE is to isolate components from a complex matrix, so work in the correct matrix.
- Run spiked samples in triplicate, at least, to make sure only the variable under investigation is changing.
- Do not mix SPE cartridges from different vendors. Although cartridges from multiple vendors may have the same functionality, they may not have the same performance characteristics. All cartridges used should be validated for the specific SPE method.

The process of automating a manual SPE method has five steps: 1) initial experiment; 2) minimizing interferences; 3) optimizing recovery; 4) reducing carryover; and 5) optimizing throughput to achieve the final method. Each step of the process should therefore accomplish a goal, moving closer to the final, optimized automated method. The five steps and their goals are outlined in Figure 2.

1. Initial Experiment

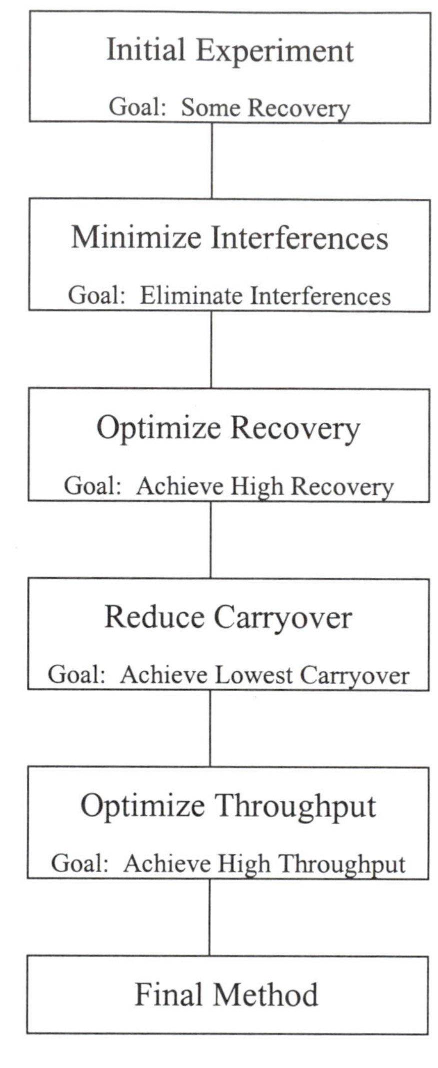

Figure 2. The process of automating a manual SPE method.

The initial experiment begins the transfer of a proven manual SPE method to an automated SPE workstation. The goal of this step is to achieve some recovery. The recovery of the initial experiment may be higher, lower, or the same as what you achieve in the manual SPE method. Additional experiments are performed to maximize recovery and optimize throughput.

Individual flow rates (or pressures) are typically selected for each step of an SPE method: Condition, load, rinse and elute. Many methods may be flow rate sensitive in the load and elute steps, but less sensitive to flow rates in the condition or rinse steps. Flow rates affect purity, recovery, and sample throughput. Manual method steps may be timed to determine the flow rates. If the automated SPE workstation allows only one flow rate for all steps, use the flow rate which yields the best recoveries — usually the slowest flow rate needed for any of the steps.

Most automated SPE workstations are not designed to handle small sample volumes. Diluting samples with a compatible reagent is the best option. This makes it easier for the workstation to handle the sample and it is easier to get the sample out of the test tube. Additionally, a more dilute sample will have better binding to the retention sites on the SPE cartridge because the mass transfer onto the cartridge will be improved. When working with plasma samples, do not dilute the initial sample unless the dilution has been tested and validated manually. This is important because dilution of a plasma sample can dilute the anti-coagulant and cause clotting. Dilution of a plasma sample with the incorrect reagent could also cause protein precipitation.

2. Minimizing Interferences

Interferences are defined as components from your sample that inhibit the ability to accurately quantitate the component(s) of interest. Interferences should be minimized prior to further method optimization to allow accurate quantitation of the analyte of interest because it is impossible to optimize recovery if we cannot measure it accurately. Interferences are removed from the SPE cartridge by rinsing and selective elution. Parameters to investigate are rinse volume and rinse flow rate (or pressure).

Rinse volumes that are too small may not wash the SPE cartridge adequately and may lead to interferences in the elution. If interferences are detected, increase the rinse volume or increase the wash solvent strength and see if the interferences decrease. If the rinse flow rate is too fast, interferences may not be adequately removed. Diffusion between the packing material and the rinse reagent may be enhanced at slower flow rates. Verify that the rinse flow rate is set correctly by decreasing it and determine if the interferences decrease.

3. Optimizing Recovery

Our goal is to get the highest recovery in as few steps as possible. A recommendation that is applicable to any SPE method development, not just automated method development, is to determine a parameter's effect on the recovery by making large changes to that parameter. This allows rapid evaluation of the effect of that variable on the results. Most automated SPE workstations should yield recoveries equivalent to or better than the manual method. Of course, only one variable should be changed in any single experiment. The following areas should be considered when optimizing recovery: load and elute flow rates (or pressures) and volumes and cartridge drying.

Load or elute flow rates that are too fast (because the vacuum or positive pressure is set too high, for example) may not allow the analyte of interest enough time to bind or be removed from the cartridge. Try decreasing the load or elute flow rate and see if there is an increase in the recovery. The optimum load or elute flow rate can be determined by running samples where the load or elute rates are varied. Observe the relationship between the load or elute flow rate and percent recovery. The goal is to determine the fastest flow rate that still yields good recovery. Figure 3 shows a graphical representation of the trade-off between recovery and flow rate for a hypothetical extraction.

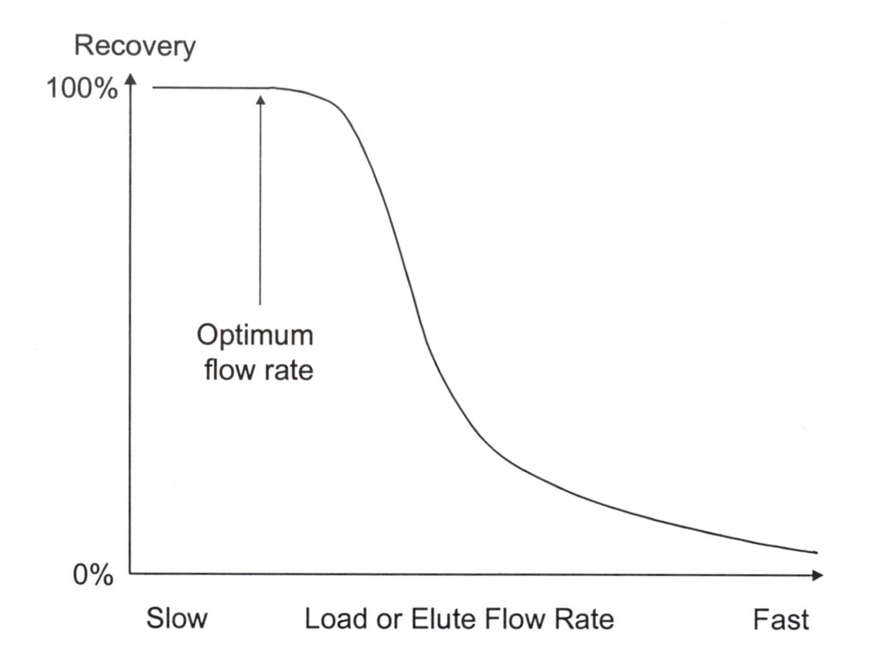

Figure 3. A graphical representation of the relationship between load or elute flow rates and recovery.

Some SPE methods may require cartridge drying to ensure optimum recovery. Drying is necessary, for example, when two immiscible reagents are used in sequence or when water from an aqueous wash step could contaminate an anhydrous eluent. Evaluate drying time by observing changes in recovery with drying time in a time-course experiment. Compare the relation between drying time and percent recovery and determine the shortest drying time that gives acceptable recovery and precision.

4. Reducing Carryover

Carryover is a type of contamination, and occurs when some of one sample remains in the system and can be "carried over" into the next sample. It is most serious when a sample with a high concentration of analyte carries over into a subsequent sample that contains a low or non-existent level of the analyte. It will be most severe in these cases when the method yields highest recovery of the analyte. Therefore, we should check for carryover after recovery has been optimized.

All automated SPE workstations have some form of a fluid path which transports reagents, samples or both reagents, and samples. Hence, checking for carryover should always be part of automating a manual SPE method. To determine carryover, first run a blank sample (one that contains no analyte), followed by a sample with a very high concentration of analyte. Follow this with two additional blank samples and measure the amount of analyte found in these samples. Do not analyze the sample containing the high concentration of analyte — that sample was used only to check for carryover. Remember that the carryover you measure is the carryover from the entire process of preparing the sample, not just the automation part.

For most applications, the carryover that is measured could come from any manual sample preparation steps, the automated SPE method, any sample evaporation or derivatization, and the final sample analysis. Determine the contribution from each portion of the sample preparation to carryover. If carryover is detected in the automated portion of the sample preparation, try to find a wash reagent to clean the fluid path that dissolves the analyte of interest well and that will also remove anything that may be remaining from the sample matrix. Keep in mind the need for wash solvent/method solvent compatibility. For example, if the fluid path last contained hexane for an elution reagent, and the beginning of the method uses methanol to condition the cartridge for the next sample, wash the fluid path with an intermediate reagent such as isopropanol or acetone before running the next sample. A general rule on cleaning any fluid path is that multiple small volume washes are more effective than a single, high-volume wash.

5. Optimizing Throughput

Up to this point during method development, we have worked to optimize the recovery of the analyte and minimize any carryover in the method. Now we need to determine if throughput needs to be optimized and, if so, how this can be done without sacrificing method performance. Not every method needs throughput optimization so before undertaking this final stage of method validation, ask yourself if the bottle neck in sample handling lies with collection, preparation, analysis, or data analysis of the sample. Consider the number of samples, the sample capacity of the automated SPE workstation, the speed of the analytical method, whether the extraction is performed on-line or off-line or is coupled to a dedicated analytical instrument, and how fast the sample results are needed.

If the automated SPE workstation is connected to an analytical workstation and direct injections are made, consider the run time for the analysis. For example, if sample preparation time is 12 minutes, and the run time is 14 minutes, the overall throughput will not be increased by further optimization of the preparation step. However, if the automated SPE workstation does not make direct injections, and the method requires all samples be evaporated and reconstituted before the final analysis, a faster sample throughput may be desired.

There are many parameters that may be adjusted to increase throughput. Adjust those steps that are most time-consuming first. To determine the most time-consuming steps, review the automated method. Those method steps with large volumes and/or slow flow rates or low pressures will be the most time-consuming steps in the automated method. Consider optimizing those steps for throughput first, as they may yield the largest increases in throughput.

Load and elute flow rates and cartridge drying steps are often the most time consuming steps in any automated SPE method. Because a slow flow rate for load and elute steps often gives the highest recoveries, these steps can be time consuming. Often even small increases in the flow rates will provide a better sample throughput, without sacrificing the recovery and you should still have data from optimization of these steps to refer to. The same rationale applies to the modification of the drying step. Always verify how recovery and carryover are affected by a change made to increase throughput.

The final result of the five-step process is an automated SPE method that has few interferences, zero carryover, and is optimized for recovery and throughput.

B. AUTOMATED SPE METHOD DEVELOPMENT

The above method transfer process is the most common introduction analysts have to automated SPE — an existing method must be converted to an automated form. As knowledge of automated SPE grows and the number of active automated workstation increases, more method development will be done directly on the workstation. There are three major advantages in automating SPE methods development:

- The ability to control the parameters that affect the separation, because the variables are held constant in an automated method. This can result in a very reproducible, rugged method.
- If the method is developed directly on the automated SPE workstation, there is no need to transfer a manual method to an automated one. This can improve the method and save time, because the method is developed as it is intended to be used.
- When method development is manual, a "working" method is developed, but there is rarely time to optimize or the method or make it rugged.

With an automated SPE workstation, it is easier to find the time to optimize and make the method rugged the method, because the labor-intensive

Table 1. Components of an automated extraction method and tasks required to optimize each component.

Component	Tasks Required
Sorbents	Write a single automated method, run with numerous different sorbents to determine which are the most favorable for the separation.
Reagents	Determine the best reagent or combination of reagents to dilute the sample matrix, to rinse the cartridge and to elute the analytes of interest.
Volumes	Determine the optimum volumes to use for rinsing or eluting fractions.
Flow Rates or Pressures	Control flow rates or pressures through the SPE cartridge to optimize recovery and build ruggedness into the method.
Analyte Class Fractionation Ability	Multiple fractions can be collected from one sample, allowing for the collection of the load step (cartridge effluent), rinse step and elution steps during method development.

process of "manual" SPE development is reduced to setting up a workstation and transferring samples and extracts to and from the workstation.

With an automated SPE workstation, the effect of the variables in a method can be examined with very little effort during the method development process. An excellent example of this testing process is given by Ren and Witkowski (1996). A summary of the requirements is tabulated in Table 1. Workstations facilitate all these steps (which should, of course, be carried out in manual SPE application development, too). What makes automated SPE method development so desirable is that all the time-consuming manual work and the record keeping can be eliminated or left for the workstation to manage. In addition, the process of changing one variable at a time provides more reliable data when performed by a workstation, since a workstation will not be prone to lapses in concentration, variability in operator technique, and so on.

If we break the method development process into to parts — isolating the compound of interest and ruggedizing the method — we will find the process of developing an automated SPE procedure easier to accomplish.

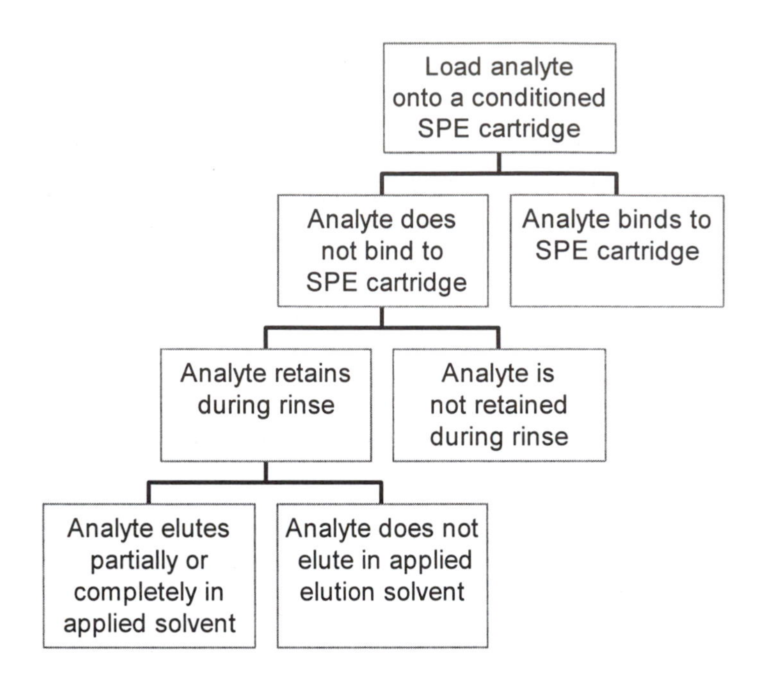

Figure 4. A Mass Balance schematic of a solid phase extraction.

1. Isolating the Compound of Interest

The goal of any SPE method is to isolate the desired analyte from the sample matrix, and to concentrate the analyte so it can be quantitated. This book applies many different practical approaches and some theory of SPE to method development. One approach that lends itself well to automated SPE method development uses the Mass Balance, a schematic of which is shown in Figure 4. Note that this has also been used in manual SPE method development (Bouvier, 1995).

The Mass Balance approach reminds us that at each stage in the process valuable information is obtained. For example, when a SPE cartridge is conditioned and a sample loaded onto the cartridge, if the sample effluent is collected and then analyzed, it will either contain large quantities of analyte, or very little if any analyte. This obvious observation is often overlooked in the author's experience. If little or no analyte is found in the analyzed fraction, the analyte is bound to the sorbent, and next the steps of rinsing and elution should be investigated. If a large quantity of analyte is found in the fraction, the analyte was not retained by the sorbent. Under these conditions the knowledge is gained that under these specific conditions, there is little or no binding of the analyte to the sorbent. Therefore, either the conditioning parameters, loading parameters, or the sorbent are not well suited

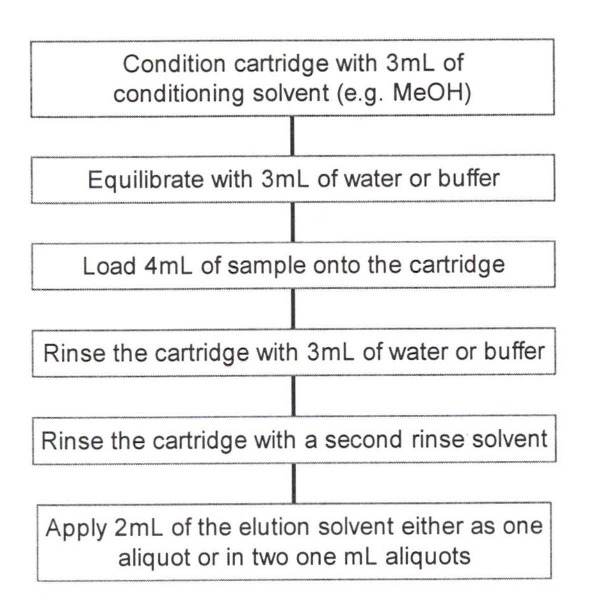

Figure 5. A typical non-polar extraction to which the mass balance will be applied

for this application. The goal is to have an analyte that binds to the sorbent, is retained during the rinse steps, and can be removed in the elution step.

Before using the mass balance we must do our "homework" and determine what mechanisms of extraction are permitted by our analyte and sample. Once the initial homework is done, a basic method can be written as a starting point. An example of a basic method for a non-polar separation example is shown in Figure 5. All flow rates or pressures for moving samples and reagents though the SPE cartridge should be set very conservatively, since they will be optimized later. The volume of sample loaded will need to be adjusted, based upon the sample size you have available and upon requirements for detection limit (LOD, LOQ). Contrary to what you would do if transferring a manual method to a workstation, you should start with the sample in a compatible buffer (for a non-polar phase application) and not in the sample matrix. Move to the sample matrix after some of the initial work has been done. All initial work should be done at a concentration in the middle of the analytical linear range.

After the basic SPE method is written, you will work down the flow chart shown in Figure 3 to determine the best sorbent, load conditions, rinse reagents and elution reagents for the SPE method. Variables of sample stability over time, temperature fluctuations and precision testing within the SPE cartridge lot and between lots should always be investigated.

2. Making the Automated Method Rugged

When developing an SPE method, there is often no time left to spend ensuring the method is rugged, once acceptable recovery has been achieved. It is not uncommon for a method to work well for a period of time and then exhibit problems of low recovery or poor reproducibility due to variables

Table 2. Ruggedization parameters for a typical SPE step.

Parameter	Value	Test
Flow Rate	12.0 mL/min	Low: 10.0 mL/min High: 14.0 mL/min
Composition	20% Methanol	Low: 15% Methanol High: 25% Methanol
Volume	3 mL	Low: 2.8 mL High: 3.2 mL

not optimized in the method development process. These problems are usually an indication that the method is not rugged.

Making the method rugged involves ensuring that all parameters used in the method are optimized. To do this, for each step of the method, multiple tests should be run. These tests examine the parameters of flow rate (or pressure), reagent composition, and reagent volume. Each of these variables is examined at a low and high range.

Table 2 illustrates an example of how the flow rate, composition and volumes could be tested to determine the ruggedness of the following step:

Rinse column with 3 mL of 20% Methanol at 12 mL/min.

To facilitate the process of testing for ruggedness, start with the method you are making rugged, then systematically edit this method to test each of the variables outlined above and save these methods to different file names. If the software permits comments to be entered when saving to a file, use this option to specify the variable under investigation in that method. Next, run these methods. If a small variation made to a variable affects the results to an appreciable degree (exactly what is "appreciable" must be determined partly by your own tolerance to recovery or precision for that assay), the method is not yet rugged. That variable must be re-examined until a value is found around which the results do not vary significantly. Finally, and this is an important point, all this data should be stored since it will prove valuable in method troubleshooting if, despite careful method ruggedization, the method should, at a later date, yield poor results.

V. CONCLUSIONS

The goal of this chapter was to explore all aspects of the automation of SPE, starting with the evaluation process, through the review of current workstation principles and capabilities, to the automation of manual method in the laboratory and finally the automation of SPE method development. Successful introduction of automation requires the laboratory to realize that an automated SPE workstation is a tool to help the laboratory personnel perform their jobs, but is not a replacement for them. Furthermore, automation that takes time to implement, and is not an instant answer to all problems in the laboratory.

There are numerous vendors offering solutions for the automation of SPE and there are many different ways of providing automation for the same basic application. This variety means that care must be taken in selecting the workstation that is the best one for your needs. Emphasis was placed throughout this chapter on the importance of assessing the needs of the project and your laboratory. When automation is implemented in the laboratory, provided the guidelines laid out in this chapter are followed, whatever workstation you use, careful SPE method development will provide rugged, reliable, and optimized automated SPE methods. In order to avoid getting lost in details of specific applications, this chapter has focused on the principles of automation. An excellent paper on one such application is given by Rossi (1999). This work puts many of the points discussed in this chapter into practice in the development of assays for a promising pharmaceutical compound in animal matrices, as applied to a high-throughput bio-analysis laboratory.

As well as looking at the automated SPE workstation, you should look at the vendor supplying it. The purchase of a workstation results in a partnership between the vendor and customer — both should be satisfied. Ask questions of the vendor to determine what validation, servicing, customer training, and support services they offer. These services, along with the capabilities of the workstation will affect the success of the automation project. Rather than providing lists of currently available workstations and options (because the lists and descriptions in this field become obsolete within months) a summary of manufacturers is provided in Appendix 4. You are urged to contact these vendors to receive up-to-date information on their product offerings.

REFERENCES

Bouvier, E.S.P., (1995) Sample Preparation Perspectives; SPE Method Development and Troubleshooting. LC.GC 13(11): 852-858.

Jordan, L., (1993) Sample Preparation Perspectives, Automating a Solid Phase Extraction Method. LC.GC 11(9): 634-638.

Jordan, L., Hansen, T.J., Zabe, N.A., (1994) Automated Mycotoxin Analysis. American Laboratory 26(5): 18-24.

Majors, R.E., (1993) Sample Preparation Perspectives; Automation of Solid-Phase Extraction. LC.GC 11(5): 336-341

Majors, R.E., (1995) Sample Preparation Perspectives; Trends in Sample Preparation and Automation — What the Experts Are Saying. LC.GC 13(9): 741-748

Paulus, M., Halvorson, M., and Paskey L., (1996) Liquid Level Detection. Laboratory Automation News, 1(5): 33-36.

Ren, X. and Witkowski, A., (1996) Use of the Rapid Trace™ SPE Workstation for Accelearted Methods Development. ISLAR '96 Proceedings (Publ. Zymark Corporation, Hopkinton, MA) 469-492.

Rossi, D.T., (1999) Automating Solid-Phase Extraction Method Development for Biological Fluids. Tends and Analysis in Bioanalysis. LC.GC. 17(45): 54-58.

15

INTEGRATION OF SPE WITH THE ANALYTICAL TECHNIQUE, PART I – LIQUID CHROMATOGRAPHY

Nigel Simpson, Sample Preparation Products, Harbor City, California

I. INTRODUCTION

Solid-phase extraction (SPE) and high-performance liquid chromatography (HPLC) are both similar and dissimilar. During the course of this chapter we will encounter aspects of SPE that show its underlying similarity and its differences to elution chromatography.

From a theoretical perspective they are at opposite ends of the separation field. Liquid chromatography is fundamentally an equilibrium-driven separation, whose effectiveness is reduced by the effect of (slow) kinetics on the partition between a solid and a liquid phase. Solid-phase extraction is ideally a kinetically driven partition from liquid phase into solid phase or vice versa (depending on whether the process is a retention or elution step). The effectiveness of this step in terms of analyte concentration is maximized when the system is at a maximum disequilibrium.

At a practical level they often appear indistinguishable. In their most common manifestations both use combinations of organic and aqueous solvents that are passed through a sorbent bed, resulting in separation of the components. Indeed, the hardware used to effect low to intermediate pressure LC separations and SPE extractions is often the same. Flash chromatography preparative systems have been modified to use SPE cartridges and, as shown in this chapter, SPE cartridges have evolved to enable them to be used as HPLC primary, concentration columns or disposable guard columns for HPLC separations.

A casual survey of the published applications of SPE reveals that roughly 60% of these combine SPE and HPLC in some way. Not surprisingly, published examples of SPE/HPLC hyphenated techniques are much more common than applications using other hyphenated SPE techniques.

II. THE DEVELOPMENT OF ON-LINE SPE/LIQUID CHROMATOGRAPHY

A. CONCENTRATION AND CLEAN-UP USING HPLC PRE-COLUMNS

The marriage of SPE and LC is an obvious one. Central to both is a sorbent/liquid partition process, and both require accurate fluid handling. Use of a guard column or "switchable" column was commonplace before the advent of SPE and the potential for using contrasting sorbent polarities to screen out or screen in analytes whose polarities themselves fall within a specified range were being explored before the first commercial silica-based SPE cartridges were introduced.

An excellent example is given by Martin et al. (1979). The authors describe how they load a sample onto a silica column which is then placed between a C18 column and a second SI column. The plumbing arrangement, shown in Figure 1, allows the sample column to be eluted in both directions using solvents supplied from two separate reservoirs and two separate pumps, and also allows the effluent from either column be sent to a detector (and the fluid lines can be flushed between uses). In this way the species in the sample that lie on either end of the polarity range will be trapped on the C18 or SI columns while the intermediate polarity compounds will pass through and be routed to the detector. This is not a true

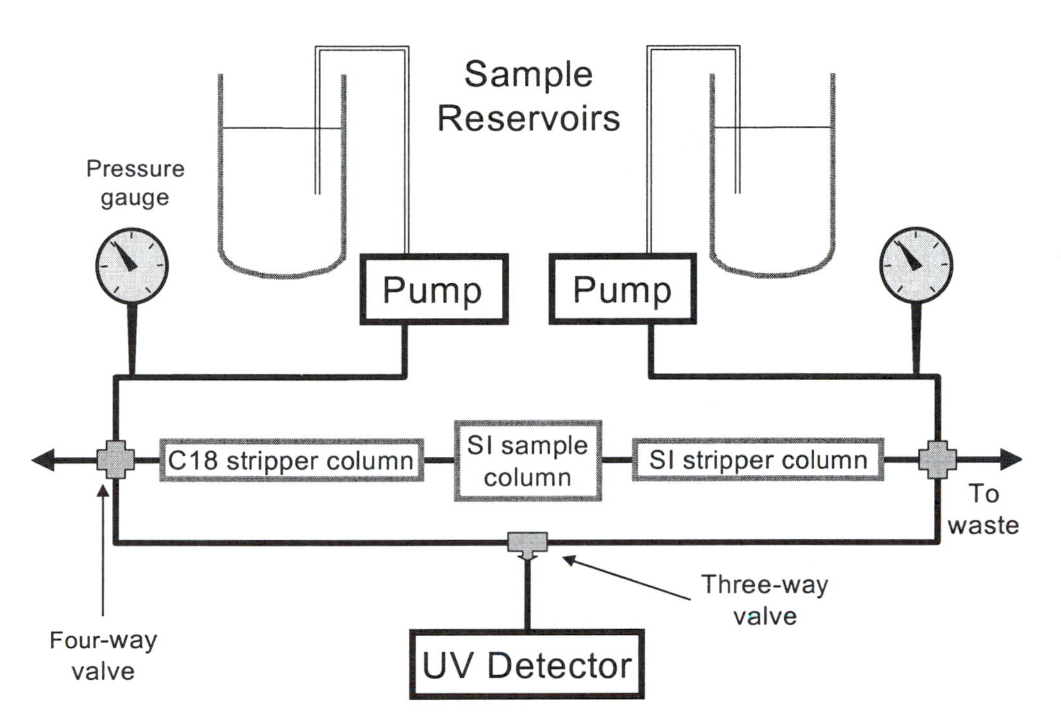

Figure 1. A plumbing diagram showing how LC column-switching has been used to perform class-fractionation and clean-up of a sample, combined with analysis.

SPE/LC application (the authors describe it as a combination of "extraction, frontal analysis, and displacement chromatography"). It is, however, clearly a demonstration of the ability of a sorbent to completely trap an analyte under one set of conditions and to release it to a detector under different set of solvent conditions.

Another illustration of how the two techniques of SPE and LC blend together is provided by the diagram shown in Figure 2 (Roos and Ögren, 1989). In this case LC hardware is coupled with a regular SPE cartridge. Use of a gradient LC pump allows an increasingly powerful eluent to sweep the SPE cartridge, onto which four compounds (an internal standard, the parent drug and two metabolites) have been deposited. This simple set up allows injection of standards onto the SPE cartridge and monitoring of their retention/elution properties under a gradient of 0% MeOH to 100% MeOH in water. It gives information on the strongest and weakest binary solvent mixtures needed for complete elution of analytes and the strongest mixtures that may be used for cartridge washing. The UV detector is placed at the far end of the SPE cartridge. It is thus possible to determine both the weakest binary elution solvent that ensures 100% elution of the weakest bound analyte (in this case, approximately 70% methanol in water) and the solvent

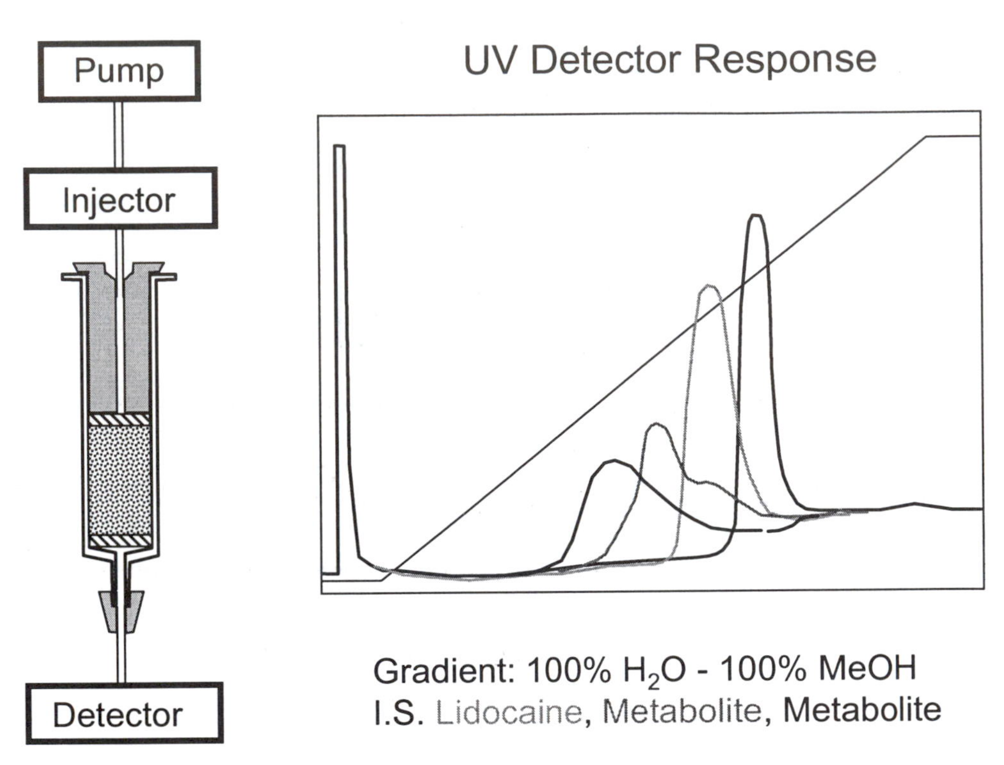

Figure 2. A demonstration of how an SPE extraction cartridge, placed in-line with a gradient LC pump and detector, can be used to optimize elution and wash conditions.

that elutes none of the strongest-bound analyte (in this case approximately 25% methanol in water). The benefits of using this experimental design are simplification of method development, automation and reduction in operator handling, leading to improved precision.

A final benefit is that a choice may be made at elution that is not always available or easy in off-line work. Either the entire eluent or a specific fraction (the front, rear, or heart of the elution bolus) may be loaded onto the top of the analytical column through the use of switchable high-pressure valves. The former enhances sensitivity, while the latter can either enhance sensitivity or clean-up or both. This idea has been examined and extended further by several reserachers (for example, Guenu and Hennion, 1994) who optimized SPE concentration by evaluating breakthrough volumes using the above system coupled to a diode array detector. Later sections of this chapter will illustrate these techniques with applications.

B. EVOLUTION OF COLUMN-SWITCHING INTO ON-LINE SPE

On-line SPE systems evolved from the practice of column-switching, which was used to solve several analytical problems. These included elimination

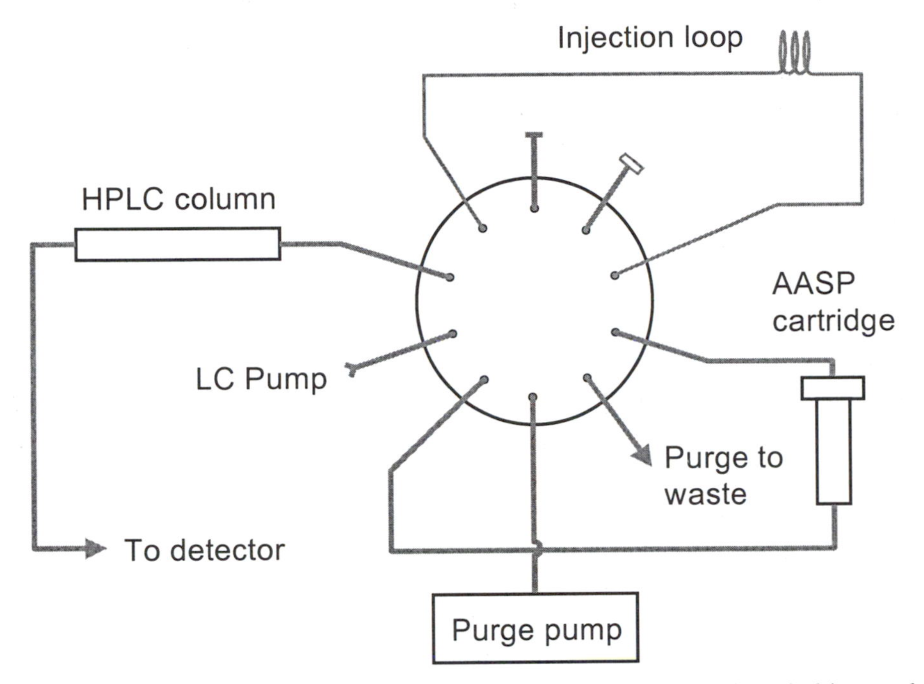

Figure 3. Analytichem Advanced Sample Processor fluid switching valve plumbing. This set-up applies to a typical extraction with on-line elution that includes a solvent purge step.

of late-eluting compounds, sample storage, and alternative to gradient elution, protection of the analytical column from blockage by sample particulates and providing a support for solid-phase derivatization. In addition, the practice evolved to include trace enrichment of analyte and removal of sample interferences — the two primary functions of SPE.

If we exclude HPLC column switching equipment, the first dedicated instrument to address on-line SPE/LC was called the AASP™[40] (Analytichem Automated Sample Processor). This device was introduced in 1984 and provided a semi-automated solution for on-line SPE/LC. The samples were loaded off-line in batches of ten, using a vacuum manifold, onto a cassette of ten steel- or titanium-fritted mini-SPE cartridges. After off-line washing of the cartridges the cassettes were inserted into the AASP unit and a fluid-switching valve (Figure 3) directed flow of a mobile phase through the SPE cartridge. The cartridge was designed so that it could be clamped into a station in the AASP unit and was made with materials that could withstand normal HPLC operating pressures.

[40] AASP is a trademark of Varian Inc., Palo Alto, California.

The purge system permitted the operator to remove early-eluting interferences that were co-retained on the sorbent and also provided a means of eliminating air bubbles that would otherwise be introduced from the SPE cartridge and lead to baseline instability (Yago, 1985). The potential to focus analytes on the head of a very non-polar HPLC column, such as a high carbon load C18, by eluting analytes off a low-capacity sorbent, like C2, using a weak SPE eluent has been exploited. This benefit partially explains the higher percentage of applications developed using C8 and C2 AASP compared to the off-line equivalents where C18 sorbents dominate.

The AASP (now discontinued) is the forerunner to more versatile, fully software-controlled units such as the Prospekt™[41] and the OSP2™[42]. These instruments use a similar mode of clamping a high-pressure SPE cartridge in-line with an analytical column but the more recent instruments also allow cartridge conditioning, liquid sample loading and washing to be handled by the extraction unit. If interlaced with a solvent delivery module or a LC autosampler, sample sizes or solvent volumes of a hundred microliters or less to more than a liter can be handled. However, the benefits of the on-line technique, notably analysis of the entire eluent and 0% analyte loss from evaporative or other effects, makes it possible to analyze much smaller samples (Brinkman, 1995) than are required for conventional off-line SPE extraction. Bed masses for the cartridges used in these systems typically fall in the range of 10 to 50 mg. The range has proven to have adequate sample extraction capacity without providing a bed volume that is too large for efficient elution.

Software control permits complex valve switching sequences to occur. Some applications have used this approach to develop the SPE extraction beyond the boundaries of what can be achieved on the SPE sorbent itself. By shunting the eluent bolus from the SPE cartridge into a switchable holding loop it is possible to cut out a fraction of the eluent (if done carefully, just that fraction of the eluent containing all the analyte). Front-cut, especially of cartridges that are back-eluted (eluted in the opposite direction to the flow of the sample at the loading step) can reduce the volume to be loaded onto the analytical column dramatically. Flushing of the sample-loading pathway by the elution solvent can be easily achieved by valve switching. This can enhance recovery and improve precision, especially for very lipophilic analytes that would otherwise be lost on the inside walls of the autosampler and SPE robot fluid handling lines. Just as with conventional LC there appears to be few limits to the inventive ways in which a separation medium, in this case an SPE cartridge, can be connected to en-

---

[41] Prospekt is a trademark of Spark Holland, B.V. Emmen, the Netherlands.

[42] OSP2 is a trademark of E. Merck GmbH, Darmstadt, Germany.

hance overall analysis. Specific examples of these on-line capabilities are given in section IV of this chapter.

III. THEORY AND PRACTICAL IMPLICATIONS

A. BREAKTHROUGH VOLUME AND OVERLOADING

The most commonly applied method for measuring breakthrough on an SPE device is to measure the UV absorbance of the effluent form the device. The curves that result, as discussed in earlier chapters, are bi-logarithmic with an inflection point defined as the retention volume of the device, V_r. It is customary to define the breakthrough volume, V_b, as that volume at which UV absorbance of the device effluent is 1% that of the sample. In other words, up to this point the device has been able to retain essentially all the analyte loaded onto it. V_b is more critical for on-line SPE than for its off-line equivalent since the typical on-line SPE sorbent bed is an order of magnitude smaller than the off-line alternatives.

We shall use these terms in the later discussion, although on a practical level, when we are trying to concentrate very low levels of analyte, it is often impossible to observe the UV absorbance of the sample, let alone 1% of it! Note, too, that in a practical experiment the capacity of the cartridge will be partly dependent on the presence of other sample components. This is because in real samples not only analyte molecules but interferences will be retained. The implication is that for compounds with a capacity factor of less than 10 on the SPE device, pre-concentration is hardly feasible. Even when the k' is much larger than this, care must be taken to determine its value in a matrix that is similar to the sample both in its bulk composition and the micro-components that will be co-extracted. The retention volume is dependent on the efficiency of the SPE device since

$$V_b = V_r - 2\sigma_v \tag{1}$$

Where σ_v is the axial dispersion of the analyte along the sorbent bed. In most cases the analyst will want to maximize the concentration effect of SPE by passing through the pre-column the maximum possible volume of sample compatible with zero breakthrough. It is possible to determine the concentration effect obtained even when the SPE device has been overloaded and breakthrough has occurred. The enrichment factor, F, is defined by the equation

$$F = C_f/C_i = (1 + k')\, V_0/V_f \qquad (2)$$

Where C_i is the initial concentration of the analyte and C_f is the final concentration, V_f is the sample volume passed through the cartridge and V_0 is the void volume of the device. Note that for SPE where, to improve clean-up, we usually wish to wash the SPE cartridge after loading the sample, we risk losing some of the analyte even when using a very gentle wash solvent. As described in earlier chapters, therefore, it is very uncommon to find SPE procedures which try to maximize the loading volume by using all of the SPE device's theoretical capacity.

We shall use a Langmuir isotherm to describe the amount of analyte that is adsorbed onto our extraction device. We begin by defining the coverage of the surface by the analyte *a*, Θ_a as

$$\Theta_a = K_a \cdot c_a /(1 + \Sigma_{i=1,n} \cdot K_i . c_i) \qquad (3)$$

In this equation K_a and K_i are adsorption equilibrium constants for component *a* and *i*, c_a and c_i are the concentrations of adsorbed analytes *a* and *i*. Analyte *a* represents the compound we are interested in; *i* represents another of the *n* compounds in the sample. The amount of analyte that can be enriched is

$$M_a = V_{SPE} \cdot \rho . S(1 - \varepsilon_T)\, \Theta_a \cdot c_{max,a} \qquad (4)$$

Where ε_T is the total bed porosity, ρ is the density of the sorbent, $c_{max,a}$ is the maximum adsorption capacity per unit area, S is the specific surface area of the sorbent and V_{SPE} is the bed volume of the SPE device. Since $V_{SPE} \cdot \rho . S(1 - \varepsilon_T)$ is independent of the nature of the analyte, according to equation (3) the amount of analyte adsorbed onto the SPE device is proportional to the analyte's concentration, c_a. This amount is not affected by other components of the system, provided $\Sigma_{i=1,n} \cdot K_i . c_i$ is much less than 1.

In practice, proteins in biological samples and humic matter in environmental samples are often present at high concentration compared to the analyte so even when K_i is small, c_i may mean that the product of the two is large.

The capacity of the SPE device is strongly influenced by the sorbent that is packed into it. We can make qualitative statements about the capacity of the sorbent in terms of the nature of the bonded phase — for example, that C18 is more retentive than C8 for many non-polar species. Literature searches yield typical capacity values of between 5 and 50 mg per gram of silica based sorbents and somewhat higher values for polymeric SPE packings, but we can seldom be quantitative. Quantitative relationships can, however, be distinguished between capacity and particle size, pore size, geometry, sample flow rate and bed shape or geometry.

1. The Effect of Sorbent Particle and Pore Size

The Van Deemter equation is a good starting point to explore sorbent performance. When the sample loading speed, v, is chosen such that pore diffusion is the largest contributor to mass transfer, the height-equivalent of theoretical plates for the device, H, is given by

$$H = A + k(d_p^2/D_p)v \qquad (5)$$

Where k is a constant, D_p is the pore diffusion term and d_p is the particle diameter. The minimum value for H (when the SPE device is operating most efficiently) occurs at approximately $2d_p$. This can be achieved by using large pore packing (when D_p is large) and by employing packings with small particle sizes (when d_p is small). Note that as pore size increases, according to equation (5), the effect of loading rate decreases as a contribution to H since A is a constant. The same behavior is found for particle size, though the resistance to flow down the sorbent bed will increase as the particle size decreases, making this a less convenient — though in practice, an easier — parameter to explore.

Materials such as PLRP-S™[43] (pore sizes of 100, 300 and 1000 Ångströms) have found routine use in on-line SPE applications, while they are barely found in applications of off-line SPE. PLRP-S is a macroporous, styrene-divinylbenzene copolymer with strong non-polar properties. Its extraction properties have been comprehensively studied and contrasted with other sorbents in several papers (for example, Guenu and Hennion, 1996; Pichon et al., 1996) with the result that we know almost as much about this rarely used sorbent as we do about the common C18 sorbents. This partly reflects, however, the fact that on-line SPE permits more easy quantitative investigation of the SPE process due to its often fully-automated nature and the ease with which breakthrough and flow rate effects can be measured or controlled.

It is not easy to explore the ramifications of equation (5) in terms only of D_p since H will affect the breakthrough curve and hence the capacity of the sorbent, while larger pores (and hence larger D_p) invariably result in a lower surface area and therefore a lower capacity. The findings of a study

---

[43] PLRP-S is a trademark of Polymer Laboratories, Church Stretton, UK.

on 23 pesticides extracted from environmental samples by various 20 μm PLRP-S sorbents **(Hennion et al. Refs)** shows a reduction in the amount of co-extracted organic matter as the pore size increases (100, 300 and 1000 Å pore sizes were investigated). The capacity towards the analytes appears to be equal for the 100 and 300 Å sorbents, however. A decline in recovery is observed on the 1000 Å sorbent. These observations imply that although H is smaller and the adsorption kinetics are faster when the pore size is large, a balance must be struck between kinetics and the capacity of the sorbent. An intermediate porosity of 300 Å appears to be best for this analysis, though it would be wrong to assume that this represents an optimum for all extractions — the sample matrix, analytes and co-extracted species will, as always, dictate the optimum sorbent specifications. In this case the interferences are relatively large species and the analytes are relatively small, so conditions can be found where H can be minimized for analytes while remaining relatively large for interferences.

Since the force driving the sample through an on-line SPE cartridge may be an HPLC pump, we can explore small particle size SPE beds. Such investigation would be outside the range of particle sizes permitted in off-line equivalents, where a maximum driving force of only one atmosphere pressure is a common limit. Thus, experimental work (Ooms et al., 1997) has shown that, as would be expected from equation (5), capacity is enhanced by the use of smaller SPE particles. A second benefit of a small d_p is that, at elution, the analyte band will be narrower than for typical SPE cartridges. If the elution is truly on-line, minimization of band-spreading on the SPE device will translate into sharper peaks on the HPLC column, without the need for "tricks" like focusing of the elution band at the head of the HPLC column.

2. The Effect of Sorbent Bed Geometry

As is the case for off-line SPE the size of the SPE device will influence the volume of eluent required for effective elution of the compound of interest. In addition to this constraint, however, on-line SPE cartridges also affect the width of the band of analytes as they enter the HPLC column. A balance must be found between the need for high capacity, low elution volume and reasonable flow rate and, practically, this has been found to result in cartridges of typical length of 10-30 mm and internal diameter of 1-3 mm. To understand how these dimensions arise we must understand that the optimal volume of the SPE cartridge is the largest volume (i.e. the volume which offers maximum capacity) that does not create band broadening on the HPLC column.

We can express this optimal (maximum) volume of the on-line SPE device, V_{SPE}, in terms of the volume of the analytical column, V_{HPLC}, by equation (6).

$$V_{SPE} \leq (1 + k').V_{HPLC}/2\sqrt{N_{HPLC}} \qquad (6)$$

Where N_{HPLC} is the number of theoretical plates on the analytical column and k' is the capacity factor for the analyte. Note that equation (6) assumes that the packing material in the SPE device/pre-column is the same as that in the analytical column. When k' is greater than 1 and for a typical HPLC column of 250 mm length and 4.0 mm internal diameter, equation (6) predicts that V_{SPE} should be approximately 40 mm^3.

A convenient practical solution has been found through the use of commercially available 2 mm I.D. polyvinylidene difluoride tubing cut in 10 mm lengths. The wall thickness of these short tubes is such that they can withstand several thousand pounds per square inch pressure. The resulting devices offer an internal volume of 30 cubic millimeters, which falls ideally between the values predicted by equation (6) for the most common analytical columns — 150x4.6 mm and 250x4.6 mm. Using the rule of a minimum elution volume of two bed volumes, it is possible to elute such a cartridge with much less than 100 μL of solvent. In practice, it is desirable to use a weak elution solvent to prevent the problems of injection of strong solvents onto an analytical column. Even so, concentration factors of 10,000 or more are clearly achievable if the analyst is patient enough to allow a 1 L sample to be processed. Note that screens, rather than frits, are desirable for the SPE device, because of the tendency of the latter to be blocked by particulates in the sample.

Exploration of larger diameter on-line SPE cartridges supports the case for 2 mm I.D. devices. Faster flow rates and higher capacities are clearly possible for 3 mm I.D. cartridges (Halmingh et al., 1998; Landrieux et al., 1998), but while this higher capacity translates into larger peak areas, band broadening may often lead to smaller peak heights. The broadening will be most acute for late-eluting compounds. Recent developments of short analytical columns and narrow-bore columns for LC/MS, in particular, have resulted in the growth of interest in narrow-bore (1 mm I.D.) on-line SPE cartridges, though the bulk of applications reported in the literature refer to the 2 mm I.D. cartridges.

3. The Effect of Flow Rate

An HPLC column can accept only small volumes of strong solvents before the efficiency of its chromatography is compromised. This means that on-line SPE devices are small compared to off-line SPE cartridges. A direct result of this is that the cross-sectional area through which the sample must be passed is small, and this will in turn, limit the speed at which samples can be passed. As discussed in the previous section typical on-line devices have bed diameters of 1-3 mm, compared to bed diameters of at least four millimeters for common off-line cartridges. Because the on-line cartridge can be subject to much larger pressure drops across its length than off-line cartridges, neither sample viscosity nor bed permeability control the flow rate. However, the height equivalent of theoretical plates, H, is clearly influenced by sample velocity (the parameter v, in equation 5).

A practical flow rate range of 2-3 mL per minute for a standard 2 mm internal diameter cartridge has been shown for a range of environmental analytes (for example, Dupas et al., 1996). This provides a relatively fast way to trace-enrich 100mL of aqueous sample — the 30 to 50 minutes this takes is often approximately the same as the time it takes to perform the analysis of the previously extracted sample. Such a flow rate is perfectly adequate for biological samples, whose volume seldom exceeds a few milliliters, but for environmental samples where1 L is a common size, such a flow rate results in preparation times of a couple of hours per sample.

To explore the trade-off between flow rate and extraction efficiency, four different flow rates (5, 10, 15, and 20 mL/minute) have been investigated by Pichon and Hennion (1994). The findings_show that satisfactory results may be obtained for flow rates up to 15 mL/minute but when this flow rate is exceeded then variable recoveries and lower recoveries result. This is partly explained by plugging of the on-line SPE device by particulates from the unfiltered samples. Even at 10 and 15 mL the peak areas of the analytes were found to be slightly lower than those from samples loaded at 5 mL/minute, though the loss of analyte is small enough to justify the convenience of faster sample processing time. As always, though, the analyst should perform tests for the sample type and analytes being extracted. Too little work has been done on the effect of flow rate to allow a set of rules to be developed that applies across all sample types and for all target compounds.

B. PACKED SORBENT BEDS VERSUS SORBENT DISCS

Discs may be inserted into on-line SPE devices and they have been used for various on-line applications. Because the properties of cartridges and discs are substantially different in off-line SPE, a few evaluations of disc-based on-line SPE have been performed

By designing a special holder for SPE discs it has been possible to place between one and ten discs on-line (Kwakman et al., 1992). One benefit of such a variable extraction bed is that for the researcher the number of discs can be selected based upon the needs of the analysis, though self-assembly of the SPE device is not a practical option for laboratories with a large, single analysis throughput.

The disc format results in an SPE device that, compared to packed-bed cartridges, offers limited but relatively high capacity and high efficiency for off-line SPE, due to the typically smaller particle size permitted by its shallow and broad sorbent bed dimensions. Unlike off-line SPE on-line SPE enables the use of much smaller particle size packings, since the pressure drop across the SPE device can be much larger than in off-line systems. In addition, alternate desorption strategies such as back-eluting (which will be discussed later) allow the operator to reduce elution volumes for on-line SPE cartridges. However, there is no reason, aside from convenience and availability of packed-bed on-line SPE devices, why on-line disc devices will not become common. From the researcher's point of view, the ability to stack different sorbent discs within a single on-line cartridge will ensure that more research will be conducted in this area. Futhermore, equation (6) applies to disc extractions, suggesting a future application for such disc-based on-line devices: they usually possess minute SPE bed volumes. This implies that discs will find use in on-line systems for micro-HPLC separations such as would be possible with HPLC columns packed with 1 μm sorbents.

IV. APPLICATIONS OF ON-LINE SOLID-PHASE EXTRACTION

On-line elution imposes some constraints on the analyst, such as volume of SPE device, choice of elution solvent and restriction to serial rather than batch SPE processing, yet it also opens up many opportunities. In this section we shall explore some of the ways the on-line elution technique has been applied and examine the benefits.

A. A SURVEY OF TECHNIQUES

The basic element of any on-line SPE or column-switching system is the switching valve. The standard format has shifted from the ten-port valve shown in Figure 3 to a six-port valve. This is illustrated in a plumbing diagram for a simple on-line SPE set up, in Figure 4.

In such an arrangement an auto-sampler may deliver the sample to the SPE cartridge (common for small volume, biological samples) via an injection loop which allows for precise loading volume. If the sample is large (typical for environmental samples) the solvent delivery unit which supplied the conditioning solvents to the SPE device, doubles its function by

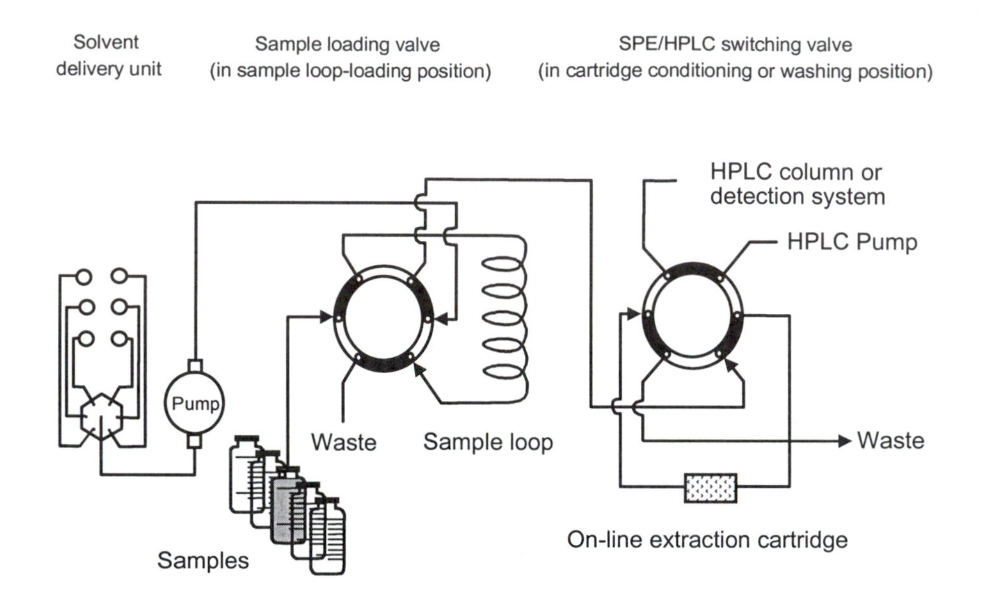

Figure 4. Fluidics of the conditioning, sample loading and wash steps for a simple on-line SPE set-up, showing how an auto-sampler can supply the sample to the cartridge, after it has been conditioned by a solvent delivery system.

also pumping the sample directly onto the cartridge. This set-up is easily configured and many applications exist which use this arrangement either in conjunction with a simple LC analysis (For example, Huen et al., 1994; Nielen et al., 1987) or as the front end for further column-switching techniques (Pastoris et al., 1995).

Improvements in reproducibility and detection are often-quoted conclusions in such references. This should not surprise us, because the total automation of the procedure and the precise control of volumes, flow rates, etc., are unique to on-line SPE. In addition this scheme, in which the entire elution bolus is injected onto the analytical column, necessarily means larger signals on the detector than would be observed in most off-line procedures, where only a fraction of the extract is usually injected onto the HPLC column. However, because the entire eluent is injected onto the analytical column, focusing of the analyte band on the head of the analytical column (which is permitted in off-line SPE extraction/HPLC analysis but is less critical to good chromatography) becomes more desirable. This helps explain the resulting preponderance of on-line methods that use less retentive sorbents than C18, which is commonly used as the analytical column.

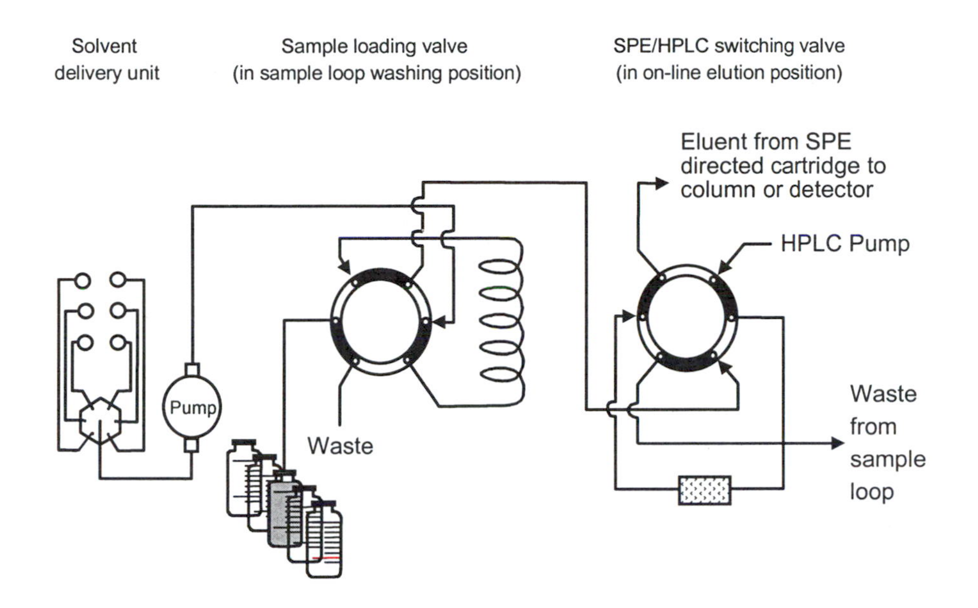

Figure 5. Fluid pathway at the elution step. By switching the valve the SPE cartridge can be brought on-line into the mobile phase stream, which will sweep the analyte onto the head of the HPLC column.

Further, mixed-mode sorbents have found few applications in on-line SPE because of the need for an elution solvent (which must be able to break up all the modes of interaction between analyte and SPE sorbent) that is often incompatible with HPLC. For this reason, too, ion-pairing agents are applied to aid retention and elution on non-polar sorbents in several applications, as an alternative to the use of ion exchange media, since the elution process again involves relatively gentle conditions. (Pastoris et al., 1995). One interesting point to note about the use of low retention sorbents in the SPE cartridge: The cartridge limits the capacity of the entire analytical system, and thus the use of high capacity sorbents such as C18 and PLRP-S is still common in the field of on-line environmental analysis.

1. Reverse Elution

The name of this practice refers to the direction of flow of sample and elution solvent through the SPE cartridge. While it has been used in off-line SPE the benefits from doing it are fewer and the practical aspects are less convenient. These benefits include minimization of injection volume and

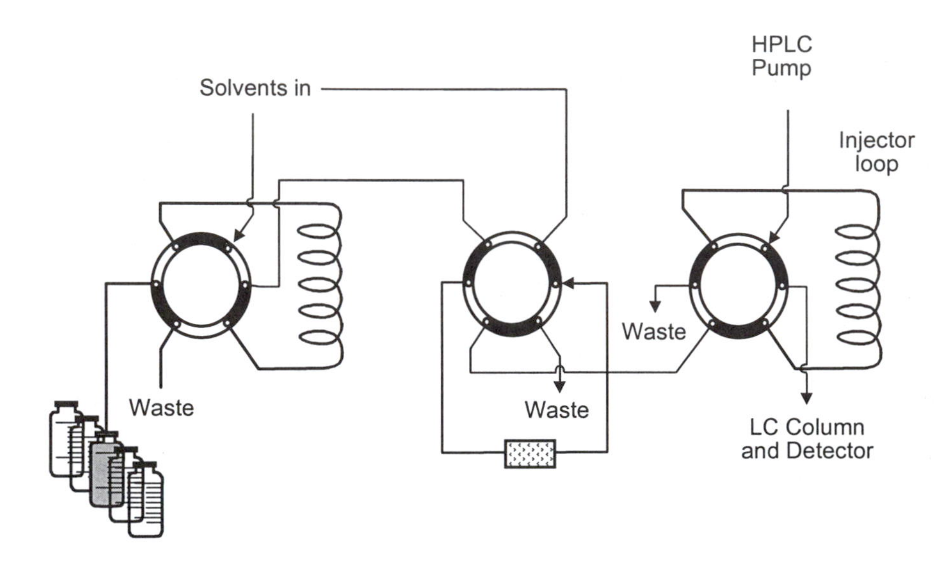

Figure 6. Plumbing diagram showing the use of a second switchable valve to perform an elution into a holding loop.

band broadening. If eluted in the opposite direction, the only part of the SPE device that can contribute to band broadening is that part on which the analytes retained.

Reverse elution on-line requires no more than a swapping of the positions of the HPLC pump and analytical column on the SPE control valve (refer to Figures 5 and 6; Figure 8 demonstrates back-flushing as part of a more complex valve-switching procedure).

In off-line solid-phase extraction, the SPE cartridge must be physically reversed (not always convenient if the cartridge has asymmetric connections such as male and female luer tips and sockets, or if the top frit is unsupported, such as in the regular syringe barrel SPE cartridge). Despite the benefits of low elution volume and band broadening, care must be taken to ensure that no particulates in the sample that deposited on the top of the SPE cartridge are fed onto the analytical column. Prefiltration of the sample using sub-micron filters is a wise precaution against this.

2. Holding Loop Elution

Such a technique is necessary if the elution solvent required to desorb the analytes from the SPE cartridge is stronger than the HPLC mobile phase. It can also help if the amount of the analyte loaded on to the HPLC column

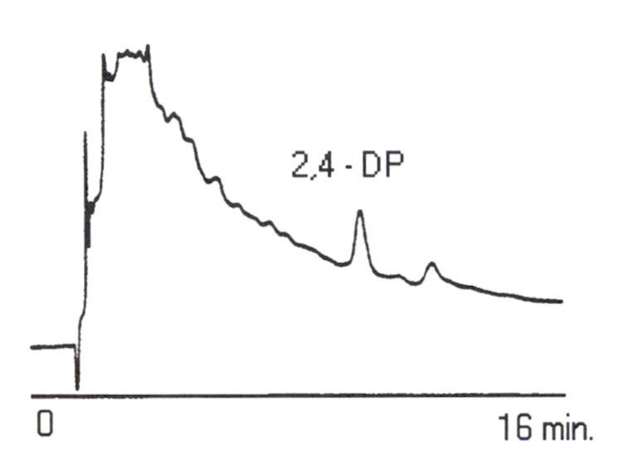

Figure 7. Chromatogram of a herbicide prepared by using a front cut of the elution bolus (Reproduced with permission from the (200 Prospekt brochure, Varian Inc., Palo Alto, California).

needs to be reduced (for that rare moment of joy when we have too much rather than too little signal!). The arrangement is shown in Figure 6.

3. Heart-cutting, Front, and Back-cutting

The idea of elution into a holding loop can be extended into another powerful technique, if we are able to control flow of the elution solvent (easy) and time the switching of the second valve (easy if the system is under computer control). With a set-up as shown in Figure 6, a fraction of the eluent can be gated out of the elution stream at the front, middle, or rear end of the elution solvent bolus and held in the loop on the second valve. The chief benefit of doing this is that the fraction of the eluent that contains the greatest concentration of the analyte may be applied to the analytical column. This is valuable if interferences that also extract onto the SPE sorbent can be encouraged to retain more or less strongly than the analytes during the elution step, thus improving the signal-to-noise ratio. The technique can also help improve signal-to-noise for the analyte peak by avoiding the dilution of the analyte into the entire volume of eluent applied to the SPE device.

An example of heart cutting is given in Figure 7 (Geerdink, 1992). In this work the elution front is routed through to the analytical system. This is the part of the eluent that was found to contain the analytes (bentazone and phenoxy acid herbicides), while more strongly retained species were found to require larger volumes of eluent to desorb and therefore did not appear in this fraction. Further manipulation of the eluent is possible; for example, the elution front may contain ionic species which could disturb the ionization process of the mass spectrometer detection system. If this was found to be a problem then the solvent front could be diverted to waste.

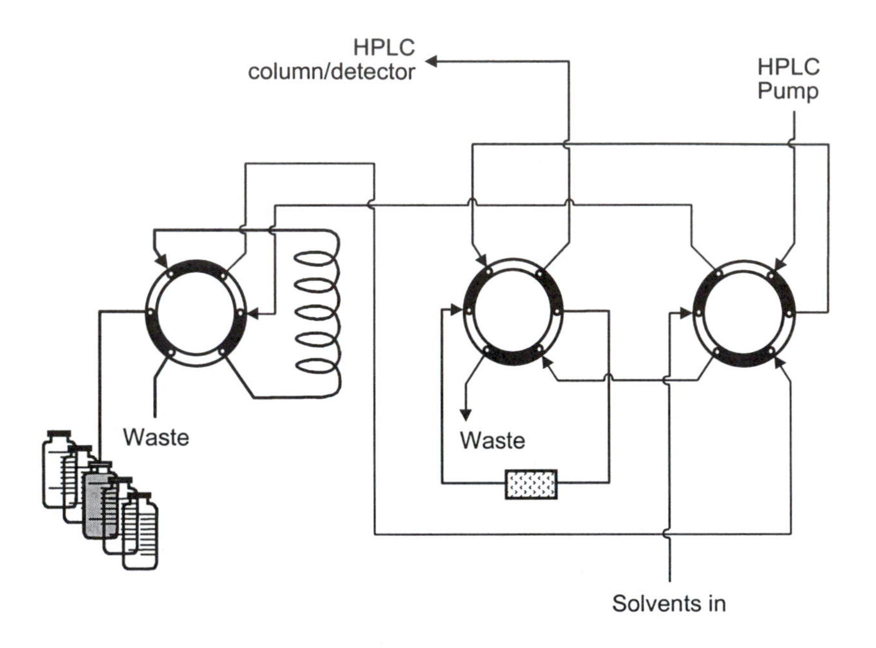

Figure 8. Plumbing diagram showing how the sample loop and SPE loading lines can be flushed during the elution process, ensuring complete transfer of non-polar analytes onto the HPLC column.

4. Sample Loop Flushing

One source of error during analysis results from loss of analyte on the walls of the extraction device and measuring equipment like pipette tips. This problem becomes acute when the analyte is a very lipopilic species. It becomes desirable in such cases to rinse the walls of the SPE device and to minimize the sample transfer steps and measuring steps in an analytical run. On-line SPE offers one way to achieve this in a fully automated way, as is illustrated in Figure 8, which has been used to analyze lipophilic indeno-indole anti-oxidants in dog plasma (Eriksson and Hedemo, 1994). Addition of a second column-switching valve allows the elution solvent to sweep the sample loading loop and the majority of the capillary tubing that takes the sample onto the SPE device. This set-up also reverse-elutes the SPE cartridge (highly desirable as the analytes, being strongly lipophilic, will be bound in a tight retention band on the head of the SPE device).

In this case the researchers not only demonstrated the value of this approach by achieving quantitative recovery and excellent precision over a

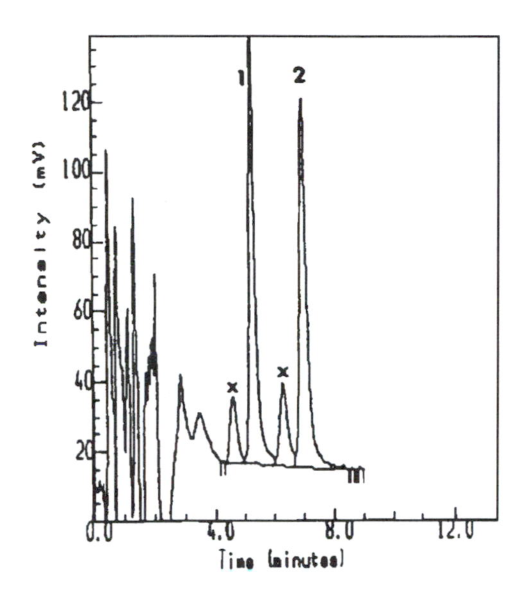

Figure 9. Chromatogram showing the effect of sample loop and sample path flushing (in this case one minute before elution of the SPE device) for two lipophilic species in a plasma sample.

large linear range. They also demonstrated, by switching the sample loop on-line, one minute before switching the SPE device on-line, that approximately 15% of the analyte was indeed retained in the sample loop rather than on the SPE sorbent bed. This is shown in Figure 9. The smaller peaks, eluting approximately one minute before the larger peaks, represent that fraction of the total sample load that failed to reach the SPE sorbent bed because of adherence to the tubing walls of the sample loop.

5. Focusing on the LC column

Several ways of improving the detectability by focusing analytes on the column head are possible. The simplest approach is to employ a more retentive HPLC column. However, band broadening can be kept small or reversed by other techniques, also. For example, if a high strength eluent is required to elute the analytes off the SPE cartridge then placing this in a holding loop until it is switched on-line with a weak mobile phase will also allow the analytical column to focus the analytes (Hartman et al., 1992). The chromatography will also benefit from use of smaller particles in the SPE column than the HPLC column or of an analytical column with larger dimensions than the SPE cartridge. Refer back to the earlier analysis of Section III for an analysis of these effects.

B. SPE FOLLOWED BY LC/MASS SPECTROSCOPIC ANALYSIS

The benefit of the MS detector lies in the high information content it produces. As delivery of information has increased through the introduction of techniques to produce molecular ions, post ionization fragmentation, through-space MS/MS and ion-trap time-controlled MS/MS or MS^3, the need for extensive clean-up to avoid contamination of the analyte signal by interferences has been reduced. What is often more critical when using these multiple mass-spectrum analyses is the need to avoid contaminating the mass spectrometer. The result has been a shift from separation and differentiation of small molecules at the sample preparation and analysis stage. The new goals are to strip out inorganic compounds that may affect the ion sources that induce ionization/fragmentation and eliminate large molecules that may foul up the interfaces between the sample introduction port and the mass spectrometer. Electrospray, thermospray, or particle beam instruments are all susceptible to overloading or clogging of the outlet of the MS chamber. We need to keep in mind this shift from conventional SPE, which is usually concerned with problems of clean-up of small molecules and concentration, when addressing this field.

An undesirable feature of atmospheric pressure ionization-MS analysts is suppression of ionization by co-extracted endogenous interferences from biofluids. To avoid false negatives, selective SPE extraction applications are required. SPE-LC/MS has been successfully demonstrated in many papers. For example, the analysis of compounds not known to be or even expected to be in a sample (one of the greatest benefits of the MS technique) has been demonstrated by Bagheri et al. (1993) and Creaser and Stygall (1993) for environmental water samples. One benefit of employing SPE for these analyses is the removal of humics and other species present. A second is the high level of concentration effected by the SPE step, permitting ultra-trace levels of pollutants to be identified positively.

1. Off-line SPE Combined with LC/MS

A key feature of the LC/MS for the analyst is the speed with which an analysis can be run. This speed derives from the ability of the MS to select out from a complex mixture only a single species and thus perform some of the functions traditionally taken by the analytical separation on, say, an HPLC column (by generation of molecular fragments, selection, and subjection of just one of these for further fragmentation). This ability to discriminate between many species in a sample and detect only a very select group of species has impacted the sample preparation step in two ways.

Firstly, the HPLC step is drastically shortened and in some cases omitted entirely. For example, Bowers et al. (1997) in the field of pharmaceutical development and Hogenboom et al. (1997) in the field of environmental monitoring, demonstrate that the HPLC separation can be eliminated entirely. The SPE cartridge alone is used to provide clean-up and solvent exchange with, perhaps, a little elution chromatography down its short length. A word of warning is appropriate at this point. When the LC/MS technology was first introduced some optimists assumed that they could eliminate the sample preparation step as well as the analytical step. The majority of early LC/MS users were researchers of pharmaceutical drug metabolism, pharmacokinetics and related disciplines. After discovering that they could indeed do this for a couple of analyses, these optimists usually encountered a rapid decline in signal intensity and quality of spectra. This was discovered to be due to components of the sample (often proteins or other macromolecules), which deposited in the transfer capillary/interface to the MS system. The apparent solution — simple protein precipitation with acetonitrile or another common solvent and application of the filtrate to the LC/MS system reduced the deleterious effect of the unobserved co-analytes. Indeed the popular 96 well plate format SPE devices have been adapted to remove precipitated proteins from biosamples that do not require SPE clean-up. However, in many cases it is still necessary to employ a typical full-scale LLE or SPE clean-up.

A second consequence of the LC/MS technology is that, with no chromatographic separation step the analysis times are much shorter — as little as one minute compared to a common range of 10 to 30 minutes for traditional HPLC/UV. This often means that classic cartridge solid-phase extractions, performed singly or in batches of 10 to 24 on vacuum manifolds, is the rate-limiting step in the total analysis. As a consequence the technology has evolved for both off-line and on-line SPE.

As instruments developed to improve sample throughput the users and manufacturers of high-throughput screening (HTS) devices standardized on a 96 well format in which collection tubes or wells in an 8 x 12 array are compactly held in a single, small footprint plate. Kaye et al. (1996) demonstrated a device in which the collection tubes were replaced by SPE cartridges, thus opening up the range of HTS robotics to the SPE stage and allowing the solid-phase extraction step to occur faster in a smaller space and with less need for human intervention. A significant trend towards this format has been observed in articles published in research journals and conference proceedings (for example, Allanson et al., 1996; Biddlecombe and Pleasance, 1996; Simpson et al., 1998) although others have shown that parallel extraction systems can achieve equally high throughput using standard SPE devices and robots (Getek et al., 1996).

2. On-Line SPE combined with LC/MS

When the SPE cartridge begins to act as a substitute for the analytical column, as it often does in the analysis of small molecules in clinical and preclinical samples during a drug's approval process, we should expect to see an increase in the use of small particle size sorbents in the SPE cartridge. In this way some separation can be achieved during the elution step and the analyte band will be tighter. The ability of the sorbent bed to filter out proteins before they elute into the MS also increases as the particle size decreases.

Fluid handling is simple: The basic plumbing diagrams shown in Figures 5 and 6 can be used to set up an on-line SPE LC/MS apparatus. The on-line approach will never be able to achieve the speed of off-line 96 well SPE plate sample preparation. However, it can still drive the LC/MS system at 100% of its capacity if the SPE extraction time is shorter than the analysis time, since one sample can be prepared at the same time as the previous one is analyzed.

The sensitivity to quenching of the ion source or other disruption of the MS fragmentation/ionization process means that it is important to eliminate proteins during the SPE stage, through the use of a buffer such as ammonium acetate. Furthermore, by washing the cartridge with pure water excess ions will be eliminated and this will, in turn, improve system performance. Elution using a pure organic solvent, without modifiers or buffer ions is desirable, just as it was for off-line LC/MS sample preparation. Consequently, polymers or sorbents which eliminate the need for modifiers or buffers (to disrupt secondary interactions) during elution are commonly encountered.

Two major claims can be made for on-line SPE/LC/MS or SPE/MS; high throughput and excellent sensitivity. Thus Beaudry et al., (1998) claim a sample throughput of between 320 and 960 samples per day for a broad-range pharmaceutical screen in human plasma (despite using a short HPLC column as well as the SPE cartridge). Marchese et al., (1998) claim a similar throughput at 5 minutes total preparation and analysis time per sample. Such a sample throughput is certainly in the same range as for off-line 96 well plate SPE. This illustrates the point that there is no "best" way to perform a sample preparation for LC/MS analysis. The choice made by the analyst for on-line or off-line SPE will be dictated by many factors, but raw speed of either technique need not be a deciding factor in the final choice.

The sensitivity issue is addressed by Bowers et al. (1997). Using an ion-spray MS/MS system linked to an auto-sampler and on-line SPE robot the authors were able to develop assays with cycle times of 5 to 7 minutes per sample, which yielded sensitivities of 50 pg/mL for sample sizes of only 200 μL. Such sensitivities are of great importance in the first human-subject trials for pharmaceuticals undergoing safety testing as part of the new drug approval process, where such factors as pharmacokinetics and toxicity need to be tested using the lowest possible drug dosage. This work was performed on 30 μm sorbent packed in a 2 mm I.D. SPE cartridge. As discussed earlier, the use of smaller particle size SPE sorbents and narrow-bore SPE devices would allow even greater sensitivity (Ooms et al., 1997).

V. CONCLUSIONS

What has been presented in this chapter is a survey of the approaches to connecting SPE to HPLC or LC/MS interfaces, off-line or on-line. One chapter cannot adequately cover all the techniques that have been used in this field. The reader is urged to consult the reference lists and application bibliographies provided by SPE vendors and vendors of SPE instrumentation (see Appendix 4). Specialist reviews also exist, including one that covers on-line SPE for environmental analysis (Barcelo and Hennion, 1995), and a more recent one on general applications for on-line SPE/HPLC (Hennion, 1999).

The major differences between off-line and on-line approaches can be summarized in two words: simplicity and flexibility. Off-line SPE requires little equipment though throughput will be influenced by automation. On-line SPE requires equipment and programming skills, but offers opportunities to squeeze more power out of the sample preparation step. Some of the imaginative applications using column-switching on-line SPE have been demonstrated here, but the opportunities for clever solutions to analytical challenges (such as extreme lipophilicity of the analytes) increase with the complexity of the fluid handling.

Most of the applications described in this chapter have been based on high-pressure on-line SPE. Some misuse of the term "on-line" has been noted in the literature by the author. Many automated applications provide off-line SPE followed by elution into a sample collection tube or holding loop which is then switched on-line to the analytical column. A better description of such applications is "at-line" since at no time does the SPE device become part of the high-pressure separating system. However, "at-line" implies that the automated SPE equipment is physically connected to the HPLC system. In most other respects, at-line SPE permits the same tricks, such as heart, front or back-cutting, and requires the same skills as are used in true on-line SPE. This is important because, as the trend to shorter, faster HPLC columns continues, the difference between the sample preparation step and the separation step in an analysis will be defined increasingly by the conditions under which the sample preparation device is used and how the extract is introduced to the detector. And, as should be clear from the applications reviewed in this chapter, the researcher will continue to find creative new solutions to analytical problems by exploiting the opportunities that exist between extraction of the sample and detection of the analyte.

Many thanks to the technical staff at Spark Holland, who assisted with advice and references for this chapter.

REFERENCES

Allanson, J.P., Biddlecombe, R.A., Jones, A.E. and Pleasance, S., (1996) The Use of Automated Solid Phase Extraction in the '96 Well' Format for High Throughput Bioanalysis Using Liquid Chromatography Couple to Tandem Mass Spectroscopy. Rapid Commun.In Mass Spec. 10: 811-816.

Bagheri, H., Brouwer, E.R., Ghijsen, R.T. and Brinkman, U.A.Th., (1993) On-Line Low-Level Screening of Polar Pesticides in Drinking and Surface Waters by Liquid Chromatography-Thermospray Mass Spectrometry. J.Chromatography 647:121-129.

Barcelo, D. and Hennion, M-C., (1995) On-line Sample Handling Strategies for the Trace-level Determination of Pesticides and their Degredation Products in Environmental Waters. Anal.Chim.Acta. 318:1-41.

Beaudry, F., Le Blanc, J.C.Y., Coutu, M. and Brown, N.K., (1998) In-vivo Pharmacokinetic Screening in Cassettes Dosing Experiments: The Use of On-line Prospekt-LC-APCI/MS/MS Technology in Drug Discovery. Presentation at the American Society for Mass Spectrometry, June 1998, Orlando FL.

Biddlecombe, R.A. and Pleasance, S., (1996) Microlute, A Semi-Automated Solid Phase Extraction System in the 96 Well Format. International Symposium on Laboratory Automatino and Robotics, ISLAR '96 Proceedings (Publ. Zymark Inc, Hopkinton, Massachussetts): 445-454.

Borgerding, A.J. and Hites, R.A., (1992) Quantitative Analysis of Alkylbenzenesulfonate Surfactants Using Continuous-Flow Fast Atom Bombardment Spectrometry. Anal.Chem., 64(13); 1449-1454.

Bowers, G.D., Clegg, C.P., Hughes, A.J., Harker, A.J. and Lambert, S., (1997) Automated SPE and Tandem MS without HPLC Columns for Quantifying Drugs at the Picogram Lavel. LC.GC 15(1): 48-53.

Brinkman, U.A.Th., (1995) On-line Monitoring of Aquatic Samples. Environ.Sci. and Technol. 29(2):40-51.

Creaser, C.S. and Stygall, J.W., (1993) Particle Baem Liquid Chromatography-Mass Spectrometry: Instrumentation and Applications. A Review. Analyst, 118:1467.

Eriksson, B-M. and Hedemo, M., (1994) On-line Solid Phase Extraction for Liquid Cromatographic Determination of Lipophilic Anti-oxidants in Blood Plasma. J.Chromatogr. A. 661: 153-159.

Geerdink, R., (1992) Bentazone and Phenoxyacid Herbicides in Watersamples. Spark Holland Application Note #44, Institute for Inland Water Management The Netherlands,

Getek, T., Mei, W. and Harihan, S., (1996). Enhanced Productivity for Bioanalytical Assays: Utilization of Automated SPE Isolation and LC/MS/MS for Steroid Analysis. International Symposium on Laboratory Automatino and Robotics, ISLAR '96 Proceedings (Publ. Zymark Inc, Hopkinton, Massachussetts): 418-428.

Guenu, S. and Hennion, M.-C., (1994) On-line Sample Handling of Water-soluble Organic Pollutants in Aqueous Samples using Porous Graphitic Carbon. J.Chromatogr. A. 665: 243-251.

Guenu, S. and Hennion, M.-C., (1996) Evaluation of New Polymeric Sorbents with High Specific Surface Areas Using an On-line Solid-phase Extraction – Liquid

Chromatographic System for the Trace-level Determination of Polar Pesticides. J.Chromatogr. A. 737: 15-24.
Halmingh, O., van Gils, G.J.M., Ooms, J.A., Wolthers, B.G., Breukelman, H. and Meiborg, G., (1998) Validation of an Automated Assay for Vitamin A and E in Serum by On-line SPE-HPLC. Paper 1661P, Book of Abstracts, The Pittsburgh Conference, New Orleans, LA.
Hartman, H., Flentge, E. and Ettema, M., (1992) On-line SPE and HPLC Analysis of PAHs in Water Samples. Laboratory for Environmental Analysis "de Punt", Glimmen, The Netherlands.
Hennion, M.-C., (1999) Solid-Phase Extraction: Method Development, Sorbents, and Coupling with Liquid Chromatography. J.Chromatogr. A. 856: 3-54.
Hogenboom, A.C., Speksnijder, P., Vreeken, R.J., Niessen, W.M.A. and Brinkman, U.A.Th., (1997) Rapid Target Analysis of Micro-Contaminants in Water by On-Line Single-Short-Column Liquid Chromatography Combined with Atmospheric Pressure Chemical Ionization Tandem Mass Spectrometry. J.Chromatogr. A. Symposium Volumes.
Huen, J.M., Gillard, R., Mayer, A.G., Baltensperger, B. and Kern, H., (1994) Automatic Measurement of Pesticides in Drinking Water Using On-Line Solid-Phase Extraction, HPLC and Chemometry. Fresenius J.Anal.Chem. 348: 606-614.
Kaye, Herron, W.J., Macrae, P.V., Robinson, S., Stopher, D.A., Venn, R.F. and Wild, W., (1996) Rapid Solid Phase Extraction Technique for the High-Throughput Assay of Darifenacin in Human Plasma. Anal.Chem., 68: 1658-1660.
Kwakman, P.J.M., Vreuls, J.J., Brinkman, U.A.Th. and Ghijsen, R.T., (1992) Determination of Organophosphorus Pesticides in Aqueous Samples by On-Line Membrane Disk Extraction and Capillary Gas Chromatography. Chromatographia 34(1/2): 41-47.
Landrieux, T., Lange, F.Th., Wenz, M., Brauch, H-J., Halmingh, O. and Ooms, J.A., (1998) Improved Automated Trace Level Determination of Aromatic Sulfonates in Water by Hyphenated SPE-HPLC. Book of Abstracts, HPLC '98, Birmingham, UK.
Martin, A.J.P., Halasz, I., Engelhardt, H. and Sewell, P., (1979) Flip-Flop Chromatography. J.Chromatogr. 186:15-24.
Marchese, A., McHugh, C., Kehler, J. and Bi. H., (1998) Determination of Pranlukast and its Metabolites in Human Plasma by LC/MS/MS with Prospekt™ On-line Solid-Phase Extraction. J.Mass Spectrom. 33: 1071-1079.
Nielen, M.W.F., Volk, A.J., Frei, R.W. and Brinkman, U.A.Th., (1987) Aully Automated Sample Handling System for Liquid Chromatography Based on Pre-column Technology and Automated Cartridge Exchange. J.Chromatogr. 393: 69-83.
Ooms, J.A., Haak, G.S.J., and Halmingh, O., (1997) High sample throughput for LC-MS/MS by automated on-line SPE-MS/MS, International Laboratory News (December): 17-18.
Pastoris, A., Carutti, L., Sacco, R., Da Vecchi, L. and Shaffi, A., (1995) Automated Analysis of Urinary Catecholamines by High-performance Liquid Chromatography and On-line Sample Pretreatment. J.Chromatogr. B. 664: 287-293.
Pichon, V., Cau Dit Coumes, C., Chen, L., Guenu, S. and Hennion, M.-C., (1996) Simple Removal of Humic and Fulvic Acid Interferences Using Polymeric Sorbents for the Simultaneous Solid-phase Extraction of Polar Acidic, Neutral and Basic Pesticides. J.Chromatogr. A. 737: 25-33P

Pichon, V. and Hannion, M.-C., (1994) Determination of Pesiticdes in Environmental Waters by Automated On-Line Trace-Enrichment and Liquid Chromatography. J.Chromatogr. A. 665: 269-281.

Roos, C. and Ögren, L., (1989) Quantitative Determination of Lidocaine in Human Plasma Using Solid Phase Extraction and Glass Capillary Gas Chromatography with Nitrogen-Phosphorus Detection.

Simpson, H., Berthemy, A., Buhrman, D., Burton, R., Newton, J., Kealy, M., Well, D. and Wu, D., (1998) High Throughput Liquid Chromatographic/Mass Spectroscopic Bioanalysis Using 96-Well Disk Solid phase Extraction Plate for the Sample Preparation. Rapid Commun.In Mass Spec. 12: 75-82.

Yago, L., (1985) Automated Sample Preparation Using Sorbent Extraction. Int.Lab. Nov/Dec. 85:40-51.

16

INTEGRATION OF SPE WITH THE ANALYTICAL TECHNIQUE, PART II

Nigel Simpson, Varian Sample Preparation Products, Harbor City, California

I. INTRODUCTION

Thus far in this book the analysis of samples following SPE extraction has been ignored or barely discussed. Obviously, however, the sample preparation and separation steps are pointless unless we can identify and quantify the extracted species. A common problem is that the detectors that are most information-rich (giving unambiguous assignment of a detected molecule) are often not the most sensitive — for example, electron capture detection is much more sensitive than GC/MS, but yields little information about the detected molecule except its electronegativity. The concentration factor potential of SPE should therefore lend itself to this. However, other features of SPE such as its ability, when used correctly to do things such as deproteinize samples, desalt them, or remove the entire sample matrix and replace it with a volatile solvent, also make it a desirable "front end" to an analysis. We address these issues in this chapter.

In this chapter we shall also explore how SPE may be coupled with the analytical system to perform sample preparation and analysis without the need for operator intervention. What is presented here are just some of the many examples where creative coupling of solid-phase extraction and analytical techniques such as GC and SPE or the combining of supercritical fluid chromatography (SFE) with SPE. The results of this research have led to expansion of the utility of both techniques, dramatically improved overall analysis time, detection limit, or some other analysis parameter.

Such techniques will be referred to as “hyphenated SPE” because the acronyms are usually hyphenated; thus, SPE-GC, SFE-SPE, and so on. In some cases, clever use of SPE has eliminated the need for a sample preparation technique that would otherwise have to be performed. In addition, examples will also be shown where, through the coupling of SPE with highly information-rich detectors such as MS, separation by chromatography prior to detection is rendered unnecessary.

A word of caution — ability to separate a molecule from a mass of other molecules and detect it via mass spectroscopy is a desirable goal. But great care must be taken to measure the effect of other species that are allowed to enter the spectrometer, since these species may not be observable directly or may be screened out of the window of detection by selective ion monitoring or some other descriminating process. These same species may affect the ionization process for the compound of interest or may have a dileterious effect upon the operation of the spectrometer over a period of time. In many conversations with users of LC/MS or GC/MS equipment the author has heard stories of attempts to eliminate the sample preparation or separation step of an assay, which began to fail when converted from a research laboratory into a high-throughput production lab.

Finally, this chapter will discuss the coupling of SPE to a variety of analytical instruments but not to its most natural partner, HPLC. This topic is covered in Chapter 15, since it represents the largest body of research and applications in the field of hyphenated SPE.

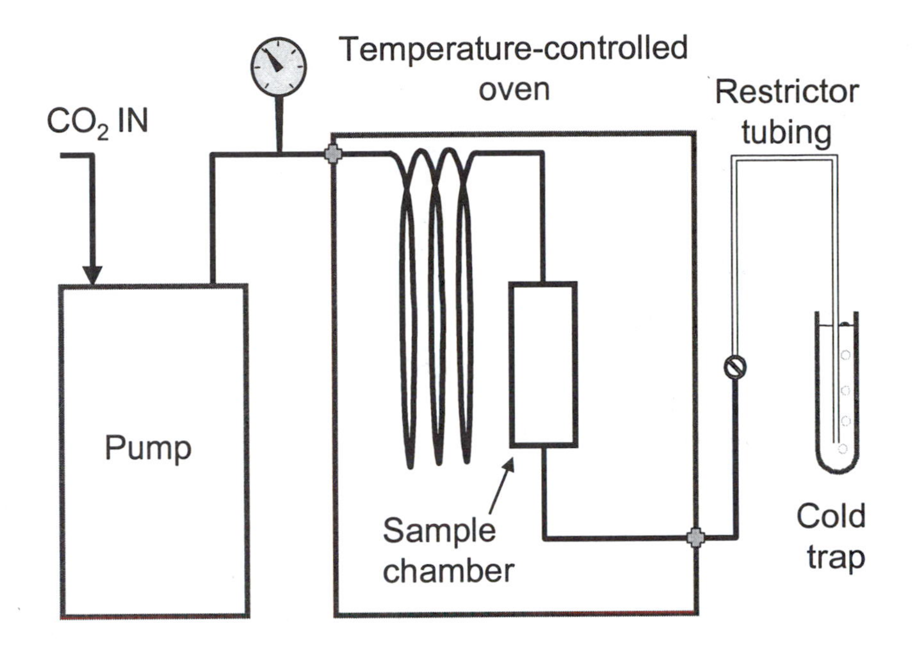

Figure 1. A standard SFE system in schematic. The sample for such a system may be a granular solid or a solid/liquid dispersed within another solid, e.g. blended in a slurry with diatomaceous earth.

II. ALTERNATIVE DESORPTION/TRAPPING STRATEGIES

A. USING SPE AS A TRAP FOR SUPERCRITICAL FLUID EXTRACTION

Supercritical Fluid Extraction (SFE) has received much attention in recent years for its apparent potential to eliminate the use of harmful solvents in the sample extraction step. It is also a readily automatable technique. Assuming that no "modifiers" (typically organic solvents of intermediate to high polarity, introduced into the supercritical fluid stream to enhance extraction) are required, then the sole waste product from the SFE process is carbon dioxide gas. Carbon dioxide is fed through a pump and oven to achieve a supercritical state. It then passes through the sample, leaching the analytes out with high efficiency, due to the supercritical stream possessing

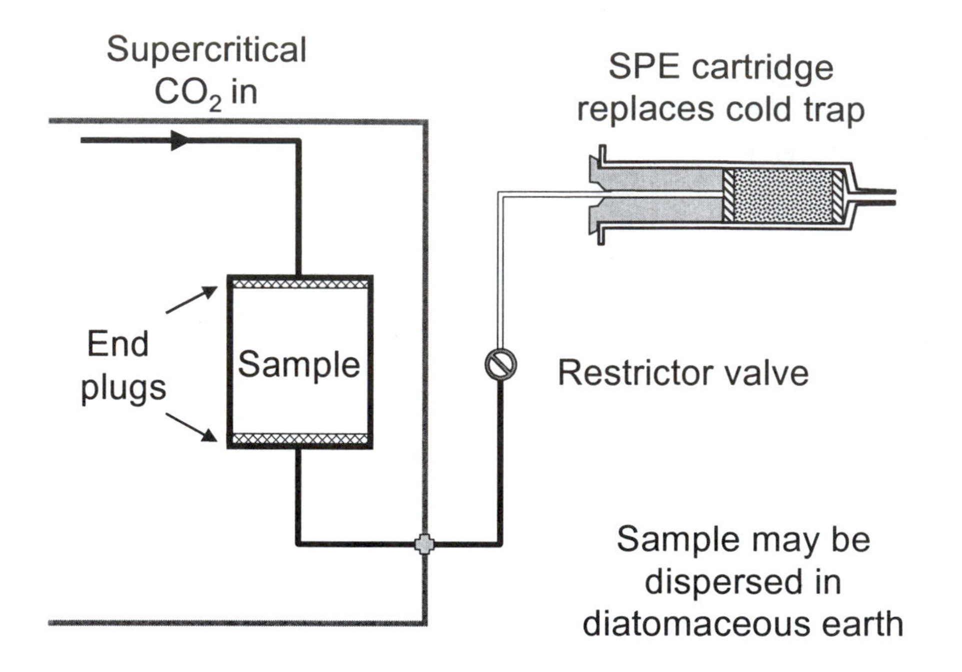

Figure 2. A typical set-up for supercritical fluid extraction/solid-phase extraction, showing how the SPE cartridge, placed after the restrictor valve, can replace a cold trap.

both high penetrative ability (like a gas) and the solubilizing capabilities of a fluid.

In order for SFE to be used for trace analytical work, it is necessary to perform a trapping step to catch analytes as they exit the sample fluid path. A common technique involves a cold trap to condense the analytes. The trap can then be transferred to an off-line sample preparation station for further work-up. SPE may be used at this stage. However, the sample handling steps inherent in off-line preparation reduce the potential benefits of the SFE process. Further, analytes may be lost during this handling process, solvents used may not be as benign as CO_2 and analytes that did not make the journey to the trap will be lost (it is not always easy to purge the fluid line properly when using a cold trap). Lastly, a cold trap is not always an ideal device to collect analytes because the stream of carbon dioxide that continues to pass through the trap will continually sparge the trap fluid that is supposed to be collecting the analytes. Hence, analytes may be swept out of the trap as the extraction proceeds.

A solution to these problems is to replace the cold trap with a SPE tube. Appropriate selection of sorbent type and bed mass will help ensure quantitative trapping of the analytes while minimizing the solvent needed to elute the SPE cartridge once the extraction is complete. A significant benefit arising from the use of SPE trapping is that a variety of solvents may be used to desorb analytes, eliminating the need for further clean-up. Comparable techniques for analysis of solids, such as Soxhlet extraction, can be very effective at exhaustively extracting a sample. However, because of restrictions to the solvent types that can be used for Soxhlet work and the conditions used, much undesirable material is co-extracted from the sample. The resulting extract usually requires additional clean-up such as Gel Permeation Chromatography or silica gel column chromatography.

Examples of this technique have demonstrated quantitative extraction of mutagenic species from food samples (Kaziunas, 1997), in which a silica SPE cartridge is used to trap a range of 10 moderately polar nitrosamines present at low ppb levels in food. Elution from the silica cartridge using pentane/methylene chloride yielded near-quantitative recoveries and good precision.

A survey, including a description of the use of SPE sorbents in SFE trapping experiments, has been provided by Levy et al. (1998).

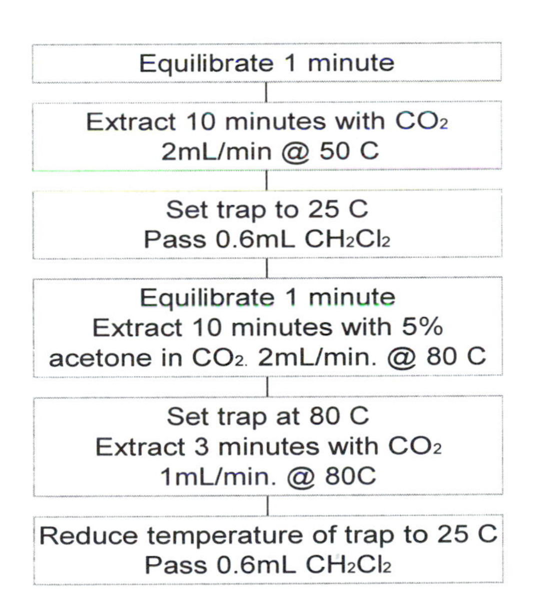

Figure 3. An illustrative SFE/SPE extraction process (Details taken from Messer et al., 1996). In this case the SPE device has been used to trap analytes from the matrix (a large volume water sample) and SFE has been used as a "solvent-free" method for desorbing or eluting the analytes.

B. USING SFE/SFC TO ELUTE FROM A SOLID-PHASE EXTRACTION CARTRIDGE

Many more examples exist of attempts to elute an SPE cartridge or disc using a supercritical fluid stream. Ezzell and Richter (1992) and Alzaga and Bayona (1993) demonstrate the technique. Wolfe et al., (1995) provide data showing that SFE elution of SPE devices can provide better elution of analytes than is given by typical liquid eluents. These applications typically use a standard nonpolar SPE cartridge to trap semivolatile compounds from water. The use of SPE is necessitated by the difficulty of performing SFE on a water sample. While it is not impossible — Hendrick and Taylor (1990) extracted phenols from a small volume (< 10 mL) for example — it is not easy and the difficulties grow as the sample volume increases. Practical constraints are imposed by several factors, including dissolution of CO_2 in the sample, which changes sample pH, and which simultaneously increases usage of CO_2.

In addition the water sample dissolves/volatilizes into the CO_2 stream, despite the low solubility of water in CO_2. This low solubility, combined with hydrogen-bonding between matrix and analyte results in poor extraction kinetics, which in turn limits the practical size of sample which may be extracted. Finally, the "mechanical mobility" of the sample often results in "splash-over" into the trap.

Replacement of the water sample by an SPE device onto which the analytes have been adsorbed resolves all these problems. The resulting SPE/SFC desorption has two advantages over the regular SPE desorption step: no organic solvent is needed and there is no subsequent solvent reduction step. Tang et al. (1993) demonstrated such a set-up in the extraction of a number of semi-volatile pollutants, using an extraction disc or a cartridge as the means of extracting analytes from a large volume water sample. Placing this disk into the path of the supercritical CO_2 stream completed the analysis. Key variables identified in this experiment were the temperature of the CO_2, the extraction mode (a non-polar mechanism was used in this case) and the presence of a modifier in the supercritical CO_2.

Messer et al. (1996) applied this idea to the analysis of 39 EPA method 525.1 analytes (a suite of semivolatile pollutants, specified to be analyzed by GC/MS) and 21 other semivolatile species. They used the same extraction scheme, which uses a C18 extraction disc, as is specified in the original EPA method. After extraction, the disc was dried thoroughly before removal from the SPE extraction manifold and insertion into the SFE chamber. The schematic for this experiment is shown in Figure 3. Elution with pure CO_2 gave reasonable recovery of the more volatile compounds such as the smaller condensed ring polyaromatic molecules. A second elution using 5% acetone to increase polarity of the desorbing supercritical fluid permitted recovery of the less volatile compounds. A final purge of the system with pure CO_2 was needed to remove the acetone modifier from the extractor set-up. Desorption was still not complete for all analytes and a further 0.6 mL of methylene chloride was used to rinse the disc after the first and third elutions. Problems of erratic recoveries for the less volatile compounds resulted in the need for longer desorption times for the second elution (23 minutes) and the consequent lengthening of the extraction procedure. The effect of a higher concentration of acetone (8%) was investigated but, despite a shorter eution time, the total volume of acetone used to achieve comparable results was greater.

Despite these method optimizations, the least volatile polyaromatic compounds still recovered poorly. A final method exploited a higher extraction chamber temperature of 110°C at 5500 psi, and used a ten minute first elution with the trap set at -20°C. This was followed by application of 0.6 mL of methylene chloride with the trap set at 25°C, a 23 minute second elution of 8% acetone in CO_2 at a flow rate of 2 mL/minute with the trap set at 80°C. A final purge was conducted with pure CO_2 for 2 minutes at 4 mL/minute with the trap at 80°C. Lowering of the trap temperature to 25°C for the final methylene chloride rinse completed the extraction. This final method gave a recovery of > 80% and relative standard deviations of less than 10% for a range of phthalates, PAHs, chlorinated compounds, and Atrazine and Simazine. However, the organic solvent consumption was higher than the original goal and the final method was not fast. Howard and Taylor (1992) also demonstrate extraction of the herbicides, sulfometuron methyl and chlorsulfuron, either onto diatomaceous earth or onto an extraction disc, followed by SFE desorption. In this case a supercritical stream modified by addition of 2% methanol was used to desorb the analytes. The modifier is essential because of the compounds' polarities and hence their lack of solubility in pure CO_2.

An observation from this study is that celite requires a shorter equilibration time than SPE sorbents (5 minutes versus 24 minutes). However, because the mechanism for celite extraction is a solid-supported liquid/liquid extraction rather than a true solid-phase extraction, this medium cannot provide preconcentration of analytes at trace levels, while the SPE sorbent can. The relationships discovered in this work are shown in Figure 4. The conclusions reached were that the analytes needed to be neutral (pH 3) for good SPE retention but recoveries were not quantitative at this pH because of secondary interactions (SiOH/analyte hydrogen bonding) which could not be disrupted by the presence of only 2% methanol in the desorbing fluid. This idea was tested by adjusting the pH of the SPE sorbent to 9 just before the SFE step. Recovery was observed to increase from approximately 80% to 95%.

Polar SPE has been applied to the trapping of pesticides in a SFE stream from a food sample of variable water and fat content (Hopper and King, 1991). The wide variety of sample composition would have resulted in variable results and extract contamination. To counteract the effect of variable water content in the sample the authors used diatomaceous earth to dehydrate and stabilize the sample during the SFE process. A polar cartridge was then used to trap the analytes and circumvent the problem of variable fat content. The florisil cartridge could then be eluted with acetone to collect the analytes.

Fractionation of analytes off an SPE device by selection of appropriate supercritical fluid parameters is given by Kane et al., (1993). In this work

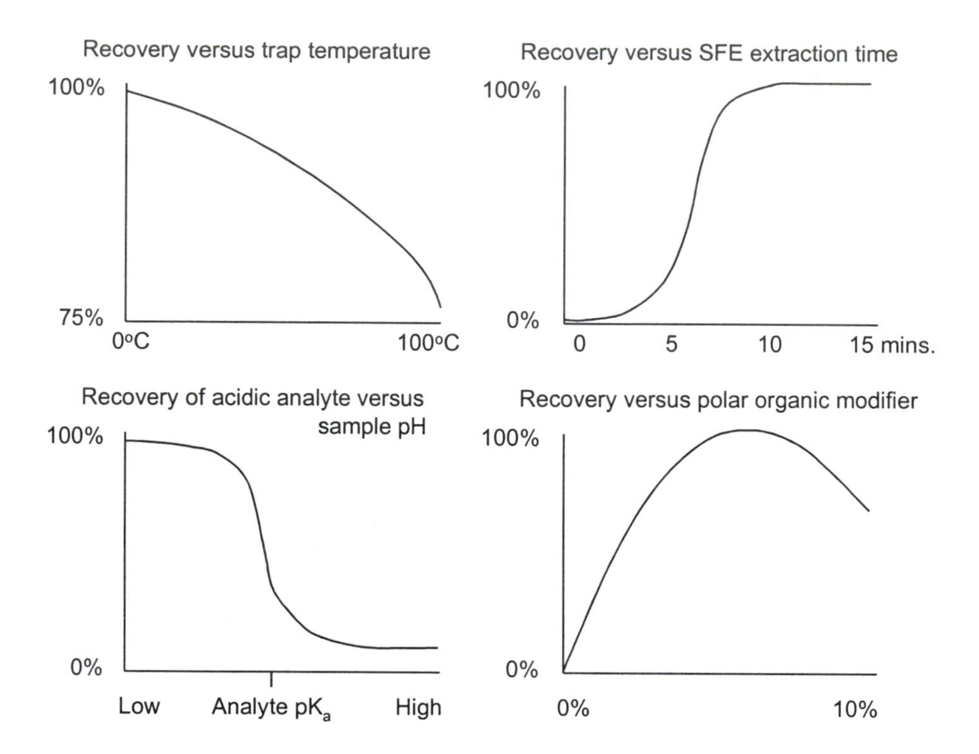

Figure 4. General trends observed in the results of research into SFE desorption and total recovery of analytes (for example, from Ho and Budde, 1994; Howard and Taylor, 1992). The observed behavior can be rationalized using the same arguments of polarity, solubility, etc., that were used to explain simple SPE extractions.

the authors successfully demonstrated the separation of two non-ionic surfactants during SFE desorption from a C18 disc. For these thermally labile analytes GC analysis was not possible so direct analysis by supercritical fluid chromatography was an obvious choice for separation. In this work the "strength" of the elution solvent was shown to be simply a function of its density. Solubility of a compound in a supercritical fluid is strongly correlated with the fluid density — the density at which significant solubility appears is referred to as the threshold density. By determining the threshold densities for the two surfactant classes it was therefore possible to effect good separation at the elution, without resorting to supercritical fluid modifiers.

Use of reverse micelles to permit polar molecules to be solubilized in supercritical CO_2 has not, to the author's knowledge, been used as a separation/SPE desorption technique, though it has been used for liquid-liquid extraction (Cooper et al., 1997)

C. SPE AND MICROWAVE OR THERMAL DESORPTION TECHNIQUES

In the same way that SPE has allowed the trapping and concentration of analytes from a SFE stream, it has been shown to be effective for the trapping and concentration of the effluent from a microwave-extracted sample (Conte et al., 1996). It has also been used to clean up the extract resulting from a liquid-liquid extraction that has been assisted by microwaves. An example of this is given by Moye et al. (1998) who apply microwaves to a finely chopped vegetable matrix immersed in an aqueous acetone mixture and placed in a microwave extraction chamber. The sample is subjected to between six and twelve minutes heating up to 200°C and the liquid extract is then applied to SPE cartridge for the extraction and clean-up of fungicide residues before GC analysis.

Thermal desorption is an appropriate technique for the analysis of volatiles and headspace and purge-and-trap are commonly applied in this area to take the analytes from the sample through to the GC. When concentration is needed then an adsorbent trap can be added and while the nature of the adsorbent is not likely to be familiar to SPE users, the process of extracting the volatile species will be. A large volume of air is drawn through the cartridge by vacuum pump and the concentrated analytes are desorbed using a small volume of liquid. The traditional sorbents such as polyurethane foam require larger desorption volumes than small SPE cartridges and thus the familiar SPE tube is seeing increased use (Woolfendedn et al., 1993). Conceivably, the SPE process could be used to extract and concentrate metabolites from exhaled air just as they are routinely used in the extraction and concentration of metabolites in urine.

III. SPE AND SPECTROSCOPY

When SPE is combined with a spectroscopic detection technique that is "information-rich" or highly selective, such as MS/MS, FT-IR or NMR, the emphasis of the SPE step shifts from clean-up and concentration to matrix removal and interference separation. Apart from the combination of direct SPE/MS, which is considered in the next chapter, there are disappointingly few papers that demonstrate SPE clean-up followed by a spectroscopic analysis. Partly this would be due to the fact that the vast majority of organic spectroscopic techniques can yield unambiguous assignment of molecular species only if the sample is pure or if it consists of a limited range of analytes or known contaminants. It is nevertheless still possible to apply

SPE to clean up complex samples for spectroscopic analysis. Three illustrative examples are described.

A. ULTRA-VIOLET/VISIBLE SPECTRA OF SPE EXTRACTS

Ultra-violet (UV) spectra do not show enough fine structure to provide absolute certainty about the components of a sample. Moderate sensitivity and the changes in the spectrum, depending on the molecule's surroundings (solvent polarity can shift the UV spectrum absorbances to much higher or lower wavelengths) limits use of UV for quantitative analysis of non-trace level analytes. Instrumental advances and software techniques that enhance data gathering through deconvolution and Fourier techniques have reduced these limitations considerably. However, spectroscopic techniques do have the benefit in that they can be applied directly to an extract without the time-consuming chromatographic separation step.

Use of second derivative UV spectroscopy in application to a sample extract (Secchieri et al., 1991) has been demonstrated to be an effective and simple way to analyze pentachlorophenol in sample types such as leather or wood shavings. Application of the spectroscopic technique directly to the samples or their n-hexane Soxhlet extracts was not possible because of the diversity of co-extracted species. Application of a simple SPE extraction based on a cyanopropyl sorbent solved this problem. Clean-up occurs while the n-hexane extract is loaded on the SPE cartridge and the cartridge is subsequently washed with n-hexane and eluted with acetonitrile. A fivefold concentration was also produced during the extraction. The resulting spectra showed good consistency between differing sample types and a linear calibration curve from 0.5 to 20 μg/mL for pentachlorophenol was demonstrated.

UV-visible quantitation has been used in the previously referenced work by Kane et al., (1993) to perform the analysis of non-ionic surfactants in a supercritical fluid extract stream from a C18 SPE disc. In this case it was found that SFC, which was used to determine the optimum elution conditions on a qualitative level, was not a reliable technique for the measurement of the analytes. Instead, having demonstrated 100% separation of the two surfactant groups during the SFE elution stage, a simple UV/visible detector gave accurate analyte measurement.

SPE brings one very significant advantage to UV/visible analysis of compounds — its ability to concentrate. This has been used to advantage by Eastwood et al. (1994) who realized that the concentration effect would allow direct detection of the analyte on the sorbent surface. Disc SPE devices offer advantages over cartridges for this type of work since the sorbent surface can be directly irradiated without the need to disgorge the sorbent out of a container, to avoid the masking of signal by UV-active components in the SPE device housing. The authors demonstrate that analysis using front surface fluorescence detection permits the quantitation of a fuel oil and various individual poly-aromatic hydrocarbons on the disc surface, though they indicate that individual components in a mixture could not be well differentiated using this technique. The benefits of the resulting quick analysis and elimination of the elution step are considerable, especially when analysis must occur in the field at the point of sampling.

B. INFRA-RED SPECTROSCOPIC ANALYSIS OF SPE EXTRACTS

Infra-red spectroscopy (IR), just like UV, has found few uses in combination with SPE. One of its largest applications — continuous process monitoring — is incompatible with techniques, like SPE, that require discreet sampling. However, some applications do exist which show use of SPE either as a method of matrix removal or matrix simplification. For example, Barber (1985) demonstrated the effectiveness of SPE when applied to simplifying mixtures such as hand cleaners and other industrial products. A general feature of such samples is the high concentrations of species present. Therefore, unlike typical SPE extractions, small sample sizes are usually loaded on to the SPE cartridge and even then, the sample may still require surfactant precipitation to avoid overloading of the SPE sorbent. Analysis of a petrolatum-containing a hand cleaner by LC or GC is precluded by the non-volatile nature of the petrolatum components and their large size. Dehydration of the hand cleaner to eliminate volatiles, followed by hexane dissolution and fractionation on a silica SPE cartridge, yielded a hexane fraction, a methylene chloride fraction and a methanol fraction, each of which was pure enough to analyze separately by IR. According to the IR spectrum of each fraction, the hexane fraction contained the petrolatum components, isopropyl myristate eluted in the methylene chloride fraction and the methanol fraction contained pure cetyl alcohol.

Infra-red spectroscopy has been applied to the analysis of oil and grease or total petroleum hydrocarbons from wastewater. A typical technique (for example, ASTM method 413.2) required the use of Freon™[44] liquid-liquid extraction of a 1 L water sample acidified to pH 2. The extract was subjected to infra-red analysis and its absorbance monitored at 2930 cm^{-1} (a C-H stretching vibration for typical hydrocarbon C-H bonds), as an estimate of the total amount of "oil and grease" hydrocarbons extracted. This technique complemented the equally arbitrary gravimetric method (arbitrary because "oil and grease" is defined as "what a certain volume of Freon extracts from a certain volume of water under certain conditions of extraction

[44] Freon is a registered trademark of the DuPont Corporation, Midland, Michigan.

and analysis") but gave greater sensitivity for low-end samples. Wells et al., (1995) who used gravimetry, and Bui and Dirksen (1994) who used IR, showed that comparable results could be obtained between the Freon extraction and an extraction based on an SPE system developed exclusively for the purpose of oil and grease testing. These papers evaluate extraction of various oil and grease "standards" ("surrogates" would be a more accurate description). Provided the SPE apparatus was sufficiently clean, it was shown that sub-part-per-million detection levels could be achieved.

Direct analysis of surfaces is possible using diffuse reflectance infra-red Fourier transform spectroscopy (DRIFT). This tool may some day be sensitive and discriminating enough to permit direct detection of trapped analytes on the SPE sorbent itself. One drawback of IR is that it does not lend itself well to low-level detection of species in a complex aqueous environment. The idea of direct detection of adsorbed analytes is explored in the next section.

C. SPE COUPLED WITH RAMAN OR ATOMIC SPECTROSCOPY

Raman spectroscopy has not been very widely applied to chemical analysis compared to chromatographic techniques or rival spectroscopic tools. However, some work combining SPE and Raman spectrocopy (SPERS) has been attempted. The Raman technique relies upon a laser, operating usually in the IR region of the spectrum. Pumping laser radiation is absorbed by a species in the sample and thrown out again at a series of wavelengths that differ from the pumping laser, the differences revealing energy levels of rotation and vibration in the molecule, and hence structures. The differences, called Raman shifts, are typically in the range of 500 to 2000 cm^{-1}. Just as with IR spectroscopy, which also probes a molecule's vibrational motion as the molecule distorts in shape, Raman spectroscopy can provide positive identification of a compound. Comparison with a pure standard spectrum confirms the identity of the sample (though it is not an effective quantitative tool).

One study (Courbariaux et al., 1997) investigates a waste stream from a process that manufactures dyes. The authors had attempted earlier extraction/elution schemes using SPE to monitor this waste stream. However, they noted that elution from the SPE cartridge had proved problematic. The benefit of using Raman detection is that elution is not necessary because SI and water have low Raman intensities. Coloration of the dye molecules made optimization of the retention step simple. The sorbent particles were then emptied from the SPE cartridge and placed in a Raman spectrometer sample chamber (the actual instrument used was a Fourier transform Raman microscope). SPERS was used successfully in this study to identify a blue dye that had appeared in the effluent stream, followed by tracing of the source of that contamination.

The role of SPE to atomic spectroscopy sample preparation is largely one of concentration, though elimination of interfering metal ions is desirable where it may be easily accomplished. An example of trace enrichment is given by Sooksamiti et al., (1996) who use a complexing agent (a crown ether) to selectively extract Pb (II) from an aqueous sample containing other metal ions. Elution was achieved using an oxalate buffer.

While some complexing products have been developed by manufacturers and commercialized (for example, the Empore™ Rad disk[45]), most literature references describe the use of either complexation in the bulk sample before extraction (Frenzel, 1998), or immobilization of complexing agent on the surface of the SPE sorbent during the conditioning step. The latter often uses a generic sorbent like C18. In such cases retention is usually, though not always, obtained by non-polar mechanisms and hence elution off the SPE sorbent may occur with solvents like acetone. Van Elteren et al. (1990) demonstrate an approach for extraction of As (III) from aqueous samples in which two different schemes were developed. One utilizes a C18 cartridge with an ion pairing mechanism between cetyl trimethyl ammonium and pyrrolidene dithiocarbamate ions; the other employs a combination of SAX and C18, with the pyrrolidene dithiocarbamate species retaining on the SAX until it complexes with the As (III). Analysis of acetone eluents off the C18 sorbent in each scheme were performed by hydride generation Atomic Absorption spectroscopy. This paper presents studies of recovery against various experimental parameters like flow rate, loading of the complexing ion during conditioning, sample pH and presence of competing ions in the sample.

D. INTRODUCTION OF SPE EXTRACTS DIRECTLY INTO MASS SPECTROMETERS

Mass spectroscopy (MS) offers the advantage of selective detection, meaning that provided care is taken not to introduce species into the MS that could degrade its performance over time or influence the ionization processes taking place during each analysis, less emphasis on sample clean-up is required. While dry-down, reconstitution, and extract introduction to the MS is still very common, other introduction techniques such as thermal desorption or laser desorption from the sorbent are possible. Krier et al. (1994) demonstrate the use of laser desorption as an alternative to solvent desorption, by inserting a sorbent disc into the cell of a Fourier transform mass spectrometer. The experiment is shown to be successful for one of the

[45] Empore is a trademark of 3M Corporation, Minneapolis-St Paul, Minnesota.

analytes (hydroxyatrazine) though volatility issues (and hence evaporation of the analyte off the solid phase in a fashion that is not correlated with the desorbing laser pulse) reduces the application's effectiveness for a second analyte (atrazine).

Barshick and Buchanan (1994) use thermal desorption as a means of separating extracts of milk containing drug residues from the solid phase sorbents to which they retained. Using a specially built thermal desorption unit into which SPE cartridges or discs could be placed, the analytes are introduced into the MS via a capillary transfer restrictor. The bulk of the desorbed sample is vented via a split valve, from where it could be trapped for further analysis.

A closely-related detection system, employing ion mobility spectroscopy, has been demonstrated for the extraction of phthalate esters from water (Poziomek and Eicmann, 1992). In this case thermal desorption from the SPE device was employed to drive analytes off the sorbent and into the ionization zone. One interesting feature of this analysis was that the extraction of phthalates was performed by dipping an extraction disc fragment into the sample and allowing diffusion to bring the analytes into contact with the sorbent, rather than drawing the sample through.

Borgerding and Jites (1992) demonstrate an alternative technique for desorption and introduction into a MS. The authors were looking for a replacement for an old complexometric method. This older technique did not give specific compound information and suffered from cross-reactivity with naturally occurring anions. In addition the chromatographic methods often failed to differentiate non-linear homologues of these surfactants. Tandem mass spectrometry can resolve both concerns and an enhancement called continuous flow fast atom bombardment (FAB) is more sensitive than the static FAB technique according to sources quoted in this paper. However, concentration of the analytes would be required, if the technique is to be applicable to analysis at the levels found in typical environmental samples.

Actual river water samples were extracted onto a SAX cartridge, then the eluent, 2M HCL, was adjusted to pH of 6.0 with NaOH and this was re-extracted onto a C18 disk. This disk was eluted with MeOH. Eluents were dried down with dry N_2 to achieve analyte concentration (up to 4000-fold), depending on the initial sample volume. These extracts were filtered and injected into a MeOH/H_2O/Glycerol (10:9:1) carrier stream which fed through a 75 μm silica capillary into a triple quadrupole MS with a modified FAB ion source.

Water samples taken from sources other than rivers could be extracted simply by the latter C18 extraction method. The benefit of using FAB-MS with tandem MS for specificity was that a simple C18 SPE procedure could be used. The authors speculate on the need for the SAX extraction of river water samples and propose two likely reasons: low extraction efficiency from river water samples and suppression of surfactant ionization by co-extracted material from this matrix. The technique also appeared to be applicable to sludge samples.

E. SPE/NUCLEAR MAGNETIC RESONANCE

Nuclear magnetic resonance (NMR) has not been a mainstay of the analytical laboratory. Its low sensitivity and requirements for deuterated solvents to reconstitute the sample before introduction reduce its convenience. Furthermore, it is not a technique that differentiates well between species. Everything in the sample is detected at the same time. Hence, NMR is not a routine analytical finish to a SPE extraction. However, the property of non-discrimination between components has been used to advantage. For example, a GC analysis can yield a flat baseline for an extracted sample yet the sample may still foul the instrument. This is especially common for biological sample extracts where many of the endogenous components of those samples — proteins, citrate, hippurate, creatinine, pigments and so on — deposit on the injector or the first few centimeters of GC column.

Wilson and Nicholson (1987) used combined SPE/NMR to identify various drug metabolites including sulfates and glucuronides, but they also detected alpha-ketoglutarate, citrate, creatinine, dimethyl amine, and other endogenous materials. The total extraction properties of several non-polar sorbents and bed masses could thus be evaluated in a semi-quantitative way by using signal-to-noise ratios of resonance peaks against the baseline noise. The non-retained fractions could also be easily analyzed. One unexpected and important finding for the authors was that the commonly selected 500 mg C18 bed mass was far larger than needed for retention of the target compounds. The excess sorbent extracted undesirable components of the sample, such as citrate and creatinine.

A further benefit of the use of NMR to monitor this extraction, because of the structural data provided by NMR spectra, was the ease of identification. This enabled quick identification of a late-eluting peak in the chromatographed (using SFC) extract of these samples. With the knowledge of the structure of the interference the authors were able to eliminate it quickly and easily by applying a 20% methanol/80% water wash to the cartridge before elution.

A review of the use of NMR in combination with SPE is given by Wilson and Nicholson (1994).

IV. SPE AND GAS CHROMATOGRAPHY

A. ANALYTICAL REQUIREMENTS FOR DIRECTLY ANALYZABLE COMPOUNDS

The advantages of combining GC with SPE are primarily the high sensitivity and resolution permitted by the GC detector. A variety of detectors enable further analyte differentiation through discrimination within the detector between different species. These benefits are balanced by one major disadvantage — the sensitivity of the GC system to water — and by one minor one — the low volume of injection allowed by the modern GC instrument (though recent advances in injector technology have somewhat alleviated this problem). By applying the MS as the detector, new opportunities are opened up in trace level screening. The two constraints mentioned above make SPE a natural choice for sample preparation for GC/MS because SPE is, by its nature, a trace enrichment step and because it is a convenient way of achieving solvent exchange, stripping away the aqueous matrix as the SPE extraction proceeds.

B. TECHNIQUES FOR COMPOUNDS REQUIRING DERIVATIZATION

If the compounds of interest cannot be directly analyzed, then an additional step, usually involving derivatization, is required. This section explores two approaches to derivatization, which use, directly or indirectly, the nature of a SPE extraction to eliminate a time-consuming and often hazardous manual derivatization step.

1. On-Cartridge Derivatization

Solid-phase derivatization has been demonstrated in many systems where the analytes are adsorbed onto the surface of sorbent particles either suspended in a reaction tube (Rosenfeld et al., 1991) or packed in a pipette or syringe barrel container (Chatfield et al., 1995). During the derivatization the sorbent surface either plays the role of indirect catalyst by concentrating the reactants at its surface, or of direct catalyst by facilitating bond breaking and making. The resulting derivatized species can then be desorbed and analyzed.

Realizing that these techniques lent themselves to standard SPE procedures very well, some researchers deliberately forced the derivatization by subjecting the SPE cartridge, after sample loading, to heat in the presence of a suitable reagent/catalyst mixture. Thus, He (1995) first demonstrated that efficient extraction of a pair of herbicides could occur on a standard C18 SPE cartridge. He then followed this step with exhaustive drying of the sorbent bed (30 minutes of airflow was found to be sufficient) and added boron trifluoride in methanol to the sorbent bed. The cartridge was then capped at both ends and placed in a sand bath at 100°C for 20 minutes. Cooling, uncapping, and elution revealed a 98% conversion rate for the analyte 2,4-dichlorophenoxy acetic acid to the corresponding methyl ester.

This research actually follows on from earlier observation of the methylation occurring on an unendcapped mixed-mode sorbent (Westwood and Dumasia, 1994). The authors were unable to extract and detect the derivatized extract of fentanyl from a horse urine sample. However, they had determined that the analyte was not being lost at any stage of the procedure. They realized that a peak they had been observing by GC/MS was methylated fentanyl. The methylation in this case occurred simply under the conditions of the extraction (application of methanol as a wash solvent to an analyte retained upon an activated surface).

Deliberate on-column derivatization of drug residues has been demonstrated by Campins-Falco et al., (1996). In this work highly efficient naphthoquinone sulphonate derivatization of amphetamines adsorbed onto C18 is investigated. The authors also summarize other papers which apply on-column derivatization of amphetamines. Further examples of on-SPE cartridge derivatization are provided by the extraction and simultaneous derivatization of aldehydes and ketones during the extraction of air samples (Zhang et al., 1994; Druzik et al., 1990). The extraction cartridge is prepared by loading 2,4-DiNitro Phenyl Hydrazine (2,4-DNPH) onto a C18 substrate. On encountering airborne aldehydes and ketones, a condensation reaction occurs, yielding a hydrazone derivative which may be eluted and analyzed. The usefulness of this approach has resulted in a commercial product, the ExPosure™[46] cartridge, which was developed for this purpose. We can expect to see more such sorbents where solid-phase

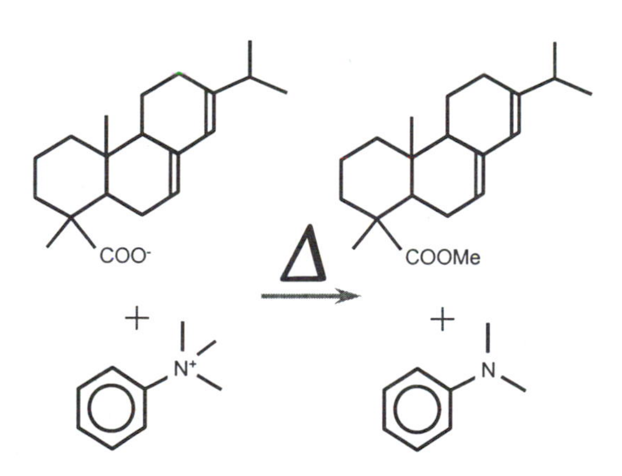

Figure 5. A reaction scheme that can occur under the conditions found in a hot GC injection port. This pyrolytic reaction occurs rapidly and completely, yielding products that can be easily chromatographed.

[46] ExPosure is a trademark of Waters Inc. Milford, MA.

reaction can eliminate "wet chemistry" preparation steps such as the methylation and condensation reactions described in this section.

2. On-GC Derivatization

The hot injection port of a GC offers many opportunities for pyrolysis reactions or other heat-assisted reactions. The difficulty lies in controlling these reactions and limiting the build-up of reaction and extraction residues that will ultimately impair chromatographic performance. An elegant example of this is given by Backa et al., (1989). In this case fatty acids discharged by paper mills were target analytes. The traditional approach involved large-volume liquid-liquid extraction with diethyl ether followed by diazomethane methylation.

It was realized that these hazardous and wasteful processes could be eliminated by the reaction scheme shown in Figure 5. This reaction scheme was known to occur at the temperatures encountered in a hot GC injection port. The tertiary amine, which is produced as a by-product, passes cleanly through the chromatograph without obscuring vital parts of the chromatogram. This technique is referred to as flash-heater esterification or ion-pair derivatization. However, if excess amine salt is used, the residue would very quickly impair performance by fouling up the injection sleeve. If too little is added then the derivatization and hence quantitation would not be accurate.

This dilemma was elegantly resolved by the use of an extraction that amounted to ion-pairing on the SPE device. The sample was loaded at a high pH onto a C18 cartridge. The trimethylphenyl ammonium salt was next loaded under conditions that ensured that it only retained when it paired with a carboxylate ion retained from the previous passage of the sample through the cartridge (the fatty acid analytes were all in their anionic form). The resulting ion-pair could then be eluted with ethanol (other analytes such as terpenoids were eluted with an aliquot of methylene chloride and the fractions were later combined) and the resulting solution with the required 1:1 correspondence of ammonium and carboxylate ions could then be injected into the GC. A similar application, employing tetrabutylammonium ion pariring with alkyl sulfonate surfactants extracted onto C18, has been demonstrated by Field et al. (1994).

C. OFF-LINE SPE/GAS CHROMATOGRAPHY

The recent introduction of large volume injection systems has helped the interfacing of SPE to GC off-line. These injectors free the elution step of the SPE extraction from the rigorous constraints of volume and dryness. Use of retention gaps also help, as do temperature-programmable injectors. Ollers et al. (1997) demonstrated a typical procedure for analysis of caffeine

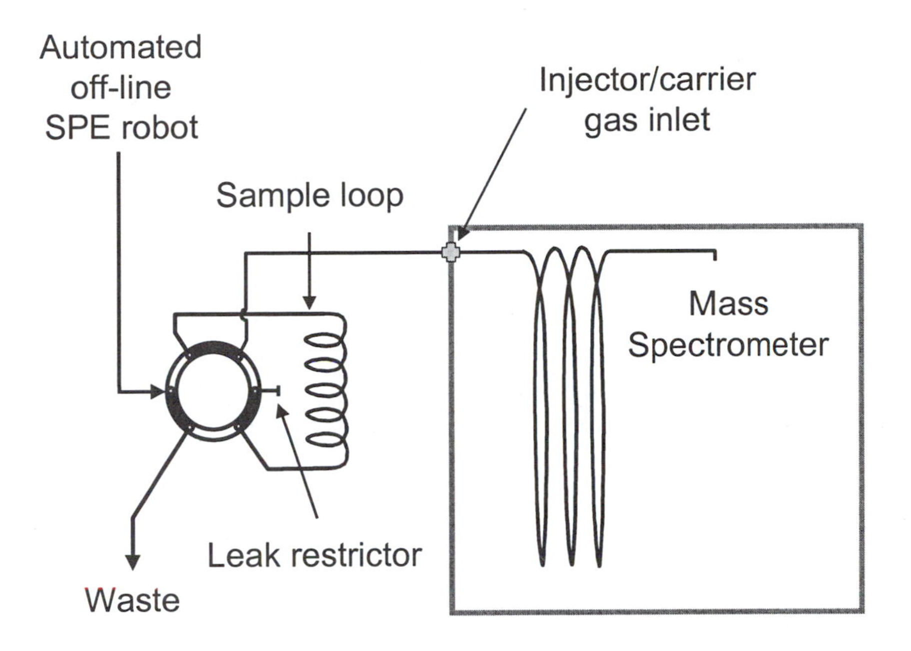

Figure 6. A simple fluidics diagram for off-line GC or GC/MS analysis combined with a SPE extraction.

in urine and triazine herbicides in drinking water. The authors claim the temperature-programmable injection technique is more reproducible and provides better results than the traditional retention gap approach with on-column injection.

The conditions stated — initial temperature of injector is 30°C below the SPE eluent's boiling point — require the injection liner to be packed with a material, similar to glass wool, which can provide pockets that hold the eluent and prevent it from percolating onto the GC injector base. By timing a solvent vent opening and high carrier gas flow the elution solvent can be swept out of the injector. Using this set-up the authors were able to load as much as 160 μL[47] of an ethyl acetate eluent onto the GC/MS system and close the solvent vent in time to ensure that qualitative transfer of triazines to the head of the GC column would occur. This experimental work used off-line SPE cartridges and a schematic is shown in Figure 6. This system is basically the same as that shown for on-line SPE/GC in the following schematic. The major differences are:

---

[47] Such a volume no longer seems large: as much as 1 mL of sample has been loaded onto a GC (Michel Bertraut, University of Montreal — personal communication).

1) A different injection condition is used.
2) Regular low-pressure SPE cartridges can be used (in this specific example, 200 mg C18 sorbent beds packed in 3 mL syringe barrels).
3) The eluent is collected in a vial and an aliquot of this is injected into the GC system.

The term "on-line" as used in this article is a little misleading: the whole system is connected on-line through a computer which synchronizes the timing events of the SPE, injection into the valve and in the GC injector itself. It does not mean that the SPE eluent is eluted directly into the GC. The authors claim that this set-up is easier to optimize because the two events, SPE extraction and GC analysis, can be uncoupled during method development and optimized separately before combining in the final, integrated procedure.

D. ON-LINE SPE/GC

A general scheme that allows a SPE extraction device to be coupled to a GC on-line is shown in Figure 7. A system like this has been used by Kwakman et al. (1992) with a GC and Louter et al. (1994) with a GC/MS. These papers use C18 and polymer SPE cartridges and discs to trace-enrich a few milliliters of aqueous sample. The cartridge must be dried before elution because of the sensitivity of the analytical system to water. Minimization of elution volume is also critical since, despite the presence of a retention gap and a retaining pre-column, it is easy to overload the GC system with elution solvent. Syringe pumps or HPLC injection loops are convenient tools to supply elution solvent to the cartridge. Provided a flow rate of no more than a couple of hundred microliters per minute is used, the eluent can be introduced directly into a capillary GC system as shown.

A "heart cut" of the eluent is still desirable because the first few drops of any commonly used solvent (ethyl acetate was the preferred solvent in this study) will push any water residing on the SPE cartridge onto the GC column. To accomplish this, the third valve in the figure is used, switching the front of the eluent out to "waste" before the heart of the eluent is switched on-line to the GC. The authors identified an additional benefit of ethyl acetate, besides its good broad-spectrum elution ability: residual water, found in the eluent, despite extensive drying using N_2 under pressure and the heart-cutting precautions, was azeotroped off in the retention gap. This azeotroped solvent/water mixture is swept out of the system through a switchable exit valve while the analytes take their time passing through the

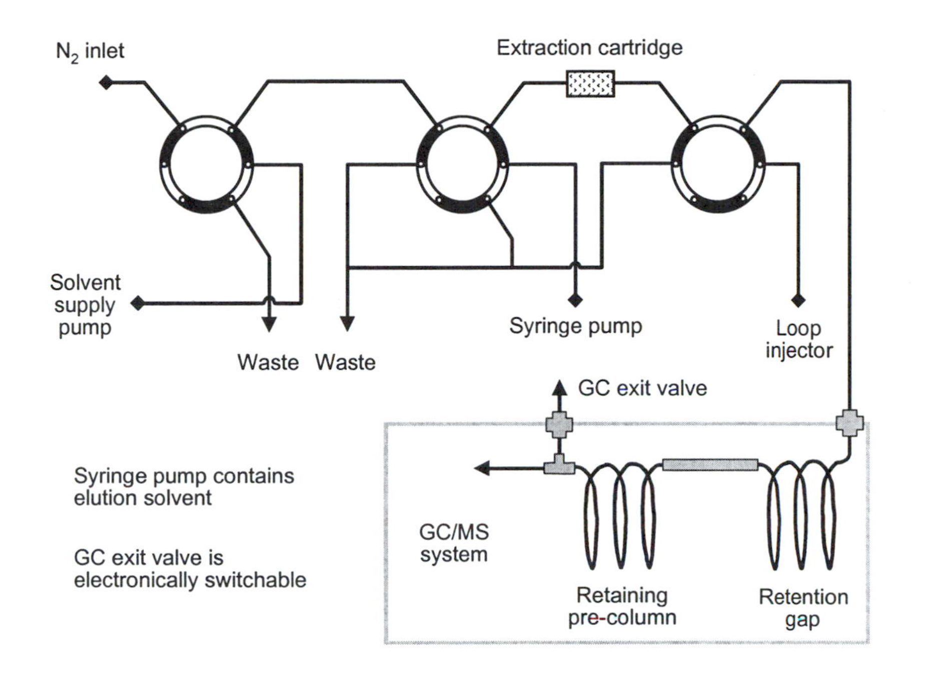

Figure 7. Valve and plumbing diagram for an on-line GC/MS solid-phase extraction system.

retention gap and pre-column. The valve can be closed in time to ensure that all the analytes pass on to the analytical column.

The difficulties resulting from water carry-over to the GC are less severe with polymer-based sorbents than with the silica-based ones. This is likely to be due to the presence of silanols on the silica sorbents, which bind the water chemically as well as providing physical space in the pores of the sorbent in which water can reside. An in-line drying cartridge of anhydrous salt, celite or silica could be used to further reduce water carry-over, as could the use of the smaller bed mass provided by extraction discs (Pico et al., 1994; Kwakman et al., 1992). Sub-10ppt levels of detection have been reported using the on-line SPE/GC technique.

Recently, an on-line SPE-GC instrument has been introduced[48], whose performance has been demonstrated by the extraction of 21 pesticides in river water at ppt levels.

---

[48] DualChrom 2000, Brechbuhler AG, Schlieren, Switzerland.

V. OTHER ANALYTICAL TECHNIQUES

A. SPE/ELECTROPHORESIS OR ELECTROKINETIC CHROMATOGRAPHY

The high efficiency of electrophoretic separations should make capillary electrophoresis a natural choice for clean-up of biological fluids, where the presence of many other components in a sample such as urine complicate the analysis. However, sensitivity towards some of the components of the sample means that a clean-up is required. The general features of extractions for these types of analytical finishes are reduction in ionic strength of the sample and interference removal, especially of ionic species and large molecules like proteins. Thus the requirements are not dissimilar from those for clean-up prior to MS detection or LC/MS analysis. However, the receiving device for the extract is a capillary and hence a large portion of the extract may have to be discarded. Thus, concentration is also a feature that must be maximized in order to get most benefit form the combination of SPE and CE.

The need for analyte concentration because of low injection may be compounded by the sensitivity afforded by the technique. Thus, when using UV detection sensitivity is not good and drug levels of no lower than 100 ng/mL are to be expected (Wernly and Thormann, 1991, 1992). These toxicology applications, using capillary electro-chromatography for the analysis, apparently needed no modification of the SPE procedure from that which was optimized for GC or HPLC finishes.

Many urine components retain and co-elute off a C18 sorbent at pH 6 (Magueregui, 1998) and interanalysis wash cycles of the Capillary Electrophoresis system with high pH solutions like aqueous potassium hydroxide are required to prevent build-up of interferences that would otherwise spoil the assay precision and limit of quantitation. If careful method development is performed it seems likely that the frequency of washes can be reduced, as is demonstrated by Veraart et al. (1998) who state that only weekly system washes were needed.

Veraart et al. (1998) demonstrate an "at-line" automated SPE system which delivers an extract to the head of a capillary electrophoresis column via a home-made interface. The system is applied to urine and serum. Direct interfacing of the extraction and the separation steps restricts the choice of solvents (buffer conditions) used for the elution and the separation steps. The extract must be particle-free and protein-free and the analytes must be desorbed in a small volume of solvent since no eluent concentration is possible in the at-line set-up. Despite these restrictions the authors show how it is possible to develop the automated extraction and analysis of non-steroidal anti-imflammatory compounds. This paper also examines the variables involved in optimization of the CE separation, sample loop loading, injection conditions and the SPE clean-up. The result is a very clean electropherogram which permits detection down to 40 ng/mL with good reproducibility.

B. ELECTROCHEMICAL SENSORS

While electrochemical sensing with ion-specific electrodes offers speed and simplicity, the risk of interference from other sample components is high. For example, analysis of nitrate ions in pure water is easily achieved using an ion-selective electrode. However, readings from this device are utterly unreliable if it is taken into the field to monitor waters rich in biological debris. Humic and fulvic acids poison the sensitive electrode surface and the resulting readout can be thousands of times higher than the true level of nitrate. A simple procedure (Csiky et al., 1985) in which a sample aliquot is passed through a NH2 cartridge removes these interferences, when used in the weak anion exchange mode. Preconditioning of the sorbent with sulfate ion (more selective than nitrate, and presumably less selective then the humic acids that must be removed) ensures that no nitrate ion is extracted with the humic acids. Subsequent analysis of samples treated this way gives accurate values of the inorganic ions.

An interesting application of solid-phase extraction combined with voltammetric detection is given by Cox et al. (1995), who use a persulfonated styrene polymer deposited over a ruthenium oxide layer to extract a range of compounds including aromatic amines and N-nitrosamines. This polymer did not affect the behavior of the catalytic ruthenium oxide layer, allowing the combined unit to extract the analytes and permit quantitation. The authors speculate on the utility of this electrode system to perform in the gas phase, as well as in the aqueous samples they investigated.

C. IMMUNOASSAY TECHNIQUES APPLIED TO SPE ELUENTS

Analysis by immunoassay of toxicant or micronutrient metabolites in unprepared samples can give spurious results. Cross-reactivity, for example, between metabolites or closely related molecular species will confound the efforts of analysts to perform quantitation or identify the presence of a species unambiguously. Further, if the samples are solid or viscous they will need to be homogenized or diluted so as to free the analytes. The solvent to be used is selected based in part upon its ability to solubilize the analytes, but the resulting solution may be incompatible with the immunoassay technique.

Here SPE can help by fractionating the sample into single metabolite groups, each of which can then be accurately quantified by immunoassay techniques. The sensitivity of immunoassays, on the other hand, means that concentration is seldom a concern, and their high specificity means that removal of matrix components of cross-reacting species may be sufficient, though the presence of compounds that may poison the assay should also be considered. These issues are considered in a review by Calverley et al. (1987).

A survey of published literature reveals that the majority of applications of SPE to this field apply to clinical analysis, although some examples of food/beverage or toxicology research have been published (Jackman, 1985; Colbert et al., 1994 — actually a solid-supported liquid-liquid extraction rather than a true solid-phase extraction). EMIT, ELISA, RIA (Jackman, 1985; Hudgins and Stromberg, 1987) and various other variants on the immunoassay technique have been employed on SPE extracts. The majority of these use a simple C18 clean-up. We may speculate that the great selectivity of the immunoassay renders more delicate clean-up unnecessary. However, this generalization can not apply to all the applications.

For example, exquisite selectivity has been achieved in the clean-up and fractionation of some closely related, and hence cross-reacting compounds (for example vitamin D metabolites). The result (Hollis, 1986; McCraw and Hug, 1990) is that fast, accurate immunoassays may be used instead of the slower TLC or HPLC separation/analysis. In both cases the use of SPE as a fractionating tool exploits the secondary interactions on C18 sorbents to separate, for example, the various hydroxylated metabolites of Vitamin D3 from plasma or serum samples. Each elution fraction is tested for activity. These two papers describe different extraction protocols to achieve the same end, though it is not stated whether the different protocols resulted from the differing requirements of the immunoassay techniques used.

VI. CONCLUSIONS

Since SPE can provide a cleaned-up and concentrated sample extract for almost any sample there is no reason why any detection system should not be coupled to a solid-phase extraction system. However, as we move away from the classic sample preparation problems of clean-up and concentration of low-level analytes in complex samples, the benefits of SPE become less significant and the convenience of a direct analytical analysis becomes more important. Direct detection involving discrimination by the detection system has been more widely used as parallel technologies like ion-specific electrodes, colorimetric affinity and other separation media are developed. However, even in these cases, SPE has a role to play.

For some of the techniques covered in this chapter it is difficult to say exactly where the preparation stage stops and the analysis begins. In some cases the SPE step involves much more than a clean-up and concentration. In others the SPE step merges with the analytical stage. Ironically, SPE also permits a greater separation of the sample preparation and analytical stages, due to the ability of SPE sorbents to stabilize, sequester and preserve the analyte. Thus, Senseman et al. (1995) and Martinez and Barcelo (1996) demonstrate that the sample extraction may be performed "in the field." The analysis may be carried out several days later, in an entirely different location, because of the stabilizing effect on the analytes, caused by the removal of the analytes from the aqueous sample matrix.

SPME has not been discussed because it is covered in other texts (Wercinski, 1999). Furthermore, from a theoretical point of view it is not a Solid Phase Extraction process. For the same reason, the use of diatomaceous earth as a means of dispersing a solid or liquid sample in a SFE extraction chamber has not been covered. From a practical perspective, though, such techniques are becoming difficult to distinguish from their true SPE counterparts, as the applications of each are being expanded and as detector versatility, especially in MS, advances.

Special thanks goes to Dr. Francis Beaudry of Phoenix International Life Sciences, Inc. (St. Laurent, Quebec) and to Dr. Danlin Wu of Purdue Pharma Inc. (Ardsley, New York) for their advice in the preparation of this text and for their generously donated time in proofing the chapter.

REFERENCES

Alzaga, R. and Bayona, J.M., (1993) Supercritical Fluid Extraction of Tributyltin and Its Degradation Products from Seawater Via Liquid-Solid Phase Extraction. J.Chromatogr. A. 655: 51-56.

Backa, S., Brolin, A. and Nilvebrant, N-O., (1989) Analyzing Wood Extractives from Process Streams by Using Alkaline Reversed-Phase Extraction Followed by *in situ* ion piring.Tappi Journal, 72(8); 139-143.

Barber, T., (1985) The use of Bonded and Unbonded Silica for the Rapid Separation of Additives from Complex Mixtures. The 2nd Annual Symposium "Sample Preparation and Isolation Using Bonded Silicas" Jan. 14-15, 1985. Philadelphia, PA. Proceedings (Varian, Harbor City, CA).

Barshick, S-A. and Buchanan, M.V., (1994) Rapid Analysis of Animal Drug Residues by Microcolumn Solid-Phase Extraction and Thermal Desorption Ion-trap Mass Spectrometry. J.Assoc.Off.Anal.Chem. 77(6):1428-1434.

Borgerding, A.J. and Hites, R.A., (1992) Quantitative Analysis of Alkylbenzenesulfonate Surfactants Using Continuous-Flow Fast Atom Bombardment Spectrometry. Anal.Chem., 64(13); 1449-1454.

Bui, V.T. and Dirksen, T., (1994) Approaches to FTIR Analysis of Oil and Grease. 3M Application Note, 3M Corporation, Minneapolis St. Paul, Minnesota.

Calverley, R.A., Jackson, R. and Pembroke-Hattersley, J.J., (1987) Solid-Phase Sample Preparation Techniques for Immunoassays. Chapter 8: 93-107.

Campins-Falco, P, Sevillano-Cabeza, A, Molins-Legua, C. and Kohlmann, M., (1996) Amphetamine and Methamphetamine Detection in Urine by Reversed-phase High-performance Liquid Chromatography with Simultaneous Sample Clean-up and Derivatization with 1,2-Naphthoquinone 4-sulphonate on Solid-phase Cartridges. J.Chromatogr. B. Biomed.Apps. 687: 239-246.

Chatfield, S.N., Croft, M.Y., Dang, T., Murby, E.J., Yu, G.Y.F. and Wells, R.J., (1995) Simultaneous Extraction and Methylation of Acid Analytes Adsorbed onto Ion Exchange Resins Using Supercritical Carbon Dioxide Containing Methyl Iodide. Anal.Chem. 67: 945-951.

Colbert, D.L., Smith, D.S., Landon, J. and Sidki, A.M., (1984) Single-reagent Polarization, Fluoroimmunoassay for Barbiturates in Urine. Clin.Chem., 30(11): 1765-1769.

Conte, D.E., Chun-Yi Shen, Perschbacher, P.W., Miller, D.W., (1996) Determination of Geosmin and Methylisoborneol in Catfish Tissue (*Ictalurus Punctatus*) by Microwave-Assisted Distillation-Solid Phase Adsorbent Trapping. J.Agric.Food Chem. 44: 829-835.

Cooper, A.I., Londono, F.D., Wignall, G., Mcclain, J.B. Samulski, E.T., Lin, J.S., Dobrynin, A., Rubinstein, M., Burke, A.L.C., Fréchet, J.M.J. and Desimone, J.M., (1997) Extraction of a Hydrophilic Compound from Water into Liquid CO2 using Dendritic Surfactants. Nature 389: 368-371.

Courbariaux, Y., Bristow, A. and Strawn, A., (1997) The Identification of Dyes in Effluent Streams Using Solid Phase Extraction — FT-Raman Spectroscopy (SPERS) U. of New Brunswick, Fredricton NB Canada. Others at Kodak Anal Labs, UK.

Cox, J.A., Alber, K.S., Brockway, C.A., Tess, M.E. and Gorski, W., (1995) Solid Phase Extraction in Conjunction with Solution or Solid State Voltammetry as a Strategy for the Determination of Neutral Organic Compounds. Anal.Chem. 67(5): 993-998.

Cziky, I., Marko-Varga, G and Jönsson, J.A., (1985) Use of Disposable Clean-up Columns for Selective Removal of Humic Substances Prior to Measurements With a Nitrate Ion-selective Electrode. Anal.Chim.Acta 178: 307-312.

Druzik, C.M., Grosjean, D., Van Neste, A. and Parmar, S.S., (1990) Sampling of Atmospheric Carbonyls with Small DNPH-coated C18 Cartridges and Liquid Chromatography Analysis with Diode Array Detection. Int.J.Environ. Anal.Chem. 38: 495-512.

Eastwood, D., Dominguez, M.E., Lidberg, R.L. and Poziomek, E.J., (1994) A Solid Phase Extraction/Solid-state Luminescence Approach for Monitoring PAHs in Water. Analusis, 22: 305-310.

Ezzell, J.L. and Richter, B.E., (1992) Supercritical Fluid Extraction of Pesticides and Phthalate Esters Following Solid Phase Extraction from Water. J.Microcol.Sep. 4: 319-323.

Field, J.A., Field, T.M., Poiger, T. and Giger, W., (1994) Detection of Secondary Alkane Sulfonates in Sewage Wastewaters by Solid Phase Extraction and Injection Port Derivatization Gas Chromatography/Mass Spectrometry. Environ.Sci. & Technol. 28(3): 497-503.

Frenzel, W., (1998) Highly Selective, Semi-quantitative Field Test for the Determination of Chromium (VI) in Aqueous Samples. Frezenius J.Anal.Chem., 1455: 1-6.

He, X., (1995) Chemical Derivatization by Coupled Solid Phase Extraction/Solid Phase Reaction For Determination of 2,4-dichlorophenoxy Acetic Acid and Haloacetic Acids by Gas and Liquid Chromatography. M.Sc. Thesis, Tennessee Technological University, Cookeville, TN.

Hendrick, J.L. and Taylor, L.T., (1990) Supercritical Fluid Extraction Strategies of Aqueous Based Matrices. J.High Resol.Chromatogr. 13: 312-316.

Hollis, B.W., (1986) Assay of Circulating 1,25-dihydroxyvitamin D Involving a Novel Single-Cartridge Extraction and Purification Procedure. Clin.Chem. 32(11): 2060-2063.

Hopper, M.L. and King, J.W., (1991) Enhanced Supercritical Fluid Extraction of Pesticides from Foods Using Pelletized Diatomaceous Earth. J.Assoc.Off.Anal.Chem. 74(4): 661-666.

Howard, A.L. and Taylor, L.T., (1992) Quantitative Supercritical Fluid Extraction of Sulfonyl Urea Herbicides from Aqueous Matrices via Solid Phase Extraction. J.Chromatogr.Sci. 30: 374-382.

Hudgins, W.R. and Stromberg, K., (1987) Solid Phase Extraction of Urinary Polypeptide Growth Factors Using Methyl Bonded Silica in Free Suspension. J.Liq.Chromatogr. 10(15): 3329-3346.

Jackman, R., (1985) Determination of Aflatoxins by Enzyme-linked Immunosorbent Assay with Special Reference to Aflatoxin M, in Raw Milk. J.Sci.Food Agric. 36: 685-698.

Kane, M., Dean, J.R., Hitchen, S.M., Dowle, C.J. and Tranter, R.L., (1993) Analysis of Non-ionic Surfactants Using Solid Phase Extraction Combined with Supercritical Fluid Extraction and Chromatography. Anal.Proceedings 30 (October): 399-400.

Kaziunas, A., (1997) SFE Plus SPE Facilitates Analysis of Nitrosamines in Cured Meats. Sample Preparation 2(11): 1-2.

Krier, G., Masselon, C., Muller, J.F., Nelieu, S. and Einhorn, J., (1994) Laser-desorption Fourier-transform Mass Spectrometry of Triazines Adsorbed on Solid-phase Extraction Membranes. Rapid Commun.Mass Spectrom. 8: 22-25.

Kwakman, P.J.M., Vreuls, J.J., Brinkman, U.A.Th. and Ghijsen, R.T., (1992) Determination of Organophosphorus Pesticides in Aqueous Samples by On-line Membrane Disk Extraction and Capillary Gas Chromatography. Chromatographia, 34(1/2): 41-47.

Lee, H.K., Wong, M.K. and Chee, K.K., (1997) Membrane Solid-Phase Extraction with Closed Vessel Microwave Elution for the Determination of Phenolic Compounds in Aqueous Matrices. Mikrochimica Acta, 126: 97-104.

Levy, J.M., Ravey, R.M. and Panella, R., (1998) The Use of Solid Phases for the Off-Line Collection in Supercritical Fluid Extraction with Modifiers. LC.GC 16(6): 570-584.

Louter, A.J.H., Rinkema, F.D., Ghijsen, R.T. and Brinkman, U.A.Th., (1994) Rapid Identification of Benzothiazole in River Water with On-line SPE-GC-MS. Int.J.Environ.Anal.Chem. 56: 49-51.

Maguregui. M., Jimenez, R.M. and Alonso. R.M., (1998) Simultaneous Detection of the β-blocker Atenolol and Several Complementary Hypertensive Agents in Pharmaceutical Formulations and Urine by Capillary Zone Electrophoresis. J.Chromatogr.Sci. 36: 516-522.

Martinez, E. and Barcelo, D., (1996) The Stability of Selected Herbicides Preconcentrated from Estuarine River Waters on Solid-Phase Extraction Disks. Chromatographia 42(1/2): 72-76.

Messer, D.C. and Taylor, L.T., (1996) Recovery of Trace Semivolatile Analytes from Reagent Water Using Solid-Phase Deposition and Supercritical Fluid Extraction. LC.GC 14(2): 134-142.

Mc.Graw, C.A. and Hug, G., (1990) Simultaneous Measurement of 25-hydroxy-, 24,25-dihydroxy- and 1,25-dihydroxyvitamin D Without Use of HPLC. Med.Lab.Sciences, 47: 17-25.

Moye, H.A., Gangadharan, M.K.P., Yoh, J. and Estevez, S., (1998) Microwave Assisted Extraction of Pesticides from Plant Tissue: Temperature, Power, Solvent, Tissue Type and Microwave Effect. Paper 7A-0074, 9th Int. Congress of Pesticide Chemistry, London, Aug. 1998.

Ollers, S., van Lieshout, M., Janssen, H.G. and Cramers, C.A., (1997) Development of an Interface for Directly Coupled Solid-Phase Extraction and GC-MS Analysis. LC.GC 15(9): 846-852.

Pico, Y., Louter, A.J.H., Vreuls, J.J. and Brinkman, U.A.Th., (1994) On-line Trace Level Enrichment Gas Chromatography of Triazine H Analyst 119: 2025-2031.

Poziomek, E.J. and Eicmann, G.A., (1992) Solid-Phase Enrichment, Thermal Desorption and Ion Mobility Spectrometry for Field Screening of Organic Pollutants in Water. Environ. Sci. Technol. 26: 1313-1318.

Rosenfeld, J.M., Moharir, Y. and Hill, R., (1991) Direct Solid Phase Extraction and Oximation of Prostaglandin E2 from Plasma and Quantitation by Gas Chromatography with Mass Spectrometric Detection in the Negative Ion Chemical Ionization Mode. Anal.Chem. 63: 1536-1541.

Secchieri, M., Benassi, C.A., Pastore, S., Semenzato, A., Bettero, A., Levorato, M and Guerrato, A., (1991) Rapid Pentachlorophenol Evaluation in Solid Matrixes by Second Derivative UV Spectroscopy for Applications to Wood and Leather Samples. J.Assoc.Off.Anal.Chem. 74(4): 674-678.

Senseman, S.A., Levy, T.L. and Mattice, J.D., (1995) Desiccation Effects on Stability of Pesticides Stored on Solid-Phase Extraction Disks. Anal.Chem. 67(17): 3064-3068.

Sooksamiti, P., Geckeis, H. and Grudpan, K., (1996) Determination of Lead in Soil Samples by In-Valve Solid-Phase Extraction-Flow Injection Flame Atomic Absorption Spectrometry. Analyst (Cambridge, U. K.) 121(10): 1413-1417.

Tang, P.H., Ho, J.S. and Eichelberger, J.W., (1993) Determination of Organic Pollutants in Reagent Water by Liquid-Solid Extraction Followed by Supercritical Fluid Elution. J.Assoc.Off.Anal.Chem.Int. 76: 72-82.

Van Elteren, J.T., Gruter, G.J.M., Das, H.A. and Brinkman, U.A.Th., (1990) Solid-Phase Extraction of As(III) from Aqueous Samples Using On-Column Formation of As(III)-trispyrrolidenedithiocarbamate. Intern.J.Environ.Anal.Chem. 43: 41-54.

Veraart, J.R., Gooijer, C., Lingemann, H., Velthorst, N.H. and Brinkman, U.A.Th., (1998) At-Line Solid-Phase Extraction for Capillary Electrophoresis: Application to Negatively Charged Solutes. J.Chromatogr. B. 719: 199-208.

Wells, M.J.M., Ferguson, D.M. and Green, J.C., (1995) Determination of Oil and Grease in Waste Water by Solid-Phase Extraction, Analyst (June) 120: 1715-1721.

Wercinski, S-A.S., (1999) Solid Phase Microextraction — A Practical Guide. (Ed. Wercinski) Marcel Dekker, New York, NY.

Wernly, P. and Thormann, W., (1991) Analysis of Illicit Drugs in Urine by Micellar Electrokinetic Chromatography with On-column Fast Scanning Polychrome Absorption Detection. Anal.Chem. 63(24): 2878-2882.

Wernly, P. and Thormann, W., (1992) Drug of Abuse Confirmation in Human Urine Using Stepwise Solid Phase Extraction and Micellar Electrokinetic Capillary Chromatography. Anal.Chem., 64(18): 2155-2159.

Westwood, S.A and Dumasia, M.C., (1994) A Note on Mixed Mode Solid Phase Extraction of Basic Drugs and Their Metabolites from Horse Urine. Sample Preparation for Biomedical and Environmental Analysis (Eds. Stevenson, D. and Wilson, I.D.) 163-166.

Wilson, I.D. and Nicholson, J.K., (1987) Solid Phase Extraction Chromatography and Nuclear Magnetic Resonance Spectroscopy for the Identification and Isolation of Drug Metabolites. Anal.Chem. 59: 2830-2832.

Wilson, I.D., and Nicholson, J.K., (1994) Proton Nuclear Magnetic Resonance Spectroscopy: A Novel Method for the Study of Solid Phase Extraction. Sample Preparation for Biomedical and Environmental Analysis (eds. Stevenson, D., and Wilson, I.D.,) Plenum Press, NY: 37-52.

Wolfe, MF., Hinton, D.E. and Seiber, J.N., (1995) Aqueous Sample Preparation for Bioassay Using Supercritical Fluid Extraction. Environ.Tox. And Chem. 14(6): 1001-1009.

Zhang, J., He, Q. and Lioy, P.J., (1994) Characteristics of Aldehydes: Concentrations, Sources and Exposures for Indoor and Outdoor Residential Microenvironments. Environ.Sci. & Technol. 28: 146-152.

17

CONCLUDING THOUGHTS — NEW DIRECTIONS FOR SPE

Nigel J. K. Simpson Varian Associates Inc., Harbor City, California

I. WHERE DOES THE FUTURE OF SPE LIE?

It seems reasonable to predict a bright future for the technique of solid-phase extraction (SPE). For over two decades applications, SPE sorbents, and engineering technology have kept pace with the increasing demands placed on the analytical laboratory and with the ever-tougher requirements of chemical analysis itself. There is every indication that it will continue to do so.

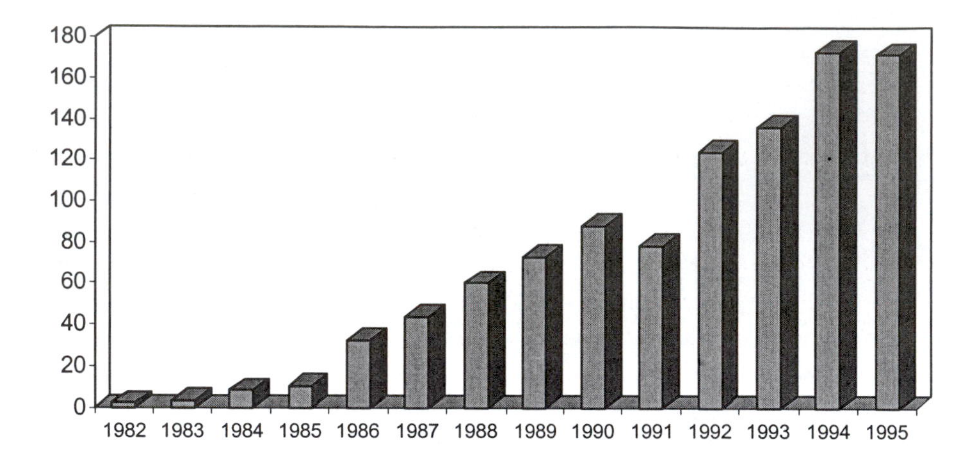

Figure 1. The number of citations having the term "solid-phase extraction" in the title of the article during the time period 1982 through 1995.

Projecting the trend in SPE citations (Figure 1) forward to the year 2000 suggests that by that year about 250 (linear fit based on 1986 through 1995; $r^2=0.95$) to 325 (quadratic fit based on 1982 through 1995; $r^2=0.98$) publications will cite this technique in the title. A search of Chemical Abstracts databases conducted in 1996 indicates that the number of papers published that use the technique even though it is not explicitly stated in the title, will be at least four times greater.

The trends provided from these literature searches, predict a continued growth in published environmental applications. Government ministries around the world are increasingly addressing the paradoxical situation that, in order to clean up a sample prior to analysis of organic pollutants, large quantities of air and water must be polluted with undesirable organic solvents when using liquid/liquid extraction techniques. The drive to improve safety in the laboratory through reducing exposure of analysts to solvent vapors will play a part in this, as will SPE automation.

II. WHAT WILL PERMIT FUTURE SPE DEVELOPMENT?

Introduction and deployment of techniques such as matrix solid phase dispersion (MSPD) will very likely expand the list of articles published on food and feed analysis. However, pharmaceutical development and clinical monitoring will continue to be a major market for SPE manufacturers. While the researchers in pharmaceutical companies were among the first to recognize the benefits of using SPE, these users show no sign of slowing their research, based on the increasing number of published articles and the increasing number of products offered for this field. This growth is catalyzed by the growth of high-speed sample preparation robots that can handle formats like the 96-well SPE plate — a format that standard in the immunoassay screening industry and combinatorial chemistry lab.

That is not all, however. Ninety-six well plate SPE "blocks" have been employed for combinatorial chemistry synthesis (Breitenbucher et al., 1998; Johnson et al., 1998), as has the use of medium to large size SPE cartridges for automated synthesis and clean-up of structurally related compounds (Lawrence et al., 1997). On-line SPE devices are being connected to increasingly diverse analytical instruments, (Chapter 16), opening up the possibilities of simpler inorganic analysis from complex matrixes, or of rapid, sensitive in-situ air and water quality monitoring; for example, SPE/gas chromatography (Vreuls et al., 1994) and SPE/capillary electrophoresis (Petersson et al., 1995).

Introduction of hyphenated instruments such as on-line SPE/GC and SPE/mass spectrometry will help ensure continued growth in this market. The previous chapter covered SFE coupled to SPE, demonstrating that while each technique can provide a solution for some sample types, the combined technique greatly expands the range of samples that can be tested. Some experimenters have already demonstrated the worth of microwave extraction/solid-phase extraction (Conte, et al., 1996). In this technique the microwave energy is used to vaporize components of the sample, from which the analytes are extracted, when the resulting vapor passes over a solid phase adsorbent trap.

III. WHAT WILL DRIVE FUTURE SPE DEVELOPMENT?

Regulatory changes and increased biological and medical knowledge will play a part in keeping SPE at the forefront of analytical method development. The changes in analytical requirements forced by the increasing potency of drugs and hence the lower detection limits required have taxed the analyst's ingenuity.

The value for a manufacturer in quickly introducing a drug to market continues to grow. Consequently, laboratories of pharmaceutical companies, as well as the clinical contract laboratories have embraced first SPE/LC/MS, then SPE/MS (Bowers et al., 1997), and even SPE/MS/MS in their quest to reduce time per assay. A further consequence of these advances has been improved sensitivity. Low picogram levels of detection from sub-three-minute extraction/analyses are not uncommon.

These trends are driven by general industry factors. We could also cite specific drivers that arise from societal problems: the need to identify explosive residues in suspected terrorist attacks; increasing concern about presence of carcinogens, identified by new medical findings, in smoked foods or distilled spirits; the appearance and abuse of a new recreational drug. For example, increased awareness of a specific drug, Rohypnol (also known as "Roofies"), combined with chilling and graphic news reports of its use on unsuspecting victims of "date rape," have catapaulted the drug into a class of federally regulated substances in the United States. Suddenly, a sensitive and specific extraction and analysis are required. SPE was able to provide such an extraction within a couple of months of the problem being identified (Varian, 1997). Such specificity and ease of method optimization will also play a part in the sustained growth of solid-phase extraction.

IV. WHAT EFFECT WILL THESE CHANGES HAVE ON THE SPE DEVICE ITSELF?

Current trends in regulations and analytical instrumentation suggest a divergence of requirements for different users. The desire to monitor pollutants at ever lower levels in water samples suggests two trends:

1) larger SPE devices or discs that can handle larger samples while maintaining small elution volumes (Hinckley and Bidleman, 1989).
2) very small cartridges that can be saturated with the pollutants and eluted with a minimal solvent volume (Brinkmann, 1995).

Chapter 13 demonstrated that the SPE column itself may be a barrier to more convenient use of SPE techniques. The recent development of solid-phase microextraction (SPME), where an optical fiber coated with a bonded phase is used to establish an equilibrium with a liquid sample by simply sitting in contact with the sample for a period of time, is another approach to introducing the sample to the sorbent. In both cases the sorbent is permitted to contact the sample directly without prior sample manipulation. What follows are some ideas for alternative formats of SPE devices that will do the same thing, as described by Barker and Floyd (1996) in their discussion of the MSPD technique to a symposium on chemically modified surfaces.

A. NEW HOUSINGS

1. Surface Phase Extractor (SUPE)

This approach would involve the derivatization of the surfaces that are to be contacted with a liquid sample. An extraction and or cleanup will occur as it passes over the device. Thus, glassware, plasticware or specially designed materials could be chemically prepared and derivatized to provide a surface chemistry capable of performing these needs. Theoretically, the interaction is identical to that in SPE. Practically, the sample does not need to squeeze into the micrometer-sized pores or interstitial spaces of a typical packed sorbent bed.

2. Web-Integrated-Phase Extractor (WIPE)

Similarly, polymeric materials in the form of a cloth that possess derivatizable functional groups could be prepared and contacted with the sample. Mixing of the sample with the "wipe" by agitation would permit rapid establishment of an adsorption equilibrium. Efficient extractions onto these surfaces could be obtained. This example could be a reality today, since the Empore disc is such a cloth or "wipe," though I am not aware of its use in this way.

3. Solid Phase On Netted Granule Extractor (SPONGE)

By enclosing SPE materials in a "bag" that is permeable to sample components one can extract the target analytes from the sample by dipping it in the sample and providing adequate time and agitation to effect efficient extraction — a reverse teabag, if you like, in which analytes diffuse from the outside in. Further, if the material of the bag could be chosen to be permeable only to lower molecular weight species (dialysis membranes, for example), proteins could be excluded.

4. Matted Organic Phase (MOP)

By derivatizing etched glass wool and attaching this material to an inert holder and handle one can perform extraction by "spinning" the device, moving it up and down in the sample or by simple agitation as described above.

Were a sequence of varying sorbent polarity "mops" to be applied to the sample, a class fractionation of the sample could be achieved while leaving the sample itself relatively intact. These acronyms are humorous, but the ideas they express are serious and any one of them could be a starting point for future SPE development.

B. MINIATURIZATION

The benefits of reduction in volume of the SPE device — smaller elution volumes required, hence smaller samples and wash volumes and smaller total extraction waste — have driven the development of miniaturized solid-phase extraction devices. This is hardly "the future" since some of these devices (discs in cartridges, sorbent plugs in pipette tips for example) already exist. However, it is one direction in which the industry has shifted. In fact, observation of the trends in products promoted by the industry over the last few years reveals a shift away from the traditional 500 mg or 1 g sorbent bed toward both larger and smaller sorbent devices. The larger ones are increasingly being applied for combinatorial clean-up (Lawrence et al., 1997) and the smaller ones for pharmaceutical development where the high throughput screening robots have automated the SPE process and allowed improvements in liquid handling that permit extremely accurate submilliliter volume dispensing (Simpson et al., 1998).

Toxicological and clinical sample sizes are getting smaller as a result of economical considerations of storage and transport and other costs, as well as the practical limitations of the size of the animal producing that sample. Reduction in size of SPE bed mass used in these extractions has paralleled sample size reduction; the smallest SPE cartridge available in 1986 contained 100 mg of sorbent. This minimum has dropped to 50 mg, then 25 mg, then 10 mg in packed cartridge format (Plumb and Jones, 1996), while the introduction of discs in cartridge form has allowed bed masses of less than 10 mg total sorbent to be used (Lensmeyer et al., 1995). It is conceivable that further miniaturization through automated handling and novel sorbent formats will continue.

A brief survey of offerings of miniaturized samples from the period 1994 to 1999 is given in table 1. Starting with reduced bed mass cartridges (25 mg or less per cartridge) and disc cartridges (with as little as 4 mg of sorbent per disc), this trend has extended through to the on-line SPE devices like the Prospekt cartridge. Initially these devices were developed around tubes with a 2 mm internal diameter. The requirements of low elution volume, minimal on-cartridge band broadening and increasing sensitivity of the LC/MS systems has allowed extractions to be performed on smaller samples. The consequence is a 1 mm I.D. cartridge, containing less than 10 mg of sorbent.

We have already seen how the desire to reduce solvent and sample volumes while maintaining assay sensitivity has led to reduced bed mass sorbents and, ultimately to the disc format. When the disc format is combined with 96-well plate robotics that can dispense microliter volumes with speed and precision this permits exceptional efficiency. For example, Simpson et al. (1998) demonstrate a high throughput assay using 96-well plate discs

Table 1. Varieties of miniaturized SPE devices

Miniature SPE device idea	Application area
Disc cartridges	Empore™[1], Spec™[2]
96 well plates (10 mg, 15 mg, 25 mg)	Microlute™[3], Versaplate™[4], Array™[5], Empore, Oasis™[6], Spec
Low Bed Mass cartridges (25 mg)	Spe-disc™[7], Isolute™[8], Bond Elut™[9]
Narrow-bore on-line SPE cartridges	Prospekt™[10]
Pipette tip plug	Zip Tip™[11]

The following are trademarked names: 1. 3M Corporation, Minneapolis-St Paul, Minnesota; 2. Ansys Corporation, Lake Forest, California; 3. Porvair Sciences, Walton, UK; 4. Varian Inc. Harbor City, California; 5. IST, Hengoed, UK; 6. Waters Corp. Milford, Massachusetts; 7. Mallinckrodt Baker, St. Louis, Missouri; 8. IST, Hengoed, UK; 9. Varian Inc. Harbor City, California; Spark Holland, Emmen, The Netherlands; Millipore Corp. Bedford, Massachusetts.

and robotic sample preparation which uses 75 μL elution volumes to give optimal recovery and maximum limits of detection.

Carrying miniaturization to extremes, a device (Zip Tip™) containing a sorbent plug of only a few milligrams, has been introduced, expressly for the extraction of femtomolar quantities of peptides. As skills in micro-fabrication and handling of smaller samples increase we should expect to see the continued miniaturization of SPE devices. AS robotic sample preparation grows in importance, the 96-well plate could give way to 384 well plate SPE devices (very unlikely to be a suitable format for manual SPE preparation) and perhaps even a 1536-well SPE plate. The only forseeable obstacle to this trend is the problem of sample homogeneity, since as we look at a smaller and smaller portion of the sample it ceases to be representative of the whole.

V. HOW WILL THE SORBENTS EVOLVE?

And the sorbents themselves will continue to evolve (Majors, 1992, 1998). Comparison of usage of common sorbents in two different years does not suggest or reveal radical changes in sorbent selection. We can expect to see modifications to existing sorbents — not so much reflecting revolutionary new ideas, but rather, incremental improvements in our knowledge of bonding chemistry or polymer functionalization (for example, Schmidt et al., 1993; Bouvier et al., 1998; Gan et al., 1999). Often these modifications and improvements will trickle down from the more easily characterizable HPLC sorbent technology. Polymer chemists have refined their craft over the last few years to produce highly uniform, clean, semi-rigid, or rigid polymers (Svec et al., 1996) at a cost that makes them feasible as the basis for SPE sorbents. Of these styrene divinylbenzene (SDVB, SDB) is the most common but polymethacrylates, celluloses, functionalized SDVB, and mixed-polarity sorbents are appearing on the lab bench (Purosep™[49], Oasis™[50], Bond Elut™ PPL[51]). We will see greater use of these sorbents for screening purposes, where the selectivity afforded by secondary interactions is not always desirable.

The more adventurous scientist has begun stretching the connections between a product and its intended application. For example, mode sequencing, commonplace in clinical toxicology has also been applied to environmental sample preparation (Sutherland, 1994). Techniques for producing chelating sorbents for metals extraction and immuno-affinity sorbents (such as were described in Chapter 12) are being developed or improved, and it is likely that a significant move toward analyte-specific products will occur as a result (Dixon et al., 1996). Scientists have even developed restricted access media or "non-fouling" sorbents, or internally reversed phases, in which the non-polar part of the sorbent resides underneath a polar surface. A model for this type of sorbent predicts that larger molecules such as proteins, which are assumed not to retain on polar phases from biological fluids, will pass through the cartridge unretained. Meanwhile the smaller molecules such as the typical pharmacologically active drug, will be able to experience the non-polar part of the sorbent and hence retain. It is certainly an intriguing possibility, though as yet very few commercial publications of this technique for SPE exist.

Review of the use of SPE sorbents (Majors, 1992) suggests that despite the availability of a huge range of sorbents, a few sorbents account for the majority of applications. According to the survey, over two-thirds of all solid-phase extractions performed during that year (1992) were developed on just four sorbents: C18, C8, CN, and SI. Indeed, a later survey (Majors, 1998) indicated that, of those using SPE, nearly nine out of ten used C18 and over 40% used SI.

A. BROAD SPECTRUM SORBENTS

With the introduction in the early 1990s of very high surface area sorbents, designed to offer maximum retention to highly polar environmental con-

---

[49] Purosep is a trademark of the Purolite Company, Pontyclun, Wales.
[50] Oasis is a trademark of Waters Corporation, Milford, Massachusetts.
[51] Bond Elut PPL is a trademark of Varian Associates, Palo Alto, California.

taminants (the first of these was Lichrolut™[52] EN), the concept of a universal sorbent, meaning one that can extract any analyte with good recovery, was born. This was followed by the commercialization of several other high-surface area sorbents — mostly polymers — and high retention silica-based sorbents — mostly based on C18.

This development was assisted not so much by new sorbent chemistry (as discussed in the first two chapters of this book, polymers had been employed in SPE from the beginning) as by the development of new analytical instrumentation. Specifically, the LC/MS system, in which the MS allowed the differentiation of smaller molecules, changed the sample preparation paradigm. Until the early 1990s the typical means of analysis for therapeutic drug metabolites was LC/UV. The majority of apparent interferences in a biological sample were therefore small molecules that were UV-active. Elimination of these small molecules was critical to good results; proteins and inorganic species were not important except insofar as they clogged the top frit of the HPLC column. LC/MS demanded, however, that the proteins and ionic species be removed — within limits the presence of other small organic molecules was not a problem because the MS detector could "select them out" of the effluent from the chromatographic column.

For this reason a class of new sorbents, the most common of which are based on polymers that use divinylbenzene combined with styrene, n-vinlpyrrolidone[53] or methacrylate[54], have been taken up enthusiastically by the research community. Not only do these sorbents offer a high recovery for many analytes, combined with a simple method. Some of them are more tolerant of method variables such as variable ionic strength of the sample and drying out (and hence de-conditioning) of the cartridge prior to loading of the sample (Bouvier et al., 1998).

B. HIGH THROUGHPUT SORBENTS

One benefit of a simple retention and elution mechanism, is that the final method is likely to require few steps and therefore be fast. This aspect lends itself to high-throughput sample preparation and the sorbents mentioned in the previous section share a common feature in that they are suitable for high throughput solid phase extraction. Low selectivity/poor clean-up is a drawback, of course. However, if the analytical technique is LC/MS, for which cycle times may be between one and two minutes per analysis, then provided co-extracted species do not interfere with ionization processes or degrade the performance of the equipment rapidly, poor clean-up is not a problem.

---

[52] Lichrolut EN, E. Merck, Darmstadt, Germany

[53] Oasis™, Waters Corp. Milford, Massachusetts

[54] NEXUS™, Varian Inc. Harbor City, California

Speed of sample extraction is now the bottleneck to analysis. For this reason fast flow sorbents (Bond Elut HF, for example) were introduced in the mid-1990s. The Oasis sorbent, designed to be resistant to drying out after conditioning and to have broad utility to lipophilic and hydrophilic species was introduced in the mid-to-late 1990s. A very recent addition (Gan et al., 1999) to the sorbent line is a sorbent called NEXUS™[55], designed such that it does not need any conditioning (termed non-conditioned SPE or NC-SPE), thus eliminating two out of a typical five steps in a solid-phase extraction method. For both these sorbents the speed of method development is increased, since fewer method parameters need to be explored. Again, it should be stressed that such sorbents cannot provide the clean-up that is provided by the more complicated yet powerful mixed-mode sorbents or those applications that utilize secondary interactions on a silica-based sorbent. However, for the needs of many high throughput labs the level of clean-up is perfectly acceptable.

Finally, it should be noted that while polymers have undergone a resurgence in the last few years, this does not imply that the future of SPE lies only in polymer chemistry. Some elegant bonding chemistry has proved that a silica-based sorbent (Gan et al., 1999) can also provide broad-spectrum selectivity without the need to perform cartridge conditioning.

VI. WILL SPE EVER BECOME A REDUNDANT TECHNOLOGY?

Developments of alternative or rival preparation technologies will, of course, continue. Solid-phase microextraction (SPME), for example, is proving of great utility in analysis of small samples for volatile or semi-volatile components. It has a bright future as a complementary technique to SPE (Penton, 1995; Zhang and Pawliszyn, 1993). Thus, where SPME is not suitable (for example, in the analysis of non-volatile components in a large sample volume) SPE is; where the analytes are too volatile to be managed by a conventional retention/elution using a SPE cartridge and a vacuum manifold, SPME has an advantage.

Supercritical fluid extraction (SFE), as demonstrated in Chapter 16, may also be a complementary technique and has not begun to displace SPE in any of the areas the latter technique dominates, namely completely aqueous samples and complex, polar samples. It is possible that microwave extraction and other accelerated extraction techniques could be coupled with SPE in the same way as has supercritical fluid extraction. The need for

[55] NEXUS is a trademark of Varian Inc. Harbor City, California.

post-extraction concentration, for example, may provide impetus for development in this field.

A common feature of these coupled techniques is that the SPE device acts as a trap for analytes which are freed from the matrix by a physical stimulation. It is not a far journey from here to a device that allows analyte detection on the solid phase itself through desorption into the detector and even direct detection of the analyte on the solid phase. Changes in surface conductivity, fluorescence, Raman, or diffuse reflectance Fourier-transform spectra could each potentially provide us with on-sorbent detection. Combine this with the microfluidic devices known as lab-on-a-chip, and one could envisage the concentration, clean-up, and detection/quantitation of minute quantities of compounds on a thumbnail-sized sliver of silicon.

The majority of separation and sample preparation techniques have been given more detailed theoretical investigation than the "low-tech" and apparently simple solid-phase extraction. Yet despite this, SPE is still by far the most commonly used of the sample preparation techniques introduced in the last twenty years. Moreover, review of surveys carried out over all sample preparation techniques (Majors, 1991; 1995, 1996, 1998) shows that it is becoming more commonly used. Labs that do not use SPE are now in the minority.

During the writing and editing of this book I have had many interesting conversations colleagues (and a few arguments!) with contributing authors and about where SPE will be after its second twenty years. The one thing nobody has disagreed on is that it will still be there, appearing in formats and ways that we cannot yet realize. Perhaps the strict definition of solid-phase extraction may not apply. After all, it has been used loosely in this book and we have included at times such unrelated (but apparently similar) techniques as solid-supported liquid-liquid extraction. But whatever the theoretical distinctions, we can expect that the partition of a compound between a solid phase or surface and a liquid phase will be as relevant an analytical tool as it is today, and certainly even more widely used.

REFERENCES

Breitenbucher, J.G., Johnson, C.A., Haight, M. and Phelan, J.C., (1998) Generation of a Piperazine-2-carboxamide Library: A Practical Application of the Phenol-Sulfide React and Release Linker. Tetrahadron Letts. 39: 1295-1298.

Bouvier, E.S.P., Iraneta, P.C., Neue, U.D., McDonald, P.D., Phillips, D.J., Capparella, M. and Cheng, Y-F., (1998) Polymeric Reversed-Phase SPE Sorbents – Characterization of a Hydrophilic-Lippophilic Balanced SPE Sorbent. Current Trends and Development in Sample Preparation, LC.GC. May: S53-S58.

Bowers G.D., Clegg C.P., Hughes S.C., Harker A.J., Lambert A., (1997) Automated SPE and Tandem MS Without HPLC Columns, for Quantifying Drugs at the Picogram Level. LC.GC 15(1): 48-53

Brinkman U.A. Th., (1995) On-line Monitoring of Aquatic Samples, Automated Procedures that Increase the Speed of Analysis and Improve Analyte Detectability. Env.Sci. & Technology, 29(2): 79-84.

Conte, D.E., Chun-Yi Shen, Perschbacher, P.W., Miller, D.W., (1996) Determination of Geosmin and Methylisoborneol in Catfish Tissue (*Ictalurus Punctatus*) by Microwave-Assisted Distillation-Solid Phase Adsorbent Trapping. J.Agric.Food Chem. 44: 829-835.

Dixon, A., Solomon, K., Pocci, R., Constantine, F., Aguilar, C. and Nau, D.R., (1996) IMAC SPE for the Clean-up and Concentration of Heavy Metals Prior to Analysis. Poster number 1291, The Pittsburgh Conference, McCormack Place, Chicago, March 1996.

Gan, K., Nguyen, H., Pocci, R. and Pippen, D., (1999) A New Sample Preparation Technology: Non-Conditioned Solid Phase Extraction for the Analysis of Pharmaceuticals in Biomatrices. Poster 1666P, The 1999 Pittsburgh Conference, Orlando, FL.

Hinckley, D.A., Bidleman, T.F., (1989) Analysis of Pesticides in Seawater After Enrichment Onto C-8 Bonded-Phase Cartridges. Environ.Sci. & Technology 23: 995-1000.

Johnson, C.R., Zhang, B., Fantauzzi, P., Hocker, M. and Yager, K.M., (1998) Libraries of N-Alkylaminoheterocycles from Nucleophilic Aromatic Substitution with Purification by Solid Supported Liquid Extraction. Tetrahedron, 54: 4097-4106.

Lawrence, R.M., Biller, S.A., Fryszman, A.M. and Poss, M.A., (1997) Automated Synthesis and Purification of Amides: Exploitation of Automated Solid Phase Extraction in Organic Synthesis. Synthesis, May: 553 – 558.

Lensmeyer, G.L. Onsager, C., Carlson I.H., Wiebe D.A., (1995) Use of a Particle-loaded Membranes to Extract Steroids for High-performance Liquid Chromatographic Analysis, Improved Analyte Stability and Detection. J.Chromatogr. A, 691: 239-246.

Majors, R.E., (1991) Sample Preparation Perspectives - An Overview of Sample Preparation,. LC.GC 9(1): 16.

Majors, R.E., (1992) Sample Preparation Perspectives – Trends in Sample Preparation. LC.GC 10(12): 912-918.

Majors, R.E., (1995) Sample Preparation Perspectives, Special Report: "Trends in Sample Preparation and Automation What the Experts Are Saying" LC.GC 13(9): 742-748.

Majors, R.E., (1996) Sample Prep Perspectives: Trends in Sample Preparation, LC.GC 14(9): 754.

Majors, R.E., (1998) A Review of Modern Solid-Phase Extraction. May LC.GC special edition: S8-S15.

Penton Z.E., (1997) Sample Preparation for GC with SPE and SPME. Advances in Chromatography Vol. 37, Chapter 5. Eds. Brown and Grushka, Marcel Dekker Inc., New York.

Petersson M., Wahlund K.-G., (1995) Nilsson S., Proceedings of the 7th Symposium on Handling of Environmental and Biological Samples in Chromatography, Lund, Sweden, 7-10 May, 1995.

Plumb, R. and Jones, C., (1996) Micro-Solid Phase Extraction in the Pharmaceutical Industry using Porvair Sciences' Microlute™, Porvair Sciences Application Note PFAP2, Shepperton, UK.

Schmidt, L., Sun, J.J., Fritz, J.S., Hagen, D.F., Markell, C.G. and Wisted, E.E., (1998) Solid-Phase Extraction of Phenols Using Membranes Loaded with Modified Polymeric Resins. J.Chromatogr. 641: 57-61.

Simpson, H., Barthemy, A., Buhrman, D., Burton, R., Newton, J., Kealy, M., Wells, D. and Wu, D., (1998) High Throughput Liquid Chromatography/Mass Spectrometry Bioanalysis Using the 96-Well Disk Solid Phase Extraction Plate for Sample Preparation. Rapid Commun.Mass Spectrom. 12: 75-82.

Sutherland, D., (1994) Method Development using Solid-Phase Extraction of Agricultural Soil and Runoff Water Samples Containing Simazine and 2,4-D for Determination by Gas and Liquid Chromatography. Thesis. Tennessee Technological University, Cookeville, TN.

Svec. F., Frechet J.M.J., (1996) New Designs of Macroporous Polymers and Supports: From Separation to Biocatalysis. Science, 273: 205-212.

Vreuls J.J., Gerhardus J. De J., Ghijsen R.T., Brinkman U.A.Th., (1994) Liquid Chromatography Coupled On-line with GC: State of the Art. J.Assoc.Off.Anal.Chem.Int, 77 (2): 306.

Zhang Z., Pawliszyn J., (1993) Headspace Solid Phase Microextraction. Anal.Chem., 6: 1843-1852.

APPENDIX 1

A SUMMARY OF COMMON MATRIX PROPERTIES AND COMPONENTS

The following guide was developed to assist the reader applying the information presented in Chapter 3. These matrix descriptions are not intended to be exhaustive but they convey the important properties and variables of each matrix, as they affect a solid-phase extraction. The text is largely reproduced, with permission, *from The Handbook of Sorbent Extraction Technology* 2nd *Edition*, (1993), Varian Inc., Palo Alto, California.

BIOMEDICAL MATRICES

Serum, Plasma

Common Extraction Modes: Non-polar, Ion-exchange
Sample Pretreatment: Usually diluted with an equal volume of water or a suitable buffer before applying to extraction sorbent. The choice of buffer and pH depends on the isolate to be extracted. TRIS buffer is generally preferable to inorganic buffers when using a non-polar extraction mechanism. Proteins may be a precipitated if necessary. Protein binding of isolates may be a problem and should be considered if recoveries of isolate standards are high but recoveries from sample are low.

Blood

Common Extraction Modes: Non-polar, Ion-exchange
Sample Pre-treatment: Blood should be handled much like serum and plasma. The difference is the presence of whole red blood cells. Red blood cells are chemically active in that many drugs and natural products uptake into the cells and are therefore not available to the sorbent surface unless the cell is disrupted by addition of an organic solvent or dilution with buffers. Also, the problem of protein binding common in serum and plasma is even greater in whole blood. If the isolate does not exist free in solution, the red blood cells should be disrupted and an effort made to also disrupt protein binding using the techniques described in Chapter 3.

Urine

Common Extraction Modes: Non-polar, Ion-exchange

Urine should first be diluted with at least an equal volume of water or appropriate buffer before applying the sample to the sorbent. As with serum and plasma, much of the protein content can be reduced by retaining the isolate on a non-polar sorbent, then washing the sorbent with water or a buffer. The protein can also be precipitated as described earlier.

Another major consideration when working with urine is the high and variable salt content. This characteristic of the urine matrix often precludes the use of ion-exchange as a first extraction mode. An excellent approach for circumventing this problem is to first retain the isolate on a non-polar sorbent, allowing facile removal of the salt through an aqueous wash, followed by isolate elution and subsequent use of an ion exchanger.

Urinary pigments are another class of compounds that are present at high levels in the matrix. Some of these retain on ion-exchangers; some on non-polar sorbents. Removal of the pigments from the rest of the sample can be accomplished through judicious use of these two extraction mechanisms.

Tissue

Common Extraction Modes: Non-polar, Polar or Ion-exchange, Depending on Sample Pretreatment

Sample Pretreatment: Tissue is homogenized as the first step in treatment (assuming matrix solid phase dispersion — see Chapter 13 — is not being employed); the choice of homogenizing solvent is largely dependent on the nature of the specific tissue sample. Fatty tissue is often homogenized in non-polar or medium-polar organic solvents, while tissue that is more strictly proteinaceous may be homogenized in more polar solvents such as methanol or alcohol/buffer mixtures. After homogenization the remaining solids are separated by filtration or centrifugation. At this point the supernatant may be diluted to optimize the environment for retention of the isolate.based on the selected mechanism.

DAIRY PRODUCTS

Milk

Common Extraction Modes: Non-polar, Ion-exchange
Sample Pretreatment: Milk is commonly diluted with the appropriate buffer or water and applied directly to the sorbent. If desired, proteins can be precipitated first using any of the techniques described earlier.

Cheese

Common Extraction Modes: Non-polar, Polar, Ion-exchange
Sample Pretreatment: Cheese is similar chemically to milk but with a higher relative fat content. Cheese can be homogenized in a variety of solvents; the choice should be based on the extraction mechanism to be used.

Note that matrix solid phase dispersion (MSPD) has been used to prepare milk, cheese, and other dairy products, and should be automatically examined during method development to see if it can be applied, since its application will save considerable time and solvent use.

PRODUCE

Tomatoes

Common Extraction Modes: Non-polar, Ion-exchange
Sample Pretreatment: Tomatoes are a highly aqueous matrix and are usually homogenized with a buffer, water, or combination of a polar organic solvent with water or a buffer. Tomatoes can also be digested in acid prior to extraction; the appropriate pH adjustment should be made before passing the digested sample through a sorbent.

Corn, Soybeans

Common Extraction Modes: Non-polar, Ion-exchange
Corn can be homogenized in a polar solvent and diluted with water of buffer as necessary, to optimize retention.

Citrus Fruits

Common Extraction Modes: Non-polar, Polar, Ion-exchange
These matrices can be homogenized in either polar solvents or non-polar ones, depending on the isolate and the extraction mechanism to be used.

Vegetable Oils

Common Extraction Modes: Polar
Sample Pretreatment: Because of their very non-polar character, oils are often treated by diluting with a non-polar organic solvent such as hexane.

GRAIN PRODUCTS

Corn Meal, Bread, Feeds

Common Extraction Modes: Non-polar, Polar, Ion-exchange, Depending on Sample Pretreatment
Sample Pre-treatment: Homogenize sample in the solvent appropriate for the extraction mechanism desired. If the sample contains high amounts of fats, a fat extraction (e.g. Goldfisch) may be performed first.

MEATS

Pork, Chicken, Beef Tissue

Common Extraction Modes: Polar
Sample Pretreatment: Due to the high fat content of these matrices, the simplest sample treatment is homogenization in a non-polar organic solvent. MSPD is an excellent choice for meat tissues when extracting non-polar or weakly polar isolates.

Liver, Kidney, Brain

Common Extraction Modes: Non-polar, Polar, Ion-exchange, Possibly in Combination, Due to the High Levels of Co-Extracted Species
Sample Pretreatment: One of the major difficulties with these matrices is the high degree of cross-linking in the sample, and the propensity for emulsion formation in the case of liver samples. MSPD is a perfect solution for both problems, but homogenization of the sample in saturated sodium chlo-

ride solution will also help avoid emulsions, if standard SPE is chosen for sample clean-up. This homogenate can be diluted with the solvent desired for extraction. The other choice is homogenization in a non-polar solvent.

CONFECTIONERY

Molasses

Common Extraction Modes: Non-polar

Sample Pretreatment: Molasses should be diluted with water or an aqueous buffer.

Chocolate

Common Extraction Modes: Polar

Sample Pretreatment: Chocolate has a high fat content and is best dealt with as a non-polar matrix. The chocolate can be homogenized in an organic solvent and the fats saponified, if desired, with methanolic KOH.

BEVERAGES

Wine, Beer, Soft Drinks

Common Extraction Modes: Non-polar, Ion-exchange

Sample Pre-treatment: The usual treatment for these very polar and largely aqueous matrices is dilution with water or a buffer, and pH adjustment when appropriate.

Coffee, Tea, Broth

Common Extraction Modes: Non-polar, Polar, Ion-exchange

Sample Pretreatment: Instant coffee, tea, and broth or soup mix can be dissolved in water directly, filtered, and the pH adjusted as necessary. Coffee beans or tea leaves can be ground and extracted directly with an organic solvent, depending on isolate chemistries.

COSMETICS

Ointments

Common Extraction Modes: Non-polar, Polar
Sample Pretreatment: Ointments are typically either water-based or oil-based. Water-based ointments may be treated by dissolving in methanol or another polar organic solvent, followed by dilution with water or a buffer. Oil-based ointments are treated by dissolving in a non-polar organic solvent.

Shampoos

Common Extraction Modes: Non-polar, Polar
Sample Pretreatment: The major issue with shampoos is the high surfactant content. Surfactants can often be removed by passing the sample through a depth filter of diatomaceous earth.

ENVIRONMENTAL

Soil

Common Extraction Modes: Non-polar, Polar, Ion-exchange
Sample Pretreatment: The major problem with soil is isolate adsorption onto the soil particles (though also note that for some isolates high humic and fulvic acid content may also complicate the sample analysis). For this reason, soil is often treated by refluxing in acid, base or an organic solvent, depending on the isolate. This extract is then diluted with a suitable solvent for the extraction mechanism to be used.

PETROLEUM PRODUCTS

Fats, Oils

Common Extraction Modes: Polar
Sample Pretreatment: These very non-polar matrices are usually treated by dilution with non-polar organic solvents.

Surfactants

Common Extraction Modes: Non-polar, Polar, Ion-exchange
Sample Pretreatments: Surfactants exhibit multiple characteristics simultaneously and thus they may often be extracted by any of the major mechanisms. They are therefore easy to extract but may prove difficult to elute from the sorbent. They may be prepared by simple dilution in a solvent that is compatible with the selected extraction mechanism.

More commonly surfactants are a co-extractant and interference in the analysis of a different isolate. In these cases the use of a depth filter packed with diatomaceous earth can help to remove the surfactant. Alternatively, a retention mechanism that is not appropriate to retain the compound of interest can be applied.

APPENDIX 2

A COMPARISON BETWEEN SOLID-PHASE EXTRACTION AND OTHER SAMPLE PROCESSING TECHNIQUES

The following glossary and guide was developed to assist the reader in understanding the differences between equilibrium and non-equilibrium extraction conditions and to relate these differences to sample size. It may prompt some users to return to the material of Chapters 4 and 5 which covered situations where sample sizes are often large and the translation from solvent extraction techniques to SPE is sometimes difficult to understand.

There are three principal types of liquid-liquid extraction (LLE), each of which may be replaced in certain applications by solid-phase extraction (SPE). It is important in deciding upon the suitability of a particular SPE method or sorbent to also understand the solvent based extraction system that it is replacing. By considering the properties of a liquid-liquid extraction system prior to developing an SPE method to replace it, the partitioning properties of the target analyte or analytes in the real sample matrix and at the working concentration may also be examined.

When considering liquid-liquid extraction, the concepts of the distribution constant (an equilibrium expression) and the extraction efficiency are of particular importance. These terms are described thus:

$$\text{Distribution Constant} = k'[\text{analyte}]_{\text{extractant}} / [\text{analyte}]_{\text{matrix}} \qquad (1)$$

$$\text{Efficiency} = [\text{analyte}]_{\text{extractant}} / ([\text{analyte}]_{\text{matrix}} + [\text{analyte}]_{\text{extractant}}) \qquad (2)$$

In the context of both liquid-liquid extraction and SPE, the equilibrium condition described by equation (1) is independent of the rate of extraction of the analyte from the sample matrix into the extractant (or onto the sorbent). Clearly however, the equilibrium condition is influenced by variables such as pH and ionic strength that alter the chemical nature of either the sample or the extraction medium. The completeness of extraction, or the extraction efficiency (often confusingly called the "extraction rate" though it is not a kinetic parameter), is also influenced by the mechanical aspects of the mixing process. These mechanical aspects will affect the significance of many variables related to the extraction; for example, phase boundary surface area, and the diffusion path length required to refresh the analyte-depleted sample matrix at the phase boundary.

Thus, while the equilibrium constant described by equation (1) is a measure of the likely efficiency of a particular extraction chemistry it does not predict the likely success of the method in allowing that equilibrium to be reached. The terms in equations (1) and (2) only become interchangeable as an extraction system reaches equilibrium. In liquid-liquid extraction the combined effect of these constraints is a mixing time required to achieve acceptable partitioning of the target analyte.

By analogy, the SPE parameter most influential over efficiency for a particular combination of sample and sorbent is the flow rate used for the passage of the sample through the sorbent. Poor wetting of the sorbent or excessive channelling also reduces the effectiveness of bringing the sorbent into contact with all parts of the sample matrix. Because these problems reduce the available surface area for retention of analytes to occur, extraction then becomes overly dependent on the diffusion of the analytes through the sample to the sorbent surface and also on a very large distribution constant.

1. BASIC LIQUID-LIQUID EXTRACTION AND BATCH ADSORPTION

Most simple applications of liquid-liquid extraction make use of solvents with a large distribution constant for the application, which leads to high extraction efficiencies with only one or two partitioning steps. In these cases, it is often considered useful to employ a back-extraction to assist with sample clean-up without the risk of significant losses of the extractant in the back-extraction solvent.

Similar methodology may be developed for SPE where target extractants are strongly retained by a sorbent. The sorbent may be washed thoroughly with the same solvent as the sample matrix to wash out weakly retained species and thereby yield a cleaner extract. Examples of this type have been given throughout the preceding chapters. In most examples, the sample volumes have been of the order of 1-10 sorbent bed volumes. Such a situation is typical of most simple SPE applications where a sample volume of 1-10 mL is extracted on a sorbent bed of 100-300 mg (which will have a total volume of approximately 0.5-1 mL).

In most of the examples given, the distribution constant is large and the amount of substance extracted from the sample is not so great as to exceed the capacity of the sorbent. The extractions can therefore be safely considered as analogous to a simple liquid-liquid extraction in which all portions of the sample are brought into contact with an extraction medium with sufficient capacity that the extracted species are efficiently stripped from the sample matrix.

As for liquid-liquid extraction, the smaller the distribution constant, the greater will be the risk of loss of the extracted species during any subsequent washing steps. However, as long as the capacity of the sorbent is not exceeded and the distribution constant is still large each portion of the added sample may be considered to be brought into contact with an extractive sorbent surface. In such a situation, breakthrough is unlikely.

Where the target compound for extraction is not retained efficiently we may consider changing sorbents to increase the retention just as we would consider changing extraction solvents to increase recovery. For example, the extraction of opiates from biological samples is usually achieved with mixtures of solvents such as chloroform-isopropanol because the use of chloroform alone gives poor recoveries. Similarly for SPE, polar compounds in an aqueous sample may be better retained on a C2 silica-bonded sorbent than by a similarly based C18 sorbent.

Multiple liquid-liquid extractions with a weaker solvent may be considered as akin to using a larger bed mass of a SPE sorbent that has a small distribution constant. Up to a point, such cases may be regarded empirically: an increase in the sample pathlength through the sorbent is equivalent to exposure of the sample to several cycles of extracting solvent. However, as the distribution constant becomes smaller and the sample volume increases, such approximations become invalid and the analogy to countercurrent extraction and frontal chromatography (described in chapter 5) become more appropriate.

2. EXHAUSTIVE OR CONTINUOUS EXTRACTION: SOXHLET AND OTHER RECYCLING EXTRACTION TECHNIQUES

Most chemists will be familiar with soxhlet extraction as an example of an exhaustive sample extraction technique. While many different apparatus may achieve the mechanical aspects of continuous extraction, it is of interest that exhaustive extraction makes use of repeated partitioning of the sample matrix with fresh volumes of extraction solvent. The techniques are particularly useful for extractants that have a small distribution coefficient and are therefore not effectively extracted by a single partitioning between sample and extractant.

SPE methods invariably make use of a single portion of sorbent for the extraction of a sample whether the sorbent is contained within a device such as a syringe barrel or is added to the sample in its free form (matrix solid-phase dispersion – see Chapter 13). To achieve more effective solid-phase extraction of a species from a sample matrix in which the distribution constant between the sample matrix and the sorbent is small, we should select an alternative sorbent with a more favourable distribution constant. This is preferable to repeated extraction of the sample with many portions of fresh sorbent. Such versatility of sorbent chemistries is one of the significant advantages that SPE presents in devising an extraction scheme.

With the exception of the arguments relating to theoretical plates within a column of sorbent to which the sample is applied, there is no clear analogy between exhaustive extraction and SPE. Examples of SPE are more closely related to either the batch type liquid-liquid extraction described above or to countercurrent extraction techniques described below.

The special case of solid-phase microextraction (SPME), which is mentioned but not treated in this text is the only solid-phase technique that may be considered to be analogous to soxhlet extraction. Rigourous mathematical and practical treatments of continuous extraction techniques are given in many analytical chemistry texts and so will not be considered here any further.

3. COUNTERCURRENT EXTRACTION

The use of countercurrent techniques proves particularly useful for compounds that have a low distribution constant, or for the separation of analytes from other matrix components with similar distribution constants.

The description of the countercurrent process provides a useful analogy for understanding the passage of a small volume sample containing a target extractant where the distribution coefficient on the particular sorbent is small. The model demonstrates that the analyte will be distributed throughout all parts of the sample and the sorbent and significant breakthrough will occur within the void volume of the sorbent.

In general, the SPE sorbent volume is small and the use of the technique aims for either total retention or total exclusion of a particular matrix component. Thus, while the countercurrent process may be used as a model for certain SPE processes, those examples are generally of applications with little practical application as they do not yield significant sample concentration or extraction.

The special case of large volume samples (particularly for environmental applications where the sample volume may be hundreds or even thousands of times the bed volume), regardless of the distribution coefficient between sample and sorbent, is best described as frontal chromatography which is covered in Chapter 4.

APPENDIX 3

ION EXCHANGE EXTRACTION

Ion exchange merits a separate discussion from the other SPE mechanisms for while it is not a unique system (in contrast to PBA extractions, for example) it does have some subtleties that set it apart from non-polar and polar extraction. These two interactions receive detailed treatment throughout this book and do not therefore require a separate appendix.

Retention

Ion exchange functions most reliably when the samples are aqueous: While it may be possible to extract an ionic species from a sample of low polarity (low dielectric constant), most ionization of organic species is suppressed. Further, the pKa values of analytes, which may be used to guide us in selecting an appropriate pH (a pH at which the analytes are ionized), cannot be applied to non-aqueous environments. It is common to apply a dilute buffer to the sample prior to performing ion exchange extraction. The buffer serves two purposes:

1) Control the pH of the sample and hence the ionic state of the analyte and potential interferences.
2) Modify the ionic strength of the sample.

Both effects will improve reproducibility of the extraction and the former will always improve recovery, provided the pH is selected correctly.

The kinetics of ion exchange are slower than those of non-polar or polar mechanisms. Several causes could be cited but the most important are the slow diffusion of ionic species (often because they also drag with them a sheath of water molecules, resulting in a diffusion radius that is much larger than the ion itself) and the process of displacement of the counter-ion that is associated with the ion exchanger. The analogous process in non-polar extraction would be the displacement of a conditioning solvent molecule by the analyte — easy and very fast. This both has consequences for retention and elution. Sample loading rates are more commonly specified when ion exchange is the mode of extraction and it is not easy to elute analytes in very small volumes.

Retention can only occur if the analyte is in an ionic state. If the analyte is a weak acid or base then this condition is obtained if the sample is two pH units above pK_a of the analyte (if it is an acid) or two units below the pK_a of the protonated form of the analyte (if the analyte is a base) respectively.

Counter-Ion Selectivity

The idea of displacing an ion that is already associated with the sorbent leads into the concept of counter-ion selectivity. The counter-ion a group or ionic species that is either present on the sorbent before the sample is loaded or that is introduced to the sorbent at the elution stage. Thus, for the most common anion exchange sorbent, SAX (timethylsilylpropyl ammonium chloride) the counter-ion is the chloride anion. Just as different organic structures lead to differing strengths of retention on non-polar sorbents, so ionic species show different affinities for ion exchange sorbents. These different affinities can be expressed relative to each other as a selectivity.

A table is given below that shows relative selectivities for several common ions, as determined for retention on SCX (the cations) and SAX (the anions). Remember that selectivity is also a function of the sorbent itself. Two general trends can be noted:

1) Selectivity increases as the charge of the ion increases (divalent ions are generally more retentive than monovalent ones)
2) Aqueous solubility and selectivity vary aproximately inversely. Thus the well-solvated F^- is a low selectivity ion while the poorly solvated I^- has high-selectivity.

In addition to this secondary interactions must be considered (see Chapter 7). Thus benzene sulfonate has a very high selectivity for SAX, partly due to the non-polar interactions that can occur between the analyte and the methyl and propyl groups present on the SAX sorbent. The ability to chelate or form multiple points of interaction between analyte and sorbent also increases selectivity. This accounts for the very high selectivity of the citrate group.

Counter-Ion Exchange Rules

The selectivity values of counter-ions can be used to determine the volume and concentration of buffer required to displace one ion for another. Such calculations are valuable if it is necessary to replace a moderately high selectivity ion (for example, chloride) by a low selectivity one (formate, acetate) in order to use the sorbent to extract a similarly low selectivity analyte (e.g. malonic acid).

In order to switch to a counter-ion of higher selectivity, 100% conversion can be achieved if you pass 2-5 bed volumes of a 1M buffer containing the higher selectivity ion.

In order to switch to a counter-ion of lower selectivity, calculate the ratio of the selectivities of the higher selectivity ion and the lower selectivity ion. Multiply this figure by 5 and pass the resulting number of bed volumes of a 1M buffer through the sorbent.

Elution

Two distinct elution regimes can be identified:

1) Displacement of the analyte by another ion of higher selectivity or by one that is present in much higher concentrations.
2) Neutralization of the sorbent or the analyte by use of a suitably acidic or basic solution.

Note that it may often in practice be impossible to distinguish between these two effects. Thus, a common elution solvent in the area of sample preparation for drugs of abuse, is 2% ammonium hydroxide in ethyl acetate. Does the hydroxyl ion neutralize the weakly basic analyte by abstracting a proton? Or does the ammonium ion, present in high numbers, displace the analyte?

It is pointless to try to use pK_a values to determine if the neutralization mechanism is feasible, because these pK_as apply to aqueous solutions of the acidic species at standard conditions and unit activity. They certainly do not apply to organic mixtures at uncontrolled temperatures and ionic activities that differ hugely from the unit activity. Consider, too, the fact that this ion exchange process occurs a few atomic diameters form a highly inhomogeneous surface of (mostly) polar substrate and non-polar tethering groups that hold the ion exchange site onto the sorbent substrate.

Keep in mind that ion exchange is a slow process compared to non-polar and polar mechanisms. This may make it very difficult to elute the analyte in the desired volume of liquid (a phenomenon that is often misinterpreted as irreversible retention). Use of high-mobility ions like OH^- and H^+ may alleviate this problem, as may the use of purely organic solutions with low levels of strongly acidic or basic additives, such as methanol/HCl (99:1 v/v) or $EtOAc/NH_4OH$ (98:2 v/v). Use of a weaker sorbent (a sorbent based on a weak acid or base rather than on a strong acid or quarternary ammonium compound) will facilitate the elution process.

Finally, note that the neutralization mechanism, if it is a cationic sorbent that is being neutralized, is indistinguishable from the displacement mechanism. The ion is simply being displaced by a proton, creating a species that we usually write as a free acid rather than as a combination of an acid anion and H^+.

Conclusions

Ion exchange is both more complex and potentially more powerful than simple non-polar or polar extractions. If ion exchange is used carefully it may, acting alone or in conjunction with other mechanisms such as non-polar retention, provide extreme levels of clean-up.

To assist in your use of this technique, a table of selectivities of some common counter ions (reproduced with permission from T*he Handbook of Sorbent Extraction Technology 2nd Edition* (1993), Varian Inc., Palo Alto, California) is provided in Figures 1 and 2.

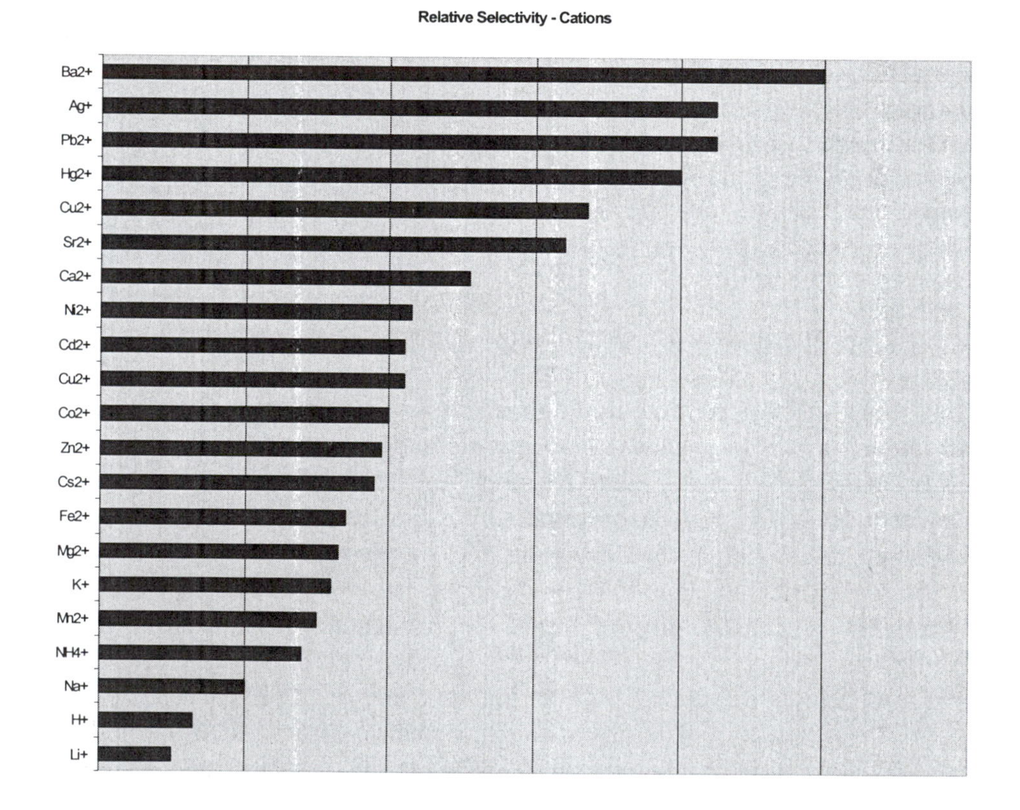

Figure 1: Cation counter-ion selectivity table, presented relative to the most strongly retained cation.

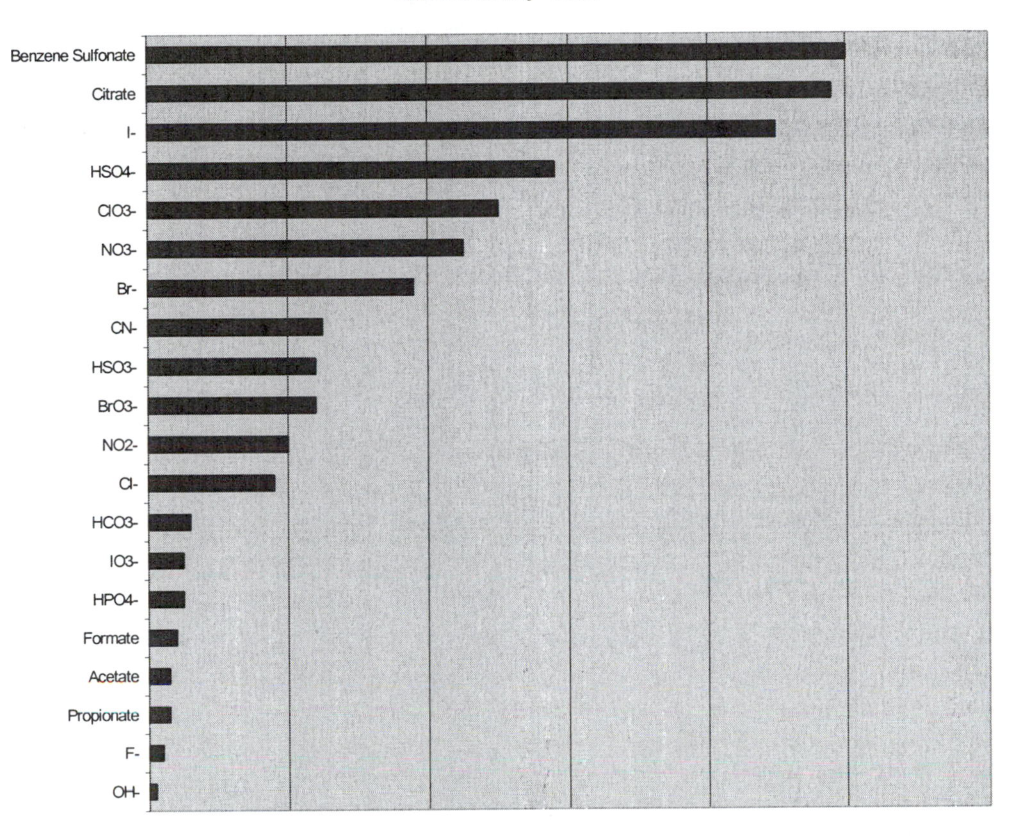

Figure 2: Anion counter-ion selectivity table, presented relative to the most strongly retained anion.

APPENDIX 4

A SUMMARY OF CURRENT SUPPLIERS OF SPE AUTOMATION EQUIPMENT

This summary was prepared in December of 1999. Since the solid-phase extraction (SPE) industry is rapidly changing, and new suppliers of consumables and automation join the market each year, the reader is advised to also review the listings of suppliers at tradeshows or through the pages of exhibitors or sponsors of conferences, obtained through the World Wide Web. An additional and increasingly useful source of up-to-date information is to be found through the links created by SPE consumables suppliers, connecting their Web site with those of SPE automation manufacturers.

U.S. Offices

Applied Separations
930 Hamilton Street
Allentown, PA 18101-1137
Phone: (610) 770-0900
Fax: (610) 770-5520
E-mail: appsep@fastnet.com

Bohdan Automation, Inc.
1500 McCormick Boulevard
Mundelein, IL 60060
Phone: (847) 680-3939
Fax: (847) 680-1199
E-mail: sales@bohdaninc.com
WWW: http://www.bohdaninc.com

Cardinal Workstation Company
901 Canal Street, Suite 1205
Bristol, PA 19007
Phone: (215) 781-2232
Fax: (215) 781-2247
E-mail: glhawk@voicenet.com

International Offices

Applied Separations Europe SarL
P.O. Box 334
CH-1000 Lausanne 9
Switzerland
Phone: +41 (21) 647 74 00
Fax: +41 (21) 647 74 00

Bohdan Europe
BP1
89630 St. Leger Vauban
France
Phone: 33-3-86-330099
Fax: 33-3-86-330106

U.S. Offices (continued)

Gilson, Inc.
3000 West Beltline Highway
P.O. Box 620027
Middleton, WI 53562-0027
Phone: (608) 836-1551
Phone 2: (800) 445-7661
Fax: (608) 831-4451
E-mail: sales@gilson.com
WWW: http://www.gilson.com

Hamilton Company
4970 Energy Way
P.O. Box 10030
Reno, Nevada 89502
Phone: (702) 858-3000
Phone 2: (800) 648-5950
Fax: (702) 856-7259
E-mail: sales@hamiltoncompany.com

Hewlett-Packard Company
2850 Centerville Road
Wilmington, DE 19808
Phone: (302) 633-8000
Phone2: (800) 227-9770
Fax: (302) 633-8901
WWW: http://www.hp.com.go/analytical

Horizon Technology, Inc.
8 Commerce Drive
Atkinson, NH 03811
Phone: (603) 893-3663
Fax: (603) 893-4994
E-mail: spe@horizontechinc.com

Jones Chromatography USA INC
P.O. Box 280329
Lakewood, CO 80228-0329
Phone: (303) 989-9200
Phone2: (800) 988-9478
Fax: (303) 988-9478
E-mail: sales@joneschrom.com

International Offices (continued)

Gilson Medical Electronics
BP 45
72 rue Gambetta
Villiers le Bel F95400
France
Phone: +33 1 3429 5000
Fax: +33 1 3429 5080

Hamilton Bonaduz AG
P.O. Box 26
CH-7402 Bonaduz
Switzerland
Phone: +41-81-37-01-01
Fax: +41-81-37-25-63

Jones Chromatography LTD
New Road
Hengoed, Mid Glamorgan CF82 8AU
United Kingdom
Phone: +44 (1443) 816991
Fax: +44 (1443) 816552

U.S. Offices (continued)

Packard Instrument Company
800 Research Parkway
Meriden, CT 06450
Phone: (800) 323-1891
Phone 2: (203) 238-2351
Fax: (203) 639-2172
E-mail: webmaster@packardinst.com
WWW: http://www. packardinst.com

SAGIAN, Inc.
5775 West 74th Street
Indianapolis, IN 46278
Phone: (800) 352-4975
Fax: (317) 328-3589

Tecan U.S.
P.O. Box 13953
Research Triangle Park, NC 27709
Phone: (919) 361-5200
Phone 2: (800) 832-2687
Fax: (919) 361-5201

Tekmar-Dohrmann
P.O. Box 429576
Cincinnati, OH 45242-9576
Phone: (513) 247-7000
Phone2: (800) 543-4461
Fax: (513) 247-7050
E-mail: info@tekmar.com
WWW: http://www.tekmar.com

Tomtec
1000 Sherman Avenue
Hamden, CT 06514
Phone: (203) 281-6790
Fax: (203) 248-5724
WWW: http://www.tomtec.com

International Offices (continued)

Packard Instrument Company (UK)
Brook House
14 Station Road
Pangbourne, Berkshire RG8 7AN
Phone: 44 (0) 118 984 4981
Fax: 44 (0) 118 984 4059
E-mail: pico_uk@packardinst.com

Spark Holland BV
Pieter de Keyserstraat 8
Emmen NL7825 VE
The Netherlands
Phone: +31 591 631700
Fax: +31 591 630035

Tecan AG
Feldbachstrasse 80
CH-8634 Hombrechtikon
Switzerland
Phone: +41 552 548 111
Fax: +41 552 443 883

U.S. Offices (continued)

Varian Instruments
2700 Mitchell Drive
Walnut Creek, CA 94598
Phone: (510) 939-2400
Fax: (510)
E-mail: helpdesk.us@varianinc.com
WWW: http://www.varian.com

Zymark Corporation
Zymark Center
Hopkinton, MA 01748
Phone: (508) 435-9500
Fax: (508) 435-3439
E-mail: solutions@zymark.com
WWW: http://www.zymark.com

International Offices (continued)

Varian Instruments
Guterstrasse 86
CH-4008 Basel, Switzerland
Phone: +41 61-295-8000
E-mail: helpdesk.eu@varianinc.com

Zymark Limited
1 Wellfield
Preston Brook
Runcorn
Cheshire WA7 3AZ
Phone: 1928-711448
Fax: 1928-791228

INDEX

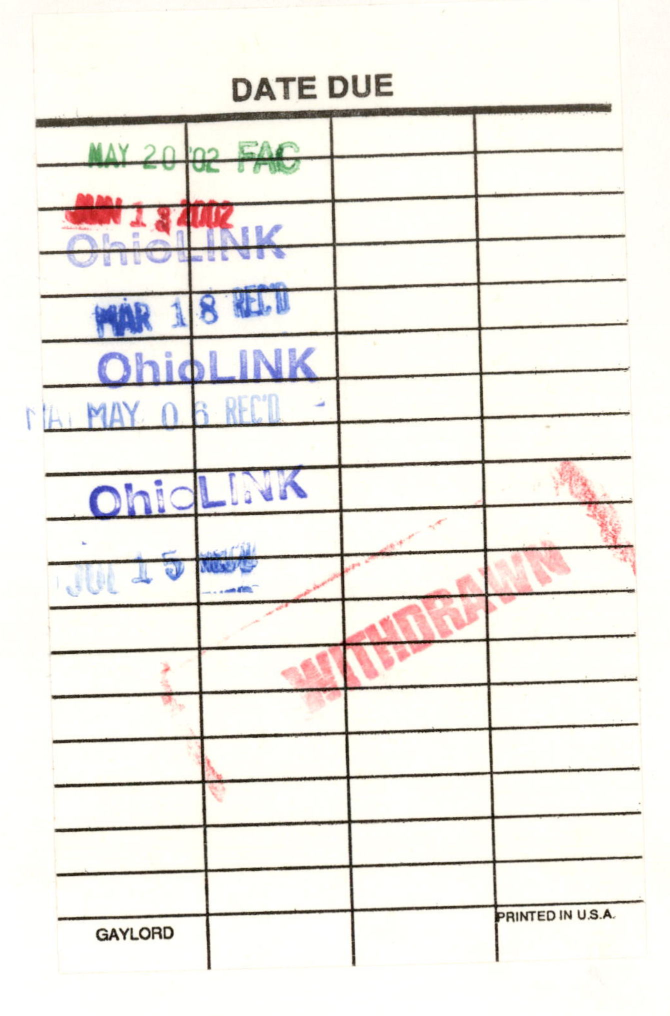
DATE DUE
MAY 20 '02 FAC
OhioLINK
MAR 18 REC'D
OhioLINK
MAY 06 REC'D
OhioLINK
WITHDRAWN
GAYLORD
PRINTED IN U.S.A.